STATISTICAL DESIGN AND ANALYSIS OF
ENGINEERING EXPERIMENTS

Statistical Design and Analysis of Engineering Experiments

Charles Lipson
Professor of Mechanical Engineering
University of Michigan

Narendra J. Sheth
Senior Research Engineer
Product Development
Ford Motor Company

McGraw-Hill Book Company

New York   St. Louis   San Francisco   Düsseldorf   Johannesburg
Kuala Lumpur   London   Mexico   Montreal   New Delhi   Panama
Rio de Janeiro   Singapore   Sydney   Toronto

17 DODO 8 9 8 7 6
This book was set in Times Roman, and was printed and bound by R. R. Donnelley & Sons Company. The designer was Merrill Haber; the drawings were done by F. W. Taylor Company. The editors were B. J. Clark and J. W. Maisel. John A. Sabella supervised production.

We Dedicate This Book
to our Wives

Contents

Preface

The purpose of this book is to present the fundamental concepts involved in the design and analysis of engineering experiments, with a strong emphasis on practical applications. For this reason, each concept in the book is followed by a solved engineering example, for a total of 100 examples. In addition, there are 99 problems listed at the ends of the chapters. The examples and problems cover a broad range of engineering fields, such as mechanical, civil, electrical, and metallurgical, and applications involving propulsions, household appliances, computers, structures, automatic control systems, chemical reactions, etc. The techniques presented may be extended to other fields. Charts, graphs, and tables are provided in the text and in the Appendix to minimize computations in solving problems. Included are a list of symbols, a glossary, and references and extensive bibliographies at the ends of the chapters.

The first two chapters cover the basic considerations in the design of experiments, Chapter 1 presenting statistical tools and Chapter 2 the principal statistical distributions which the test data may follow. Chapters 3 through 9 are devoted to the organization of experiments so that the resultant data will be statistically significant. Along with the experimental techiques, methods are presented for the analysis of these data. Chapters 10 through 14 are concerned specifically with the analysis of data.

The book is written in a relatively simple style so that a reader with a moderate knowledge of mathematics may follow the subject matter. No prior knowledge of statistics is necessary. Until now no single book has treated the broad area of statistical considerations in engineering problems, as related to the design and analysis of experiments. The key feature of this book is a balanced treatment and integrated presentation of the various phases of design and analysis of engineering experiments.

The book is at present used at the senior-graduate level at the University of Michigan and for engineering summer conferences conducted for government and industry. In notes form, the book has also been used for in-plant courses and seminars conducted by the senior author for over 20 major industrial firms.

The authors wish to acknowledge the contribution of R. Disney, Professor of Industrial Engineering, University of Michigan, to the preparation of the chapter on interference theory; the help of Professor H. Schenck of The University of Rhode Island for the review of the manuscripts; the help of Mr. S. Bussa, Principal Staff Engineer, Ford Motor Company, in the review and suggestions in the preparation of the chapter on fatigue experiments; and the assistance of the following students in the preparation of these notes: A. Krafve, L. Mitchell, J. Kerawalla, R. Pittman, and R. Bharti.

CHARLES LIPSON
NARENDRA J. SHETH

List of Symbols

α	Probability that a variable will exceed a certain value
b	Weibull slope of sample (shape parameter); also, specific value of x
β	Weibull slope of population
B_q	Life at which q percent of the items in a sample would fail
$\hat{B}_q$	Life at which q percent of the items in a population would fail
C	Confidence level
χ^2	Random variable of x^2 distribution
$\chi^2_{\alpha;\nu}$	Specific value of variable x^2
e	Base of natural logarithm ($e = 2.718$)
$E(x)$	Expected value of x
$E(\bar{x})$	Expected value of mean
F	Random variable of F distribution
$F_{\alpha;\nu_1;\nu_2}$	Specific value of variable F
$F(x)$	Cumulative distribution function
$f(x)$	Density function
g	Number of unsuccessful occurrences
H_A	Alternate hypothesis
H_D	Design hypothesis
H_o	Null hypothesis
j	Order number
k	Number of defectives (failures)
K_f	Fatigue strength reduction factor
K_t	Stress concentration factor
λ	Hazard rate
$m(t)$	Average number of failures or replacements by time t
μ	Population mean
μ_l	Population log mean
N	Statistical (total) sample size ($N = n + k$)
n	Sample size

$N(t)$	Number of failures or replacements by time t for a system with more than one component
$n(t)$	Number of failures or replacements by time t
ν	Degrees of freedom
$1 - \alpha$	Confidence level
p	Probability of finding a defective item
p_L	Lower limit of p in binomial distribution
p_U	Upper limit of p in binomial distribution
$p(x)$	Probability of occurrence of variable x
$p(x_i)$	Probability of occurrence of event x_i
$P(x \leq a)$	Probability that x is less than or equal to a
q	Notch sensitivity; also, probability of finding an OK item in binomial distribution
q_L	Lower limit of q in binomial distribution
q_U	Upper limit of q in binomial distribution
R	Range of sample; also, reliability based on sample
$\hat{R}$	Reliability of population
r	Number of successful occurrences; also correlation coefficient
R_C	Reliability at confidence level C
S	Stress or strength
s or s_x	Sample standard deviation
s_l	Sample log standard deviation
s_l^2	Sample log variance
S_n	Fatigue strength of fabricated part
S_n'	Fatigue strength of standard laboratory specimen
$s_{\bar{x}}^2$	Sample variance of mean
σ or σ_x	Population standard deviation
σ_l	Population log standard deviation
σ_l^2	Population log variance
$\sigma_{\bar{x}}^2$	Population variance of mean (standard error of mean)
T	Operating (test) time
t	Independent time variable; also, random variable of t distribution
$t_{\alpha;\nu}$	Specific value of variable t.
T_D	Desired mean time
T_0	Total test time
θ	Characteristic value of sample (scale parameter)
$\hat{\theta}$	Characteristic value of population
u	Random variable of u distribution

W_D	Desired statistical mean value (load, voltage, temperature, etc.)
W_0	Test value
X	Specific value of x
x	Continuous random variable
$\bar{x}$	Sample mean
x_c	Cut off (critical) point for variable x
x_i	Discrete random variable
x_l	Log x
$\bar{x}_l$	Sample log mean
x_0	Expected minimum value of population (location parameter)
$(x - C_{\alpha/2}), (x + C_{\alpha/2})$	Confidence limits
$[(x - C_{\alpha/2})$ to $(x + C_{\alpha/2})]$	Confidence interval
z	Random variable of z distribution (standardized normal variate)
z_α	Specific value of z, $[z_\alpha = (b - \mu)/\sigma]$

Introduction

Manufacturers of industrial products, universities, and other nonprofit organizations frequently spend much time and effort conducting laboratory and field tests. Sometimes, just because the test is run, one gets a feeling of security, without assuring himself of what the test really means. A few years ago, in connection with a testing program, we went through several manufacturers' files containing test data accumulated over a period of 20 years. The purpose of this search was to reexamine past data in the light of new statistical methods for data interpretation. We found that fully 75 percent of these data were statistically not significant. What this really meant was that the conclusions drawn from these data, and some of the decisions based on these conclusions, may not have been valid at all.

What saved those people was one of three things: Either they operated with such large factors of safety that their products were grossly overdesigned in the first place; or the real criterion of these designs was not structural integrity but rigidity, wear, bearing area, or ease of manufacture; or, what we consider most likely, the Good Lord looked out for them.

What, then, is a good experiment? A good experiment is one which provides the required information with the minimum amount of time and effort. Anything else means you are wasting somebody else's money.

To achieve a good experiment, the program must be well prepared; that is, the experiment must be designed. Specifically, the following steps should be taken:

Step 1. Decide whether the test is necessary in the first place, since it is not economically feasible to test every part. If you suspect during the design stage that the part is weak, better redesign it.

Step 2. Define the scope of the test:
Should the test be destructive or nondestructive?
Should the test be terminal or run to failure?
Should the test duplicate the operating conditions, or should it be accelerated?
Samples should be taken so that they are really representative of the parts being tested.
The number of items to be tested should be such that the final data are statistically significant.

Step 3. The test should be carefully designed so that statistically significant data are obtained. This is particularly true if the improvement you are seeking is small, of the order of 5 to 10 percent.

Step 4. The test must be related to a real (we emphasize real) problem, that is, to a problem that exists and not to a problem that you think exists. This observation is so obvious that it would be superfluous except for the fact that it often crops up to plague us. Take, for example, the problem of a gas turbine wherein most of the testing effort was spent on those parts which differentiate the turbine from the piston engine, that is, turbine and compressor rotors, liners, and vanes. The company thought this was the problem. When they finally collected service records, that is, records of premature engine removals, they found that all those elements accounted for only 5 percent of causes for engine removals, while the greatest culprit, bearings, with 47 percent, received no testing attention at all. This was not a well-organized test program.

Step 5. Lastly, what is probably most important, the final data should be properly systematized and analyzed. With the advent of statistical tools we can now do the job much more effectively than in the past. The way test data were handled in the past can be best summarized by quoting from Moroney's book "Facts from Figures:" " Really, the slipshod way we deal with data is a disgrace to civilization. Never have so many data been collected in files and left unanalyzed. Never have so many data been taken out of files and misread." Proper interpretation of data will bring you great rewards. By this we mean proper interpretation of good data. If the test data are bad, even the best statisticians will not be able to do much with it. This means we must have good data and good interpretation of these data: in short, we must have a *good* experiment.

Thus, it can truly be said that any test, regardless of its nature, can only be as good as the care taken in its preparation.

This book is an attempt to provide means for conducting a good experiment and for the analysis of the experimental data.

1
Statistical Tools

The purpose of this chapter is to introduce the basic concepts of statistical tools necessary for designing experiments.

No two parts produced are exactly the same, no matter how carefully they are made. Although the differences may be small, they nevertheless exist, and, in turn, they have an important bearing on the design of experiments. We recognize this by employing statistical tools.

This variability, or scatter, faces the test engineer, and it complicates his task. If the scatter did not exist, that is, if all parts produced were exactly the same, his task would be simple. He would take only one item, and the conclusions drawn from this single test would be applicable to the whole production lot, that is, to the population.

Unfortunately, the parts are not the same. He therefore takes several items and conducts the experiment, and the conclusions drawn from this test he applies to the whole population.

Now how valid is this application? We intuitively feel that if the sample were large, say 100 gears or 100 bearings, we would be reasonably certain of the conclusions drawn. But in engineering, generally for economic reasons, we are faced with small samples, say 5 to 10 gears, or only one prototype housing. How valid is a conclusion drawn from such a small sample? This is illustrated by the following example.

Example 1.1 Sixty bearings were picked at random from a lot of one thousand bearings. They were divided into six groups of 10 bearings and tested to failure.

Sample number	Number of bearings tested	Average life of sample, 10^6 cycles
1	10	2.9
2	10	8.1
3	10	0.7
4	10	0.9
5	10	10.0
6	10	4.5

In this example the following will be noted:

1. The lives of so-called identical bearings under identical testing conditions show considerable scatter.
2. The sample averages vary considerably, and it would be difficult to predict which of the sample averages is representative of the average population life.

Consider another example:

Example 1.2 Two different gear designs A and B are to be compared. A sample of 10 gears from each design is tested with the following results:

Number of hours to failure

Gear design A	Gear design B
120	110
150	160
210	350
250	400
260	410
270	430
510	500
870	520
980	530
1,140	570
Average life = 476 h	*Average life = 398 h*

Comparing the average lives of the two designs, we find that A appears to be substantially better than B. However, from the raw data, it is noted that six gears, or 60 percent of gear design A, failed with a life less than the average of 476 h. In design B only three gears, or 30 percent of the sample, failed with a life less than the average of 398 h. Thus, the average values are not necessarily a good index of the quality of the two sets of data.

These examples lead to the fundamental problem with which the test engineer is faced: Given a small random sample out of a much larger set (known as the population or universe), how may the information derived from the sample be used to obtain valid conclusions about the population? This kind of problem can be treated only with the aid of statistical tools. The following sections deal with some of the basic concepts.

1.1 THE CONCEPT OF RANDOM VARIABLE

In Example 1.1 it was noted that even though identical samples were tested under identical conditions, inherent scatter was observed. The samples were chosen at random, and the average lives obtained can be considered also random.

Since average lives of the samples cannot be less than zero, then of all possible outcomes the values range from zero to infinity. If, on the other hand, one were to experiment with the numbers on the upturned face of a fair die, all possible outcomes would consist of the numbers 1, 2, 3, 4, 5, 6. Thus, to each face of the die, a number has been assigned. This number is known as a random variable. In the examples on bearing lives, the random variable was the life expressed in number of cycles.

In the case of the upturned fair die, the faces correspond to the sample points or set of outcomes $A_1, A_2, \ldots, A_6$, and the numbers on these faces correspond to the numbers x_i assigned to each A_i by a random-variable rule. Hence, the numbers 1, 2, $\ldots$, 6 are the values of the random variables.

This example should not lead one to believe that a random variable can be obtained only from experimentation. In fact, any function of a random variable is also a random variable. For instance, in the case of the bearings, the random variable was the life in cycles. The sample average life obtained is another random variable since it was derived from the random variable, the bearing life.

Thus, random variables can be classified into two categories: (1) continuous and (2) discrete. The random variable associated with the bearings is a continuous random variable, since bearing life can attain any value continuously between zero and infinity. The random variable in the case of the die experiment is discrete because it can attain only one of the six discrete values 1, 2, 3, 4, 5, or 6.

1.2 PROBABILITY

So far, all possible outcomes have been discussed without regard to the fact that they may not occur with the same frequency. That is, some of them may occur more frequently than others. This concept can be expressed as follows:

$$\text{Probability of occurrence} = \frac{\text{number of successful occurrences}}{\text{total number of trials}}$$

$$= \lim_{n \to \infty} \frac{m}{n}$$

$$= p(x) \qquad \text{for continuous random variable}$$
$$= p(x_i) \qquad \text{for discrete random variable}$$

$$(1.1)$$

The following are the properties associated with probability:

1. Probability is a nonnegative number. This is because neither m nor n can be negative.
2. If an event is certain, then m and n are equal and the probability is unity.
3. If the events A and B are mutually exclusive, then the probability of occurrence of the event A or B is the probability of A plus the probability of B. Mathematically,

$$p(A + B) = p(A) + p(B) \qquad (1.2)$$

In the case of the number on the upturned face of a fair die, these numbers are mutually exclusive, since the occurrence of one automatically excludes the possibility of the occurrence of any other. In such a case, the probability of occurrence of number 1 or $4 = p(1) + p(4) = \frac{1}{6} + \frac{1}{6} = \frac{1}{3}$.
4. If events A and B are independent of each other (i.e., the occurrence of one does not affect the occurrence of the other), then the probability that both A and B will occur is the probability of A times the probability of B, or

$$p(AB) = p(A) \times p(B) \qquad (1.3)$$

Thus, the probability that 2 and 5 will occur in two dice simultaneously is

$$p(2,5) = p(2) \times p(5) \times 2$$
$$= \tfrac{1}{6} \times \tfrac{1}{6} \times 2$$
$$= \tfrac{1}{36} \times 2$$
$$= \tfrac{1}{18}$$

Similar examples can be offered in the case of continuous variables.

1.3 DENSITY FUNCTION

The density function can be defined as the function which yields the probability that the random variable takes on any one of its admissible values. Consider the data in Example 1.3, which represents the lives of 35 compression springs. The table below gives the number of failures obtained within each life interval.

Example 1.3

Life interval, 10^6 cycles	Number of failures
0–1	0
1.1–2	2
2.1–3	5
3.1–4	7
4.1–5	5
5.1–6	5
6.1–7	4
7.1–8	4
8.1–9	2
9.1–10	1
	Total 35

These results are reproduced in Fig. 1.1. In this histogram the base of each vertical rectangle represents the interval of life, and the height the number of failures observed during that life. It will be noted that the frequency of failure

$$\frac{\text{Number of failures during an interval}}{\text{Total number of items on test}}$$

is not the same for all life intervals. There is a high concentration of failures between 3.0×10^6 and 4.0×10^6 cycles. If a much larger number of springs were tested, a smooth curve could be drawn through the peaks of the rectangles. This curve would then represent probability, that is, frequency of failure. The curve would thus be a density function of the compression springs. In general, a density function can be represented as in Fig. 1.2.

It will be noted in both Figs. 1.1 and 1.2 that (1) the probability of occurrence is not constant over the whole range of values attained by the variable and (2) even though the value of probability is not a constant, there seems to be a definite trend involved.

All the above pertained to the continuous random variable. Next, consider the example of the number on the upturned face of a fair die. The probability of occurrence in this case is always constant and equals $\frac{1}{6}$, as shown in Fig. 1.3. In either case, there is an advantage in having an analytical

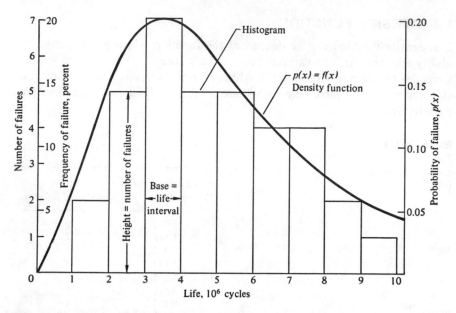

Fig. 1.1   Histogram of compression-springs failure data.

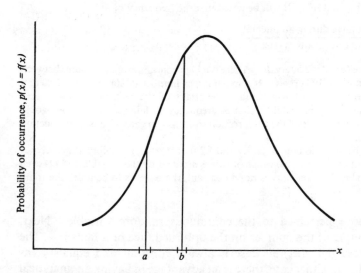

Fig. 1.2   Density function for a continuous random variable.

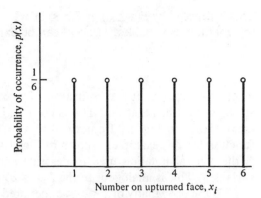

Fig. 1.3 Density function for a discrete random variable of a fair die.

expression which relates the value of $f(x)$ (the probability of obtaining the value x) with x (the random variable). Such a relation has the form $p(x) = f(x)$, and it is called the density function.

There are many well-known theoretically established density functions. These functions may have a wide variety of shapes. This is because in actual practice the random variable arises from a variety of causes, and these causes govern the probability of occurrence of the random variable. The principal functions are discussed in Chap. 2.

1.4 CUMULATIVE DISTRIBUTION FUNCTION

Plotting the density function, as in Fig. 1.2, is not the only method of presenting the experimental data. There is another method of plotting data which is easier to evaluate. This method involves plotting of the cumulative distribution function versus the random variable.

The cumulative distribution function $F(x)$ is defined as

$$F(x \leq a) = \sum_{i=-\infty}^{a} p(x_i) \qquad \text{where } x_i \leq a \tag{1.4}$$

If $p(x)$ is continuous, then

$$F(x \leq a) = F(a) = \int_{-\infty}^{a} p(x)\, dx = \int_{-\infty}^{a} f(x)\, dx = P(x \leq a) \tag{1.5}$$

For a discrete variable,

$$F(x \leq a) = p(x_1) + p(x_2) + \cdots + p(a)$$

$$= \text{probability that } x \text{ is less than or equal to } a \tag{1.6}$$

This is denoted by $P(x \leq a)$ or $F(x \leq a)$.   In the case of Example 1.3, involving continuous variables, $F(x \leq a)$ can be expressed as

$$\int_0^a f(x) \ dx$$

where the lower limit is zero instead of $-\infty$ since the life of a compressor spring cannot be negative.   This is nothing but the area under the frequency curve shown in Fig. 1.2 from $x = 0$ to $x = a$.   The cumulative distribution function for this example is shown in Fig. 1.4.   It should be noted that $F(x)$ is the general notation used to describe such a function for a continuous variable.   This is analogous to the $F(x \leq a) = P(x \leq a)$ in the discrete case.

If the probability of $a \leq x \leq b$ [denoted by $P(a \leq x \leq b)$] is to be found, then

$$P(a \leq x \leq b) = P(x \leq b) - P(x \leq a)$$
$$= F(b) - F(a)$$

In the case of a discrete variable, as in the case of the example of the number on the face of an upturned fair die (Fig. 1.3),

$$F(x \leq 4) = p_1 + p_2 + p_3 + p_4$$
$$= \tfrac{1}{6} + \tfrac{1}{6} + \tfrac{1}{6} + \tfrac{1}{6}$$
$$= \tfrac{2}{3}$$

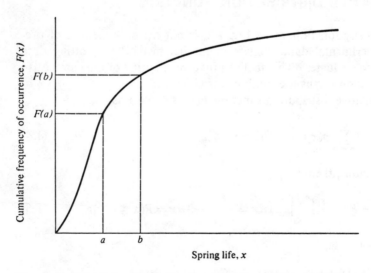

Fig. 1.4   Cumulative distribution function for a continuous random variable.

In Fig. 1.5, the cumulative distribution function for this example is plotted.

Some of the properties of the CDF are:

1. Since $p(x_i) \geq 0$ and

$$F(a) = \sum_{i=-\infty}^{a} p(x_i)$$

then $F(x)$ is a nondecreasing function of x.

2. $F(+\infty) = \sum_{i=-\infty}^{+\infty} p(x_i) = 1$

In the case of the spring this suggests that the area under the frequency distribution of Fig. 1.2 is unity. In the case of the fair die (Fig. 1.3):

$$F(+\infty) = \sum_{i=0}^{6} p(x_i) = p(x_1) + p(x_2) + p(x_3) + p(x_4) + p(x_5) + p(x_6)$$
$$= \tfrac{1}{6} + \tfrac{1}{6} + \tfrac{1}{6} + \tfrac{1}{6} + \tfrac{1}{6} + \tfrac{1}{6}$$
$$= 1$$

3. $F(-\infty) = \lim_{b \to -\infty} F(b) = 0$

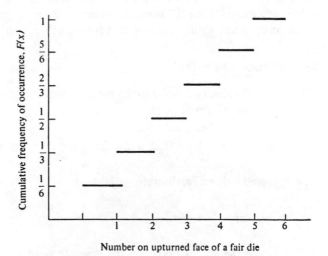

Number on upturned face of a fair die

Fig. 1.5 Cumulative distribution function for a discrete random variable.

1.5 SAMPLE AND POPULATION

A sample is defined as a group of items under test, selected randomly from a lot of similar items of infinite size. (See also definition in the Glossary.) This lot is known as the population. In most engineering situations the problem is to predict the behavior of the population from the knowledge of a sample whose characteristics are experimentally determined.

The concept of sample and population can be illustrated by the data in Example 1.3.   Comparison of the density-function curve with the histogram in Fig. 1.1 shows that the histogram matches the density curve quite well in most places.   Yet there are some areas where the histogram, derived from the sample of 35 springs, does not truly represent the behavior of all the springs (population).   "How much confidence can one have in stating that the sample is truly representative of the population?" is a question which shall be answered in later chapters.   At this time it is sufficient to recognize the basic difference between the population and the sample.

1.6 MEASURES OF CENTRAL TENDENCY OF A POPULATION

In the case of a random variable, whether discrete or continuous, one frequently searches for a "typical value."   In Fig. 1.1 it was noted that the random variable, life, could take any value from zero to infinity, but it is obvious that some values are more likely to occur than others.   That is, some values are more "typical" than the others.   These typical values, known as mean or expected value, median, and mode, are defined below.

1.6.1 POPULATION MEAN

The expected value of a continuous random variable x with density function $f(x)$, such that $p(x) = f(x)$, is

$$\text{Mean or expected value} = \mu = E(x) = \int_{-\infty}^{\infty} xf(x) \ dx \tag{1.7}$$

The expected value of a discrete variable is

$$\mu = E(x) = \sum_{i=1}^{n} x_i p(x_i) \tag{1.8}$$

where x_i are the values taken by the discrete random variable, and $p(x_i)$ is the probability of occurrence associated with x_i.   The expected value μ is generally referred to as the mean of the population.

As an example, consider the numbers on the upturned face of a fair die. In this case the expected value is

$$E(x) = \sum_{i=1}^{6} x_i p(x_i)$$
$$= (1)(\tfrac{1}{6}) + (2)(\tfrac{1}{6}) + (3)(\tfrac{1}{6}) + (4)(\tfrac{1}{6}) + (5)(\tfrac{1}{6}) + (6)(\tfrac{1}{6})$$
$$= 3.5$$

1.6.2 POPULATION MEDIAN

In the case of a continuous random variable x, the median or the x_{50} value of a variable where the density function is $p(x) = f(x)$ is

$$0.5 = \int_{-\infty}^{x_{50}} f(x) \, dx \tag{1.9}$$

That is, the population median is the value of x at which the cumulative distribution function $F(x)$ has a value of 0.5.

1.6.3 POPULATION MODE

In the case of a continuous random variable x with distribution $p(x) = f(x)$, the population mode is that value of x at which

$$\frac{dp(x)}{dx} = \frac{df(x)}{dx} = 0 \tag{1.10}$$

That is, the mode is the value of x corresponding to the peak value of the probability of occurrence of any continuous or discrete distribution.

1.7 MEASURES OF CENTRAL TENDENCY OF A SAMPLE

Consider Example 1.1 on bearing lives. Even though the lives occurred over a wide band, one would like to find out the single value which best represents these lives. Thus, a numerical value of the central tendency in this distribution is desired. As in the case of population, there are three different and well-established measures of central tendency for a given sample.

1.7.1 SAMPLE MEAN

The mean, or arithmetic average, is defined as

$$\bar{x} = \frac{\sum_{i=1}^{n} x_i}{n} \tag{1.11}$$

This is a well-known measure of central tendency. Thus, the average of 10, 12, 8, 9, 10, 11, 7, 5 is

$$\frac{10 + 12 + 8 + 9 + 10 + 11 + 7 + 5}{8} = 9$$

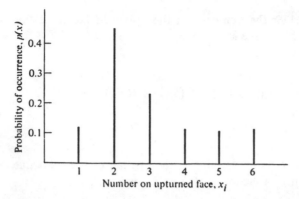

Fig. 1.6 Density function for a discrete random variable
of an unfair die (compare with Fig. 1.3).

1.7.2 SAMPLE MEDIAN

The sample median is the number in the middle when all the observations
are ranked in their order of magnitude.   For example:

The median of 2, 7, 11, 15, 18, 20, 21, is 15.
The median of 6, 10, 14, 18, 19, 20 is $(14 + 18)/2 = 16$.

1.7.3 SAMPLE MODE

In the case of a continuous variable, as in Fig. 1.1, the mode is the value
of the variable corresponding to the maximum probability of occurrence
(about 3.5×10^6 cycles in Fig. 1.1).   In the case of the discrete distribution,
as in Fig. 1.6, the mode is 2.

1.8 MEASURES OF VARIABILITY OF A POPULATION

A population can generally be described in terms of two parameters: central
tendency (generally, mean) and variation about that mean.   Variation is
usually expressed by a quantity known as the variance or the standard deviation, where the standard deviation is the square root of the variance.

In the case of a continuous random variable, the variance is

$$\sigma^2 = E[(x - \mu)^2] = \int_{-\infty}^{\infty} (x - \mu)^2 f(x)\, dx \qquad (1.12)$$

where x is the random variable and $p(x) = f(x)$.   The population standard
deviation $\sigma = \sqrt{\sigma^2}$, where σ^2 is as defined above.

In the case of a discrete random variable, the variance is

$$\sigma^2 = E[(x - \mu)^2] = \sum_{i=1}^{n} (x_i - \mu)^2 p(x_i) \tag{1.13}$$

where x_i is the value of the random variable and $p(x_i)$ is the probability associated with it.

1.9 MEASURES OF VARIABILITY OF A SAMPLE

The simplest way to specify variability would be to state the range which encompasses all the observations. In the case depicted in Fig. 1.7, this range is 0 to 140 s. Where there appears to be no set central tendency, this method is as good as any. In cases which do exhibit central tendency, there are two ways to measure the variability: (1) through the estimate of the population standard deviation; (2) by the variance, which is the square of the standard deviation.

The standard deviation of a sample is

$$s = \sqrt{\frac{\sum_{i=1}^{n} (x_i - \bar{x})^2}{n - 1}} \tag{1.14}$$

s is known as the sample standard deviation, and s^2 the sample variance. This is generally referred to as an unbiased estimate of the population standard deviation.

Three forms of the above equation are

$$s = \sqrt{\frac{\sum_{i=1}^{n} (x_i - \bar{x})^2}{n - 1}} \tag{1.15}$$

$$s = \sqrt{\frac{n \sum_{i=1}^{n} x_i^2 - \left(\sum_{i=1}^{n} x_i\right)^2}{n(n - 1)}} \tag{1.16}$$

$$s = \sqrt{\frac{\sum_{i=1}^{n} x_i^2 - n\bar{x}^2}{n - 1}} \tag{1.17}$$

Equations (1.14) to (1.17) are generally referred to as unbiased forms of the standard deviation. In some cases the root-mean-square (rms) value, known as the biased estimate of standard deviation, is used instead. In this case, $n - 1$ in the above equations is replaced by n.

If a fixed quantity is added to or subtracted from each of the observations, the mean will be increased or decreased by that amount, but the variance and the standard deviation will not be affected. If each of the observations is multiplied by a fixed quantity m, the new mean is m times the old mean, the new variance is m^2 times the old variance, and the new standard deviation is m times the old standard deviation.

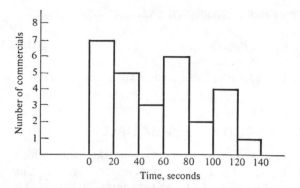

Fig. 1.7 Frequency distribution of duration times of television commercials.

1.10   THE CONCEPT OF CONFIDENCE LEVEL

A concept of considerable importance in the design of experiments is one of confidence levels or confidence intervals. Consider the lot of the 1,000 bearings in Example 1.1. What is the mean life μ of these bearings? The bearings may have a distribution such as in Fig. 1.8. However, the true shape of this distribution is not known unless all 1,000 bearings are run to failure. This is frequently an impossible task. The next best thing is to obtain a small sample of n items and find $\bar{x}$, the sample mean. Depending on the type of the distribution function, the value of the other parameters of this distribution, the number of items involved, etc., the value of the sample mean $\bar{x}$ may fall somewhere near the value of the population mean μ. How-

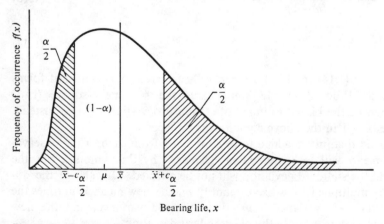

Fig. 1.8 The concept of confidence level and confidence interval.

ever, the chances of finding an $\bar{x}$ which exactly equals μ are very small. Therefore, an interval $(\bar{x} - C_{\alpha/2})$ to $(\bar{x} + C_{\alpha/2})$ is picked which will contain μ. The degree of confidence or the probability that this interval will contain the population mean μ is $1 - \alpha$. This is known as the confidence level (Fig. 1.8). The interval $[(\bar{x} - C_{\alpha/2})$ to $(\bar{x} + C_{\alpha/2})]$ is the confidence interval.

It should be noted that a parameter such as μ is a fixed constant which depends only on $f(x)$. It is the values of the confidence intervals that fluctuate, depending on the sample chosen.

Similarly, the probability (confidence level) that x will be larger than X is α, and smaller than or equal to X is $1 - \alpha$ (Fig. 1.9). In all the above, it is understood that the confidence levels are the areas under the density curve; the total area under the curve is always equal to 1.0.

1.11   THE CONCEPT OF RELIABILITY

In this text confidence and reliability are interrelated, particularly in the solved examples. Confidence was defined as the probability that a given interval determined from the test data will contain the population parameters (such as μ and σ). Reliability, as defined here, is the probability that a part, system, or assembly will perform the intended function, in a given environment, for a predetermined period of time, without failure. Numerically, reliability is the percent survivors.

Reliability and confidence are related when one attempts to project reliability determined from a test sample into the population, since reliability applies to the part, and confidence to the test itself. Suppose a part has 90 percent reliability at 80 percent confidence. This means that when, say, a sample of 20 parts is tested, 2 parts will fail (10 percent failure, or 90 percent reliability). When, say, 10 such samples of 20 parts each are tested, 8 out

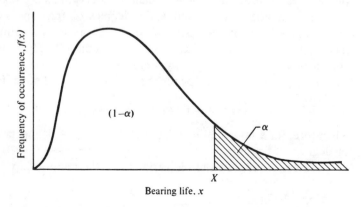

Fig. 1.9   Confidence level.

of 10 samples will have two failures or less (80 percent confidence). The relationship between reliability and confidence is generally expressed by Bayes' theorem [Eq. (5.30)].

1.12   DEGREES OF FREEDOM

Degrees of freedom v can be defined as the number of observations made in excess of the minimum theoretically necessary to estimate a statistical parameter or any unknown quantity. For example, one measurement is theoretically sufficient to measure the diameter of a shaft. If eight measurements are made instead, then the system of eight measurements has seven degrees of freedom, because there are seven more measurements made than are required to determine the shaft diameter. In general, if n measurements are made on m unknown quantities, then the degrees of freedom are $n - m$. For example, the degrees of freedom in Eqs. (13.30) and (13.16) are $n - 3$ and $n - 2$ since the number of parameters to be estimated is three and two, respectively. In Eqs. (3.5) and (3.16), $v = n - 1$ since only one parameter is to be estimated.

Another way of defining degrees of freedom is as follows: Say n observations are made: $x_1, x_2, x_3, \ldots, x_n$. The mean, from Eq. (1.11), is

$$\bar{x} = \frac{\sum_{i=1}^{n} x_i}{n}$$

Degrees of freedom are then defined as the number of independent observations available to measure the deviation of x_i from the mean $\bar{x}$. When the number of observations is n, the total number of deviations is n ($x_1 - \bar{x}$, $x_2 - \bar{x}, x_3 - \bar{x}, \ldots, x_n - \bar{x}$), out of which only $n - 1$ deviations are independent since all n deviations should sum to zero by definition of the mean. Hence the degrees of freedom are $n - 1$ when n observations are taken.

Still another way to define the degrees of freedom is as $(n - 1)$ independent contrasts out of n observations. Suppose in a sample with $n = 5$, measurements x_1, x_2, x_3, x_4, and x_5 are made. Then the independent contrasts are $x_1 - x_2, x_2 - x_3, x_3 - x_4$, and $x_4 - x_5$. The additional contrast $x_1 - x_5$ is not independent since its value is known from

$$(x_1 - x_2) + (x_2 - x_3) + (x_3 - x_4) + (x_4 - x_5) = x_1 - x_5$$

Therefore, for a sample of $n = 5$, there are four $(n - 1)$ independent contrasts or degrees of freedom.

In calculating the sample variance s^2, the relationship

$$\sum_{i=1}^{n} \frac{(x_i - \bar{x})^2}{n - 1}$$

is used, since $\bar{x}$ is taken as an estimate of μ; therefore, $m = 1$. Hence degrees of freedom equal $n - m = n - 1$. However, if the population mean μ is known, every measurement provides a deviation (measured value − mean), and the degrees of freedom available to estimate the variance s^2 equal the number of measurements n.

In the case of regression analysis (Sec. 13.1), if it takes a multiparameter model to express the observed test results, the degrees of freedom to establish correlation are $n - m$, where m is the number of regression coefficients a_i :

$$y = a_1 + a_2 x_1 + a_3 x_2{}^2 + a_4 x_1 x_2 + a_5 x_3{}^4 + \cdots$$

The degrees of freedom also enter:

1. In the calculation of the confidence intervals of the regression coefficients a_i to determine whether the coefficients are significantly different from zero at some level of confidence
2. In testing the correlation coefficient r to conclude with some level of confidence whether r is significant
3. In calculating the confidence interval containing the true value of y so as to compare the predicted with experimental values

1.13 THE CONCEPT OF RANKING

When a large sample is available, the usual method of estimating population from the sample is to reduce the data into a frequency-histogram form. However, in most engineering situations only a small sample is available; therefore, the shape of the histogram varies considerably with change in class interval. Hence, for these situations a cumulative distribution plot is preferred. This involves plotting the observations on the abscissa against the rankings of these observations on the ordinate. It would appear that if a sample of five measurements were taken, and if these measurements were arranged in an increasing order, then the rank of the first (lowest) measurement would be 20 percent, the second 40 percent, etc. This, however, would imply that 20 percent of the whole population would have a value less than the lowest of the five observed. This may or may not be true; therefore it is necessary to resort to a statistical estimate as to the fraction of the population that the lowest of five measurements does represent.

1.13.1 MEDIAN RANKS

Suppose a sample of five items is picked randomly from a population whose density function is known (Fig. 1.10). These items are tested, say, to failure (although the concept of ranking applies to any measurements, whether failures are involved or not), and the values of lives are recorded.

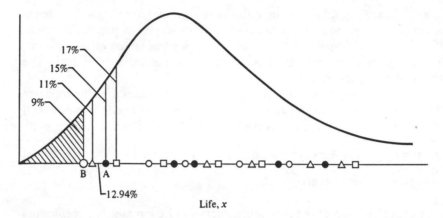

17%

15%

11%

9%

B   A

—12.94%

Life, x

Fig. 1.10 The concept of median ranks.

Thus, in sample 1 (of five items) the first failure may have occurred at A (Fig. 1.10), where 15 percent of the population has values less than A. If another sample of five were taken and the test repeated, the lowest life B might represent only 9 percent of the population.

When this is repeated many times, the data generate a series of such percentage numbers randomly distributed. The median of these numbers is then used as a true rank of the first failure out of the group tested. This is known as median, or 50 percent rank. In Fig. 1.10 a median rank of 12.94 percent is indicated for the first failure out of five.

In the same manner, median ranks for the second, third, fourth, and the fifth failure out of five can be estimated. These values are given in Tables A-11 to A-13 for various sample sizes n and various failure orders j. Thus, for example, for the sample size $n = 5$, the first failure represents 12.94 percent of the population, the second 31.47 percent, and so on. To obtain a cumulative-distribution-function graph, these median ranks are plotted on the ordinate against lives (or any other measurements) arranged in an increasing order on the abscissa.

The following approximate formula for the calculation of median ranks may be used when the tables are not available, or when sample size is beyond the range covered by the table:

$$\text{Median or } 50\% \text{ rank} = \frac{j - 0.3}{n + 0.4}$$

where j = failure order number
$\quad n$ = sample size

Another method of ranking involves mean ranks instead of median ranks. The latter, however, have found much wider application in engineering practice.

1.13.2 OTHER RANKS

The concept of ranking can be extended to any other ranks such as 5, 10, 90, or 95 percent. Thus, for example, the 5 percent rank of the lowest of seven measurements is 0.74 percent (Table A-14). This means that only in 5 percent of the cases the lowest of seven measurements would represent less than 0.74 percent of the population; in 95 percent of cases the lowest of seven would obviously represent more than 0.74 percent of the population. Similarly, for the 95 percent rank the lowest of seven represents 34.82 percent of the population. This means that in 95 percent of the cases the lowest of seven would represent as much as 34.82 percent of the population. Only in 5 percent of cases the lowest of seven would represent even more than 34.82 percent of population. The 0.5, 2.5, 5, 50 (median), 95, 97.5, and 99.5 percent ranks are given in Tables A-11 to A-16. The mathematical derivations of these ranks can be found in Ref. 1.

1.13.3 SUSPENDED ITEMS

The concept of suspended items is very useful in engineering experiments involving the lives of components or systems, although it can be applied to any set of measurements. The word "suspended" describes items which have not failed or, in general, which have not been measured or whose measurements have not been completed.

Consider a test on four gears, three of the four gears having failed with the following results:

Gear 1. Failure at 3,500 h
Gear 2. Test suspended at 4,000 h
Gear 3. Failure at 5,100 h
Gear 4. Failure at 6,700 h

Failure at 3,500 h represents the lowest life in this set of measurements and therefore can be described as failure order 1. However, failure at 5,100 h may be failure order 2 or 3, depending on item 2 under the test. That is, if the test were allowed to continue and failure occurred at, say, 5,400 h, failure at 5,100 h would be failure order 2. If, however, the failure occurred at 4,800 h, the 5,100 h failure would be failure order 3. Thus, the failure of gear 3 would be somewhere between failure orders 2 and 3.

This, then, indicates that, in general, the order numbers of all failed gears following a suspended item will have noninteger order numbers. In order to determine an order number of the first failure following any suspension, it is convenient to compute an increment to be added to the previous mean order number. This increment is given by the general formula

$$\text{New increment} = \frac{(n + 1) - \text{previous failure order number}}{1 + \text{number of items following present suspended set}}$$

where n is the total sample size on test.   This increment is used on all failures following a suspended item until another suspended item is encountered. Then a new increment must be computed corresponding to this new suspended item.   An example using this technique is given in Chap. 5.

SUMMARY

Basic statistical tools necessary for the design and analysis of engineering experiments are presented in this chapter.   These tools, in various degrees, are used in all the chapters.   For a quick reference they are symbolically noted in the List of Symbols and are defined in the Glossary.   Their use and location in the book can be found in the Index.

PROBLEMS

1.1.   A sample of 10 items was picked at random from a given population, and the following observations were made:

    5, 8.5, 7, 6.5, 5, 7.5, 8, 6, 5.5, 6

Compute sample mean $\bar{x}$, sample standard deviation s, and sample variance s^2.

1.2.   Using the data of Prob. 1.1:

    (a) Subtract 5 from each of the observations and compute $\bar{x}$, s, and s^2.   Compare the results with the results of Prob. 1.1.

    (b) Multiply each of the observations by 10 and find $\bar{x}$, s, and s^2.   Compare the results with those of Prob. 1.1.

1.3.   Select in any sequence 50 or more random numbers from Table A-48 and construct the frequency histogram, using the abscissa interval of magnitude 0.4.   Determine which number occurs most frequently.

1.4.   Consider three fair dice.   When these dice are thrown simultaneously, what is the probability that the following numbers will appear on their upturned faces: 1, 1, 1; 6, 6, 6; 2, 4, 5; and 4, 4, 5.

1.5.   Assign 2.5, 5, 50, 95, and 97.5 percent ranks to the following life data in cycles: 1.7×10^5, 2×10^6, 3.2×10^4, 7×10^5, 1.5×10^4, 5×10^5, 4×10^6, 3×10^5, 2.1×10^4, and 8×10^4.

1.6.   Assign 5, 50, and 95 percent ranks to the following data:

Item 1. Failure at 400 h
Item 2. Test suspended at 420 h
Item 3. Test suspended at 450 h
Item 4. Failure at 500 h
Item 5. Failure at 510 h
Item 6. Test suspended at 515 h
Item 7. Failure at 520 h

1.7.   In a certain city the law requires that the waste water must be purified before it is dumped in a river.   One of the water pollutants is oil and grease, and its concentration in the waste water can be 100 ppm (parts per million) or more.   A purification process was introduced that brings the concentration level down to an acceptable level.   The oil and grease concentration in the purified water was measured periodically and the following was found, in ppm: 14.0, 15.0, 14.7, 15.2, 15.5, 14.5, 14.8, 15.1, 15.2, 15.3, 15.7.

(*a*) Compute the average concentration level of oil and grease.

(*b*) Compute the standard deviation of these measurements. (This standard deviation can be considered as a measure of the variation in the purification efficiency of the process.)

1.8. A water purification process reduced the phenol concentration level of 400 ppb (parts per billion) to the following level: 20, 17, 18, 19, 17, 17.3, 18, 21, 20.5, 19.5. Compute the average and the standard deviation of these levels.

REFERENCE

1. Johnson, L. G.; The Median Ranks of Sample Values in Their Population with an Application to Certain Fatigue Studies, *Ind. Math.*, vol. 2, 1951.

BIBLIOGRAPHY

Bowker, A. H., and G. J. Lieberman: "Engineering Statistics," Prentice-Hall, Inc., Englewood Cliffs, 1963.

Dixon, W. J., and F. J. Massey, Jr.: "Introduction to Statistical Analysis," 3d ed., McGraw-Hill Book Company, New York, 1969.

Guttman, I., and S. S. Wilks: "Introductory Engineering Statistics," John Wiley & Sons, Inc., New York, 1965.

Mood, A. M., and F. A. Graybill: "Introduction to the Theory of Statistics," 2d ed., McGraw-Hill Book Company, New York, 1963.

Moroney, M. J.: "Facts from Figures," Penguin Books, Inc., Baltimore, 1965.

Natrella, M. G.: "Experimental Statistics," *Nat. Bur. Stand. Hand.* 91, Aug. 1, 1963.

Wallis, W. A., and H. V. Roberts: "Statistics: A New Approach," The Free Press, New York, 1956.

2
Statistical Distributions

In Chap. 1 a density function was defined as a function which relates the random variable x to its probability of occurrence $f(x) = p(x)$. One way to arrive at such a function is to plot experimental data in the form of a histogram. However, this procedure has two drawbacks: (1) Experimental data give good information regarding the distribution of the sample. However, when this information is extrapolated to the distribution of the population, only a limited amount of confidence can be placed in the result. (2) The distribution function, derived experimentally, is usually awkward and difficult to handle in subsequent manipulations.

These drawbacks are overcome by the use of standard or "ready-made" distributions. These distributions are known mathematically, and their values are usually available in tabulated forms. Then, instead of determining a distribution function from experimental data, one finds a standard distribution which best fits or represents his observed data.

2.1 NORMAL DISTRIBUTION

The normal distribution is the best-known and most widely used distribution. It was discovered in the eighteenth century by Gauss, Laplace, and De Moivre. Although all these discoveries were independent of each other, the normal distribution is generally identified as the gaussian distribution or the gaussian error law.

Consider Fig. 2.1. The abscissa in this case is the stress at failure. The left ordinate is the observed number of failures m in a given stress range. This number is converted to the frequency of failure m/n, where n is the total number of trials $= \sum m$. This frequency is plotted on the right-hand ordinate. Such a diagram is called a histogram. Now, if the class intervals are made of smaller width, the jumps in the histogram from one class to the next will become less perceptible, until the histogram has the appearance of a smooth bell-shaped curve, as shown by the dotted line. This curve can best be represented by a mathematical distribution function known as the normal distribution function which is given mathematically by

$$f(x) = \frac{1}{\sigma\sqrt{2\pi}} \exp\left[-\frac{(x-\mu)^2}{2\sigma^2}\right] \qquad -\infty \leq x \leq \infty \qquad (2.1)$$

where x is the random variable, $f(x)$ is the probability of occurrence of x, and μ and σ are constant parameters:

μ = measure of central tendency of population, which is the mean, the median, and the mode

σ = standard deviation of population

Distributions having the same standard deviation but different means are represented in Fig. 2.2; those with the same mean but different standard deviations, in Fig. 2.3.

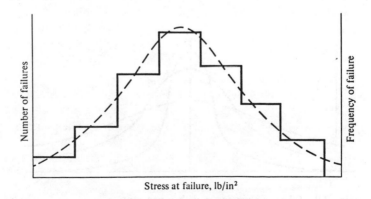

Fig. 2.1 Distribution of the stress at failure.

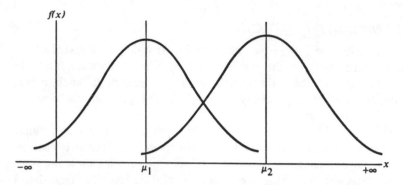

Fig. 2.2 Two normal distributions with the same σ but different means.

2.1.1 THE STANDARDIZED NORMAL VARIATE

Consider the problem of determining the probabilities of occurrence for a certain event. Given a random variable x which has a normal distribution with mean μ and variance σ^2, the probability that x is greater than b is given by the shaded area α under the curve in Fig. 2.4. Mathematically, the probability of $x > b$, or $P(x > b)$, is found by integrating the normal distribution function from b to ∞. That is,

$$P(x > b) = \int_b^\infty \frac{1}{\sigma\sqrt{2\pi}} \exp\left[-\frac{(x-\mu)^2}{2\sigma^2}\right] dx$$

This is not an elementary function, and its integral is not available in a simple form. Hence, it is necessary to resort to a tabulation in order to avoid duplication of labor involved in evaluating this integral. But in this particular form the integral would need a separate table for each (μ, σ) pair chosen.

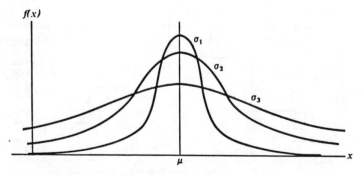

Fig. 2.3 Effect of changing σ, with μ constant.

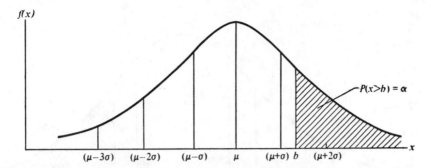

Fig. 2.4 A normal distribution with mean μ and variance σ^2.

This is a difficult task, but fortunately, by a simple transformation, the problem can be reduced to computing a single table for use with all choices of μ and σ. Let

$$\frac{x - \mu}{\sigma} = z \tag{2.2}$$

where z is called the standardized normal variate. Then $dx = \sigma\,dz$.

$$P(x > b) = \int_{(b-\mu)/\sigma}^{\infty} \frac{1}{\sigma\sqrt{2\pi}} \exp\left(\frac{-z^2}{2}\right)(\sigma\,dz)$$

$$= \int_{(b-\mu)/\sigma}^{\infty} \frac{1}{\sqrt{2\pi}} \exp\left(\frac{-z^2}{2}\right) dz$$

$$= \int_{(b-\mu)/\sigma}^{\infty} f(z)\,dz = \alpha \tag{2.3}$$

Then

$$P(x > b) = P\left(z > \frac{b - \mu}{\sigma}\right) = \alpha \tag{2.4}$$

It will be noted that z also has a normal distribution with $\mu = 0$ and $\sigma = 1$. This is known as a standardized normal distribution, or the z distribution. The integral equation (2.3) has been determined, and its values are tabulated against values of $z_\alpha = (b - \mu)/\sigma$, where z_α is a specific value of z.

z_α is known as the upper α percentage point (the cutoff point) or the normal deviate corresponding to α, and it is defined by

$$P(z > z_\alpha) = \int_{z_\alpha}^{\infty} \frac{1}{\sqrt{2\pi}} \exp\left(\frac{-z^2}{2}\right) dz = \alpha$$

$$= \text{shaded area in Fig. 2.5}$$

This area α is tabulated versus z in Table A-1.

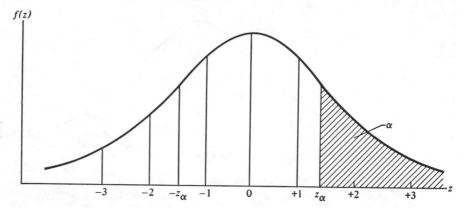

Fig. 2.5 Standardized normal distribution.

It will be noted that

$$P(x > b) = \text{shaded area in Fig. 2.4}$$
$$= P(z > z_\alpha)$$
$$= \text{shaded area in Fig. 2.5} \tag{2.5}$$

Hence, to evaluate $P(x > b)$, one evaluates the shaded area in Fig. 2.5, since this value is already given in Table A-1.

Example 2.1 Consider a normal distribution with $\mu = 20$, $\sigma = 30$, $b = 80$. Compute the following:

1. $P(x > 80)$
2. $P(x \leq 80)$
3. $P(x \leq -80)$
4. $P(50 \leq x \leq 80)$

Solution

1. The probability that $x > 80 = P(x > 80) = P(z > z_\alpha)$ from Eq. (2.5), where

$$z_\alpha = \frac{b - \mu}{\sigma} = \frac{80 - 20}{30} = \frac{60}{30} = 2.0$$

From Table A-1, $\alpha = 0.0228$ for $z_\alpha = 2.0$. Therefore
$$P(x > 80) = 0.0228$$
$$= 2.28\% \quad \text{(Fig. 2.6}a)$$

2. The probability that $x \leq 80$ can be found from the relation
$$P(x \leq b) + P(x > b) = 1$$
$$P(x \leq b) \ = 1 - P(x > b)$$
$$P(x \leq 80) = 1 - P(x > 80)$$
$$= 1 - 0.0228$$
$$= 0.9772$$
$$= 97.72\% \quad \text{(Fig. 2.6}a)$$

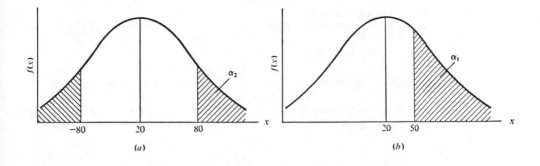

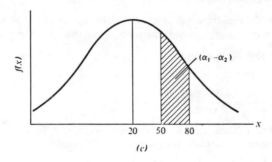

Fig. 2.6 Frequency distribution for Example 2.1.

3. The probability that $x \leq -80$ is obtained from the symmetry of Fig. 2.6a is

$$P\left(z \leq \frac{-80 - 20}{30}\right) = P(z \leq -3.33)$$

$$
\begin{aligned}
P(z \leq -3.33) &= P(z > 3.33) \\
&= 0.000434 \\
P(x \leq -80) &= 0.0434\% \qquad \text{(Fig. 2.6a)}
\end{aligned}
$$

4. The probability that $50 \leq x \leq 80$ is given in Fig. 2.6c as the area under the curve between 50 and 80. This is the same as the area under the curve from 50 to ∞ (Fig. 2.6b) minus the area from 80 to ∞ (Fig. 2.6a). Mathematically,

$$
\begin{aligned}
P(50 \leq x \leq 80) &= P(x \geq 50) - P(x > 80) \\
&= P(z \geq z_{\alpha_1}) - P(z > z_{\alpha_2})
\end{aligned}
$$

$$z_{\alpha_1} = \frac{50 - \mu}{\sigma} = \frac{50 - 20}{30} = 1.0 \qquad \text{thus } \alpha_1 = 0.1587$$

$$z_{\alpha_2} = \frac{80 - \mu}{\sigma} = \frac{80 - 20}{30} = 2.0 \qquad \text{thus } \alpha_2 = 0.0228$$

Thus

$$P(50 \leq x \leq 80) = \alpha_1 - \alpha_2$$
$$= 0.1587 - 0.0228$$
$$= 0.1359$$
$$= 13.59\%$$

Example 2.2 In one manufacturing process, shafts of less than 1.5 in in diameter are usable in production. This dimension is normally distributed with mean 1.490 in and standard deviation 0.005 in. What percent of the shafts are usable for production?

Solution

$$\mu = 1.490 \text{ in} \qquad \sigma = 0.005 \text{ in}$$
$$b = x_c = \text{critical diameter} = 1.5 \text{ in}$$

where x_c denotes the cutoff point of the distribution of x. Figure 2.7 illustrates the situation graphically. If x is the random variable, then the probability of scrap is

$$P(x > 1.5) = P(z > z_a) = \alpha$$

but

$$z_a = \frac{x_c - \mu}{\sigma} = \frac{1.5 - 1.490}{0.005} = 2.0$$

From Table A-1 for $z_a = 2.0$, $\alpha = 0.0228$. Thus, the probable scrap is 2.28 percent. Therefore, the percent of usable shafts is

$$P(x \leq 1.5) = 100 - 2.28$$
$$= 97.72\%$$

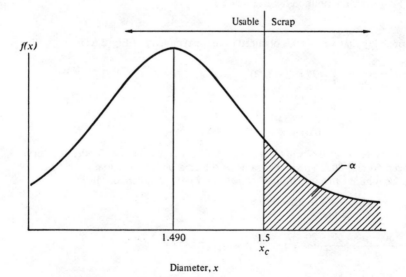

Fig. 2.7 Frequency distribution for Example 2.2.

2.1.2 THE CUMULATIVE DISTRIBUTION FUNCTION

In the above examples, the probability of occurrence was found as the area under the frequency distribution curve. But the probability that $x > x_i$, which is $P(x > x_i) = P(z > z_a)$, also equals the ordinate of the cumulative distribution function (CDF) at $x = x_{c_a}$. Hence, one can obtain $P(x > x_c)$ by plotting a CDF and reading the ordinate. This is illustrated in Fig. 2.8. In Fig. 2.8a the density function for the variable x is plotted, and in this case the shaded area is the area α, to be found. This area is equal to 1 minus the ordinate corresponding to x_{c_a} in Fig. 2.8b, which is the CDF of Fig. 2.8a. When this CDF is plotted on probability paper, the resulting curve is a straight line. Hence, instead of reading the ordinate of Fig. 2.8b, one can read the ordinate of Fig. 2.8c to find $1 - \alpha$ and then the required area α. Moreover, any of these three curves can be used to find the probability α; but, since straight lines are easiest to plot, Fig. 2.8c is commonly used. To illustrate this approach, Example 2.2 is solved graphically in Example 2.3.

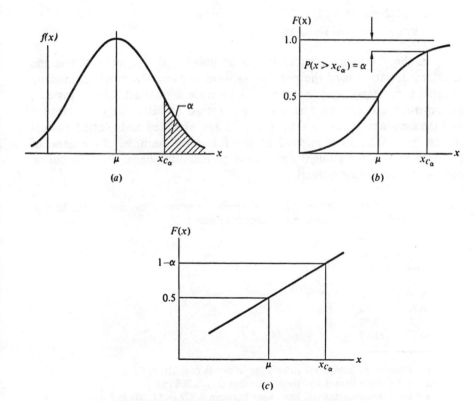

Fig. 2.8 Cumulative distribution function.

Example 2.3   (Refer to Example 2.2.)

$\mu = 1.490$ in $\sigma = 0.005$ in $x_c = 1.5$ in
$\alpha = ?$

Solution   On the CDF plot in Fig. 2.9, when $x = \mu = 1.490$,

$F(x) = F(1.49) = 0.5$

When $x = \mu + \sigma = 1.490 + 0.005 = 1.495$ in,

$F(1.495) = 0.8413$,

since when $x = \mu + \sigma$,

$$z = \frac{(\mu + \sigma) - \mu}{\sigma} = 1$$

and for $z_\alpha = 1$, $\alpha = 0.1587$ (from Table A-1),

$F(x_c) = 1 - \alpha = 1 - 0.1587 = 0.8413$

The values $[x = 1.49, F(1.49) = 0.5]$ and $[x = 1.495, F(1.495) = 0.8413]$ are plotted on probability paper as shown in Fig. 2.9 and a straight line is drawn through these points.   From this figure the value of $F(1.5) = 0.978 = 1 - \alpha$,

$\alpha = 0.022 = $ scrap $= 2.2\%$

or 97.8 percent parts are usable.

In the previous examples it was assumed that the mean μ and the standard deviation σ of the population were known.   In most engineering situations the characteristics of the population are seldom available.   Instead, an experiment is run on a sample, and instead of calculating the mean $\bar{x}$ and the standard deviation s, the test data are tabulated and plotted against the median ranks, as illustrated by the following example.   In general, if the data plot as the straight line on the probability paper, the distribution can be assumed to be normal.

Example 2.4   Friction linings from 10 wheels, removed after certain application, were measured for wear with the following results:

Wear, 0.001 *in*	
8.05	5.00
9.10	8.53
6.20	6.80
7.50	9.60
10.30	11.46

1. How many linings will show wear in excess of 0.010 in?
2. How many linings will show wear less than 0.008 in?
3. How many linings will have wear between 0.007 and 0.009 in?

Provide the answer with the aid of median ranks.

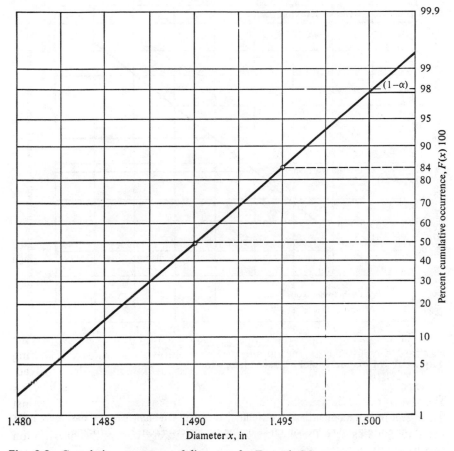

Fig. 2.9 Cumulative occurrence of diameters for Example 2.3.

Solution Arrange the data in ascending order and assign median ranks, using Table A-11
for $n = 10$.

Wear, 0.001 *in*	Median ranks, %
5.00	6.70
6.20	16.32
6.80	25.94
7.50	35.57
8.05	45.19
8.53	54.81
9.10	64.43
9.60	74.06
10.30	83.68
11.46	93.30

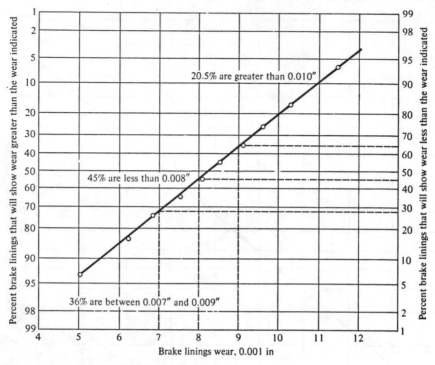

Fig. 2.10 Median-rank plot of brake lining wear.

Plot these data on normal probability paper with the lining wear on the abscissa and the corresponding median ranks on the ordinate. From Fig. 2.10:

1. 20.5 percent will show wear in excess of 0.010 in.
2. 45 percent will show wear less than 0.008 in.
3. 63 percent will show wear less than 0.009 in, and 27 percent less than 0.007 in. Therefore, 63 − 27 percent, or 36 percent, will show wear between 0.007 and 0.009 in.

2.2 LOG NORMAL DISTRIBUTION

In the normal distribution, discussed in the previous section, the distribution is symmetrically disposed around the mean value.

There are many distributions which are not symmetrical, and one of the most important is log normal distribution, as it describes life or durability phenomena. Such a distribution is shown in Fig. 2.11.

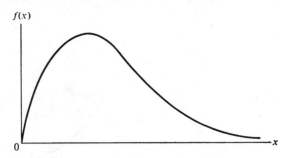

Fig. 2.11 Log normal distribution.

In log normal distribution the following nomenclature is used:

$$x_l = \log x$$

$$\mu_l = \int_{-\infty}^{\infty} x_l f(x_l)\, dx_l = \text{log population mean}$$

$$\sigma_l^2 = \int_{-\infty}^{\infty} (x_l - \mu_l)^2 f(x_l)\, dx_l = \text{log population variance}$$

$$\bar{x}_l = \frac{\sum_{i=1}^{n} x_{l_i}}{n} = \text{log sample mean}$$

$$s_l = \sqrt{\frac{\sum_{i=1}^{n} (x_{l_i} - \bar{x}_l)^2}{n-1}} = \text{log sample standard deviation}$$

If x is a random variable which has a log normal distribution, then x_l will have a normal distribution. Mathematically, a variable x has a log normal distribution if

$$f(x_l) = \frac{1}{\sigma_l\sqrt{2\pi}} \exp\left[\frac{-(x_l - \mu_l)^2}{2\sigma_l^2}\right] \qquad 0 \le x_l \le \infty \qquad (2.6)$$

Since x_l has a normal distribution (Fig. 2.12), everything that has been said about the normal distribution will be applicable here. Thus, the probability of occurrence of x_l can be found in the same manner as the probability of occurrence of a normal variable in Sec. 2.1.

Example 2.5 A manufacturer receives a contract to supply a 0.2-in-diameter steel wire which has an endurance strength in shear of 30×10^3 psi. This wire is used for springs holding the control rods of a nuclear reactor. Replacement of these springs is made after they have received 10^6 cycles of load at the operating stress. It

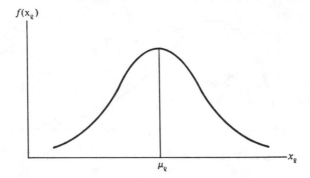

Fig. 2.12 Log normal distribution.

is known from previous experience that the failure distribution of lives at a constant stress is log normal with the following parameters:

$$\mu_l = \log(1.38 \times 10^6) = 6.1399$$
$$\sigma_l = \log 1.269 = 0.1035$$

What are the chances of a failure?

Solution   Expected life $= x_c = 10^6$ cycles. Therefore, $x_{l_c} = \log 10^6 = 6.000$. The probability of failure is equal to the probability of having a life less than 10^6 cycles, or $P(x \le 10^6)$.

$$P(x \le 10^6) = P(x_l \le 6.00)$$
$$= P(z \le z_\alpha)$$

where

$$z_\alpha = \frac{x_{l_c} - \mu_l}{\sigma_l}$$

$$= \frac{6.000 - 6.1399}{0.1035}$$

$$= -1.36$$

since $P(z \le -1.36) = P(z > 1.36)$ from symmetry. Table A-1 gives $\alpha = 0.0869$. That is, the chance of a failure is 8.69 out of 100.

Example 2.5 can also be solved graphically. In Sec. 2.1 it was pointed out that, if a variable has normal distribution, its CDF is a straight line on a probability paper. In case of a log normal variable x, if the CDF of x_l were plotted on probability paper, the result would be a straight line. To avoid the trouble of taking logarithms of x and·then plotting x_l, a special-purpose paper is constructed. This paper is called log probability paper, and the CDF of a log normal variable x plots out to be a straight line on this paper. When plotting the CDF on a log probability paper, if the resulting plot is a straight line, it can be assumed that the distribution is log normal.

Since spring lives are known to be distributed in a log normal manner, the CDF will be a straight line on log probability paper (Fig. 2.13). To determine this

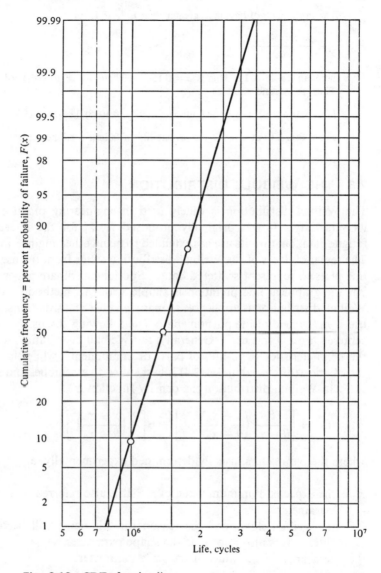

Fig. 2.13 CDF of spring lives.

line, one has to locate two points on this paper. The first point is derived from the fact that when

$$x_l = \mu_l, \qquad F(\mu_l) = 0.5$$

That is, the first point is

$$[\text{antilog}(\mu_l), 50] = [1.38 \times 10^6, 50]$$

The probability of having x_i greater than $\mu_i + \sigma_i$ is the

$$z_\alpha = \frac{\mu_i + \sigma_i - \mu_i}{\sigma_i} = 1.0$$

point in Table A-1. This is $\alpha = 0.1587$. Therefore, $F(\mu_i + \sigma_i) = 1 - \alpha = 0.8413$
Thus, the second point is

[antilog($\mu_i + \sigma_i$), $1 - \alpha$] or [antilog(6.1399 + 0.1035),0.8413] = [1.755×10^6,0.8413]

From Fig. 2.13 at $x = 10^6$ the probability of failure = 8.7 percent.

2.3 THE WEIBULL DISTRIBUTION

The Weibull distribution is widely used in engineering practice because of its versatility. It was originally proposed [1, 2] for the interpretation of fatigue data, but now its use has extended to many other engineering problems. The real usefulness of the Weibull distribution stems from the use of straight-line plots to represent scattered data. Special coordinate paper is required, but the graphical interpretation is simple with this method. Although the Weibull distribution is applicable to many engineering situations, its principal use is in the field of life phenomena. The following discussion, therefore, is centered around this use. Generally, the Weibull distribution well describes the characteristics (e.g., life) of parts or components, whiie the exponential distribution (discussed in Sec. 2.4) is best suited to assemblies and systems.

In Weibull distribution the density function is

$$f(x) = \left[\frac{b}{\theta - x_0} \left(\frac{x - x_0}{\theta - x_0} \right)^{b-1} \right] \left\{ \exp\left[-\left(\frac{x - x_0}{\theta - x_0} \right)^b \right] \right\} \qquad (2.7)$$

where the parameters, usually determined experimentally, are x_0, b, and θ.

x_0 is the expected minimum value of x. It is also referred to as the location
   parameter.
b is the Weibull slope, and the reason for this name will become obvious
   later. It is also known as the shape parameter.
θ is the characteristic value, or the scale parameter.

A plot of this function is given in Fig. 2.14. The cumulative distribution function is

$$F(x) = \int_{-\infty}^{x} f(x)\, dx = \int_{x_0}^{x} f(x)\, dx$$

$$= \int_{x_0}^{x} \frac{b}{\theta - x_0} \left(\frac{x - x_0}{\theta - x_0} \right)^{b-1} \exp\left[-\left(\frac{x - x_0}{\theta - x_0} \right)^b \right] dx$$

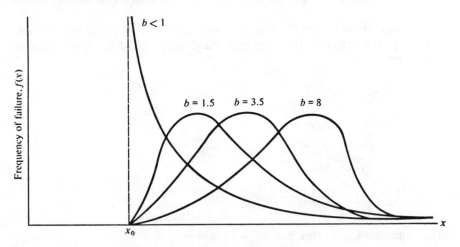

Fig. 2.14 The density function of the Weibull distribution.

Let

$$\left(\frac{x - x_0}{\theta - x_0}\right)^b = y$$

Therefore

$$b\left(\frac{x - x_0}{\theta - x_0}\right)^{b-1} \frac{1}{\theta - x_0} dx = dy$$

or

$$F(x) = \int e^{-y} \, dy$$

$$= -e^{-y}\Big|_{y_0}^{y}$$

$$= -\exp\left(\frac{x - x_0}{\theta - x_0}\right)^b\Big|_{x_0}^{x}$$

Therefore

$$F(x) = 1 - \exp\left[-\left(\frac{x - x_0}{\theta - x_0}\right)^b\right] \qquad (2.8)$$

2.3.1 TWO-PARAMETER (θ, b) WEIBULL FUNCTION

In dealing with certain life phenomena it is reasonable to assume that the lower bound of life x_0, that is, the expected minimum of the population, is equal to zero. This reduces the Weibull cumulative distribution function [Eq. (2.8)] to

$$F(x) = 1 - \exp\left[-\left(\frac{x}{\theta}\right)^b\right] \qquad (2.9)$$

This is a two-parameter form of the Weibull CDF, which is considerably simpler to use than the three-parameter function [Eq. (2.8)]. Equation (2.9) can be rewritten

$$\frac{1}{1 - F(x)} = \exp\left(\frac{x}{\theta}\right)^b$$

Taking natural logarithms,

$$\ln \frac{1}{1 - F(x)} = \left(\frac{x}{\theta}\right)^b$$

$$\ln \ln \frac{1}{1 - F(x)} = b(\ln x) - (b \ln \theta)$$

This equation has a form $Y = bX + C$, where

$$Y = \ln \ln \frac{1}{1 - F(x)}$$

$$X = \ln x$$

$$C = -b \ln \theta$$

The equation $Y = bX + C$ represents a straight line with a slope b and intercept C on the cartesian X, Y coordinates. Hence, a plot of

$$\ln \ln \frac{1}{1 - F(x)}$$

against $\ln x$ will also be a straight line with slope b. The reason for calling b the Weibull slope is therefore apparent (Fig. 2.15).

The characteristic life θ can be related to the median life. The median life, that is, 50 percent life, or B_{50} life, is the life corresponding to 50 percent of the population. That is, 50 percent of the population is expected to have life less than or equal to B_{50} life. In general, q percent of the population is expected to have life less than or equal to B_q life. Therefore, if B_q life is the required life, q is the probability of failure, which is commonly expressed in percent. Since $F(x) = 0.5 =$ probability of failure, Eq. (2.9) gives

$$0.5 = 1 - \exp\left[-\left(\frac{B_{50}}{\theta}\right)^b\right]$$

or

$$\frac{1}{2} = \frac{1}{\exp(B_{50}/\theta)^b}$$

or

$$\ln 2 = \left(\frac{B_{50}}{\theta}\right)^b$$

and

$$\theta = \frac{B_{50}}{(\ln 2)^{1/b}} \tag{2.10}$$

To determine the probability that a part will fail at the characteristic life or less, from Eq. (2.9) for $x = \theta$,

$$F(x) = 1 - \exp\left[-\left(\frac{x}{\theta}\right)^b\right]$$

$$= 1 - \exp\left[-\left(\frac{\theta}{\theta}\right)^b\right] = 1 - e^{-1} = 1 - \frac{1}{e}$$

$$= 1 - \frac{1}{2.718} = 0.632$$

$$= 63.2\% \quad \text{for } x = \theta \tag{2.11}$$

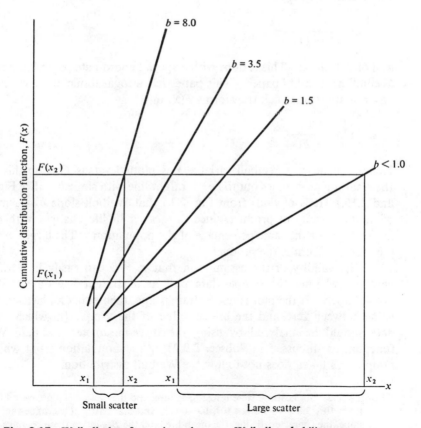

Fig. 2.15 Weibull plots for various slopes on Weibull probability paper.

Therefore, θ is the life by which 63.2 percent of the parts will have failed. This is similar to the median, which is the life by which 50 percent of the parts have failed.

Since Weibull is generally not a symmetrical distribution, the mean and the median values will not be the same, as was the case in the normal distribution. In order to determine mean life, locate the Weibull slope b on the abscissa of Table A-17 and read off percentage failed at the mean on the ordinate. The life on the Weibull plot corresponding to this percentage will be the mean life.

It was stated above that a plot of

$$Y = \ln \ln \frac{1}{1 - F(x)} \qquad \text{vs} \qquad X = \ln x$$

is a straight line with a slope b. It is cumbersome to calculate X and Y for each application. Hence there is a need for some method of expediting the conversion of $F(x)$ to

$$\ln \ln \frac{1}{1 - F(x)}$$

and of X to $\ln x$. This is done with a special coordinate paper known as the Weibull probability paper. This paper has a logarithmic abscissa scale and an ordinate scale which transforms $F(x)$ into

$$\ln \ln \frac{1}{1 - F(x)}$$

Hence, whenever a Weibull variable x is plotted versus $F(x)$ on this paper, the resulting plot comes out to be a straight line with slope b. (See Figs. 2.15 and 2.16.) It is obvious from Fig. 2.15 that Weibull slope b is a measure of the uniformity of product, since the scatter in life changes with change in the slope for the same percentage of the population. The larger the slope, the more uniform is the product.

The validity of the assumption that $x_0 = 0$ can easily be verified by checking whether the sample data plot as a straight line on the Weibull graph paper. If the plot is not a straight line, then either x_0 has some finite value between zero and the lowest value of the sample [in which case the data should be analyzed by using the three-parameter (x_0, θ, b) Weibull function, as discussed in Subsec. 2.3.2] or the population from which the sample was taken does not follow the Weibull distribution.

Example 2.6 Six bearings were tested to failure, and the cycles to failure were 4.0×10^5, 1.3×10^5, 9.8×10^5, 2.7×10^5, 6.6×10^5, and 5.2×10^5. Determine the Weibull slope, the characteristic life, the mean life, and the B_{10} life of these bearings.

Solution Arrange the life values in an increasing order. Assign median ranks from Table A-11 for $n = 6$.

Life to failure, 10^5 cycles	Median ranks, %
1.3	10.91
2.7	26.55
4.0	42.18
5.2	57.82
6.6	73.45
9.8	89.09

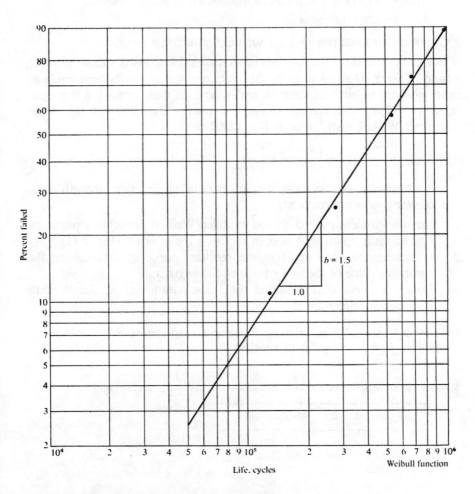

Fig. 2.16 Weibull plot for bearing lives (Example 2.6).

Plot these data on the Weibull probability paper with the life on the abscissa and the corresponding median ranks on the ordinate. Draw the best-fitting line through these points (Fig. 2.16). For higher accuracy this line should be fitted by the least-squares method, which is discussed in Chap. 13.

Weibull slope b: The Weibull probability paper was constructed in such a manner that the Weibull slope of any line on it is the arithmetic slope. The Weibull slope b, therefore, is 1.5.

Characteristic life θ: By definition, characteristic life θ is the life corresponding to 63.2 percent failure. Therefore, $\theta = 5.7 \times 10^5$ cycles.

Mean life: From Table A-17, the percent failed at mean $= 57.5$ percent for $b = 1.5$. From Fig. 2.16, the life corresponding to 57.5 percent is

Mean life $= 5.1 \times 10^5$ cycles

B_{10} life: From Fig. 2.16, the life corresponding to 10 percent failure is

B_{10} life $= 1.26 \times 10^5$ cycles

2.3.2 THREE-PARAMETER (θ, b, x_0) WEIBULL FUNCTION

While in the analysis of life data it is often reasonable to assume that the lower bound of life x_0 is equal to zero, this assumption may not be applicable to other phenomena such as strength and wear. In these cases the three-parameter Weibull distribution should be used. The CDF for the three-parameter Weibull distribution is [Eq. (2.8)]

$$F(x) = 1 - \exp\left[-\left(\frac{x - x_0}{\theta - x_0}\right)^b\right]$$

In order to determine the Weibull parameters for, say, strength data, the following steps are required:

1. The strengths are plotted on the modified Weibull probability paper on the abscissa against the median ranks on the ordinate (Fig. 2.17).
2. A correction is then made to the resultant curve by determining the probable value of the lower bound of strength x_0.
3. From the curve thus modified the three parameters of the Weibull function are determined.

Example 2.7 The following represents the scatter in fatigue strengths in ksi of a certain component. Determine the Weibull parameters.

57.3, 59.2, 62.5, 55.3, 61.4

Solution Rearrange the data in an increasing order, and assign median ranks.

Fatigue strength x, ksi	Median ranks, %
55.3	12.94
57.3	31.47
59.2	50.00
61.4	68.53
62.5	87.06

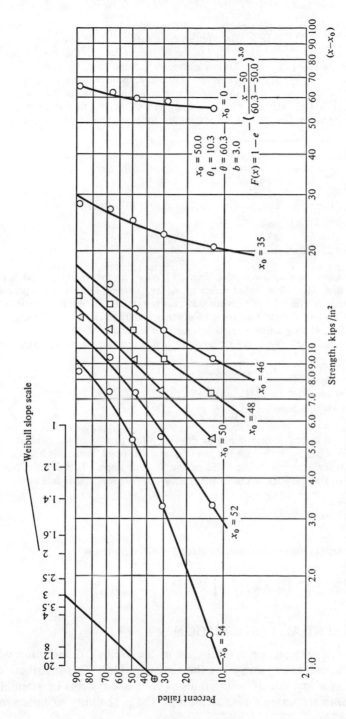

Fig. 2.17 Modified Weibull plot for determination of Weibull parameters.

These data are then plotted on the modified Weibull probability paper as shown in Fig. 2.17, curve A. In plotting these data, an assumption was made that the lower bound of strength x_0 (that is, the minimum strength that can be expected in the population) is zero. This is obviously not the case, as mechanical parts must have a strength greater than zero. Therefore, the next step is to determine the probable value of x_0. This value should be somewhere between the lowest value of the sample (55.3 ksi) and zero. As the first trial, assume that x_0 is 35 ksi. By subtracting x_0 from the original set of data, the following new set of data is obtained:

$x - x_0$, ksi	Median ranks, %
20.3	12.94
22.3	31.47
24.2	50.00
26.4	68.53
27.5	87.06

When these data are plotted (Fig. 2.17, curve B), the resultant curve is not a straight line. Therefore, other values of x_0 are assumed, and the same procedure is repeated until, for a certain assumed x_0, the test points can best be linearized. In this case the best straight line corresponds to $x_0 = 50$ ksi, curve C. Through these points, then, a straight line is fitted by using the least-squares method.

The value of $x - x_0$ at 63.2 percent is read off to determine the characteristic strength θ:

$\theta = x$ at 63.2%
$(x - x_0)_{63.2\%} = \theta_1 = 10.3$ ksi
$\theta = (x)_{63.2\%} = \theta_1 + x_0 = 10.3 + 50$
$\qquad\qquad\qquad = 60.3$ ksi

The Weibull slope b is determined by drawing a line parallel to the straight line of $x_0 = 50$ and passing it through the pivot point. The point where this line intersects the Weibull slope scale is the value of the Weibull slope. In this case, $b = 3.0$. Hence, the Weibull parameters for the given set of fatigue-strength data are

$x_0 = 50$ ksi
$\theta = 60.3$ ksi
$b = 3.0$

The analytical form for the corresponding Weibull equation is

$$F(x) = 1 - \exp\left[-\left(\frac{x - 50}{60.3 - 50}\right)^{3.0}\right]$$

2.4 EXPONENTIAL DISTRIBUTION

Exponential distribution is a special case of the Weibull distribution when $b = 1$, $x_0 = 0$, and $x = t$, where t is the time variable. This distribution is very useful in the analysis of failure rates of complete systems or assemblies. For example, in the case of aircraft pumps, the probability of failure of a component, such as a shaft or a bearing, can be described by a log normal or Weibull distribution. However, for the pump as a whole (as a system) the exponential distribution is a better choice. Exponential distribution may also apply to components, provided that their failures result from chance occurrence alone, such as a stone hitting an automobile windshield. This failure is clearly not a function of time, as the event may occur whether the automobile has 1,000 or 100,000 m on it.

The density function for an exponential distribution is

$$f(t) = \frac{1}{\theta} e^{-t/\theta} \tag{2.12}$$

where $f(t)$ = probability of failure
$\quad\quad t$ = operating time (independent variable)
$\quad\quad \theta$ = characteristic life = $B_{50}/\ln 2$ [see Eq. (2.10)]
$\quad\quad e$ = 2.718 = base of natural logarithms

Equation (2.12) is plotted in Fig. 2.18. The expected value

$$E(t) = \text{mean} = \int xf(x)\, dx = \int t\, \frac{1}{\theta} e^{-t/\theta}\, dt$$

Hence

$$E(t) = \theta = \text{mean} \tag{2.13}$$

Thus, exponential distribution has only one parameter, θ.

The cumulative distribution function is

$$F(t) = \int_0^t f(t)\, dt = \int_0^t \frac{1}{\theta} e^{-t/\theta}\, dt$$

$$= 1 - e^{-t/\theta} \tag{2.14}$$

Equation (2.14) is plotted in Fig. 2.19.

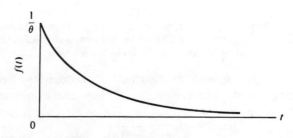

Fig. 2.18 Exponential density function.

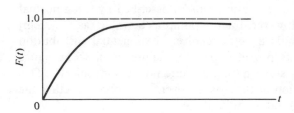

Fig. 2.19 Exponential cumulative distribution function.

Example 2.8 Determine the B_{50} and B_{90} lives of an electric generator. From previous tests it is known that under highly overloaded conditions 20 percent of these generators will fail at the end of 50 h of operation.

Solution The variable is time t. At the end of 50 h

$$F(t) = 1 - e^{-t/\theta}$$
$$0.2 = 1 - e^{-50/\theta}$$
$$e^{-50/\theta} = 0.8$$
$$e^{50/\theta} = 1.25$$

$$\frac{50}{\theta} = \ln 1.25$$

$$\theta = \frac{50}{\ln 1.25} = \frac{50}{0.223}$$

$$= 223 \text{ h}$$

B_{50} life: by definition,

$$B_{50} = \theta \ln 2.0$$
$$= (223)(0.693)$$
$$= 155 \text{ h}$$

B_{90} life:

$$F(t) = 1 - e^{-t/\theta}$$
$$0.9 = 1 - e^{-B_{90}/223}$$
$$e^{-B_{90}/223} = 0.1$$
$$e^{B_{90}/223} = 10$$
$$B_{90} = 223 \ln 10$$
$$= 513 \text{ h}$$

An alternative solution to this problem is obtained by means of the Weibull probability paper, where a line with the slope b equal to 1.0 is drawn through a point corresponding to 223 h (θ) on the abscissa and 63.2 percent on the ordinate (Fig. 2.20). Read on the abscissa B_{50} life = 155 h and B_{90} life = 513 h.

In exponential distribution a quantity frequently of considerable importance is hazard rate λ. The general expression is

$$\lambda = \text{hazard rate} = \frac{f(x)}{1 - F(x)}$$

$$= P(X \leq x \leq X + dX \,|\, x \geq X) \tag{2.15}$$

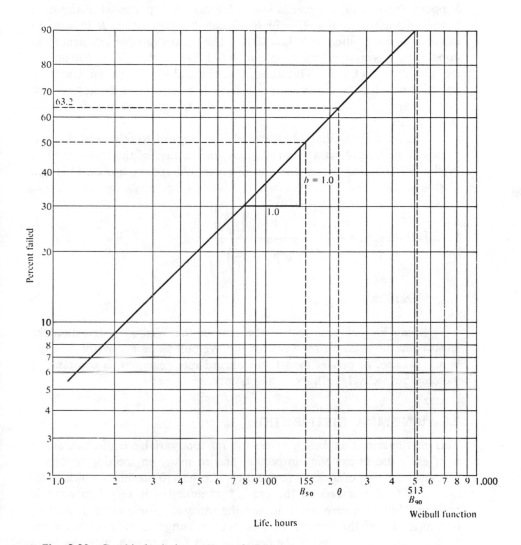

Fig. 2.20 Graphical solution to Example 2.8.

This means that the hazard rate is a probability that a given item on test will fail between X and $X + dX$ time when it has already survived up to the time X.

By substituting expressions in Eqs. (2.12) and (2.14) into Eq. (2.15), the hazard rate for the exponential distribution becomes

$$\text{Hazard rate } \lambda = \frac{(1/\theta)e^{-t/\theta}}{1 - (1 - e^{-t/\theta})} = \frac{1}{\theta} = \text{const} \qquad (2.16)$$

Suppose, for example, a population A follows an exponential distribution having characteristic life $\theta_A = 20$ h. Similarly, population B follows an exponential distribution with $\theta_B = 40$ h. The hazard rate for any item from population A is constant and equal to $1/\theta = 1/20 = 0.05$. For population B, the hazard rate is 0.025. This means that the probability that an item from population A will fail during any one hour is 5 percent regardless of how many hours the item has survived. Similarly, the probability of failure for B is 2.5 percent.

The general equation (2.15) can also be applied to other distributions to determine hazard rates. For example, in the case of the two-parameter Weibull distribution [when x is the time variable and x_0 in Eqs. (2.7) and (2.8) is zero],

$$\text{Hazard rate} = \frac{\frac{b}{\theta}\left(\frac{x}{\theta}\right)^{b-1}\exp\left[-\left(\frac{x}{\theta}\right)^b\right]}{1 - \left\{1 - \exp\left[-\left(\frac{x}{\theta}\right)^b\right]\right\}}$$

$$\text{Hazard rate } \lambda = \frac{b}{\theta^b}(x)^{b-1} \tag{2.17}$$

It can be seen here that in the case of the Weibull distribution the hazard rate depends on the value of the Weibull slope b. When b is less than, equal to, or greater than 1, the hazard rate decreases, is constant, or increases respectively with increasing time x.

2.5 BINOMIAL DISTRIBUTION

This distribution is applicable where the random variable is discrete, and it is of great use in sampling inspection and in many engineering problems. Consider the case of a tire manufacturer who has 10 machines which apply adhesives to the carcass of the tire. This adhesive is very important in making a bond between the tread and the carcass. There arose a situation in which one of these machines was not applying the adhesive correctly. This machine was producing the defective tires for a month, with the result that out of a total of 100,000 tires produced during that month, 10 percent were defective. These tires got thoroughly mixed when they were sent to a buyer.

If a retail buyer were to buy one single tire, the probability of his getting a defective tire would be 10 percent or 0.1. If he were to purchase two tires, the probability of receiving two defectives would be $0.1 \times 0.1 = 0.01$. This follows from the fact that the probability of the simultaneous occurrence of independent events equals the product of their individual probabilities of occurrence.

Since there were 10 percent defectives produced, the probability of finding a defective tire is 0.1. This is called p. The probability of finding an OK tire is $q = 1 - p = 1 - 0.1 = 0.9$. Hence, the probability of finding two OK tires will be $q^2 = 0.9^2 = 0.81$.

Similarly, the probability of finding one defective tire in purchasing two tires can be determined by considering the following possible occurrences: (1) The first tire is defective and the second is OK. (2) The first tire is OK and the second is defective. In that case the total probability of occurrence will be $pq + qp = 2pq = 0.18$. In this manner, Table 2.1 was constructed, including the cases of $n = 3$ and $n = 4$.

The results obtained by this procedure are accurate and interesting, but as n increases, the tabulation becomes very involved. This can be avoided if one notes that the values obtained in the last column in each case

Table 2.1   Binomial distribution

Number of trials n (number of tires purchased)	Result r	Ways of occurring	Probability	Probability of type result $P(r = r_i)$
	2 def.*	Def., def.	pp	$p^2 = 0.01$
	1 def., 1 OK	OK, def.	qp	$2pq = 0.18$
2		Def., OK	pq	
	0 def., 2 OK	OK, OK	qq	$q^2 = 0.81$
				1.00
	3 def.	Def., def., def.	p^3	$p^3 = 0.001$
	2 def., 1 OK	OK, def., def.	qp^2	
		Def., OK, def.	pqp	$3qp^2 = 0.027$
3		Def., def., OK	p^2q	
	1 def., 2 OK	OK, OK, def.	q^2p	
		OK, def., OK	qpq	$3pq^2 = 0.243$
		Def., OK, OK,	pq^2	
	0 def., 3 OK	OK, OK, OK	q^3	$q^3 = 0.729$
				1.000
	4 def.	Derived as above		$p^4 = 0.0001$
	3 def., 1 OK			$4p^3q = 0.0036$
4	2 def., 2 OK			$6p^2q^2 = 0.0486$
	1 def., 3 OK			$4pq^3 = 0.2916$
	0 def., 4 OK			$q^4 = 0.6561$
				1.0000

* Defective.

follow the same pattern. This is an expansion of $(p + q)^n$. That is, when $n = 2$, probabilities of $r = 2, 1$, and 0 are p^2, $2pq$, and q^2, respectively, which can be obtained from $(p + q)^2 = p^2 + 2pq + q^2$. Thus, the probabilities of $n, n - 1, \ldots, 3, 2, 1, 0$ defectives in a sample of n items drawn at random from a population whose proportion defective is p, and whose proportion OK is q, are given by the successive terms of the expansion of $(p + q)^n$, reading from left to right.

The above rule can be stated in more general terms. Consider an experiment with the following characteristics: (1) The experiment consists of n independent trials. (2) The overall probability of a successful occurrence is p, and $q = 1 - p$. Then the probability that the number of successful occurrences r during the n trials is equal to r_1 is given by the discrete distribution function

$$p(r = r_1) = \binom{n}{r_1} p^{r_1} q^{n-r_1} \quad \text{for} \quad r_1 = 0, 1, 2, \ldots, n \tag{2.18}$$

where

$$\binom{n}{r_1} = \frac{n!}{r_1!(n - r_1)!} \tag{2.19}$$

and $n!$, $r_1!$, and $(n - r_1)!$ are factorials of the respective quantities.

This distribution function [Eq. (2.18)] is known as the binomial distribution, having mean

$$E(r) = np \tag{2.20}$$

and standard deviation

$$\sigma(r) = \sqrt{npq} \tag{2.21}$$

All the values obtained in Table 2.1 can be reproduced by using Eq. (2.18), as indicated in the following example.

Example 2.9 In a population containing 10 percent defectives, what is the probability of finding four, three, two, one, and zero defectives in a sample of four items?

Solution The number of trials $n = 4$, the value of $p = 0.1$, and $q = 0.9$. The probability that the number of defectives r equals four, three, two, one, zero successively is given by [Eq. (2.18)]:

$$p(r = 4) = \binom{4}{4}(0.1^4)(0.9^{4-4}) = \frac{4!}{4!\,0!}(0.1^4)(0.9^0) = 0.1^4 \qquad = 0.0001$$

$$p(r = 3) = \binom{4}{3}(0.1^3)(0.9^{4-3}) = \frac{4!}{3!\,1!}(0.1^3)(0.9^1) = 4(0.1^3)(0.9) = 0.0036$$

$$p(r = 2) = \binom{4}{2}(0.1^2)(0.9^{4-2}) = \frac{4!}{2!\,2!}(0.1^2)(0.9^2) = 6(0.1^2)(0.9^2) = 0.0486$$

$$p(r=1) = \binom{4}{1}(0.1^1)(0.9^{4-1}) = \frac{4!}{1!3!}(0.1^1)(0.9^3) = 4(0.1^1)(0.9^3) = 0.2916$$

$$p(r=0) = \binom{4}{0}(0.1^0)(0.9^4) = \frac{4!}{0!4!}(0.1^0)(0.9^4) = 0.9^4 \qquad = 0.6561$$

$$\text{Total probability, containing all possible results} = \overline{1.0000}$$

Note that $0! = 1$ for the following reason:

$$n! = n(n-1)(n-2)\cdots 3, 2, 1$$
$$= n(n-1)!$$

If

$$n = 1$$
$$1! = 1 \times (1-1)!$$
$$1! = 1 \times 0!$$

$$0! = \frac{1!}{1} = 1$$

So, $0!$ must equal 1 to be consistent with normal notation.

On comparing the above values with values from Table 2.1, it is found that they are identical.

In the case of $n = 4$ and $p = 0.1$, it will be noted that one small sample could give very misleading results. This is illustrated in histogram in Fig. 2.21. That is, in a population containing 10 percent defectives, if four items are drawn, the following will be found: 65.6 percent of the time all four items will contain no defectives, suggesting that the population from which they come is 100 percent OK, where actually it contains 10 percent defectives; 29.2 percent of the time there will be one defective in a sample of four, suggesting that the population contains 25 percent defectives, where actually it has only 10 percent defectives. Similarly, in nearly 5 percent of the cases, two out of four tires, that is, 50 percent, will be found defective while the population has only 10 percent defectives.

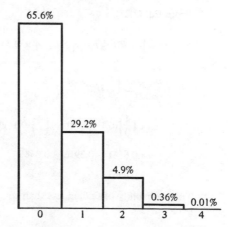

Fig. 2.21 Number of defectives in sample of four items.

In some engineering situations one is interested not in the probability itself but in a probability that the certain event will occur less than a specified number of times. That is, the function of interest is

$$P\left(\frac{r}{n} \leq g\right) \tag{2.22}$$

Since

$$P\left(\frac{r}{n} \leq g\right) = \sum_{i=0}^{gn} p(r = r_i)$$

it follows that

$$P\left(\frac{r}{n} \leq g\right) = P(r \leq ng) = \sum_{i=0}^{gn} \binom{n}{r_i} p^{r_i} q^{n-r_i} \tag{2.23}$$

where gn is the largest integer less than or equal to $g \cdot n$.

The expected value μ of this variable is

$$\mu = E\left(\frac{r}{n}\right) = p \tag{2.24}$$

and the standard deviation is

$$\sigma\left(\frac{r}{n}\right) = \sqrt{\frac{pq}{n}} \tag{2.25}$$

Example 2.10 In a population containing 10 percent defectives, what is the probability of finding up to and including half defectives in a sample of four items?

Solution

$$n = 4 \qquad \frac{r}{n} = g = 0.5$$

From Eq. (2.23),

$$P\left(\frac{r}{n} \leq g\right) = P(r \leq ng) = P(r \leq 4 \times 0.5)$$

$$P(r \leq 2) = \sum_{i=0}^{2} \binom{4}{r_i} p^r q^{4-r_i}$$

$$= \binom{4}{0}(0.1^0)(0.9^4) + \binom{4}{1}(0.1^1)(0.9^3) + \binom{4}{2}(0.1^2)(0.9^2)$$

$$= 0.6561 + 0.2916 + 0.0486$$

$$= 0.9963$$

The probability of finding less than or equal to two defectives out of four is 99.63 percent.

2.6 HYPERGEOMETRIC DISTRIBUTION

The hypergeometric distribution is used for the same situations as the binomial distribution, with one important exception. In the binomial distribution the percent defectives (and, therefore, the percent OKs) is assumed to be constant throughout the test. This is true if the lot from which the sample is taken is so large that withdrawal of the sample does not affect the proportion of defectives. Or, if the lot is small, it is assumed that when the sample is taken, each item in the sample is replaced before the next one is drawn. If the proportion of defectives cannot be assumed to be constant, a hypergeometric distribution should be used instead.

Suppose a lot contains N items of which Np are defective and Nq are OK, and where $p + q = 1$. If n of these items are withdrawn from this population without replacement, then

$$p(\text{number of defectives} = d) = \frac{\binom{Np}{d}\binom{Nq}{n-d}}{\binom{N}{n}} \tag{2.26}$$

where, in general [see, for example, Eq. (2.19)], the symbol

$$\binom{a}{b} = \frac{a!}{b!(a-b)!} \tag{2.27}$$

and

$N =$ lot size

$n =$ sample size

$Np =$ number of defectives in a lot

$Nq =$ number of OKs in a lot

$d =$ number of defectives in a sample

Example 2.11 A supplier receives a lot of 15 bulbs and he sells 10 of them. Later he is informed that 5 of these 15 bulbs were defective. What is the probability that his customer received 3 defectives?

Solution

$$N = 15 \qquad p = \text{fraction defective} = \frac{5}{15} = 0.333$$

$q = 0.667$

$Np = 5 \qquad Nq = 10 \qquad n = 10 \qquad d = 3$

From Eq. (2.26),

$$p(d=3) = \frac{\binom{5}{3}\binom{10}{7}}{\binom{15}{10}}$$

$$= \frac{5!}{3!(5-3)!} \times \frac{10!}{7!(10-7)!} \times \frac{10!(15-10)!}{15!}$$

$$= \frac{5!\,10!\,10!\,5!}{3!\,2!\,7!\,3!\,15!}$$

Evaluation can be simplified by taking logarithms of the factorials and using Table A-54. Hence,

$$
\begin{aligned}
p(d=3) &= \text{antilog}[(\log 5! + \log 10! + \log 10! + \log 5!) - (\log 3! \\
&\quad + \log 2! + \log 7! + \log 3! + \log 15!)] \\
&= \text{antilog}[(2.0792 + 6.5598 + 6.5598 + 2.0792) - (0.7781 \\
&\quad + 0.3010 + 3.7024 + 0.7781 + 12.1165)] \\
&= \text{antilog}(17.2780 - 17.6761) \\
&= \text{antilog}[-(0.3981)] \\
&= 0.398
\end{aligned}
$$

Hence, the probability that the customer received 3 defective bulbs is 39.8 percent.

2.7 POISSON DISTRIBUTION

In the case of binomial and hypergeometric distributions a sample is taken and the number of occurrences is determined. In both cases the number of times the event did occur and did not occur is known. There are situations, however, where the number of times an event occurs can be observed, but it would not be meaningful to inquire as to the number of times the event did not occur. If one travels down a highway every day for a number of years, he can tell how many stones hit his windshield during that period of time or how many nails punctured his tires. But he cannot say how many stones did not hit his windshield or how many nails did not puncture his tire. In such cases the binomial and hypergeometric distributions are not applicable.

The following distribution describes very well the occurrence of isolated events in a continuum. If y represents the expected or average number of occurrences of an event, the probability of observing the occurrence of r_i events is given by

$$p(r=r_i) = \frac{e^{-y}y^{r_i}}{r_i!} \qquad r_i = 0, 1, 2, 3, \ldots \tag{2.28}$$

This is known as the Poisson distribution, where the expected value μ is

$$\mu = E(r) = y \tag{2.29}$$

and the standard deviation

$$\sigma(r) = \sqrt{y} \tag{2.30}$$

That is, all one needs to know is the average number of occurrences of the event y, and he can work out the probabilities of observing the various possible occurrences.

Example 2.12 Consider the following tests conducted on a single-cylinder engine running at part throttle with a certain kind of fuel. The number of detonations during each consecutive minute, for 30 min, was recorded. Ten such engines were tested. The results are given below. Determine the probability of occurrence of zero, one, two, three, and four detonations per minute per engine.

Number of detonations during a minute	Number of times this number of detonations occurred during the test	Number of detonations occurring during the test
0	150	$0 \times 150 = 0$
1	100	$1 \times 100 = 100$
2	36	$2 \times 36 = 72$
3	10	$3 \times 10 = 30$
4	3	$4 \times 3 = 12$
5	1	$5 \times 1 = 5$
6	0	$6 \times 0 = 0$
	300	219

Solution The total number of detonations is 219, so that the average number per minute per engine is

$$\frac{219}{10 \times 30} = 0.73$$

Therefore, the expected value $y = 0.73$, and e^{-y} is 0.482. From Eq. (2.28),

For $r_i = 0$:

$$p(r = 0) = \frac{0.482(0.73^0)}{0!} = 0.482$$

For $r_i = 1$:

$$p(r = 1) = \frac{0.482(0.73^1)}{1!} = 0.352$$

For $r_i = 2$:

$$p(r = 2) = \frac{0.482(0.73^2)}{2!} = 0.128$$

Hence

Number of detonations per minute per engine, r_i	0	1	2	3	4
Probability $(r = r_i)$	0.482	0.352	0.128	0.0313	0.0057
Predicted frequency	145	106	38	9.4	1.7
Observed frequency	150	100	36	10	3

The Poisson distribution has one stipulation which must be met: The expectation y must be constant from trial to trial. The probability of having assembly-line shutdown because one of the stamping machines is out of order may be affected by several factors, such as worker fatigue, labor relations, and number of repairmen at hand. If these factors vary considerably, the expectation y may vary, too.

Example 2.13 The following data were obtained for the number of assembly-line breakdowns in a day due to a particular stamping machine. The data were taken for 360 consecutive days during which records were kept. Find the probability of having zero, one, two, three, four, and five breakdowns per day.

Number of breakdowns per day	Number of days these breakdowns occurred, in a given machine	Number of breakdowns
0	92	$0 \times 92 = 0$
1	110	$1 \times 110 = 110$
2	68	$2 \times 68 = 136$
3	48	$3 \times 48 = 144$
4	29	$4 \times 29 = 116$
5	7	$5 \times 7 = 35$
6	4	$6 \times 4 = 24$
7	2	$7 \times 2 = 14$
	360	579

Solution In 360 days 579 breakdowns occurred. This gives an average of 1.60 breakdowns per day. Therefore,

$$y = 1.6 \quad \text{and} \quad e^{-y} = 0.2$$

Thus, from Eq. (2.28)

$$p(r = 0) = 0.2 \frac{1.6^0}{0!} = 0.2 \qquad p(r = 1) = 0.2 \frac{1.6^1}{1!} = 0.32$$

$$p(r = 2) = 0.2 \frac{1.6^2}{2!} = 0.256 \qquad p(r = 3) = 0.2 \frac{1.6^3}{3!} = 0.136$$

By working out the successive terms of this distribution, the following is found:

Breakdowns per day:	0	1	2	3	4	5
Probability $(r = r_i)$	0.2	0.32	0.256	0.136	0.0546	0.0018
Predicted frequency	72	115	92	49	20	6.5
Observed frequency	92	110	68	48	29	7

This agreement is reasonably satisfactory. It suggests that the disturbing factors of labor relations and worker fatigue do not exert an appreciable effect.

In some engineering situations it is of interest to determine the probability that a certain event occurred at least r_i number of times. This can be done graphically with the aid of the Poisson probability paper, which is standard log probability paper where the abscissa represents expectation y and the ordinate the probability of occurrence,

$$P(r \geq r_a) = \sum_{i=a}^{\infty} p(r = r_i) \tag{2.31}$$

The curved lines drawn represent the values of occurrence r (Fig. 2.22).

Example 2.14 With reference to Example 2.13, find the probability of having at least four breakdowns per day.

Solution The probability of at least four breakdowns per day $= P(r \geq 4)$.

$$P(r \geq 4) = \sum_{i=4}^{\infty} p(r = r_i)$$

$$= p(r = 4) + p(r = 5) + p(r = 6) + p(r = 7) + \cdots$$

In this case the expectation y is 1.6.

For $y = 1.6$ and $r = 4$, $P = 0.085$ is found from Fig. 2.22. Therefore, out of 360 days one would expect $360 \times 0.085 = 30.6$ days when there will be four or more breakdowns. The observed data show that there were $29 + 7 + 4 + 2 = 42$ days when there were four or more breakdowns.

A second use of this paper is to determine whether an actual distribution of observed data may be described by the Poisson law and, if so, what is the expectation y. Consider, for example, the data in Example 2.12:

Number of detonations	0	1	2	3	4
Number of times this number of detonations occurred during the test	150	100	36	10	3

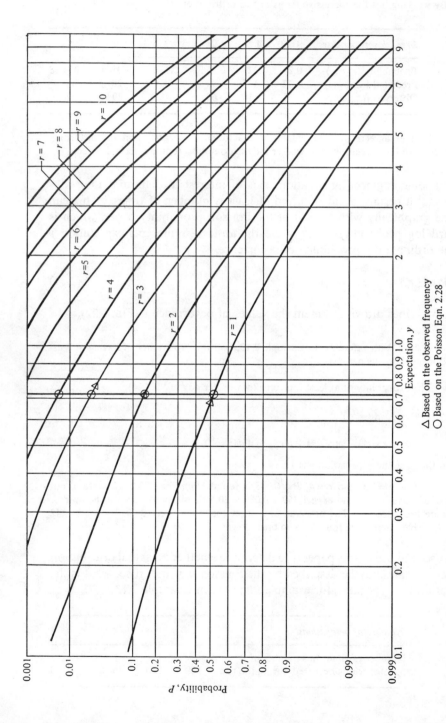

Fig. 2.22 Poisson probability paper, showing the probability P that at least r_i number of times an event will occur (that is, r times or more) when the expected number of occurrences has the value of y.

△ Based on the observed frequency
○ Based on the Poisson Eqn. 2.28

The probability of each number of detonations can be derived from the frequencies corresponding to each group, by dividing these frequencies by the total number of readings, $30 \times 10 = 300$. Thus

Number of detonations		0	1	2	3	4
Probability of occurrence	Based on the Poisson Eq. (2.28)	0.482	0.352	0.128	0.0313	0.0057
	Based on the observed frequency	0.500	0.333	0.120	0.033	0.010

From this table the following is obtained:

Number of detonations, r	Probability of occurrence of at least r number of detonations	
	Based on the Poisson Eq. (2.28)	Based on the observed frequency
1	$(1 - 0.482) = 0.518$	0.500
2	$(1 - 0.834) = 0.166$	0.167
3	$(1 - 0.962) = 0.038$	0.047
4	$(1 - 0.9933) = 0.0067$	0.014

When these data are plotted on the Poisson paper, by placing each point on its proper line, $r = 1$, $r = 2$, etc., opposite the corresponding value of the probability, it will be found that they lie on a vertical straight line, at the value $y = 0.73$. Thus one may test whether a given distribution follows a Poisson-type law by plotting it as described above. If the points plot on a vertical straight line, the data follow the Poisson distribution with expectation y equal to the abscissa for the vertical line.

2.8 DETERMINATION AND APPLICATION OF STATISTICAL DISTRIBUTIONS

Before a meaningful experiment can be designed or a set of data analyzed, it is essential to determine what statistical distribution the test data will follow. In the planning stage of the experiment, prior to obtaining the test data, there is no foolproof method which would guide the engineer in the choice of the distribution. In the absence of any evidence to the contrary, as the first try, one should assume that the data will follow a normal distribution (or, in the case of life, log normal or Weibull). For more specific guides, reference should be made to Table 2.2, which is based on experience and actual test cases.

Table 2.2   Fields of application of statistical distributions

Statistical distribution	Fields of application	Test cases
Normal	Various physical, mechanical, electrical, chemical, etc., properties	Capacity variation of electrical condensers; tensile strength of aluminum alloy sheet; monthly temperature variation; penetration depth of steel specimens; rivet-head diameters; electrical power consumption in a given area; electrical resistance; gas molecules velocities; wear; noise generator output voltage; wind velocity; hardness; chamber pressure from firing ammunition
Log normal	Life phenomena; asymmetric situations where occurrences are concentrated at the tail end of the range, where differences in observations are of a large order of magnitude	Automotive mileage accumulation by different customers; amount of electricity used by different customers; downtime of large number of electrical systems; light intensities of bulbs; concentration of chemical process residues
Weibull (two-parameter)	Same as log normal cases. Also situations where the percent occurrences (say failure rates) may decrease, increase, or remain constant with increase in the characteristic measured, for parts at debug, wear-out, and chance failure stages of product's life	Life of electronic tubes, antifriction bearings, transmission gears, and many other mechanical and electrical components; corrosion life; wear-out life
Weibull (three-parameter)	Same as two-parameter Weibull and, in addition, various physical, mechanical, electrical, chemical, etc., properties, except less common than in the case of normal distribution	Same cases as above two-parameter Weibull. In addition, electrical resistance, capacitance, fatigue strength, etc.
Exponential	The life of systems, assemblies, etc. For components, situations where failures occur by chance alone and do not depend on time in service; frequently applied when the design is completely debugged	Vacuum-tube failure life; expected cost to detect bad equipment during reliability testing; expected life of indicator tubes used in radar sets; life to failure of light bulbs, dishwashers, water heaters,

Table 2.2 (*continued*)

Statistical distribution	*Fields of application*	*Test cases*
	for production errors	clothes washers, aircraft pumps, electric generators, automobile transmissions
Binomial	Number of defectives in n sample size drawn from a large lot having p fraction defectives; probability of x occurrences in a group of y occurrences, that is, situations involving "go–no-go," "OK-defective," "good-bad" types of observations. Proportion of lot does not change appreciably as a result of sample drawn	Inspection for defectives in a shipment of steel parts; inspection of defective tires in a production lot; determination of defective weld joints; probability of obtaining electrical power of a certain wattage from a source; probability that a production machine will perform its function
Hypergeometric	Inspection of mechanical, electrical, etc., parts from a small lot having known percent defectives. Same as in binomial cases, except the proportion of lot may change as a result of sample drawn	Probability of obtaining 10 satisfactory resistors from a lot of 100 resistors having 2% defectives; similar cases involving light bulbs, bolts, piston rings, transistors
Poisson	Situations where the number of times an event occurs can be observed but not the number of times the event does not occur. Applies to events randomly distributed in time	Number of machine breakdowns in a plant; automobiles arriving simultaneously at an intersection; number of times dust particles found in atmosphere in some number of spot checks; industrial plant personnel injury accidents; dimensional errors in engineering drawings; automotive accidents in a given location per unit time; automotive traffic; hospital emergencies; telephone circuit traffic; a defect along a long tape, wire, chain, bar, etc.; tire punctures; stones hitting windshield; number of defective rivets in an airplane wing; radioactive decay; number of engine detonations; number of flaws per yard in sheet metal

Table 2.3   Methods for determining whether test data follow given distribution

Statistical distribution	*How to determine whether a set of data follow given distribution*
Normal	Data should plot as a straight line on normal probability paper. See Examples 2.4 and 10.3 for the procedure.
Log normal	Data should plot as a straight line on log normal probability paper. See Example 2.5 for the procedure.
Weibull (two-parameter)	Data should plot as a straight line on Weibull probability paper. See Examples 2.6 and 13.9 for the procedure.
Weibull (three-parameter)	Modified Weibull plot of the test data should result in a straight line. See Example 2.7 for the procedure.
Exponential	Data should plot as a straight line with Weibull slope equal to 1 on Weibull probability paper. See Example 2.4 for the procedure.
Binomial	See Example 2.9 for the procedure.
Hypergeometric	See Example 2.11 for the procedure.
Poisson	Compute probabilities of occurrence for at least one, two, three, etc., events. Plot these on Fig. 2.22, using the appropriate curve for each point. The resulting plot should be a vertical line at a given value of y (the expected average of occurrences).

Once the test data are obtained, the analysis is greatly facilitated by establishing the distribution these data will follow. A widely used procedure is the *chi-square test*. However, this method requires a great deal of test data; therefore, its application to most engineering problems is limited. As an alternative the following method is suggested (also see Table 2.3):

Arrange the set of data in an increasing order and assign median ranks (Table A-11) to each test point, as shown in Example 2.4. Plot these data on various probability papers (normal, log normal, Weibull, etc.), and the plot which gives the best straight line is an indication of the distribution which the test data will follow. The drawing of the best-fitting line can be done reasonably well by "eyeballing." For a more sophisticated approach a correlation coefficient r is computed (see Chap. 13) for each case. The maximum value of r denotes the distribution which fits the test data best.

SUMMARY

In designing an experiment, the single most important problem facing the engineer is how to determine what distribution the test data will follow. Thus a decision whether the test will be terminal or run to failure, whether it will be a predetermined or a sequential experiment, what should be the sample size and the confidence levels required depends very much on the type of distribution. The statistical distribution depends also on the variable being measured: an electrical conductivity is likely to follow a different distribution than a fatigue life. Even for a given variable the distribution may vary, depending on the circumstances. Thus, dimensions of a machine part may follow normal, rectangular, triangular, or circular distribution, depending on the method of machining and the manner of specifying the dimensions.

In the planning stage of the experiment, before the actual test data are obtained, consult Sec. 2.8 and specifically Table 2.2. Once the data are taken, see Table 2.3. The distributions discussed in this chapter are normal, log normal, Weibull, exponential, binomial, hypergeometric, and Poisson. These are the major distributions which test data may follow; therefore, they were grouped together in one chapter. There are several other distributions discussed in this book, such as z, t, χ^2, F, but these are used principally for the interpretation of these data; therefore, they are presented in pertinent chapters.

PROBLEMS

2.1. In a given mechanical system five pump assemblies are used. The system will operate satisfactorily if any three out of the five pumps are functioning. What are the chances for a system to malfunction if 15 percent of the total pump production is defective?

2.2. A manufacturer tests five steel specimens for impact strength and obtains the following results (in ksi):

31.0, 26.5, 36.0, 20.0, 42.5

(a) Using the Weibull probability paper, find the Weibull slope and the mean impact strength.

(b) What other statistical distribution function does this population approximately follow besides the Weibull?

2.3. In what way does hypergeometric distribution differ from the binomial distribution as far as the application to engineering is concerned?

2.4. The acceptable width of a slot in a certain forging is 0.900 ± 0.005 in. From past experience it is known that $\sigma = 0.003$ in. What is the percent of scrap forgings? Solve the problem

(a) with the aid of Table A-1;

(b) with the aid of a CDF plot.

2.5. Determine the percent of the population that lies within $\pm 0.675\sigma$; $\pm 1.0\sigma$; $\pm 2.0\sigma$; $\pm 3.0\sigma$; $\pm 4.0\sigma$.

2.6. In order to evaluate the fatigue life of transmission gears, six gears were picked at random out of a production run and fatigue tested at 50,000 psi, with the following results (life cycles in 10^6): 1.2, 1.0, 1.6, 1.3, 1.4, 2.0.

(a) How many gears in the gear population do you expect to have a life lower than the lowest life in the above sample?

(b) What is the B_{10} life? Provide the answer with the aid of median ranks.

2.7. An engineer is testing a lot of 100 transmissions. At the end of 1,000 h, the total number of failures is five.

    (a) Determine the expected characteristic life θ.

    (b) Determine the expected B_{50} life.

    (c) Determine the expected B_{10} life.

    (d) What are the chances for any transmission failure at any hour?

2.8. (a) How is the exponential distribution function related to the Weibull distribution function?

    (b) What are their parameters?

2.9. What is the quickest way of determining whether a given set of test data follow a log normal distribution?

2.10. Weibull is basically a three-parameter (x_0, b, θ) distribution. In what kind of problems, situations, mechanisms, etc., can we generally assume that $x_0 = 0$; and in what kind of problems, etc., can we not assume that $x_0 = 0$? In the latter case, describe briefly how the problem should be handled.

2.11. Suppose 15 percent of total production of a given type of electric generator is defective. What is the probability that when five generators are picked randomly from this lot the following are found:

    (a) no defectives;

    (b) only two defectives;

    (c) no more than two defectives;

    (d) only three defectives;

    (e) no more than three defectives.

2.12. The following observations on traffic flow were made, where the number of times two vehicles approached an intersection simultaneously in any one minute was recorded. The frequency of such events per minute was observed for a total of 2 h:

Number of events in any one minute	0	1	2	3	4	5
Number of minutes in which the above events occurred	25	26	20	27	15	7

    (a) Determine which statistical distribution these data follow. Justify by making appropriate calculations and plots.

    (b) Determine the probability of vehicles' approaching simultaneously two, three, and four times in any one minute.

2.13. The following lives to failure (in hours) of a certain component were obtained:

    100; 245; 425; 620; 850; 1,180; 1,450; 1,900; 2,600; 3,500

    (a) Determine which distribution these data follow and compute the parameters of this distribution.

    (b) If the distribution and its parameters are known, what can be said about the failure rate of this component?

2.14. Flow rates of 12 identical pumps were measured (gallons per minute):

    7.8, 8.4, 9.0, 9.2, 9.5, 9.8, 10.2, 10.3, 10.7, 11.0, 11.6, 12.2

Determine the statistical distribution and its parameters that can best express the random characteristics of flow rates, from pump to pump.

2.15. Consider Prob. 1.7, and assume that (a) the average and the standard deviation computed from the sample is the same as the population mean μ and the population standard deviation σ; (b) the measurement is a random variable and follows a normal distribution. Estimate the percent of the time the oil and grease concentration level in the purified water will not exceed 16 ppm. Repeat this for 16.5 and 17.0 ppm.

2.16. Since the efficiency of the water purification process may vary from time to time, using the information from Prob. 2.15 determine the average level of oil and grease concentration that must be maintained by the purification process so that 99 percent of time the level is below 15.0 ppm.

2.17. The amount of sulfur dioxide in air was measured in terms of parts per hundred million (pphm) over a period of time: 3.0, 0.7, 1.0, 0.5, 1.2, 0.6, 1.6, 0.65, 2.3, 0.3, 1.7, 0.8, 1.3, 1.5, 1.9.

(a) Determine which statistical distribution these data follow.

(b) Assuming that these data follow normal distribution, determine percent of time the pollution level would be higher than 2.5 pphm.

REFERENCES

1. Weibull, W.: A Statistical Distribution Function of Wide Applicability, *J. Appl. Mech.*, vol. 18, pp. 293–297, September, 1951.
2. Weibull, W: A Statistical Representation of Fatigue Failures in Solids, *Acta Polytech.*, *Mech. Eng. Ser.*, vol. 1, no. 9, 1949.
3. Lipson, C., and N. J. Sheth: Prediction of Percent Failures from Stress/Strength Interference, *SAE Paper* 680084, January, 1968.
4. Moroney, M. J.: "Facts from Figures," Penguin Books, Inc., Baltimore, 1965.

BIBLIOGRAPHY

Bowker, A. H., and G. J. Lieberman: "Engineering Statistics," Prentice-Hall, Inc., Englewood Cliffs, N.J., 1963.
Dixon, W. J., and F. J., Massey, Jr.: "Introduction to Statistical Analysis," 3d ed., McGraw-Hill Book Company, New York, 1969.
Guttman, I., and S. S. Wilks: "Introductory Engineering Statistics," John Wiley & Sons, Inc., New York, 1965.
Johnson, L. G.: "Theory and Technique of Variation Research," Elsevier Publishing Company, Amsterdam, 1964.
Kao, J. H. K.: The Mixed Weibull Distribution in Life-testing of Electron Tubes, *Trans. Amer. Soc. Qual. Cont.*, 1959 *Nat. Conv.*, pp. 267–289, 1959.
Lipson, C., and N. J. Sheth: Prediction of Percent Failures from Stress/Strength Interference, *SAE Paper* 680084, January, 1968.
———, ———, and R. L. Disney: Reliability Prediction—Mechanical Stress/Strength Interference, *Tech. Rep.* RADC-TR-66-710, Rome Air Development Center, Air Force Systems Command, Griffiss Air Force Base, New York, March, 1967.
———, ———, ———, and A. Mehmet: Reliability Prediction—Mechanical Stress/Strength Interference (Nonferrous), *Tech. Rep.* RADC-TR-68-403, Rome Air Development Center, Air Force Systems Command, Griffiss Air Force Base, New York, February, 1969.

Mood, A. M., and F. A. Graybill: "Introduction to the Theory of Statistics," 2d ed., McGraw-Hill Book Company, New York, 1963.

Moroney, M. J.: "Facts from Figures," Penguin Books, Inc., Baltimore, 1965.

Natrella, M. G.: Experimental Statistics, *Nat. Bur. Stand. Hand.* 91, Aug. 1, 1963.

Weibull, W.: A Statistical Representation of Fatigue Failures in Solids, *Acta Polytech., Mech. Eng. Ser.*, vol. 1, no. 9, 1949.

————: A Statistical Distribution Function of Wide Applicability, *J. Appl. Mech.*, vol. 18, pp. 293–297, September, 1951.

3
Experiments of Evaluation

One of the important tasks in designing engineering experiments is to draw conclusions about a population from a sample. An engineer seldom knows the characteristics of the population. In most cases such information just does not exist. Therefore, he chooses a sample of a limited number of items. Such a sample may or may not be truly representative of the whole population, and therefore it is essential to predict the characteristics of the population from the sample.

Experiments can be designed around samples which would evaluate the following characteristics: the quality of the product, the uniformity of the product, and the probability of occurrence of a particular event associated with the product. The first two are the basic parameters of the normal and Weibull distributions. In the case of the normal distribution, the quality is expressed by mean μ and the uniformity by standard deviation σ or variance σ^2. For the Weibull distribution, the corresponding two parameters are the characteristic value θ and the Weibull slope b. The probability of occurrence of a particular event p is the basic parameter of the binomial distribution.

3.1 NORMAL DISTRIBUTION

3.1.1 ESTIMATE OF THE UNIFORMITY OF A PRODUCT

In many engineering problems the proper functioning of the product is determined by the uniformity of the characteristics of the product. A test therefore is run, sample standard deviation s is determined, and from this the population standard deviation σ is predicted. In these cases, therefore, it is desirable to know the relation between s and σ. This relation can be derived from the χ^2 distribution.

The chi-square (χ^2) distribution   Consider an experiment which has the following characteristics:

1. The experiment consists of ν trials.
2. The variables $x_1, x_2, \ldots, x_\nu$ are obtained by the first, second, $\ldots$, νth trial, respectively.
3. These variables are independent and normally distributed.
4. If the normal distribution has mean $\mu = 0$ and variance $\sigma^2 = 1$, then

$$\chi^2 = \text{sum of squares of random variable}$$

$$= x_1{}^2 + x_2{}^2 + x_3{}^2 + \cdots + x_\nu{}^2 = \sum_{i=1}^{\nu} (x_i)^2$$

5. If the variable is distributed with mean μ and variance σ^2, then

$$\chi^2 = \sum_{i=1}^{\nu} \frac{(x_i - \mu)^2}{\sigma^2} \tag{3.1}$$

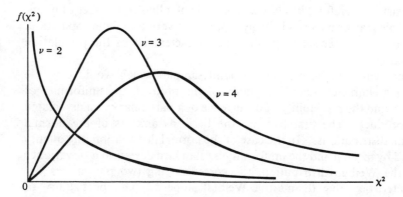

Fig. 3.1   Density function of a χ^2 random variable.

The χ^2 is another random variable which has a χ^2 distribution, and its density function is

$$f(\chi^2) = \frac{1}{2^{\nu/2}\Gamma(\nu/2)} (\chi^2)^{(\nu-2)/2} e^{-\chi^2/2} \qquad \text{for } \chi^2 \geq 0 \qquad (3.2)$$

where ν = number of trials = degrees of freedom, and Γ represents the gamma function (Table A-10). Figure 3.1 shows such a distribution. There are many physical phenomena which have a probability distribution approximated by χ^2.

The expected value $E(\chi^2)$ of a χ^2 distribution is

$$E(\chi^2) = \int_0^\infty (\chi^2) f(\chi^2) \, d\chi^2$$

$$= \int_0^\infty \frac{1(\chi^2)^{\nu/2}}{2^{\nu/2}\Gamma(\nu/2)} e^{-\chi^2/2} \, d(\chi^2)$$

$$= \frac{1}{2^{\nu/2}\Gamma(\nu/2)} \frac{\Gamma(\nu/2 + 1)}{\frac{1}{2}(\nu/2 + 1)}$$

$$= \frac{2^{\nu/2+1}}{2^{\nu/2}} \frac{(\nu/2)\Gamma(\nu/2)}{\Gamma(\nu/2)} = \nu$$

Hence

$$E(\chi^2) = \mu = \nu \qquad (3.3)$$

The variance of a χ^2 distribution is

$$\sigma_{\chi^2}^2 = \int_0^\infty (\chi^2)^2 f(\chi^2) \, d(\chi^2)$$

$$= \int_0^\infty \frac{1(\chi^2)^2(\chi^2)^{\nu/2-1}}{2^{\nu/2}\Gamma(\nu/2)} e^{-\chi^2/2} \, d(\chi^2)$$

$$= 2\nu \qquad (3.4)$$

or

$$\sigma = \sqrt{2\nu}$$

Hence, the mean of a χ^2 distribution is ν and the standard deviation is $\sqrt{2\nu}$.

The probability that $\chi^2 \geq \chi_{\alpha;\,\nu}^2$ is

$$P(\chi^2 > \chi_{\alpha;\,\nu}^2) = \int_{\chi_{\alpha;\,\nu}^2}^\infty f(\chi^2) \, d\chi^2$$

Graphically, this means that the probability that $\chi^2 > \chi_{\alpha;\,\nu}^2$ is given by the area α under the curve, with ν degrees of freedom, between $\chi_{\alpha;\,\nu}^2$ and ∞.

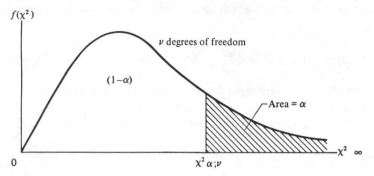

Fig. 3.2   The meaning of $\chi^2_{\alpha;\nu}$.

This is shown in Fig. 3.2.   The distribution starts at zero and therefore is skewed to the right.   These probabilities are given in Table A-2 for different values of α and ν.   For example, if $\nu = 2$ and $\chi^2_{\alpha;\,\nu} = 7.824$, then

$$P(\chi^2 > \chi^2_{\alpha;\,\nu}) = \alpha$$

From Table A-2, α is given as 0.02, or 2 percent.

Prediction of the population variance from the sample variance
The following derivation is based on a general case of a random variable when distributed normally with mean μ and variance σ^2.   From Eq. (1.14), when $\bar{x}$ is assumed to be equal to μ,

$$s^2 = \frac{\sum (x - \mu)^2}{n - 1}$$

where n is the sample size and $n - 1$ are the degrees of freedom (see Sec. 1.12) for unbiased estimate of variance.   From Eq. (3.1),

$$\chi^2 = \frac{\sum (x - \mu)^2}{\sigma^2}$$

when mean $= \mu$ and variance $= \sigma^2$.   Therefore,

$$s^2 = \frac{\chi^2(\sigma^2)}{n - 1}$$

with degrees of freedom ν equal to $n - 1$.
The two quantities s^2 and σ^2 are related by the equation

$$\chi^2 = (n - 1)\frac{s^2}{\sigma^2} \qquad \text{with } \nu = n - 1 \tag{3.5}$$

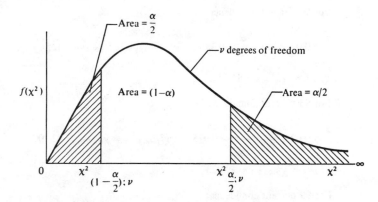

Fig. 3.3 The $(1 - \alpha)$ confidence interval for a χ^2 distribution.

From Fig. 3.3:

$$P(\chi^2_{(1-\alpha/2);\,v} \le \chi^2 \le \chi^2_{\alpha/2;\,v}) = 1 - \alpha$$

$$P\left[\chi^2_{(1-\alpha/2);\,v} \le (n-1)\frac{s^2}{\sigma^2} \le \chi^2_{\alpha/2;\,v}\right] = 1 - \alpha$$

or the confidence interval for the population standard deviation σ is

$$\sqrt{\frac{(n-1)s^2}{\chi^2_{\alpha/2;\,v}}} \le \sigma \le \sqrt{\frac{(n-1)s^2}{\chi^2_{(1-\alpha/2);\,v}}} \qquad (3.6)$$

Example 3.1 Friction linings from 10 wheels, removed after a certain use, were measured for wear with the following results:

Wear, 0.001 *in*	
8.05	5.0
9.1	8.53
6.2	6.8
7.5	9.6
10.3	11.46

Determine the standard deviation of the wear of the population from which this sample was taken, at 90, 95, and 99 percent confidence.

Solution

$\bar{x} = 8.254 \times 10^{-3}$ in $\qquad s = 1.94 \times 10^{-3}$ in [from Eq. (1.14)]

$v = n - 1 \qquad v = 10 - 1 = 9$

For 90 percent confidence:

$$\frac{\alpha}{2} = 0.05 \qquad 1 - \frac{\alpha}{2} = 0.95 \qquad \chi^2_{(1-\alpha/2);\,v} = \chi^2_{0.95;9} = 3.325$$

$$\chi^2_{\alpha/2;\,v} = \chi^2_{0.05;9} = 16.919$$

For 95 percent confidence:

$$\frac{\alpha}{2} = 0.025 \qquad 1 - \frac{\alpha}{2} = 0.975 \qquad \chi^2_{(1-\alpha/2);\,v} = \chi^2_{0.975;9} = 2.700$$

$$\chi^2_{\alpha/2;\,v} = \chi^2_{0.025;9} = 19.023$$

For 99 percent confidence:

$$\frac{\alpha}{2} = 0.005 \qquad 1 - \frac{\alpha}{2} = 0.995 \qquad \chi^2_{(1-\alpha/2);\,v} = \chi^2_{0.995;9} = 1.735$$

$$\chi^2_{\alpha/2;\,v} = \chi^2_{0.005;9} = 23.589$$

From Eq. (3.6):

$$\sqrt{\frac{(n-1)s^2}{\chi^2_{\alpha/2;\,v}}} \le \sigma \le \sqrt{\frac{(n-1)s^2}{\chi^2_{(1-\alpha/2);\,v}}}$$

For 90 percent confidence:

$$\sqrt{\frac{(10-1)(1.94^2) \times 10^{-6}}{16.919}} \le \sigma \le \sqrt{\frac{(10-1)(1.94^2) \times 10^{-6}}{3.325}}$$

$$\sqrt{10^{-6} \times \frac{33.8}{16.919}} \le \sigma \le \sqrt{\frac{33.8}{3.325} \times 10^{-6}}$$

Therefore

$$\sqrt{2.305 \times 10^{-6}} \le \sigma \le \sqrt{10.3 \times 10^{-6}}$$

or

$$1.43 \times 10^{-3} \text{ in} \le \sigma \le 3.24 \times 10^{-3} \text{ in}$$

Hence, it can be concluded with 90 percent confidence that the standard deviation of wear of the population of the lining is a minimum of 1.43×10^{-3} in and a maximum of 3.24×10^{-3} in. Similarly, for 95 percent confidence:

$$1.35 \times 10^{-3} \text{ in} \le \sigma \le 3.58 - 10^{-3} \text{ in}$$

and for 99 percent confidence:

$$1.22 \times 10^{-3} \text{ in} \le \sigma \le 4.52 \times 10^{-3} \text{ in}$$

3.1.2 ESTIMATE OF THE QUALITY OF A PRODUCT

Probably the most important characteristic of a product is its quality, which statistically can be best expressed by the mean value. Since the mean of the population μ is seldom known, a test is run on a sample taken from this population, and the sample mean $\bar{x}$ is determined. From $\bar{x}$ the value of μ is then predicted. To make this prediction, it is essential to establish the relation between $\bar{x}$ and μ. This relation is established here for the following cases:

1. If the standard deviation of the population σ is known, in which case the sample size chosen may be small; or if σ is not known, in which case sample size must be large (> 30) so that the standard deviation of the sample s is approximately equal to the standard deviation of the population σ—in these cases use the z distribution (Subsec. 2.1.1).
2. If the standard deviation of the population σ is not known and the sample size is small (< 30), use the t distribution (discussed later in this section)

In order to predict the population mean μ from the sample mean $\bar{x}$, for both cases it is necessary first to establish the relationship between the distribution of the variable x and of $\bar{x}$.

Relationship between the distributions of x **and** $\bar{x}$ Consider a test conducted with the following sample of n items, $x_1, x_2, \ldots, x_n$. Then

$$\bar{x}_n = \frac{x_1 + x_2 + x_3 + \cdots + x_n}{n}$$

$$= \frac{x_1}{n} + \frac{x_2}{n} + \cdots + \frac{x_n}{n}$$

Hence, $\bar{x}$ can be considered a new variable which is the linear combination of independent random variables $x_1, x_2, \cdots x_n$. The expected value of this variable is

$$E(\bar{x}) = E\left(\frac{1}{n} x_1 + \frac{1}{n} x_2 + \cdots + \frac{1}{n} x_n\right)$$

$$= \frac{1}{n} E(x_1) + \frac{1}{n} E(x_2) + \cdots + \frac{1}{n} E(x_n)$$

$$= \frac{1}{n} (\mu + \mu + \cdots + \mu)$$

$$= \frac{n\mu}{n} = \mu$$

In a similar manner the variance of $\bar{x}$ is given by

$$\sigma_{\bar{x}}^2 = a_1{}^2\sigma_1{}^2 + a_2{}^2\sigma_2{}^2 + a_3{}^2\sigma_3{}^2 + \cdots + a_n{}^2\sigma_n{}^2$$

$$= \left(\frac{\sigma}{n}\right)^2 + \left(\frac{\sigma}{n}\right)^2 + \cdots + \left(\frac{\sigma}{n}\right)^2$$

$$= \frac{1}{n^2}(n\sigma^2)$$

$$= \frac{\sigma^2}{n}$$

or

$$\sigma_{\bar{x}} = \frac{\sigma}{\sqrt{n}} \tag{3.7}$$

$\sigma_{\bar{x}}$, therefore, is the standard deviation of the variable $\bar{x}$. It is frequently referred to as the standard error of the mean. Thus

$$E(\bar{x}) = \mu \tag{3.8}$$

Similarly,

$$s_{\bar{x}} = \frac{s}{\sqrt{n}} \tag{3.9}$$

Figure 3.4 shows the relation between the distribution of $\bar{x}$ and the distribution of x.

The z distribution   In Eqs. (2.2) and (2.4), the variable was x and the standardized normal variate z was defined as

$$z = \frac{x - \mu}{\sigma}$$

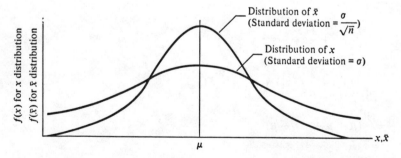

Fig. 3.4 Relation between the distributions of sample mean $\bar{x}$ and the variable x.

and the probability that x is larger than the predetermined value b was defined as

$$P(x > b) = P\left(z > \frac{b - \mu}{\sigma}\right) = \alpha$$

or, in general,

$$P(-z_{\alpha/2} \leq z \leq z_{\alpha/2}) = 1 - \alpha \tag{3.10}$$

In the present case, the variable under consideration is $\bar{x}$ instead of x; therefore z can be defined as

$$z = \frac{\bar{x} - \mu}{\sigma_{\bar{x}}} \tag{3.11}$$

With the aid of Eq. (3.7) this becomes

$$z = \frac{\bar{x} - \mu}{\sigma/\sqrt{n}} \tag{3.12}$$

This equation is used for the prediction of the population mean from the sample mean.

Prediction of the population mean from the sample mean by using z distribution As pointed out before, z distribution applies to the situations when the standard deviation of the population σ is known, in which case the sample size may be small; or if σ is not known, the sample size must be over 30.

Substituting Eq. (3.12) into Eq. (3.10),

$$P\left(-z_{\alpha/2} \leq \frac{\bar{x} - \mu}{\sigma/\sqrt{n}} \leq z_{\alpha/2}\right) = 1 - \alpha \tag{3.13}$$

It will be recalled that this is applicable when the sample size is large (over 30). For smaller sample sizes, t distribution, discussed later in this section, should be used.

Example 3.2 A sample of 36 resistors was tested, and the following data were obtained:

$$\bar{x} = 20 \text{ k}\Omega \qquad s = 4.2 \text{ k}\Omega \cong \sigma$$

With 95 percent confidence, predict the true (population) mean value of the resistance.

Solution

$$n = 36 \qquad 1 - \alpha = 0.95 \qquad \alpha = 0.05 \qquad \frac{\alpha}{2} = 0.025$$

$$z_{\alpha/2} = 1.96 \text{ from Table A-1}$$

$$P\left(-z_{\alpha/2} \leq \frac{\bar{x} - \mu}{\sigma/\sqrt{n}} \leq z_{\alpha/2}\right) = 1 - \alpha$$

$$P\left[\left(\bar{x} - z_{\alpha/2}\,\frac{\sigma}{\sqrt{n}}\right) \le \mu \le \left(\bar{x} + z_{\alpha/2}\,\frac{\sigma}{\sqrt{n}}\right)\right] = 1 - \alpha$$

$$\left(20 - 1.96\,\frac{4.2}{\sqrt{36}}\right) \le \mu \le \left(20 + 1.96\,\frac{4.2}{\sqrt{36}}\right) = 0.95$$

$$P(18.63 \le \mu \le 21.37) = 0.95$$

Therefore, with a 95 percent confidence level, the true mean value μ of the resistance is between 18.63 and 21.37 kΩ.

The t distribution t distribution is used for the prediction of the population mean μ from the sample mean $\bar{x}$ when the population standard deviation σ is not known and the sample size is small.   The t distribution is mathematically defined as follows:

1. If x_1, x_2, ..., x_n are normally distributed random variables with zero mean and unit variance, and
2. χ^2 is a random variable independent of x and having a χ^2 distribution with v degrees of freedom, then the random variable

$$\frac{x}{\sqrt{\chi^2/v}}$$

has a t distribution with v degrees of freedom.

Such a distribution has a frequency distribution function

$$f(t) = \frac{1}{\sqrt{\pi v}}\,\frac{\Gamma[(v+1)/2]}{\Gamma(v/2)}\left(1 + \frac{t^2}{v}\right)^{-(v+1)/2} \qquad -\infty < t < \infty \qquad (3.14)$$

A z distribution extends symmetrically from $-\infty$ to $+\infty$ whereas a χ^2 distribution extends from 0 to $+\infty$.   Hence, a t distribution also extends symmetrically from $-\infty$ to $+\infty$.   Figure 3.5 shows a plot of $f(t)$ versus t. From this figure it is seen that as $v \to \infty$, a t distribution tends to be a normal distribution.

The expected value of a t distribution is

$$E(t) = \int_{-\infty}^{\infty} tf(t)\,dt = 0 \qquad \text{for } v > 1$$

and the variance of a t distribution is

$$\sigma_t^2 = \int_{-\infty}^{\infty} t^2 f(t)\,dt = \frac{v}{v-2} \qquad \text{for } v = 2$$

The evaluation of these integrals is generally not made because the $E(t)$ and σ_t^2 are of little use in applications of the t distribution.

The probability α that the value of t exceeds a preset constant value $t_{\alpha;\,v}$ is given by the shaded area in Fig. 3.6.   This is denoted by $\alpha = P(t > t_{\alpha;\,v})$.

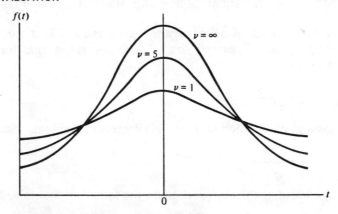

Fig. 3.5 Density plot of t distribution.

Values of α are given for various values of $t_{\alpha;\,\nu}$ in Table A-3. It is interesting to note that the values of α for $t_{\alpha;\,\nu}$ when $\nu = \infty$ are the same as the values of α given in Table A-1 for the z distribution. This follows from the fact that as $\nu \to \infty$ the t distribution asymptotically approaches a normal distribution. The tabulation of α is made only up to $\nu = 500$. This is because α does not change appreciably with changes in ν above 500. For example, to find the probability that a variable with t distribution has a value greater than or equal to 2.787, given that $\nu = 25$, $P(t > t_{\alpha;\,\nu}) = \alpha$. Then, from Table A-3 for $\nu = 25$ and $t_{\alpha;\,\nu} = 2.787$, $\alpha = 0.005 =$ required probability.

Prediction of the population mean from the sample mean by using t distribution It has been shown before that when $x_1, x_2, \ldots, x_n$ are independent and normally distributed with mean μ and variance σ^2, the variable

$$z = \frac{\bar{x} - \mu}{\sigma / \sqrt{n_x}}$$

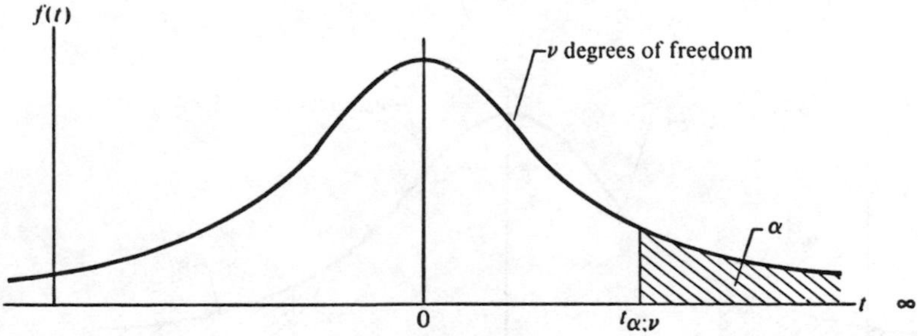

$f(t)$

ν degrees of freedom

α

0 $t_{\alpha;\nu}$ t ∞

Fig. 3.6 Meaning of $t_{\alpha;\nu}$.

is also normally distributed with zero mean and unit variance; $\bar{x}$ here was obtained from n_x sample size. It was also shown that for a sample size n_s the variable

$$\frac{(n_s - 1)s^2}{\sigma^2}$$

has a χ^2 distribution with $(n_s - 1)$ degrees of freedom. Since, by definition,

$$t = \frac{x}{\sqrt{\chi^2/v}}$$

by combining the above relations, the following is obtained:

$$\frac{[(\bar{x} - \mu)\sqrt{n_x}/\sigma]\sqrt{n_s - 1}}{\sqrt{(n_s - 1)s^2/\sigma^2}} = \frac{(\bar{x} - \mu)\sqrt{n_x}}{s} = t \tag{3.15}$$

has a t distribution with $(n_s - 1)$ degrees of freedom.

The usefulness of this variable t becomes apparent when one observes that $\bar{x}$ and s are the parameters associated with a given sample and as such could be found from a test. The relationship between the sample mean $\bar{x}$ and the population mean μ is derived from the following probability statements.

Referring to Fig. 3.7:

$$P(t \leq t_{\alpha/2;\,v}) = 1 - \frac{\alpha}{2} \quad \text{where } v = n - 1$$

$$P(-t_{\alpha/2} > t) = \frac{\alpha}{2}$$

$$P(-t_{\alpha/2;\,v} \leq t \leq +t_{\alpha/2;\,v}) = 1 - \alpha$$

where $1 - \alpha$ is the confidence, and $t = \dfrac{\bar{x} - \mu}{s/\sqrt{n}}$.

$$P\left(-t_{\alpha/2;\,v} \leq \frac{\bar{x} - \mu}{s/\sqrt{n}} \leq +t_{\alpha/2;\,v}\right) = 1 - \alpha$$

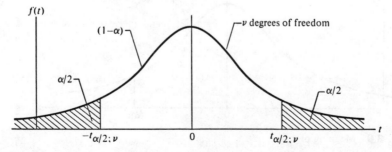

Fig. 3.7 The $(1 - \alpha)$ confidence interval for a t distribution.

or $(1 - \alpha)$ confidence interval for μ is

$$\bar{x} - \frac{s}{\sqrt{n}}\, t_{\alpha/2;\,v} \leq \mu \leq \bar{x} + \frac{s}{\sqrt{n}}\, t_{\alpha/2;\,v} \tag{3.16}$$

Example 3.3 A filling machine is set to fill packages to a certain weight x. Sixteen packages were chosen at random; their average weight was found to be 95 oz, and the standard deviation 4.00 oz. Estimate the population mean μ with 90 percent confidence.

Solution

$\bar{x} = 95$ oz $\qquad s = 4$ oz $\qquad n = 16 \qquad v = n - 1 = 15$

$t_{\alpha/2;\,v} = t_{0.05;\,15} = 1.753$ from Table A-3

Substituting the values in Eq. (3.16),

$$95 - \frac{4}{\sqrt{16}} \times 1.753 \leq \mu \leq 95 + \frac{4}{\sqrt{16}} \times 1.753$$

or

93.247 oz $\leq \mu \leq 96.753$ oz

Hence, it can be concluded with 90 percent confidence that the range 93.247 to 96.753 oz will contain the population mean μ.

3.1.3 ESTIMATE OF THE POPULATION LIMITS OF A PRODUCT

Sections 3.1.1 and 3.1.2 were concerned with the estimation of the uniformity of a product in terms of the standard deviation (σ) and of the quality of a product in terms of its mean value (μ) respectively. A pair of confidence limits were determined so that the interval between the two limits would contain the population parameters σ or μ with a certain confidence. However, in many engineering situations it is important to determine an interval that would cover a fixed proportion of the population distribution. This interval can be expressed in the form

$\bar{x} \pm Ks$

where $\bar{x}$ = sample average based on the sample size n
$\quad s$ = sample standard deviation
and K = factor that depends on sample size n, the desired proportion P of the population distribution, and the confidence at which this interval is estimated

This interval for population distribution is larger than the one that covers a single parameter of the population.

Table 3.1 gives the value of K for the two-sided cases where the estimate of both the upper limit ($\bar{x} + Ks$) and of the lower limit ($\bar{x} - Ks$) are to be made. Table 3.2 is to be used for one-sided cases when an estimate is to be made of only one of the two limits.

Table 3.1 Factors (K) for two-sided population limits for normal distributions*

n	C = 0.75					C = 0.90					C = 0.95					C = 0.99				
P	0.75	0.90	0.95	0.99	0.999	0.75	0.90	0.95	0.99	0.999	0.75	0.90	0.95	0.99	0.999	0.75	0.90	0.95	0.99	0.999
2	4.498	6.301	7.414	9.531	11.920	11.407	15.978	18.800	24.167	30.227	22.858	32.019	37.674	48.430	60.573	114.363	160.193	188.491	242.300	303.054
3	2.501	3.538	4.187	5.431	6.844	4.132	5.847	6.919	8.974	11.309	5.922	8.380	9.916	12.861	16.208	13.378	18.930	22.401	29.055	36.616
4	2.035	2.892	3.431	4.471	5.657	2.932	4.166	4.943	6.440	8.149	3.779	5.369	6.370	8.299	10.502	6.614	9.398	11.150	14.597	18.383
5	1.825	2.599	3.088	4.033	5.117	2.454	3.494	4.152	5.423	6.879	3.002	4.275	5.079	6.634	8.415	4.643	6.012	7.855	10.260	13.015
6	1.704	2.429	2.889	3.779	4.802	2.196	3.131	3.723	4.870	6.188	2.604	3.712	4.414	5.775	7.337	3.743	5.337	6.345	8.301	10.548
7	1.624	2.318	2.757	3.611	4.593	2.034	2.902	3.452	4.521	5.750	2.361	3.309	4.007	5.248	6.676	3.233	4.613	5.488	7.187	9.142
8	1.568	2.238	2.663	3.491	4.444	1.921	2.743	3.264	4.278	5.446	2.197	3.136	3.732	4.891	6.226	2.905	4.147	4.936	6.468	8.234
9	1.525	2.178	2.593	3.400	4.330	1.839	2.626	3.125	4.098	5.220	2.078	2.967	3.532	4.631	5.899	2.677	3.822	4.550	5.966	7.600
10	1.492	2.131	2.537	3.328	4.241	1.775	2.535	3.018	3.959	5.046	1.987	2.839	3.379	4.433	5.649	2.508	3.582	4.265	5.594	7.129
11	1.465	2.093	2.493	3.271	4.169	1.724	2.463	2.933	3.849	4.906	1.916	2.737	3.259	4.277	5.452	2.378	3.397	4.045	5.308	6.766
12	1.443	2.062	2.456	3.223	4.110	1.683	2.404	2.863	3.758	4.792	1.858	2.655	3.162	4.150	5.291	2.274	3.250	3.870	5.079	6.477
13	1.425	2.036	2.424	3.183	4.059	1.648	2.355	2.805	3.682	4.697	1.810	2.587	3.081	4.044	5.158	2.190	3.130	3.727	4.893	6.240
14	1.409	2.013	2.398	3.148	4.016	1.619	2.314	2.756	3.618	4.615	1.770	2.529	3.012	3.955	5.045	2.120	3.029	3.608	4.737	6.043
15	1.395	1.994	2.375	3.118	3.979	1.594	2.278	2.713	3.562	4.545	1.735	2.480	2.954	3.878	4.949	2.060	2.945	3.507	4.605	5.876
16	1.383	1.977	2.355	3.092	3.946	1.572	2.246	2.676	3.514	4.484	1.705	2.437	2.903	3.812	4.865	2.009	2.872	3.421	4.492	5.732
17	1.372	1.962	2.337	3.069	3.917	1.552	2.219	2.643	3.471	4.430	1.679	2.400	2.858	3.754	4.791	1.965	2.808	3.345	4.393	5.607
18	1.363	1.948	2.321	3.048	3.891	1.535	2.194	2.614	3.433	4.382	1.655	2.366	2.819	3.702	4.725	1.926	2.753	3.279	4.307	5.497
19	1.355	1.936	2.307	3.030	3.867	1.520	2.172	2.588	3.399	4.329	1.635	2.337	2.784	3.656	4.667	1.891	2.703	3.221	4.230	5.399
20	1.347	1.925	2.294	3.013	3.846	1.506	2.152	2.564	3.368	4.300	1.616	2.310	2.752	3.615	4.614	1.860	2.659	3.168	4.161	5.312
21	1.340	1.915	2.282	2.998	3.827	1.493	2.135	2.543	3.340	4.264	1.599	2.286	2.723	3.577	4.567	1.833	2.620	3.121	4.100	5.234
22	1.334	1.906	2.271	2.984	3.809	1.482	2.118	2.524	3.315	4.232	1.584	2.264	2.697	3.543	4.523	1.808	2.584	3.078	4.044	5.163
23	1.328	1.898	2.261	2.971	3.793	1.471	2.103	2.506	3.292	4.203	1.570	2.244	2.673	3.512	4.484	1.785	2.551	3.040	3.993	5.098
24	1.322	1.891	2.252	2.959	3.778	1.462	2.089	2.489	3.270	4.176	1.557	2.225	2.651	3.483	4.447	1.764	2.522	3.004	3.947	5.039
25	1.317	1.883	2.244	2.948	3.764	1.453	2.077	2.474	3.251	4.151	1.545	2.208	2.631	3.457	4.413	1.745	2.494	2.972	3.904	4.985
26	1.313	1.877	2.236	2.938	3.751	1.444	2.065	2.460	3.232	4.127	1.534	2.193	2.612	3.432	4.382	1.727	2.469	2.941	3.865	4.935
27	1.309	1.871	2.229	2.929	3.740	1.437	2.054	2.447	3.215	4.106	1.523	2.178	2.595	3.409	4.353	1.711	2.446	2.914	3.828	4.888
30	1.297	1.855	2.210	2.904	3.708	1.417	2.025	2.413	3.170	4.049	1.497	2.140	2.549	3.350	4.278	1.668	2.385	2.841	3.733	4.768
35	1.283	1.834	2.185	2.871	3.667	1.390	1.988	2.368	3.112	3.974	1.462	2.090	2.490	3.272	4.179	1.613	2.306	2.748	3.611	4.611
40	1.271	1.818	2.166	2.846	3.635	1.370	1.959	2.334	3.066	3.917	1.435	2.052	2.445	3.213	4.104	1.571	2.247	2.677	3.518	4.493
45	1.262	1.805	2.150	2.826	3.609	1.354	1.935	2.306	3.039	3.871	1.414	2.021	2.408	3.165	4.042	1.539	2.200	2.621	3.444	4.399

n																				
50	1.255	1.794	2.138	2.809	3.588	1.340	1.916	2.284	3.001	3.833	1.396	1.996	2.379	3.126	3.993	1.512	2.162	2.576	3.385	4.323
55	1.249	1.785	2.127	2.795	3.571	1.329	1.901	2.265	2.976	3.801	1.382	1.976	2.354	3.094	3.951	1.490	2.130	2.538	3.335	4.260
60	1.243	1.778	2.118	2.784	3.556	1.320	1.887	2.248	2.955	3.774	1.369	1.958	2.333	3.066	3.916	1.471	2.103	2.506	3.293	4.206
65	1.239	1.771	2.110	2.773	3.543	1.312	1.875	2.235	2.937	3.751	1.359	1.943	2.315	3.042	3.886	1.455	2.080	2.478	3.257	4.160
70	1.235	1.765	2.104	2.764	3.531	1.304	1.865	2.222	2.920	3.730	1.349	1.929	2.299	3.021	3.859	1.440	2.060	2.454	3.225	4.120
75	1.231	1.760	2.098	2.757	3.521	1.298	1.856	2.211	2.906	3.712	1.341	1.917	2.285	3.002	3.835	1.428	2.042	2.433	3.197	4.084
80	1.228	1.756	2.092	2.749	3.512	1.292	1.848	2.202	2.894	3.696	1.334	1.907	2.272	2.986	3.814	1.417	2.026	2.414	3.173	4.053
85	1.225	1.752	2.087	2.743	3.504	1.287	1.841	2.193	2.882	3.682	1.327	1.897	2.261	2.971	3.795	1.407	2.012	2.397	3.150	4.024
90	1.223	1.748	2.083	2.737	3.497	1.283	1.834	2.185	2.872	3.669	1.321	1.889	2.251	2.958	3.778	1.398	1.999	2.382	3.130	3.999
95	1.220	1.745	2.079	2.732	3.490	1.278	1.828	2.178	2.863	3.657	1.315	1.881	2.241	2.945	3.763	1.390	1.987	2.368	3.112	3.976
100	1.218	1.742	2.075	2.727	3.484	1.275	1.822	2.172	2.854	3.646	1.311	1.874	2.233	2.934	3.748	1.383	1.977	2.355	3.096	3.954
110	1.214	1.736	2.069	2.719	3.473	1.268	1.813	2.160	2.839	3.626	1.302	1.861	2.218	2.915	3.723	1.369	1.958	2.333	3.066	3.917
120	1.211	1.732	2.063	2.712	3.464	1.262	1.804	2.150	2.826	3.610	1.294	1.850	2.205	2.898	3.702	1.358	1.942	2.314	3.041	3.885
130	1.208	1.728	2.059	2.705	3.456	1.257	1.797	2.141	2.814	3.595	1.288	1.841	2.193	2.883	3.683	1.349	1.928	2.298	3.019	3.857
140	1.206	1.724	2.054	2.700	3.449	1.252	1.791	2.134	2.804	3.582	1.282	1.833	2.184	2.870	3.666	1.340	1.916	2.283	3.000	3.833
150	1.204	1.721	2.051	2.695	3.443	1.248	1.785	2.127	2.795	3.571	1.277	1.825	2.175	2.859	3.652	1.332	1.905	2.270	2.983	3.811
160	1.202	1.718	2.047	2.691	3.437	1.245	1.780	2.121	2.787	3.561	1.272	1.819	2.167	2.848	3.638	1.326	1.896	2.259	2.968	3.792
170	1.200	1.716	2.044	2.687	3.432	1.242	1.775	2.116	2.780	3.552	1.268	1.813	2.160	2.839	3.627	1.320	1.887	2.248	2.955	3.774
180	1.198	1.713	2.042	2.683	3.427	1.239	1.771	2.111	2.774	3.543	1.264	1.808	2.154	2.831	3.616	1.314	1.879	2.239	2.942	3.759
190	1.197	1.711	2.039	2.680	3.423	1.236	1.767	2.106	2.768	3.536	1.261	1.803	2.148	2.823	3.606	1.309	1.872	2.230	2.931	3.744
200	1.195	1.709	2.037	2.677	3.419	1.234	1.764	2.102	2.762	3.529	1.258	1.798	2.143	2.816	3.597	1.304	1.865	2.222	2.921	3.731
250	1.190	1.702	2.028	2.665	3.404	1.224	1.750	2.085	2.740	3.501	1.245	1.780	2.121	2.788	3.561	1.286	1.839	2.191	2.880	3.678
300	1.186	1.696	2.021	2.656	3.393	1.217	1.740	2.073	2.725	3.481	1.236	1.767	2.106	2.767	3.535	1.273	1.820	2.169	2.850	3.641
400	1.181	1.688	2.012	2.644	3.378	1.207	1.726	2.057	2.703	3.453	1.223	1.749	2.084	2.739	3.499	1.255	1.794	2.138	2.809	3.589
500	1.177	1.683	2.006	2.636	3.368	1.201	1.717	2.046	2.689	3.434	1.215	1.737	2.070	2.721	3.475	1.243	1.777	2.117	2.783	3.555
600	1.175	1.680	2.002	2.631	3.360	1.196	1.710	2.038	2.678	3.421	1.209	1.729	2.060	2.707	3.458	1.234	1.764	2.102	2.763	3.530
700	1.173	1.677	1.998	2.626	3.355	1.192	1.705	2.032	2.670	3.411	1.204	1.722	2.052	2.697	3.445	1.227	1.755	2.091	2.748	3.511
800	1.171	1.675	1.996	2.623	3.350	1.189	1.701	2.027	2.663	3.402	1.201	1.717	2.046	2.688	3.434	1.222	1.747	2.082	2.736	3.495
900	1.170	1.673	1.993	2.620	3.347	1.187	1.697	2.023	2.658	3.396	1.198	1.712	2.040	2.682	3.426	1.218	1.741	2.075	2.726	3.483
1000	1.169	1.671	1.992	2.617	3.344	1.185	1.695	2.019	2.654	3.390	1.195	1.709	2.036	2.676	3.418	1.214	1.736	2.068	2.718	3.472
∞	1.150	1.645	1.960	2.576	3.291	1.150	1.645	1.960	2.576	3.291	1.150	1.645	1.960	2.576	3.291	1.150	1.645	1.960	2.576	3.291

* By permission from C. Eisenhart, M. W. Hastay, and W. A. Wallis, "Techniques of Statistical Analysis," chap. 2, McGraw-Hill Book Company, New York, 1947.

Table 3.2 Factors (K) for one-sided population limits for normal distributions*

	C = 0.75					C = 0.90				
n \ P	0.75	0.90	0.95	0.99	0.999	0.75	0.90	0.95	0.99	0.999
3	1.464	2.501	3.152	4.396	5.805	2.602	4.258	5.310	7.340	9.651
4	1.256	2.134	2.680	3.726	4.910	1.972	3.187	3.957	5.437	7.128
5	1.152	1.961	2.463	3.421	4.507	1.698	2.742	3.400	4.666	6.112
6	1.087	1.860	2.336	3.243	4.273	1.540	2.494	3.091	4.242	5.556
7	1.043	1.791	2.250	3.126	4.118	1.435	2.333	2.894	3.972	5.201
8	1.010	1.740	2.190	3.042	4.008	1.360	2.219	2.755	3.783	4.955
9	0.984	1.702	2.141	2.977	3.924	1.302	2.133	2.649	3.641	4.772
10	0.964	1.671	2.103	2.927	3.858	1.257	2.065	2.568	3.532	4.629
11	0.947	1.646	2.073	2.885	3.804	1.219	2.012	2.503	3.444	4.515
12	0.933	1.624	2.048	2.851	3.760	1.188	1.966	2.448	3.370	4.420
13	0.919	1.606	2.026	2.822	3.722	1.162	1.928	2.403	3.310	4.341
14	0.909	1.591	2.007	2.796	3.690	1.139	1.895	2.363	3.257	4.274
15	0.899	1.577	1.991	2.776	3.661	1.119	1.866	2.329	3.212	4.215
16	0.891	1.566	1.977	2.756	3.637	1.101	1.842	2.299	3.172	4.164
17	0.883	1.554	1.964	2.739	3.615	1.085	1.820	2.272	3.136	4.118
18	0.876	1.544	1.951	2.723	3.595	1.071	1.800	2.249	3.106	4.078
19	0.870	1.536	1.942	2.710	3.577	1.058	1.781	2.228	3.078	4.041
20	0.865	1.528	1.933	2.697	3.561	1.046	1.765	2.208	3.052	4.009
21	0.859	1.520	1.923	2.686	3.545	1.035	1.750	2.190	3.028	3.979
22	0.854	1.514	1.916	2.675	3.532	1.025	1.736	2.174	3.007	3.952
23	0.849	1.508	1.907	2.665	3.520	1.016	1.724	2.159	2.987	3.927
24	0.845	1.502	1.901	2.656	3.509	1.007	1.712	2.145	2.969	3.904
25	0.842	1.496	1.895	2.647	3.497	0.999	1.702	2.132	2.952	3.882
30	0.825	1.475	1.869	2.613	3.454	0.966	1.657	2.080	2.884	3.794
35	0.812	1.458	1.849	2.588	3.421	0.942	1.623	2.041	2.833	3.730
40	0.803	1.445	1.834	2.568	3.395	0.923	1.598	2.010	2.793	3.679
45	0.795	1.435	1.821	2.552	3.375	0.908	1.577	1.986	2.762	3.638
50	0.788	1.426	1.811	2.538	3.358	0.894	1.560	1.965	2.735	3.604

* Adapted by permission from *Industrial Quality Control*, vol. xiv, no. 10, April, 1958, from article entitled "Tables for One-Sided Statistical Tolerance Limits," by G. J. Lieberman.

Table 3.2 (*continued*)

n \ P	C = 0.95					C = 0.99				
	0.75	0.90	0.95	0.99	0.999	0.75	0.90	0.95	0.99	0.999
3	3.804	6.158	7.655	10.552	13.857					
4	2.619	4.163	5.145	7.042	9.215					
5	2.149	3.407	4.202	5.741	7.501					
6	1.895	3.006	3.707	5.062	6.612	2.849	4.408	5.409	7.334	9.540
7	1.732	2.755	3.399	4.641	6.061	2.490	3.856	4.730	6.411	8.348
8	1.617	2.582	3.188	4.353	5.686	2.252	3.496	4.287	5.811	7.566
9	1.532	2.454	3.031	4.143	5.414	2.085	3.242	3.971	5.389	7.014
10	1.465	2.355	2.911	3.981	5.203	1.954	3.048	3.739	5.075	6.603
11	1.411	2.275	2.815	3.852	5.036	1.854	2.897	3.557	4.828	6.284
12	1.366	2.210	2.736	3.747	4.900	1.771	2.773	3.410	4.633	6.032
13	1.329	2.155	2.670	3.659	4.787	1.702	2.677	3.290	4.472	5.826
14	1.296	2.108	2.614	3.585	4.690	1.645	2.592	3.189	4.336	5.651
15	1.268	2.068	2.566	3.520	4.607	1.596	2.521	3.102	4.224	5.507
16	1.242	2.032	2.533	3.463	4.534	1.553	2.458	3.028	4.124	5.374
17	1.220	2.001	2.486	3.415	4.471	1.514	2.405	2.962	4.038	5.268
18	1.200	1.974	2.453	3.370	4.415	1.481	2.357	2.906	3.961	5.167
19	1.183	1.949	2.423	3.331	4.364	1.450	2.315	2.855	3.893	5.078
20	1.167	1.926	2.396	3.295	4.319	1.424	2.275	2.807	3.832	5.003
21	1.152	1.905	2.371	3.262	4.276	1.397	2.241	2.768	3.776	4.932
22	1.138	1.887	2.350	3.233	4.238	1.376	2.208	2.729	3.727	4.866
23	1.126	1.869	2.329	3.206	4.204	1.355	2.179	2.693	3.680	4.806
24	1.114	1.853	2.309	3.181	4.171	1.336	2.154	2.663	3.638	4.755
25	1.103	1.838	2.292	3.158	4.143	1.319	2.129	2.632	3.601	4.706
30	1.059	1.778	2.220	3.064	4.022	1.249	2.029	2.516	3.446	4.508
35	1.025	1.732	2.166	2.994	3.934	1.195	1.957	2.431	3.334	4.364
40	0.999	1.697	2.126	2.941	3.866	1.154	1.902	2.365	3.250	4.255
45	0.978	1.669	2.092	2.897	3.811	1.122	1.857	2.313	3.181	4.168
50	0.961	1.646	2.065	2.863	3.766	1.096	1.821	2.296	3.124	4.096

(handwritten annotation near row 16, column 0.95: 2.5353)

For example, the data in Example 3.3 gives the estimate only of the population mean (μ) to lie between 93.247 and 96.753 oz with a confidence of 90 percent on the basis of the sample mean ($\bar{x}$) being 95 oz, $s = 4$ oz, and $n = 16$. However, it is also possible to get an estimate of the population limits as follows:

$$\bar{x} = 95 \text{ oz}$$
$$s = 4 \text{ oz}$$
$$n = 16$$

From Table 3.1, for $C = 90$ percent confidence, $P = 95$ percent, and $n = 16$; $K = 2.676$. Therefore, the interval $\bar{x} \pm Ks = 95 \pm (2.676)$ (4), or the interval which would cover 95 percent of the population, at 90 percent confidence, is 84.3 to 105.7 oz.

3.2 WEIBULL DISTRIBUTION

Weibull distribution, because of its versatility, is widely used in engineering practice. Although it can be applied to a variety of engineering applications, its predominant use so far has been in the field of fatigue life phenomena associated with failure problems, warranties, etc. As in the case of normal distribution, two parameters are considered here, the quality of the product and its uniformity. In normal distribution the quality is expressed by μ and the uniformity by σ. In the case of the Weibull distribution the corresponding two values are μ or $\hat{\theta}$ and $\hat{b}$, where

μ is the mean value of the Weibull population.
$\hat{\theta}$ is the characteristic value of the Weibull population.
$\hat{b}$ is the slope of the Weibull population.

3.2.1 ESTIMATE OF THE UNIFORMITY OF A PRODUCT

Plot the test data on the Weibull paper with the aid of median ranks (see Subsec. 2.3.1), and determine the Weibull slope b of the test sample. To predict the true Weibull slope of the population $\hat{b}$, refer to Tables A-18 and A-19, which give the Weibull slope error in relation to the sample size for 50 and 90 percent confidence. This is illustrated in Example 3.4.

3.2.2 ESTIMATE OF THE QUALITY OF A PRODUCT

To predict population mean μ or population characteristic value $\hat{\theta}$ from the sample, the sample test data are plotted on the Weibull paper with the aid of the median ranks as before; then for, say, 90 percent confidence, these data are plotted on the same paper against 5 and 95 percent ranks (see Table A-14). The resultant plot gives a 90 percent confidence interval within which the mean and the characteristic values will lie. (For other confidence intervals such as 95 percent confidence, use the appropriate pair of ranks—in this case 2.5 and 97.5 percent.) This is illustrated by the following example.

Example 3.4 Five gears were fatigue tested to failure, and the following life cycles were recorded: 2.2×10^5, 0.51×10^5, 1.5×10^5, 3.0×10^5, 0.97×10^5. Predict at 90 percent confidence:

1. The true Weibull slope b
2. The population mean μ
3. The true characteristic life θ
4. The true $\hat{B}_{10}$ life

Solution Plot the median, 5 percent, and 95 percent ranks, from Tables A-11 and A-14, on Weibull paper (Fig. 3.8).

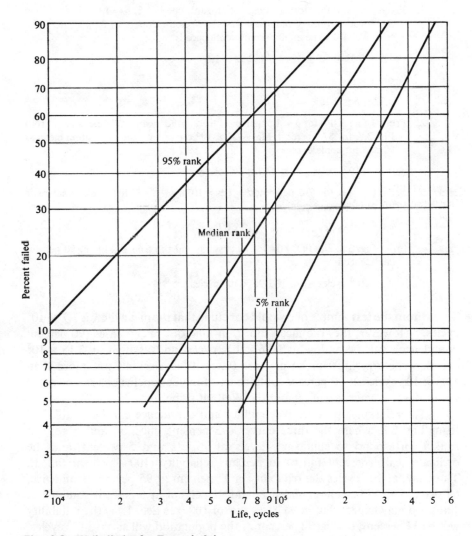

Fig. 3.8 Weibull plot for Example 3.4

Life, 10^5 cycles	Median ranks	5% ranks	95% ranks
0.51	0.1294	0.0102	0.4507
0.97	0.3147	0.0764	0.6574
1.5	0.5000	0.1893	0.8107
2.2	0.6853	0.3426	0.9236
3.0	0.8706	0.5493	0.9898

1. From Fig. 3.8, the slope of the median rank line = 1.5, $b = 1.5$. From Table A-18, for number of failures $n = 5$, the Weibull slope = ± 52 percent. The true Weibull slope at 90 percent confidence is

$$1.5 - 0.52 \times 1.5 \leq \hat{b} \leq 1.5 + 0.52 \times 1.5$$

or

$$0.72 \leq \hat{b} \leq 2.28$$

2. From Table A-17, for $b = 1.5$, percent failed at the mean is 57.5 percent. From Fig. 3.8, for 57.5 percent, life = 1.7×10^5 cycles, and the population mean at 90 percent confidence is

$$0.76 \times 10^5 \text{ cycles} \leq \mu \leq 3.15 \times 10^5 \text{ cycles}$$

3. From Fig. 3.8, for 63.2 percent, $\theta = 1.9 \times 10^5$ cycles, and the true characteristic life at 90 percent confidence is

$$8.6 \times 10^4 \text{ cycles} \leq \hat{\theta} \leq 3.4 \times 10^5 \text{ cycles}$$

4. From Fig. 3.8, B_{10} life = 0.43×10^5 cycles, and the true $\hat{B}_{10}$ life at 90 percent confidence is

$$0.9 \times 10^4 \text{ cycles} \leq \hat{B}_{10} \leq 1.05 \times 10^5 \text{ cycles}$$

From the test sample one would conclude that the mean life $\bar{x}$ is 1.7×10^5 cycles. However, the true mean life μ, in nine out of ten cases (90 percent confidence), can be as low as 0.76×10^5 cycles and as high as 3.15×10^5 cycles. The B_{10} life can be at least 0.9×10^4 cycles but not more than 1.05×10^5 cycles at 90 percent confidence. Similar conclusions can be made about the characteristic life $\hat{\theta}$ and the Weibull slope $\hat{b}$.

The relationship between reliability and confidence can be established from Fig. 3.8, where the three curves (95 percent, median, and 5 percent ranks) correspond to confidence levels of 95, 50, and 5 percent, and the ordinate scale (percent failed) to percent reliability (100 − percent failed). Thus, say, at 10^5 cycles the reliability is 32 percent at 95 percent confidence, 68 percent at 50 percent confidence, and 91 percent at 5 percent confidence. Thus, we may expect that in 95 cases out of 100 (see Sec. 1.11) the reliability will be 32 percent, that is, 32 percent of the population will survive 10^5 cycles. This, therefore, can be regarded as the minimum reliability of the product.

If in a given engineering application the high confidence of 95 percent is not required, and the judgment can be made on the basis of 50 percent confidence (as is sometimes done), the percent of population lasting 10^5 cycles will be 68 percent.

3.3 BINOMIAL DISTRIBUTION

Normal and Weibull distributions are used where the random variable x is continuous, that is, when it can assume any value within the specified limits. If the random variable x_i is discrete, as in the case of "go–no-go," "failure-success," "acceptable–not acceptable" situations, binomial distribution applies. In Sec. 2.5, the percent of defectives in the population was assumed to be known, and an attempt was made to predict the percent of defectives in the sample chosen from this population. In many engineering problems the reverse is needed: A sample is picked from an unknown population, it is tested, and from the resultant test data an attempt is made to describe (predict) the population.

Suppose, for example, 10 parts are picked at random from a large-size lot and tested. After, say, 100 h of testing, the 10 parts are inspected and two are found worn to the point where they must be considered failures. With the aid of the technique described below it should be possible to predict the percent of failure (at 100 h) in the whole lot, with a given confidence. (Although this situation involves a life problem, Weibull distribution cannot be used because life to failure is not known. The only information is how many parts have failed and how many did not fail at 100 h.) The probability of having a unit operate without malfunction is designated as q, and the probability of having a unit fail is p. Therefore

$$p + q = 1$$

The probability (confidence) of obtaining up to r_1 defective units from a randomly drawn sample of size n, as given in Chap. 2, is

$$p(r = r_1) = \binom{n}{r_1} p^{r_1} q^{n-r_1} \tag{3.17}$$

where

$$\binom{n}{r} = \frac{n!}{r!\,(n-r)!}$$

The goal is to develop a procedure for estimating the lower and upper limits of the probability of success q (which is synonymous with the term "reliability") at various confidence levels, given a sample of size n having r_1 defectives.

Some additional notation is now introduced. Let C_1 be the confidence limit that the probability of failure of the population is not lower than p_L. Let C_2 be the confidence limit that the probability of failure of the population is not larger than p_U where C_1 can equal C_2. Thus

$$p_L = \sum_{i=k}^{n} p(r = r_i) = 1 - C_1 \tag{3.18}$$

$$p_U = \sum_{i=0}^{n} p(r = r_i) = 1 - C_2 \tag{3.19}$$

The above equations are unwieldy. By making two transformations, the commonly tabulated F distribution can be used. The results are [1]

$$p_L = \frac{1}{1 + [(n - r + 1)/r]F_L} \tag{3.20}$$

where F_L = value from the F distribution tables corresponding to the degrees of freedom (discussed on p. 106)

$$v_1 = 2(n - r_1 + 1) \quad \text{and} \quad v_2 = 2r_1$$

$$p_U = \frac{1}{1 + \dfrac{n - r_1}{r_1 + 1} \dfrac{1}{F_U}} \tag{3.21}$$

where F_U = value from the F distribution tables corresponding to the degrees of freedom

$$v_1 = 2(r_1 + 1) \quad \text{and} \quad v_2 = 2(n - r_1)$$

The probability of success q can be similarly derived: $g = n - r$, where g = number of successes, n = number of trials, and r = number of failures. The relationships are

$$q_L = \frac{1}{1 + [(n - g + 1)/g]F_L} \tag{3.22}$$

where

$$v_1 = 2(n - g + 1)$$
$$v_2 = 2g$$

and

$$q_U = \frac{1}{1 + \dfrac{n - g}{g + 1} \dfrac{1}{F_U}} \tag{3.23}$$

$$v_1 = 2(g + 1)$$
$$v_2 = 2(n - g)$$

The subscripts U and L denote the upper and lower limits, respectively.

These relationships are plotted as the families of curves for 90, 95, and 99 percent confidence levels (Tables A-41 to A-46). Values for both the upper and the lower limits of the probability of success (reliability R) can be found from these tables. The example problem illustrates both the analytical and the graphical method of solving problems.

Example 3.5 A coal-mining firm is interested in purchasing a new type of drilling head for a mining machine. Each time one of these heads fails to function properly, there is a downtime at considerable expense to the company. The firm has purchased 20 new drilling heads for trial and found that 4 of the 20 failed to operate satisfactorily. What is the reliability of the new drilling heads at 95 percent confidence?

Solution

$$C = 95\% \qquad n = 20 \qquad r = 4 \qquad g = 16 \qquad R = \frac{16}{20} = 0.8$$

Lower limit for the probability of success (*reliability*):

$$q_L = \frac{1}{1 + [(n - g + 1)/g]F_L}$$

where

$$\nu_1 = 2(n - g + 1)$$
$$= 2(20 - 16 + 1)$$
$$= 10$$
$$\nu_2 = 2g$$
$$= 32$$

From F tables, Table A-5, $F_{0.05;10;32} = 2.14 = F_L$. Therefore

$$q_L = \frac{1}{1 + [(20 - 16 + 1)/16]2.14} = 60\%$$

$$= R_L = 60\%$$

Upper limit for the probability of success (*reliability*):

$$q_U = \frac{1}{1 + \dfrac{n - g}{g + 1}\dfrac{1}{F_U}}$$

where

$$\nu_1 = 2(g + 1)$$
$$= 2(16 + 1)$$
$$= 34$$

and

$$\nu_2 = 2(n - g)$$
$$= 2(20 - 16)$$
$$= 8$$

From F tables, Table A-5, $F_{0.05;\,34;\,8} \cong 3.06 = F_U$.   Therefore

$$q_U = \cfrac{1}{1 + \cfrac{20-16}{16+1}\cfrac{1}{3.06}}$$

$$= R_U = 93\%$$

Thus, combining the above results,

$$0.60 < \hat{R} < 0.93$$

where $\hat{R}$ is the true reliability of the population.   Hence, with 95 percent confidence, it can be stated that at least 60 but not more than 93 drilling heads out of 100 will function properly for the specified time.   The above answer can also be obtained graphically.   Tables A-41 through A-46 provide the lower and upper estimates of reliability for various confidence levels.   Using Table A-43 for 95 percent confidence, sample size of 20, and 80 percent survivors (or from the nomograph in Table A-40):

$$q_L = R_L = 60\%$$

From Table A-44:

$$q_U = R_U = 93\%$$

Thus, from the sample one would have concluded that the reliability (R) is 80 percent (16/20) when actually it can be as low as 60 percent and as high as 93 percent.   If sample size is increased to 40, and the percent defective is still the same (20 percent defectives, that is, 80 percent reliability), then from Table A-43, the lower estimate of reliability increases from 60 to 67 percent for 95 percent confidence.   This means that the reliability as based on the sample (80 percent) approaches the true reliability of the population as the sample size increases.

3.4 THE OUTLIERS

In engineering experimentation a situation frequently arises of obtaining a set of data in which one or more observations should be retained or rejected as being faulty.

There have been many methods proposed for rejecting experimental data, out of which the following three are given here because of their simplicity and ease of application to the engineering problems.

3.4.1 MODIFIED THREE-SIGMA TEST

The simplest criterion for rejection would be to calculate the sample mean $\bar{x}$ and sample standard deviation s by using all test data, and if the suspected observation lies outside the range $\bar{x} + 3s$, there is about 99 percent chance that the observation is an outlier.   However, as the sample size increases, the probability of getting an extreme observation increases too.   To account for this, the following modification should be made.

Calculate sample mean $\bar{x}$ and sample standard deviation s as before. Compute

$$z' = \frac{x - \bar{x}}{s} \tag{3.24}$$

where x is the suspect observation. Look up in Table A-1 the value of α corresponding to z' equal to z in the table.

If $n\alpha \le 0.1$, where n is the number of observations including the suspect one, reject x.

Example 3.6 Ten measurements were made with the following results (inches):

1.121	1.126
1.123	1.126
1.124	1.127
1.125	1.128
1.125	1.135

It was felt that the last observation (1.135) might be suspect. Determine whether this observation should be included in the analysis of the above data.

Solution By using Eqs. (1.11) and (1.14), $\bar{x}$ and s were computed:

$$\bar{x} = 1.126 \text{ in} \qquad s = 3.8 \times 10^{-3} \text{ in}$$

$$x = 1.135 \text{ (suspect observation)} \qquad n = 10$$

From Eq. (3.24),

$$z' = \frac{x - \bar{x}}{s} = \frac{1.135 - 1.126}{0.0038}$$

$$= 2.37$$

From Table A-1, the value of α corresponding to $z' = z = 2.37$ is

$$\alpha = 0.00889$$
$$n\alpha = 10 \times 0.00889$$
$$= 0.0889$$

Since $n\alpha = 0.0889 < 0.1$, the suspect observation (1.135) should be excluded from the analysis of the test data.

3.4.2 THE DIXON METHOD

This method, as described by Natrella [2], applies to situations prevalent in engineering problems where neither the mean μ nor the standard deviation σ of the population is known. In those few cases where these quantities are available, methods described in Ref. 2 can be used.

Arrange the data in an increasing order:

$$x_1 < x_2 < x_3 < \cdots < x_{n-1} < x_n$$

where n is the sample size.   If

$$3 \leq n \leq 7 \qquad \text{compute } r_{10}$$
$$8 \leq n \leq 10 \qquad \text{compute } r_{11}$$
$$11 \leq n \leq 13 \qquad \text{compute } r_{21}$$
$$14 \leq n \leq 25 \qquad \text{compute } r_{22}$$

where r_{ij} are computed as follows:

r_{ij}	*If x_n is suspect*	*If x_1 is suspect*
r_{10}	$(x_n - x_{n-1})/(x_n - x_1)$	$(x_2 - x_1)/(x_n - x_1)$
r_{11}	$(x_n - x_{n-1})/(x_n - x_2)$	$(x_2 - x_1)/(x_{n-1} - x_1)$
r_{21}	$(x_n - x_{n-2})/(x_n - x_2)$	$(x_3 - x_1)/(x_{n-1} - x_1)$
r_{22}	$(x_n - x_{n-2})/(x_n - x_3)$	$(x_3 - x_1)/(x_{n-2} - x_1)$

Table 3.3   Criteria for rejecting suspect observations [4] (the Dixon method)

Statistic	Sample size n	Critical values Probability levels $\alpha/2$						
		0.30	0.20	0.10	0.05	0.02	0.01	0.005
	3	0.684	0.781	0.886	0.941	0.976	0.988	0.994
	4	0.471	0.560	0.679	0.765	0.846	0.889	0.926
r_{10}	5	0.373	0.451	0.557	0.642	0.729	0.780	0.821
	6	0.318	0.386	0.482	0.560	0.644	0.698	0.740
	7	0.281	0.344	0.434	0.507	0.586	0.637	0.680
	8	0.318	0.385	0.479	0.554	0.631	0.683	0.725
r_{11}	9	0.288	0.352	0.441	0.512	0.587	0.635	0.677
	10	0.265	0.325	0.409	0.477	0.551	0.597	0.639
	11	0.391	0.442	0.517	0.576	0.638	0.679	0.713
r_{21}	12	0.370	0.419	0.490	0.546	0.605	0.642	0.675
	13	0.351	0.399	0.467	0.521	0.578	0.615	0.649
	14	0.370	0.421	0.492	0.547	0.602	0.641	0.674
	15	0.353	0.402	0.472	0.525	0.579	0.616	0.647
	16	0.338	0.386	0.454	0.507	0.559	0.595	0.624
	17	0.325	0.373	0.438	0.490	0.542	0.577	0.605
	18	0.314	0.361	0.424	0.475	0.527	0.561	0.589
	19	0.304	0.350	0.412	0.462	0.514	0.547	0.575
r_{22}	20	0.295	0.340	0.401	0.450	0.502	0.535	0.562
	21	0.287	0.331	0.391	0.440	0.491	0.524	0.551
	22	0.280	0.323	0.382	0.430	0.481	0.514	0.541
	23	0.274	0.316	0.374	0.421	0.472	0.505	0.532
	24	0.268	0.310	0.367	0.413	0.464	0.497	0.524
	25	0.262	0.304	0.360	0.406	0.457	0.489	0.516

Look up $r_{\alpha/2}$ for the given sample size n from Table 3.3, where $1 - \alpha$ is the confidence level desired. If $r_{ij} > r_{\alpha/2}$, reject the suspect observation with $(1 - \alpha)$ percent confidence.

The above applies when the suspect observations are present on both high and low sides of the sample. In cases where only one of the two situations arises, the value of $r_{\alpha/2}$ corresponds to the confidence level $(1 - \alpha/2)$ instead of $(1 - \alpha)$.

Example 3.7 Repeat the analysis of the test data in Example 3.6, using the Dixon method.

Solution From the test data $x_n = 1.135$ (suspect observation), $n = 10$, $x_1 = 1.121$ in, $x_9 = 1.128$ in, and $x_{10} = 1.135$ in $= x_n$. Since $n = 10$, the value r_{11} is to be computed.

$$r_{11} = \frac{x_n - x_{n-1}}{x_n - x_2} = \frac{x_{10} - x_9}{x_{10} - x_2}$$

or

$$r_{11} = \frac{1.135 - 1.128}{1.135 - 1.123} = 0.583$$

Since the suspect observation is present on only one side (high), the value of $r_{\alpha/2}$ from Table 3.3 would correspond to $(1 - \alpha/2)$ confidence level.

From Table 3.3, for $\alpha/2 = 0.02$ (98 percent confidence) and $n = 10$, $r_{\alpha/2} = 0.551$.

Since $r_{11} > r_{\alpha/2}$, it can be concluded with 98 percent confidence that the suspect observation should be excluded from the analysis. This means that the risk taken in making this decision is 2 percent.

3.4.3 THE GRUBBS METHOD [3]

Arrange the data in an increasing order: $x_1 < x_2 < x_3 < \cdots < x_{n-1} < x_n$, where n is the sample size. Let

$$v^2 = \sum_{i=1}^{n} (x_i - \bar{x})^2$$

$$v_1^2 = \sum_{i=2}^{n} (x_i - \bar{x}_1)^2$$

$$v_n^2 = \sum_{i=1}^{n-1} (x_i - \bar{x}_n)^2$$

where

$$\bar{x} = \frac{\sum_{i=1}^{n} x_i}{n}$$

$$\bar{x}_1 = \frac{\sum_{i=2}^{n} x_i}{n - 1}$$

and

$$\bar{x}_n = \frac{\sum_{i=1}^{n-1} x_i}{n-1}$$

If x_n is the suspect value, calculate v_n^2/v^2; if x_i is the suspect value, calculate v_1^2/v^2. If these values are less than those in Table 3.4, reject the suspect observations at the appropriate confidence level.

Example 3.8 Repeat the analysis of the test data in Example 3.6, using the Grubbs method.

Table 3.4 Criteria for rejecting suspect observations [3] (the Grubbs method)

Sample size n	Critical values Confidence levels			
	99%	97.5%	95%	90%
3	0.0001	0.0007	0.0027	0.0109
4	0.0100	0.0248	0.0494	0.0975
5	0.0442	0.0808	0.1270	0.1984
6	0.0928	0.1453	0.2032	0.2826
7	0.1447	0.2066	0.2696	0.3503
8	0.1948	0.2616	0.3261	0.4050
9	0.2411	0.3101	0.3742	0.4502
10	0.2831	0.3526	0.4154	0.4881
11	0.3211	0.3901	0.4511	0.5204
12	0.3554	0.4232	0.4822	0.5483
13	0.3864	0.4528	0.5097	0.5727
14	0.4145	0.4792	0.5340	0.5942
15	0.4401	0.5030	0.5559	0.6134
16	0.4634	0.5246	0.5755	0.6306
17	0.4848	0.5442	0.5933	0.6461
18	0.5044	0.5621	0.6095	0.6601
19	0.5225	0.5785	0.6243	0.6730
20	0.5393	0.5937	0.6379	0.6848
21	0.5548	0.6076	0.6504	0.6958
22	0.5692	0.6206	0.6621	0.7058
23	0.5827	0.6327	0.6728	0.7151
24	0.5953	0.6439	0.6829	0.7238
25	0.6071	0.6544	0.6923	0.7319

Solution   Arrange the data in an increasing order from x_1 to x_{10}.

$$\bar{x} = \frac{\sum_{i=1}^{10} x_i}{10} = 1.126 \text{ in}$$

$$\bar{x}_n = \frac{\sum_{i=1}^{n-1} x_i}{9} = 1.125 \text{ in}$$

$$\nu^2 = \sum_{i=1}^{10} (x_i - \bar{x}^2)^2 = 126 \times 10^{-6}$$

$$\nu_n^2 = \sum_{i=1}^{9} (x_i - \bar{x}_n)^2 = 41 \times 10^{-6}$$

Since x_n is the suspect value, calculate

$$\frac{\nu_n^2}{\nu^2} = \frac{41 \times 10^{-6}}{126 \times 10^{-6}} = 0.325$$

The value of $\nu_n^2/\nu^2 = 0.325$ is less than the value (0.3526), from Table 3.4, for $n = 10$ and confidence level 97.5 percent.

Hence, with 97.5 percent confidence the suspect observation (1.135) should be excluded from the analysis.

SUMMARY

In this chapter methods for evaluating a product (design, material, process) are presented.   This is done by predicting the characteristics of the population from the characteristics of the sample on which the test was conducted. Since the means for accomplishing this depend upon the statistical distribution which the test data follow, methods are presented for three major distributions: normal, Weibull, and binomial.   Table 3.5 gives the summary of the information.   To determine which distribution to use, see Sec. 2.8. For a review of the distributions themselves, consult Secs. 2.1, 2.3, and 2.5.

The evaluation of a product is not limited to Chap. 3.   It is the concern of the entire text, and it is covered, in various degrees, in every chapter. Specifically, Chaps. 4 and 10 describe how to evaluate a product by comparing it with another product.   Chapters 5, 7, 8, and 9 discuss ways of evaluating a product through various experimental means.   Chapter 6 evaluates a product by determining the effect of different variables on some characteristics of the product.   The uniqueness of Chap. 3 is that it deals with the very basic approach to the evaluation of a product: Given a set of data derived from a test sample, determine the characteristics of the population.

Table 3.5 Summary of information for evaluating a product

Distribution which the data follow	Parameters which can be evaluated from the sample	Population parameters to be predicted	Distributions or charts to use	Equations to use
Normal	$\bar{x}, s, n$	σ	X^2 distribution Table A-2	$\left[\dfrac{(n-1)s^2}{\chi^2_{\alpha/2;\,\nu}}\right]^{1/2} \leq \sigma \leq \left[\dfrac{(n-1)s^2}{\chi^2_{1-\alpha/2;\,\nu}}\right]^{1/2}$
		μ	z distribution Table A-1	$\bar{x} - \dfrac{\sigma}{\sqrt{n}}\,z_{\alpha/2} \leq \mu \leq \bar{x} + \dfrac{\sigma}{\sqrt{n}}\,z_{\alpha/2}$
		μ	t distribution Table A-3	$\bar{x} - \dfrac{s}{\sqrt{n}}\,t_{\alpha/2;\,\nu} \leq \mu \leq \bar{x} + \dfrac{s}{\sqrt{n}}\,t_{\alpha/2;\,\nu}$
		Population limits	Tables 3.1, 3.2	$\bar{x} \pm ks$
Weibull	b, n	b	Tables A-18, A-19	
	$\bar{x}$ θ B_{10} b	μ θ $\hat{B}_{10}$	Tables of 5 and 95% for 90% C; 2.5 and 97.5% for 95% C; etc.	
Binomial	%R n, r, g	%$\hat{R}$	Tables A-41 to A-46	$\dfrac{1}{1+\dfrac{n-g+1}{g}F_L} \leq \hat{R} \leq \dfrac{1}{1+\dfrac{n-g}{g+1}\dfrac{1}{F_U}}$

PROBLEMS

3.1. The sample mean of nine bores is 1.004 in, and the standard deviation is 0.005 in. The distribution of dimension sizes is normal.

(a) Determine the average of the population with 95 percent confidence.

(b) Determine the average of the population with 99 percent confidence.

3.2. Determine by Weibull method, with 90 percent confidence, the mean life and the characteristic life, on the basis of the following test data (life cycles in 10^5): 0.51, 0.97, 1.50, 2.20, 3.00.

3.3. Fifty items were picked at random from a lot of two thousand, and ten items were found defective. Predict with 90 percent confidence the minimum and the maximum number of items that could be defective in the remaining items of this lot.

3.4. Ten metallic pieces were chosen at random from a normal population, and their hardnesses were found to be 66, 68, 67, 69, 71, 70, 70, 71, 63, and 63 RC. Plot the mean hardnesses of the population versus the confidence levels.

3.5. Ten press-fit assemblies, involving a shaft and a bore, were picked out of a lot of one hundred thousand assemblies, and their interferences measured, with the following results:

Press-fit assembly number	Diametral interference, in
1	0.00060
2	0.00040
3	0.00050
4	0.00035
5	0.00060
6	0.00055
7	0.00050
8	0.00045
9	0.00045
10	0.00060

What is the average interference of this lot of 100,000 assemblies? Provide the answer at 95 percent confidence level.

3.6. An engineer tests a sample of 11 bearings for hardness. The sample variance is 2.850. The test data are normally distributed. Find the range of the population variance with 80 percent confidence.

3.7. A certain type of light bulb has a variance in burning time of 10,000 h^2. A sample of 20 bulbs was picked, presumably from this lot, and its variance was found to be 12,000 h^2. At 95 percent confidence level, determine whether the 20 bulbs were picked from the right lot.

3.8. Fifteen plastic linings made out of standard material were tested for wear with the following results: $\bar{x} = 0.0090$ in, $s = 0.0021$ in. What can be said about the population from which this sample was taken? Answer with 95 percent confidence.

3.9. How many observations should be made to ensure with 90 percent confidence that $0.47 \leq s^2/\sigma^2 \leq 1.70$?

3.10. A shaft is manufactured with a large number of stress raisers. There is considerable doubt as to whether the shaft can withstand the required load for 5×10^5 cycles. Ten shafts were tested to failure, with the following results: Number of cycles to failure, in 10^6: 0.37, 0.60, 0.78, 0.94, 1.1, 1.25, 1.48, 1.7, 1.9, and 2.3. Using a Weibull plot:

(*a*) What can be said about the population that this sample represents? (Show graphically.)

(*b*) Give the range of percent of shafts that will fail before 5×10^5 cycles.

3.11. Ten gears were fatigue tested to failure, and the life cycles were recorded as follows:

Life to failure, cycles	
1.6×10^6	2.3×10^7
3.8×10^6	2.7×10^6
1.2×10^7	5.4×10^6
5.8×10^5	1.6×10^7
9.2×10^6	7.2×10^6

Using Weibull distribution, determine the following:

(*a*) The median, characteristic, and $B_{2.0}$ lives.

(*b*) The mean life of the sample and the 90 percent confidence limit for the true mean.

(*c*) The B_{10} life of the sample and the 90 percent confidence limit for the true B_{10} life.

(*d*) The Weibull slope and the 90 percent confidence limit for the true slope.

3.12. An intra-vane pump was failing because of a substantial amount of smears and seizures in the cartridge of the pump. Consequently, the pump was redesigned. Ten of these new pumps were then tested, and two of these were found unsatisfactory. What can be said about the reliability (maximum and minimum) of the new pumps, with 90 percent confidence?

3.13. Using the data in Prob. 1.7, determine the 90, 95, and 99 percent confidence limits for the population mean μ.

3.14. Using the data in Prob. 1.8, determine the 90, 95, and 99 percent confidence limits on the uniformity of the process σ.

3.15. The amount of suspended particles, in air, was measured from day to day in terms of micrograms per cubic meter: 150, 170, 165, 148, 181, 160, 175, 182, 168, 172, 158, 180, 155, 180, 170.

(*a*) Determine the upper and lower limits within which 90 percent of all the future measurements are expected to fall.

(*b*) Determine the 90 percent confidence limits on the true mean pollution level.

REFERENCES

1. Pieruschka, E.: "Principles of Reliability," Prentice-Hall, Inc., Englewood Cliffs, N.J., 1963.
2. Natrella, M. G.: Experimental Statistics, *Nat. Bur. Stand. Hand.* 91, Aug. 1, 1963.
3. Grubbs, F. E.: Sample Criteria for Testing Outlying Observations, *Ann. Math. Statist.*, vol. 21, 1958.
4. Dixon, W. J., and F. J. Massey, Jr.: "Introduction to Statistical Analysis," 2d ed., McGraw-Hill Book Company, New York, 1957.

BIBLIOGRAPHY

Bowker, A. H., and G. J. Lieberman: "Engineering Statistics," Prentice-Hall, Inc., Englewood Cliffs, N.J., 1963.

Dixon, W. J., and F. J. Massey, Jr.: "Introduction to Statistical Analysis," 3d ed., McGraw-Hill Book Company, New York, 1969.

Duncan, A. J.: "Quality Control and Industrial Statistics," Richard D. Irwin, Inc., Homewood, Ill., 1959.

Grubbs, F. E.: Sample Criteria for Testing Outlying Observations, Ann. Math. Statist., Vol. 21, 1958.

Moroney, M. J.: "Facts from Figures," Penguin Books, Inc., Baltimore, 1965.

Natrella, M. G.: Experimental Statistics, *Nat. Bur. Stand. Hand.* 91, Aug. 1, 1963.

Pieruschka, E.: "Principles of Reliability," Prentice-Hall, Inc., Englewood Cliffs, N.J., 1963.

Volk, W.: "Applied Statistics for Engineers," 2d ed., McGraw-Hill Book Company, New York, 1969.

Wallis, W. A., and H. V. Roberts: "Statistics: A New Approach," The Free Press, New York, 1956.

4
Experiments of Comparison

There are many engineering situations where the qualities and the uniformities of two products are to be compared from the knowledge of the test data. The appropriate experiments are designed, and the test data are collected and analyzed.

In this chapter the design of experiments of comparison is dealt with through (1) the preliminary approach and (2) the detailed approach. The *preliminary approach*, the simpler of the two, is used to determine whether one product is better or more uniform than the other, without inquiring what the actual differences are. It can also be used to establish the confidence interval which would contain the population difference between the means or the population ratio of variances (see Examples 4.1 to 4.4). This approach employs the same type of statistics as used in Chap. 3. In effect, new variables are formed which represent the differences or ratios pertaining to the qualities or uniformities of the two products, and appropriate distributions are used for the solution of the problem. The cases where data follow normal distribution and Weibull distribution are included.

In the *detailed approach* means are provided for establishing not only whether differences between two products exist but also how much the differences are. Generally, these differences are numerically predetermined in the statement of the problem (see Examples 4.5, 4.6, 4.11, 4.12, and others). The detailed approach is divided into two broad categories: absolute comparison (Subsec. 4.2.1) and relative comparison (Subsec. 4.2.2). The first involves the type of problems where the characteristics of one product (mean or standard deviation) are known and it is desired to determine whether the other product is better or worse than the first by a predetermined amount (Examples 4.5, 4.6, and others). The second category deals with situations where the characteristics of neither product are known and it is desired to determine whether one is different from the other by a predetermined amount (Examples 4.12, 4.15, and others).

In the detailed approach the statistics of the tests of hypothesis are used. Here, sample sizes necessary for statistically significant tests are determined, and the comparison between two products is judged quantitatively against the confidence level initially established. This approach is used for cases where data follow normal distribution.

In those cases where the underlying distribution is not known or cannot be established, the comparison between products can be treated by non-parametric experiments, as described in Chap. 8.

4.1 PRELIMINARY APPROACH

Two distributions are considered here: (1) normal and (2) Weibull. In the case of the normal distribution, the qualities of two products are compared in terms of the differences between their means when their variances are known and when they are not known but equal. The uniformities of two products are compared in terms of the ratios of their variances. In the case of the Weibull distribution, the qualities of two products are compared in terms of the ratios of their means.

4.1.1 NORMAL DISTRIBUTION

Comparison of the qualities of two products when their variances are known If x_1, x_2, ..., x_{n_x} are normally distributed independent random variables with mean μ_x and variance σ_x^2, then the variable

$$\bar{x} = \frac{\sum_{i=1}^{n_x} x_i}{n_x}$$

has a normal distribution with mean μ_x and variance σ_x^2/n_x. Also, if y_1, y_2, y_3, ..., y_{ny} are normally distributed independent random variables with mean μ_y and variance σ_y^2, then

$$\bar{y} = \frac{\sum_{i=1}^{n_y} y_i}{n_y}$$

has a normal distribution with mean μ_y and variance σ_y^2/n_y. Furthermore, these two normal distributions can be subtracted to give a distribution $\bar{x} - \bar{y}$ with mean $\mu_x - \mu_y$ and variance $\sigma_x^2/n_x + \sigma_y^2/n_y$. From this it follows that the variable

$$z = \frac{(\bar{x} - \bar{y}) - (\mu_x - \mu_y)}{\sqrt{\sigma_x^2/n_x + \sigma_y^2/n_y}} \tag{4.1}$$

has a z distribution.

The following test is used to determine whether the difference between $\bar{x}$ and $\bar{y}$ is significant.

If $|z| > z_{\alpha/2}$, $\bar{x}$ is significantly different from $\bar{y}$, where $+z$ implies that $\bar{x} > \bar{y}$, and $-z$ indicates $\bar{x} < \bar{y}$. The value of $z_{\alpha/2}$ is read from Table A-1 for $(1 - \alpha)$ confidence level. The value of z is found from

$$z = \frac{\bar{x} - \bar{y}}{\sqrt{\sigma_x^2/n_x + \sigma_y^2/n_y}}$$

The concept of confidence interval is used to predict the difference between the two populations from the knowledge of their sample.

$$P(-z_{\alpha/2} \leq z \leq z_{\alpha/2}) \equiv 1 - \alpha \qquad \text{confidence}$$

The working equation is

$$(\bar{x} - \bar{y}) - \left(\frac{\sigma_x^2}{n_x} + \frac{\sigma_y^2}{n_y}\right)^{1/2} z_{\alpha/2} \leq (\mu_x - \mu_y) \leq (\bar{x} - \bar{y}) + \left(\frac{\sigma_x^2}{n_x} + \frac{\sigma_y^2}{n_y}\right)^{1/2} z_{\alpha/2} \tag{4.2}$$

Here the engineer starts out by selecting n_x and n_y items and finding their respective sample means $\bar{x}$ and $\bar{y}$. He then predicts the difference between μ_x and μ_y.

Example 4.1 Two vibratory feeders are set to give equal flow rates of powder. The operator of one feeder makes five random checks on his feeder and finds that the average flow rate for five observations is 150 g/s. The operator of the other feeder does a similar test and finds that for 20 observations the average is 120 g/s. It is known from past experience that the variance of the first machine is 100 and the variance of the second machine is 400. Determine whether the average flow rate of one feeder is significantly different from the flow rate of the other. What is the true difference between the two flow-rate means at 95% confidence level?

Solution

$$\bar{x} = 150 \text{ g/s} \qquad \bar{y} = 120 \text{ g/s} \qquad n_x = 5 \qquad n_y = 20 \qquad \sigma_x^2 = 100$$

$$\sigma_y^2 = 400 \qquad 1 - \alpha = 0.95 \qquad \frac{\alpha}{2} = 0.025$$

$$z_{\alpha/2} = 1.96 \text{ from Table A-1}$$

Also, with $\bar{x} - \bar{y} = +30$ and $\sigma_x^2/n_x = 20$ and $\sigma_y^2/n_y = 20$, then

$$\sqrt{\frac{\sigma_x^2}{n_x} + \frac{\sigma_y^2}{n_y}} = \sqrt{20 + 20} = \sqrt{40} = 6.325 \text{ g/s}$$

$z = 30/6.31 = 4.74$. Since $4.74 > 1.96$ and the value of z is positive, it can be concluded (even in the presence of the variations in the flow-rate data) that the flow rate of feeder x is significantly larger than the flow rate of feeder y at 95 percent confidence level.
Using Eq. (4.2):

$30 - 6.325 \times 1.96 \leq \mu_x - \mu_y \leq 30 + 6.325 \times 1.96$ for 95 percent confidence

$30 - 12.4 \leq \mu_x - \mu_y \leq 30 + 12.4$

or

$17.6 \text{ g/s} \leq \mu_x - \mu_y \leq 42.4 \text{ g/s}$

Hence, it can be concluded with 95 percent confidence that the true difference in the average flow rate of the two machines can be as low as 17.6 g/s and not more than 42.4 g/s.

Comparison of the qualities of two products when their variances are unknown but equal

If $x_1, x_2, \ldots, x_{n_x}$ are independent normally distributed variables with mean μ_x and variance σ^2, then the variable

$$\bar{x} = \sum_{i=1}^{n_x} \frac{x_i}{n_x}$$

is normally distributed with mean μ_x and variance σ^2/n_x. Also

$$\frac{(n_x - 1)s_x^2}{\sigma^2}$$

has a χ^2 distribution with $(n_x - 1)$ degrees of freedom. Similarly, if y_1, $y_2, \ldots, y_{n_y}$ are independent and normal variables with mean μ_y and variance σ^2, then the variable

$$\bar{y} = \sum_{i=1}^{n_y} \frac{y_i}{n_y}$$

is normally distributed with mean μ_y and variance σ^2/n_y. Also

$$\frac{(n_y - 1)s_y^2}{\sigma^2}$$

has a χ^2 distribution with $(n_y - 1)$ degrees of freedom.

By combining properties of two normal distributions, one finds that $\bar{x} - \bar{y}$ is normally distributed with mean $\mu_x - \mu_y$ and variance

$$\sigma^2 \left(\frac{1}{n_x} + \frac{1}{n_y} \right)$$

Therefore

$$\frac{(\bar{x} - \bar{y}) - (\mu_x - \mu_y)}{\sigma \sqrt{1/n_x + 1/n_y}}$$

has a z distribution.

By combining two χ^2 distributions, one finds that

$$\frac{(n_y - 1)s_y^2}{\sigma^2} + \frac{(n_x - 1)s_x^2}{\sigma^2}$$

has a χ^2 distribution with $(n_x + n_y - 2)$ degrees of freedom.

By the definition of a t distribution, a t_2 distribution is given by

$$t_2 = \frac{z \text{ distribution}}{\sqrt{\chi^2 \text{ distribution}/v}}$$

or

$$t_2 = \frac{[(\bar{x} - \bar{y}) - (\mu_x - \mu_y)]/\sigma\sqrt{1/n_x + 1/n_y}}{\sqrt{\frac{(n_x - 1)s_x^2/\sigma^2 + (n_y - 1)s_y^2/\sigma^2}{n_x + n_y - 2}}}$$

$$= \frac{(\bar{x} - \bar{y}) - (\mu_x - \mu_y)}{\sqrt{\frac{(n_x - 1)s_x^2 + (n_y - 1)s_y^2}{n_x + n_y - 2}} \sqrt{\frac{1}{n_x} + \frac{1}{n_y}}} \tag{4.3}$$

Therefore, t_2 has a t distribution with $v = n_x + n_y - 2$. As before, the following test is used to determine whether the difference between $\bar{x}$ and $\bar{y}$ is significant:

If $|t_2| > t_{\alpha/2}$, $\bar{x}$ is significantly different from $\bar{y}$, where $+t_2$ implies that $\bar{x} > \bar{y}$, and $-t_2$ that $\bar{x} < \bar{y}$. The value of $t_{\alpha/2}$ is found from Table A-3 for $(1 - \alpha)$ confidence. The value of t_2 is computed by using Eq. (4.3) and substituting $\mu_x - \mu_y = 0$.

t_2 can be used to determine the value of $\mu_x - \mu_y$, since

$$P(-t_{\alpha/2;v} < t_2 \leq t_{\alpha/2;v}) = 1 - \alpha$$

The advantage of using the variable t_2 is that one does not have to know anything about the populations, except that the two variances are equal.

The following relationship is used to predict, at a given confidence $1 - \alpha$, the difference between μ_x and μ_y of two products:

$$(\bar{x} - \bar{y}) - At_{\alpha/2;v} \leq (\mu_x - \mu_y) \leq (\bar{x} - \bar{y}) + At_{\alpha/2;v} \tag{4.4}$$

where $v = n_x + n_y - 2$, and

$$A = \left[\frac{(n_x - 1)s_x{}^2 + (n_y - 1)s_y{}^2}{n_x + n_y - 2} \frac{n_x + n_y}{n_x n_y}\right]^{1/2}$$

This is illustrated by the following example.

Example 4.2 In order to determine the relative merits of two materials x and y, 10 specimens of x and 14 of y were tested for wear, with the following results:

	Material x	Material y
Number of specimens tested	10	14
Average wear, in	0.00825	0.01320
Standard deviation, in	0.00194	0.00245

Based on the above test data, at what confidence level can one state that one material is better than the other (material with less amount of wear is better)? Determine also the true difference in average wear of the populations of material x and material y at 90, 95, and 99 percent confidence.

The two materials x and y were fabricated on the same machines, and the variances of the two populations, therefore, can be assumed to be the same.

Solution

$$\bar{x} = 8.25 \times 10^{-3} \text{ in} \qquad \bar{y} = 13.2 \times 10^{-3} \text{ in}$$
$$n_x = 10 \qquad\qquad\qquad n_y = 14$$
$$s_x = 1.94 \times 10^{-3} \text{ in} \qquad s_y = 2.45 \times 10^{-3} \text{ in}$$
$$v = n_x + n_y - 2 = 10 + 14 - 2 = 22$$

From Table A-3, read the appropriate values of t. For 90 percent confidence, $1 - \alpha = 0.9$. Therefore

$$\frac{\alpha}{2} = 0.05 \qquad \text{and} \qquad t_{0.05; 22} = 1.717$$

For 95 percent confidence, $1 - \alpha = 0.95$. Therefore

$$\frac{\alpha}{2} = 0.025 \qquad \text{and} \qquad t_{0.025; 22} = 2.074$$

For 99 percent confidence, $1 - \alpha = 0.99$. Therefore

$$\frac{\alpha}{2} = 0.005 \qquad \text{and} \qquad t_{0.005; 22} = 2.819$$

Using Eq. (4.4),

$$A = \left[\frac{(10 - 1)(1.94^2) \times 10^{-6} + (14 - 1)(2.45^2) \times 10^{-6}}{10 + 14 - 2} \times \frac{10 + 14}{10 \times 14}\right]^{1/2}$$

$$= 0.925 \times 10^{-3} \text{ in}$$

By means of Eq. (4.3),

$$t_2 = \frac{-4.95 \times 10^{-3} - 0}{0.925 \times 10^{-3}} = -5.35$$

Since $|t_2| = 5.35$ is considerably higher than the value $t_{\alpha/2} = 2.819$ for 99 percent confidence, it can be concluded with more than 99 percent confidence that the average wear rate $\bar{x}$ is significantly less than $\bar{y}$ (since t_2 is negative). Hence material x is better than y.

The next step is to determine the true difference between the two wear rates.

$$(\bar{x} - \bar{y}) - A t_{\alpha/2; \, \nu} \leq (\mu_x - \mu_y) \leq (\bar{x} - \bar{y}) + A t_{\alpha/2; \, \nu} \qquad \text{for } (1 - \alpha) \text{ confidence}$$

For 90 percent confidence:

$$(8.25 - 13.2) - 0.925 \times 1.717 \leq (\mu_x - \mu_y) \leq (8.25 - 13.2) + 0.925 \times 1.717 \text{ in terms}$$
of 10^{-3} in

$$-6.54 \times 10^{-3} \leq (\mu_x - \mu_y) \leq -3.36 \times 10^{-3} \text{ in}$$

$$3.36 \times 10^{-3} \leq (\mu_y - \mu_x) \leq 6.54 \times 10^{-3} \text{ in}$$

Hence, it can be concluded with 90 percent confidence that the true difference in average wear of materials x and y can be as little as 3.36×10^{-3} in and as much as 6.54×10^{-3} in. Similarly, for higher confidences:

$$3.03 \times 10^{-3} \leq (\mu_y - \mu_x) \leq 6.87 \times 10^{-3} \text{ in for 95 percent confidence}$$

$$2.34 \times 10^{-3} \leq (\mu_y - \mu_x) \leq 7.56 \times 10^{-3} \text{ in for 99 percent confidence}$$

Comparison of the uniformities of two products The comparison of the uniformities of two products can be made by means of the F distribution. F distribution is defined as follows:

If χ_x^2 and χ_y^2 are two independent random variables which have χ^2 distribution, with ν_x and ν_y degrees of freedom, respectively, the variable

$$F = \frac{\chi_x^2 / \nu_x}{\chi_y^2 / \nu_y} \qquad (4.5)$$

has an F distribution with ν_x and ν_y degrees of freedom. The F distribution is shown in Fig. 4.1. The probability that F is greater than $F_{\alpha; \, \nu_x; \, \nu_y}$ is as shown in Fig. 4.2.

$$P(F > F_{\alpha; \, \nu_x; \, \nu_y}) = \alpha$$

This is the area under the curve from $F_{\alpha; \, \nu_x; \, \nu_y}$ to ∞. This area is difficult to obtain analytically, but it is available in tabulated form. Since the relation between $F_{\alpha; \, \nu_x; \, \nu_y}$ and α depends on two parameters ν_x and ν_y, three-dimensional tables are required. These are given in Tables A-4 to A-9.

Because

$$\frac{\chi_x^2 / \nu_x}{\chi_y^2 / \nu_y}$$

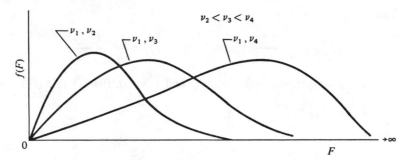

Fig. 4.1 Plot of $f(F)$ versus F.

has an F distribution with v_x and v_y degrees of freedom,

$$\frac{\chi_y^2/v_y}{\chi_x^2/v_x}$$

has an F distribution with v_y and v_x degrees of freedom. In other words, the order of v_x and v_y is important. Also

$$P\left(\frac{\chi_x^2/v_x}{\chi_y^2/v_y} > F_{\alpha; v_x; v_y}\right) = P\left(\frac{\chi_y^2/v_y}{\chi_x^2/v_x} \le \frac{1}{F_{\alpha; v_x; v_y}}\right) = \alpha$$

But

$$P\left(\frac{\chi_y^2/v_y}{\chi_x^2/v_x} \ge \frac{1}{F_{\alpha; v_x; v_y}}\right) = 1 - \alpha$$

Also

$$1 - \alpha = P\left(\frac{\chi_y^2/v_y}{\chi_x^2/v_x} \ge F_{1-\alpha; v_y; v_x}\right)$$

$f(F)$

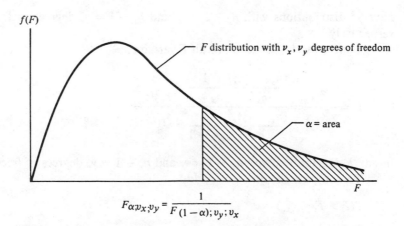

F distribution with v_x, v_y degrees of freedom

α = area

$$F_{\alpha; v_x; v_y} = \frac{1}{F_{(1-\alpha); v_y; v_x}}$$

Fig. 4.2 Meaning of $F_{\alpha; v_x; v_y}$.

Therefore

$$F_{1-\alpha;\,\nu_y;\,\nu_x} = \frac{1}{F_{\alpha;\,\nu_x;\,\nu_y}} \qquad \text{or} \qquad F_{\alpha;\,\nu_x;\,\nu_y} = \frac{1}{F_{1-\alpha;\,\nu_y;\,\nu_x}} \tag{4.6}$$

For example, if $\alpha = 0.1$, $\nu_x = 8$, and $\nu_y = 20$, from Table A-4

$$F_{0.10;\,8;\,20} = 2.0$$

Therefore

$$\frac{1}{F_{1-0.10;\,20;\,8}} = \frac{1}{F_{0.9;\,20;\,8}} = 2.0$$

or $F_{0.9;\,20;\,8} = 1/2.0 = 0.5$.   Tables A-4 to A-9 contain values for $F_{0.10;\,\nu_x;\,\nu_y}$ to $F_{0.001;\,\nu_x;\,\nu_y}$.   These correspond to 90 to 99.9 percent confidence levels.

The F distribution is very useful in the analysis-of-variance ratio of populations, that is, in the analysis of the uniformities of two products. Consider two normally distributed independent random samples

$$x_1, x_2, \ldots, x_{n_x}$$

with mean μ_x, variance σ_x^2, and a sample variance s_x^2, for n_x items; and

$$y_1, y_2, \ldots, y_{n_y}$$

with mean μ_y, variance σ_y^2, and a sample variance s_y^2, for n_y items.   Then, as shown in Chap. 3, the variables

$$\frac{(n_x - 1)s_x^2}{\sigma_x^2} = \chi_x^2 \qquad \text{and} \qquad \frac{(n_y - 1)s_y^2}{\sigma_y^2} = \chi_y^2$$

have χ^2 distributions with $n_x - 1 = \nu_x$ and $n_y - 1 = \nu_y$ degrees of freedom, respectively.

Hence, by the definition of an F variable,

$$F = \frac{\chi_x^2/\nu_x}{\chi_y^2/\nu_y} = \frac{\dfrac{(n_x - 1)s_x^2}{\sigma_x^2}\dfrac{1}{n_x - 1}}{\dfrac{(n_y - 1)s_y^2}{\sigma_y^2}\dfrac{1}{n_y - 1}} = \frac{s_x^2/\sigma_x^2}{s_y^2/\sigma_y^2} \tag{4.7}$$

has an F distribution with $n_x - 1 = \nu_x$ and $n_y - 1 = \nu_y$ degrees of freedom.

To compute the limits of σ_x^2/σ_y^2:

$$P(F > F_{\alpha;\,\nu_x;\,\nu_y}) = \alpha$$

$$P(F \leq F_{\alpha;\,\nu_x;\,\nu_y}) = 1 - \alpha$$

From Eq. (4.7),

$$F = \frac{s_x^2/\sigma_x^2}{s_y^2/\sigma_y^2}$$

$$P\left(\frac{s_x^2/\sigma_x^2}{s_y^2/\sigma_y^2} \leq F_{\alpha;\, \nu_x;\, \nu_y}\right) = 1 - \alpha$$

$$P\left(\frac{s_x^2}{s_y^2}\frac{\sigma_y^2}{\sigma_x^2} \leq F_{\alpha;\, \nu_x;\, \nu_y}\right) = 1 - \alpha \qquad (4.8)$$

Also,

$$\frac{s_y^2/\sigma_y^2}{s_x^2/\sigma_x^2}$$

has an F distribution with the value under consideration written as $F_{\alpha;\, \nu_y;\, \nu_x}$.

$$F = \frac{s_y^2/\sigma_y^2}{s_x^2/\sigma_x^2}$$

$$P\left(\frac{s_y^2/\sigma_y^2}{s_x^2/\sigma_x^2} \leq F_{\alpha;\, \nu_y;\, \nu_x}\right) = 1 - \alpha$$

$$P\left(\frac{s_y^2}{s_x^2}\frac{\sigma_x^2}{\sigma_y^2} \leq F_{\alpha;\, \nu_y;\, \nu_x}\right) = 1 - \alpha$$

By inverting the inequality

$$P\left(\frac{s_x^2}{s_y^2}\frac{\sigma_y^2}{\sigma_x^2} \geq \frac{1}{F_{\alpha;\, \nu_y;\, \nu_x}}\right) = 1 - \alpha \qquad (4.9)$$

Combining Eqs. (4.8) and (4.9):

$$P\left(\frac{1}{F_{\alpha/2;\, \nu_y;\, \nu_x}} \leq \frac{s_x^2}{s_y^2}\frac{\sigma_y^2}{\sigma_x^2} \leq F_{\alpha/2;\, \nu_x;\, \nu_y}\right) = 1 - \alpha$$

Therefore, with $(1 - \alpha)$ confidence,

$$\frac{1}{F_{\alpha/2;\, \nu_y;\, \nu_x}(s_x^2/s_y^2)} \leq \frac{\sigma_y^2}{\sigma_x^2} \leq \frac{F_{\alpha/2;\, \nu_x;\, \nu_y}}{s_x^2/s_y^2} \qquad (4.10)$$

Frequently we are not interested in the limits of σ_x^2/σ_y^2, but only whether σ_x^2 is larger or smaller than σ_y^2, that is, whether σ_x^2/σ_y^2 is larger or smaller than 1. In this case the derivation is as follows:

From Eq. (4.8):

$$P\left(\frac{s_x^2}{s_y^2}\frac{\sigma_y^2}{\sigma_x^2} \leq F_{\alpha;\, \nu_x;\, \nu_y}\right) = 1 - \alpha$$

Therefore, with $(1 - \alpha)$ confidence,

$$\frac{s_x^2}{s_y^2} \frac{\sigma_y^2}{\sigma_x^2} \leq F_{\alpha; \, \nu_x; \, \nu_y}$$

or

$$\frac{\sigma_y^2}{\sigma_x^2} \leq F_{\alpha; \, \nu_x; \, \nu_y} \frac{s_y^2}{s_x^2} \tag{4.11}$$

From the above, the following test can be used to establish whether the measurements x are more variable than the measurements y:

If $s_x^2/s_y^2 > F_{\alpha; \, \nu_x; \, \nu_y}$, the ratio of the true variances σ_x^2/σ_y^2 is greater than 1 at $(1 - \alpha)$ confidence level, where x and y can be used interchangeably.

Example 4.3   A spring manufacturer was offered a wire claimed to have more uniform diameter than his standard production wire. To check the claim, he tested 11 current production springs and 21 springs made from the new wire, with the following results:

Current production	New spring
$n_x = 11$	$n_y = 21$
$\nu_x = n - 1 = 10$	$\nu_y = n_y - 1 = 20$
$s_x = 0.048$ in	$s_y = 0.033$ in
$s_x^2 = 23.5 \times 10^{-4} \, in^2$	$s_y^2 = 10 \times 10^{-4} \, in^2$

Is the new wire really better than the current stock? Provide the answer with the highest confidence level possible. Establish the 90 and 95 percent confidence limits on the true relative uniformities of two types of wires.

Solution

$$\frac{s_x^2}{s_y^2} = \frac{23.5 \times 10^{-4}}{10.0 \times 10^{-4}} = 2.35$$

From Tables A-4 and A-5,

$$F_{\alpha; \, \nu_x; \, \nu_y} = F_{0.10; \, 10; \, 20} = 1.94 \quad \text{for 90 percent confidence level}$$

$$F_{\alpha; \, \nu_x; \, \nu_y} = F_{0.05; \, 10; \, 20} = 2.35 \quad \text{for 95 percent confidence level}$$

By applying the above test $(s_x^2/s_y^2 > F_{\alpha; \, \nu_x; \, \nu_y})$, it can be seen that s_x^2/s_y^2 is greater than the tabulated value 1.94 at 90 percent confidence level and it is equal to the tabulated value corresponding to 95 percent confidence. Hence, it can be concluded at about 94 percent confidence level that the diameter of the current wire is more variable than the diameter of the new one; therefore, the new wire is better.

At 90 percent confidence level, by means of Eq. (4.10),

$$\frac{1}{F_{\alpha/2;\,v_y;\,v_x}(s_x^2/s_y^2)} \leq \frac{\sigma_y^2}{\sigma_x^2} \leq \frac{F_{\alpha/2;\,v_x;\,v_y}}{s_x^2/s_y^2}$$

From Table A-5,

$$F_{\alpha/2;\,v_x;\,v_y} = F_{0.05;\,10;\,20} = 2.35$$

$$F_{\alpha/2;\,v_y;\,v_x} = F_{0.05;\,20;\,10} = 2.77$$

$$\frac{1}{(2.77)(2.35)} \leq \frac{\sigma_y^2}{\sigma_x^2} \leq \frac{2.35}{2.35}$$

$$0.154 \leq \frac{\sigma_y^2}{\sigma_x^2} \leq 1.0$$

Therefore, the 90 percent confidence limits on the true relative uniformities are

$$0.392 \leq \frac{\sigma_y}{\sigma_x} \leq 1.0$$

since the ratio of σ_y/σ_x is less than 1 for both the upper and the lower limits. This confirms the above conclusion with 90 percent confidence that the new wire is more uniform.

At 95 percent confidence level,

$$F_{\alpha/2;\,v_x;\,v_y} = F_{0.025;\,10;\,20} = 2.77$$

$$F_{\alpha/2;\,v_y;\,v_x} = F_{0.025;\,20;\,10} = 3.42$$

$$\frac{1}{(3.42)(2.35)} \leq \frac{\sigma_y^2}{\sigma_x^2} \leq \frac{2.77}{2.35}$$

$$0.352 \leq \frac{\sigma_y}{\sigma_x} \leq 1.086$$

Therefore, at 95 percent confidence it can no longer be stated that the new wire is more uniform than the current stock.

4.1.2 WEIBULL DISTRIBUTION

The Weibull distribution provides a useful way of comparing the mean lives (qualities) of two products. The test data of the two products are plotted on the Weibull paper with the aid of the median rank, and the straight lines are drawn as described in Example 2.4. The ratio of their mean lives is determined from the graph and compared for significance with the mean-lives ratio statistically required for a given confidence.

Example 4.4 Two types of steel were tested for life. The following test data were obtained. Which steel has a higher life?

| Lot x, standard steel | | Lot y, experimental steel | |
Failure No.	Life, 10^5 cycles	Failure No.	Life, 10^5 cycles
1	0.3	1	0.47
2	0.6	2	0.9
3	0.7	3	1.0
4	1.0	4	1.5
5	1.3	5	1.8
6	1.5	6	2.0
7	2.1	7	2.5
		8	3.2
		9	3.6
		10	4.4

Solution The above data are plotted on Weibull probability paper as shown in Fig. 4.3. The ordinate, percent failed, is in terms of median ranks, discussed previously. The experimental steel appears to be superior for the samples tested. In order to determine the relative merits of the respective populations, the following analysis is made. From the sample sizes of the two products:

Degrees of freedom of lot $x = 7 - 1 = 6$
Degrees of freedom of lot $y = 10 - 1 = 9$
Combined degrees of freedom $= 6 \times 9 = 54$

The Weibull slopes (as measured) of the two plots are both about 1.7. From Table A-17, the mean is located at about the 56 percent failed level for Weibull slope of 1.7. Hence

Mean of lot $x = 1.1 \times 10^5$ cycles (56% level in Table A-17)
Mean of lot $y = 2.2 \times 10^5$ cycles (56% level in Table A-17)

The mean-life ratio (ratio of the larger to the smaller mean life) is

Mean-life ratio (MLR) $= 2.0$

    In order to determine whether this ratio is significant for the population, Tables A-20 to A-22 are used. For a Weibull slope of 1.7 and for 54 degrees of freedom the confidence exceeds 95 percent. Hence, it can be concluded with at least 95 percent confidence that the experimental steel (lot y) is better than the standard steel (lot x).

    In cases when the slopes of the two products are not the same, first assume that they are both equal to the slope of the first product. From the table, determine the confidence level at which the MLR is significant. Repeat this process by assuming the slope is equal to the slope of the second product, and again determine the confidence level. As an estimate of the true confidence, take the average of these two confidence levels.

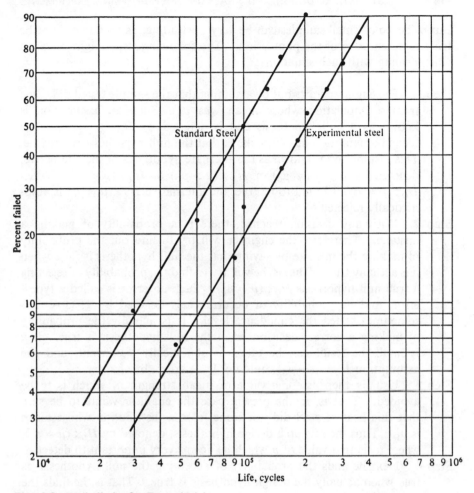

Fig. 4.3 Weibull plot for Example 4.4.

4.2 DETAILED APPROACH

An engineer is frequently confronted with situations in which he is compelled to make decisions on the basis of a small sample.   Consider a case where he is asked whether a given heat of steel comes from the same population as the standard heat.   To come to a decision, he will test a sample of n items and then find the mean strength of this sample,

$$\bar{x} = \sum_{i=1}^{n} \frac{x_i}{n}$$

But evidence of small samples can be very misleading, as they might not be truly representative of the population. The following steps might be taken when dealing with such situations:

Step 1. The engineer starts out by assuming that the sample tested did come from a population whose mean was μ_0. Such an assumption is known as a null hypothesis and is denoted by $H_O : (\mu = \mu_0)$. The second assumption he can make is that the null hypothesis is not true. This assumption is known as the alternate hypothesis or the alternative, denoted by $H_A : (\mu \neq \mu_0)$ or $H_A : (\mu > \mu_0)$ or $H_A : (\mu < \mu_0)$. When $H_O : (\mu = \mu_0)$ is found to be true, the alternative $H_A : (\mu \neq \mu_0)$ is automatically rejected.

Step 2. With any decision reached, there is a probability of making a mistake. Therefore, the engineer will try to find out the probability of making the mistake by saying that the null hypothesis is false when it is actually true. That is, he will try to find the probability of rejecting a true null hypothesis $H_O : (\mu = \mu_0)$. Such an error is called a type I error, and its probability of occurrence is denoted by α; $(1 - \alpha)$ is sometimes called the confidence level. It is actually the probability of making a correct decision when the null hypothesis is true. If α is small the result can be taken as statistically significant, and the assumption $H_O : (\mu = \mu_0)$ could not have arisen by chance.

Step 3. Usually there is some value of mean strength μ_1 which is to be avoided. That is, in the present case the engineer wants to be sure that this heat of steel did not come from a distribution whose mean is μ_1. Thus, he sets up a design hypothesis, denoted by $H_D : (\mu = \mu_1)$, where μ_1 is that value of μ which he deems very important to detect.

Step 4. Next he finds the probability of accepting the null hypothesis as true when actually the design hypothesis is true. That is, he finds the probability β of saying that $\mu = \mu_0$, when actually $\mu = \mu_1$. This error is called a type II error or sometimes the β error.

After these steps are completed, and provided that α and β are both small, the engineer has a good statistical basis for making a decision about the mean of the new heat as compared with the other heat. If α and β are not as small as desired, he can only say that there is not enough evidence to draw a conclusion which is statistically significant.

Types of alternatives Step 4 indicated that an engineer would have to choose a value of $\mu = \mu_1$ which is necessary to detect or to avoid. Depending on this value μ_1, the alternative can take two distinct forms.

In the first case, the engineer either (1) has an a priori knowledge that a certain value of strength cannot occur or (2) is not worried about values of the unknown parameter being greater than or equal to μ_0. In such a case, the alternative is known as one-sided, because he is worried only about values which are *either* above *or* below μ_0 (by an amount $|\mu_0 - \mu_1|$) but not both above and below μ_0. Some examples where one-sided alternatives are reasonable are:

1. Drinking water may be tested for bacteria count, and the hypothesis that the count is at a safe level may be rejected only if the count is too high. If the count is low, this causes no worry.
2. A new fertilizer may be tested for crop yield and the old fertilizer rejected only if a significantly higher yield results with the new one. There is no concern about the possibility of the new one's having a lower yield. That is, no change in fertilizer will result unless the new fertilizer produces a significantly better yield than the old.
3. Cores of concrete may be tested for strength and rejected only if they are unusually low in strength. High strengths are acceptable.

In the second case, the engineer may wish to avoid values which are both higher or lower than μ_0. In such a case he adopts an alternative which is called a two-sided alternative. Under such an alternative he will reject the null hypothesis if the value of μ seems too high and also if the value of μ seems too low.

Acceptance regions The above procedures can be represented graphically. Consider a test in which it is assumed that the mean of a given normal distribution has a specified value $H_O : (\mu = \mu_0)$. In this test the alternative is $\mu < \mu_0$ or $H_A : (\mu < \mu_0)$. The value of μ important to detect is μ_1, hence $H_D : (\mu = \mu_1)$. In Fig. 4.4 two distributions are shown; 1 is the

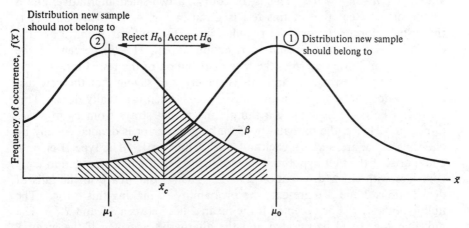

Fig. 4.4 Decision theory, one-sided.

distribution of the standard heat whose mean is μ_0. If the mean for the new heat is μ and if $\mu = \mu_0$, then the hypothesis H_O : $(\mu = \mu_0)$ will be true. In other words, distribution 1 is the distribution the new heat should belong to, in order to validate H_O : $(\mu = \mu_0)$. Distribution 2 is the distribution of the design hypothesis H_D : $(\mu = \mu_1)$. This is the distribution the new heat should not come from. If the new heat actually does come from this distribution, it must be detected with probability $1 - \beta$.

The type I error α is the error of saying that the new heat belongs to a distribution centered at some value less than μ_0 when it actually is centered at μ_0. This α error is the area under distribution 1 from $-\infty$ to $\bar{x}_c$ (cutoff point). The type II error β is the probability of saying that the new heat came from distribution 1 when it actually came from distribution 2. This β error is given by the area under distribution 2 between $\bar{x}_c$ and $+\infty$.

The cutoff point $\bar{x}_c$ is the critical value of $\bar{x}$. In order to make any decision about H_O or H_A, one will have to take a small sample out of the new heat and find its mean $\bar{x}$. If $\bar{x}_c$ is the critical value of $\bar{x}$ and if $\bar{x}$ lies to the right of $\bar{x}_c$, H_O is accepted. It will be noted from Fig. 4.4 that if $\bar{x}_c$ were moved to the left, then the area β would be larger and the chances of making a type II error would be increased. That is, the chances of accepting the null hypothesis would increase, and the chances of accepting the alternative hypothesis would reduce.

This kind of testing of hypothesis need not be done only on the basis of the sample mean $\bar{x}$. One can well test for significance by using the median or the mode. It has been found that if $\bar{x}$ is used instead of the median or mode for a given number of specimens, the errors α and β obtained are minimum. Hence, the test procedure using $\bar{x}$ is an optimum procedure and is thus preferred over all others.

Consider now a test for significance with H_O : $(\mu = \mu_0)$, H_A : $(\mu \neq \mu_0)$, and H_D : $(|\mu - \mu_0| = d)$. This is, of course, a two-sided alternative, and a graphical representation of this test is given in Fig. 4.5. Distribution 1 is the distribution of the original population. Distributions 2 and 3 are the distributions that should result in rejection of H_O, $(1 - \beta)$ percent of the time, if either 2 or 3 describes the true distribution of the tested lot.

The type I error is α, and this is the error of saying that the new distribution does not coincide with distribution 1 when it actually does. This error can be made by finding a sample mean that strays from μ_0 in *either* direction. Hence, the probability of making this error is divided equally by choosing equal areas $\alpha/2$ in each tail of distribution 1. The type II error β could arise if the null hypothesis is accepted when the new population truly coincides with either distribution 2 *or* 3. Hence, the area β in the tails of distributions 2 and 3 represents the probability of making this error. The null hypothesis will be accepted if the mean $\bar{x}$ lies between $\bar{x}_{c_1}$ and $\bar{x}_{c_2}$. The null hypothesis will be rejected and the alternative accepted if the mean $\bar{x}$

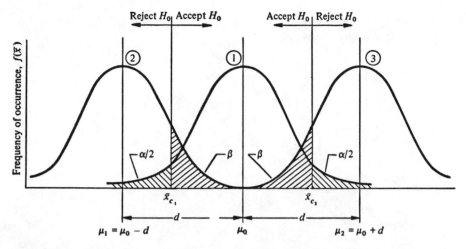

Fig. 4.5 A graphical representation of a two-sided test procedure.

lies outside the range of $\bar{x}_{c_1}$ and $\bar{x}_{c_2}$. The above represents the reasoning behind tests of hypothesis which are the basis for the experiments of comparison. In the above discussion one of the products was being evaluated in terms of a standard product whose parameters (μ_0 or σ_0^2) were assumed to be known. In this case the variables were $\bar{x}$ (see Fig. 4.4) or s^2. The same reasoning can apply to two products, neither of which is standard, provided one of them is known. This category of problems will be discussed first under Subsec. 4.2.1. The second category of problems (Subsec. 4.2.2) involves situations where neither of the two products is known. In this case, the two products are compared in terms of the differences between their means or the ratios of their variances. The variables are $\bar{x} - \bar{y}$ and s_x^2/s_y^2 (see Figs. 4.10 and 4.12).

4.2.1 ABSOLUTE COMPARISON OF TWO PRODUCTS

Several situations arise in this category, as described below:

Comparison of the means when σ^2 is known

Case 1—*the one-sided alternative* In such a case the null hypothesis is H_O : ($\mu = \mu_0$), alternative H_A : ($\mu < \mu_0$), and H_D : ($\mu = \mu_1$). If variable $\bar{x}$ has a normal distribution with mean μ and variance σ^2/n, then

$$\frac{\bar{x} - \mu}{\sigma/\sqrt{n}}$$

has a z distribution. From Fig. 4.4 it is seen that a type I error will be made if the null hypothesis is rejected when true.   This error will be made when $\bar{x} \leq \bar{x}_c$.   Hence $P(\bar{x} \leq \bar{x}_c) = \alpha$; that is,

$$P(\bar{x} \leq \bar{x}_c) = P\left(\frac{\bar{x} - \mu_0}{\sigma/\sqrt{n}} \leq \frac{\bar{x}_c - \mu_0}{\sigma/\sqrt{n}}\right) = \alpha$$

but

$$P\left(\frac{\bar{x} - \mu_0}{\sigma/\sqrt{n}} \leq -z_\alpha\right) = \alpha$$

by the definition of z_α in Subsec. 2.1.1.   Therefore

$$-z_\alpha = \frac{\bar{x}_c - \mu_0}{\sigma/\sqrt{n}}$$

Similarly,

$$z_\beta = \frac{\bar{x}_c - \mu_1}{\sigma/\sqrt{n}}$$

Hence, by subtraction,

$$z_\alpha + z_\beta = \frac{\mu_0 - \mu_1}{\sigma/\sqrt{n}}$$

and

$$n = \frac{(z_\alpha + z_\beta)^2 \sigma^2}{(\mu_0 - \mu_1)^2} = \text{sample size required to test} \qquad (4.12)$$

Also, addition gives

$$-z_\alpha + z_\beta = \frac{2\bar{x}_c - \mu_0 - \mu_1}{\sigma/\sqrt{n}}$$

$$z_\beta - z_\alpha = \frac{2\bar{x}_c - (\mu_0 + \mu_1)}{\sigma/\sqrt{n}}$$

$$\bar{x}_c = \frac{(z_\beta - z_\alpha)\sigma}{2\sqrt{n}} + \frac{\mu_0 + \mu_1}{2} = \text{cutoff point}$$

rewritten as

$$\bar{x}_c = \frac{1}{2}(z_\beta - z_\alpha)\sqrt{\frac{\sigma^2}{n}} + \frac{\mu_0 + \mu_1}{2} \qquad (4.13)$$

Hence, the following steps should be taken:

1. Decide on $H_O : (\mu = \mu_0)$, $H_A : (\mu > \mu_0)$ or $(\mu < \mu_0)$, and $H_D : (\mu = \mu_1)$.
2. Decide on the errors α and β that can be tolerated.
3. Use Eq. (4.12) to find n.
4. Test n items and find $\bar{x}$.
5. Find $\bar{x}_{c_1}$, the cutoff point, from Eq. (4.13).
6. If $H_A : (\mu > \mu_0)$, accept H_O if $\bar{x} \leq \bar{x}_c$ and reject H_O if $\bar{x} > \bar{x}_c$.
   If $H_A : (\mu < \mu_0)$, accept H_O if $\bar{x} \geq \bar{x}_c$ and reject H_O if $\bar{x} < \bar{x}_c$.

Example 4.5 A manufacturer of flashlight batteries feels that his batteries will operate for an average of 80 h. The batteries will be accepted with a confidence of 95 percent if the mean life is not less than 75 h. Set up a test for significance, if $\alpha = 0.025$ and $\sigma = 6.44$ h.

Solution

$$H_O : (\mu = 80 \text{ h}) \qquad H_A : (\mu < 80 \text{ h}) \qquad H_D : (\mu = 75 \text{ h})$$

$$\alpha = 0.025 \text{ from Table A-1} \qquad \therefore z_\alpha = 1.960$$

$$\beta = 1.00 - 0.95 = 0.05 \qquad z_\beta = 1.645$$

$$n = \frac{(z_\alpha + z_\beta)^2}{(\mu_0 - \mu_1)^2} \sigma^2$$

$$= \frac{(1.645 + 1.960)^2}{25} \, 41.5$$

$$\approx 22$$

A sample of 22 items is chosen and $\bar{x}$ is found to be 78 h. Now,

$$\bar{x}_c = \frac{(z_\beta - z_\alpha)}{2\sqrt{n}} \sigma + \frac{\mu_0 + \mu_1}{2}$$

$$= \frac{-(0.315) \times 6.44}{2 \times 4.69} + \frac{155}{2}$$

$$= -0.215 + 77.5$$

$$= 77.285 \text{ h} \qquad \text{DECISION PT.}$$

Hence, since $\bar{x} > \bar{x}_c$, $H_O : (\mu = 80 \text{ h})$ is accepted with $\alpha = 0.025$ and $\beta = 0.05$. Therefore, the batteries meet the specified requirement.

ACCEP Null Hyoth

Case 2—the two-sided alternative In such a case a decision is made on the basis of $\bar{x}$, the mean of a sample of n items. As before, it is known that $(\bar{x} - \mu)/(\sigma/\sqrt{n})$ has a z distribution.

Here, by considering Fig. 4.5, one can conclude that

$$-z_{\alpha/2} = \frac{\bar{x}_{c_1} - \mu_0}{\sigma/\sqrt{n}}$$

$$z_{\beta} = \frac{\bar{x}_{c_1} - \mu_1}{\sigma/\sqrt{n}}$$

Adding,

$$-(z_{\alpha/2} - z_{\beta}) = \frac{2\bar{x}_{c_1} - \mu_0 - \mu_1}{\sigma/\sqrt{n}}$$

Therefore

$$2\bar{x}_{c_1} = \frac{(z_{\beta} - z_{\alpha/2})\sigma}{\sqrt{n}} + \mu_0 + \mu_1$$

or

$$\bar{x}_{c_1} = \frac{\mu_0 + \mu_1}{2} + \frac{(z_{\beta} - z_{\alpha/2})\sigma}{2\sqrt{n}}$$

rewritten as

$$\bar{x}_{c_1} = \frac{1}{2}(z_{\beta} - z_{\alpha/2})\sqrt{\frac{\sigma^2}{n}} + \frac{\mu_0 + \mu_1}{2} \tag{4.14}$$

Subtracting and squaring yields

$$n = \frac{(z_{\alpha/2} + z_{\beta})^2 \sigma^2}{(\mu_0 - \mu_1)^2} \tag{4.15}$$

It can be shown also that

$$\bar{x}_{c_2} = \mu_0 + (\mu_0 - \bar{x}_{c_1})$$
$$= 2\mu_0 - \bar{x}_{c_1} \tag{4.16}$$

Hence, the following procedure may be used:

1. Decide on $H_O: (\mu = \mu_0)$, $H_A: (\mu \neq \mu_0)$, and $H_D: (|\mu - \mu_0| = d)$.
2. Choose the errors α and β that can be tolerated.
3. Use Eq. (4.15) to determine the sample size necessary to give α and β.
4. Test n random items to determine $\bar{x}$.
5. Find $\bar{x}_{c_1}$ and $\bar{x}_{c_2}$ from Eqs. (4.14) and (4.16).
6. If $\bar{x}_{c_1} \leq \bar{x} \leq \bar{x}_{c_2}$, accept H_O; otherwise, reject H_O.

Example 4.6 A material was proposed for a certain part which is to be produced by a stamping process. The process cannot handle material with the ultimate tensile strength Su greater than 85 ksi. The stress requirements make it necessary to have Su greater than 75 ksi. Hence, the strength of 80 ksi is thought to be desirable. Choosing $\beta = 0.05$ and $\alpha = 0.05$ and knowing from past experience that $\sigma = 6.44$ ksi, set up a test for significance to determine whether the material fulfills the desired requirements.

Solution

$\mu_0 = 80$ ksi $\therefore H_0 : (\mu = 80$ ksi$)$

$H_A : (\mu \neq 80)$ $H_D : (|\mu - \mu_0| = 5$ ksi$)$ $\therefore \mu_2 = 85$ ksi $\mu_1 = 75$ ksi

$\alpha = 0.05$ $\dfrac{\alpha}{2} = 0.025$ $z_{\alpha/2} = 1.960$

$\beta = 1.00 - 0.95 = 0.05$ $z_\beta = 1.645$

$$n = \frac{(z_{\alpha/2} + z_\beta)^2}{(\mu_0 - \mu_1)^2}\,\sigma^2$$

$$= \frac{(1.645 + 1.960)^2}{25}\ 41.5$$

$$= 21.5$$

$$\simeq 22$$

Suppose that you sampled 22 items made from the proposed material and $\bar{x}$ came out as 83.0 ksi.

$$\bar{x}_{c1} = \frac{(z_\beta - z_{\alpha/2})\sigma}{2\sqrt{n}} + \frac{\mu_0 + \mu_1}{2}$$

$$= \frac{-0.315 \times 6.44}{2 \times 4.69} + \frac{155}{2}$$

$$= 77.284 \text{ ksi}$$

$$\therefore \bar{x}_{c2} = 2 \times 80 - 77.284$$

$$= 82.716 \text{ ksi}$$

By comparing $\bar{x}$ with $\bar{x}_{c1}$ and $\bar{x}_{c2}$, it is found that

77.284 ksi $\leq 83.0 \nleq 82.716$ ksi

Hence $H_0 : (\mu = 80$ ksi$)$ is rejected, and it is concluded that the material does not fulfill the stated requirement.

Comparison of the means when σ^2 is unknown
In actual engineering situations the population variance σ^2 is seldom known. Hence, before one can make decisions as to the value of μ, he has to determine the variance s^2.

Case 1—the one-sided alternative   In this case the null hypothesis is
$H_O: (\mu = \mu_0)$ and the alternative is $H_A: (\mu < \mu_0)$ with $H_D: (\mu = \mu_1)$. In
order to use a procedure based on $\bar{x}$ (the sample mean of n items), s rather
than σ must be used.   From the results of Subsec. 3.1.2 it is known that

$$\frac{\bar{x} - \mu}{s} \sqrt{n}$$

has a t distribution with $(n - 1)$ degrees of freedom.   It is assumed that
$n_s = n_x$.   Observe in Fig. 4.6 that for a type I error $\bar{x} \leq \bar{x}_c$.   The probability
of having $\bar{x} \leq \bar{x}_c$ is α, if the null hypothesis is true.   In other words,

$$P(\bar{x} \leq \bar{x}_c) = \alpha$$

But

$$P(\bar{x} \leq \bar{x}_c) = P\left(\frac{\bar{x} - \mu_0}{s/\sqrt{n}} \leq \frac{\bar{x}_c - \mu_0}{s/\sqrt{n}}\right) = \alpha$$

Also, by definition of a t distribution,

$$P\left(\frac{\bar{x} - \mu_0}{s/\sqrt{n}} \leq -t_{\alpha;\,v}\right) = \alpha$$

where $v = n - 1$.   Then, at the cutoff point,

$$-t_{\alpha;\,v} = \frac{\bar{x}_c - \mu_0}{s/\sqrt{n}}$$

and also

$$t_{\beta;\,v} = \frac{\bar{x}_c - \mu_1}{s/\sqrt{n}}$$

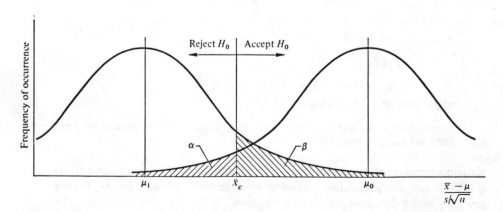

Fig. 4.6   One-sided alternative.

Subtraction gives

$$t_{\alpha;\nu} + t_{\beta;\nu} = \frac{\mu_0 - \mu_1}{s/\sqrt{n}}$$

or

$$n = \frac{(t_{\alpha;\nu} + t_{\beta;\nu})^2}{(\mu_0 - \mu_1)^2} s^2 \qquad (4.17)$$

Addition gives

$$-t_{\alpha;\nu} + t_{\beta;\nu} = \frac{2\bar{x}_c - \mu_0 - \mu_1}{s/\sqrt{n}}$$

$$\bar{x}_c = \frac{(t_{\beta;\nu} - t_{\alpha;\nu})s}{2\sqrt{n}} + \frac{\mu_0 + \mu_1}{2}$$

rewritten as

$$\bar{x}_c = \frac{1}{2}(t_{\beta;\nu} - t_{\alpha;\nu})\sqrt{\frac{s^2}{n}} + \frac{\mu_0 + \mu_1}{2} \qquad (4.18)$$

It is obvious from Eq. (4.17) that if one wants to find the sample size n, he will have to use a trial-and-error procedure. The following steps should be taken:

1. Decide on $H_O : (\mu = \mu_0)$, $H_A: (\mu < \mu_0)$ or $(\mu > \mu_0)$, and $H_D : (\mu = \mu_1)$.
2. Decide on the magnitude of α and β that can be tolerated.
3. Select at random n_1 items, conduct the test, and find s_1 and $\bar{x}_1$.
4. Use Eq. (4.17) with s_1 and $\nu = n_1 - 1$. Check to see if $n \le n_1$. If $n \le n_1$, then proceed with step 5; if not, select more items until $n \le n_1$. Then proceed with step 5.
5. Use Eq. (4.18) to find $\bar{x}_c$. Then compare $\bar{x}_1$ with $\bar{x}_c$ to make the decision about the mean. For $H_A : (\mu < \mu_0)$, if $(\bar{x}_1 < \bar{x}_c)$, reject H_O ; if $(\bar{x}_1 \ge \bar{x}_c)$, accept H_O. For $H_A : (\mu > \mu_0)$, if $(\bar{x}_1 > \bar{x}_c)$, reject H_O ; if $(\bar{x}_1 \le \bar{x}_c)$, accept H_O.

Example 4.7 Strain gauges in an internal-combustion-engine connecting rod indicate that the stress fluctuates with each revolution. Strength considerations show that a mean value of this maximum stress should be $\mu_0 = 60$ ksi. It is deemed necessary to detect $\mu_1 = 70$ ksi with a probability of 95 percent. With $\alpha = 0.05$, determine whether it is safe to assume that $\mu = 60$ ksi, if (1) $n_1 = 11$, $s_1 = 10$, and $\bar{x} = 65$; and (2) $n_2 = 21$, $s_2 = 11$, and $\bar{x} = 64$.

Solution a

$$H_O: (\mu = 60) \qquad H_D: (\mu = 70) \qquad H_A: (\mu > 60)$$

$$\alpha = 0.05 \qquad \beta = 0.05$$

$$n_1 = 11 \qquad s_1 = 10 \qquad \bar{x}_1 = 65$$

Observation of the strain 11 times gives $s_1 = 10$ and $\bar{x} = 65$. Hence Eq. (4.17) gives

$$n = \frac{(t_{\alpha;\nu} + t_{\beta;\nu})^2}{(\mu_0 - \mu_1)^2} s^2$$

$$\alpha = 0.05 \qquad \nu = 11 - 1 = 10 \qquad \therefore t_{\alpha;\nu} = 1.812 = t_{\beta;\nu}$$

$$\therefore n = \frac{(2 \times 1.812)^2}{100} \times 100 = \frac{3.28 \times 400}{100} = 13.1$$

but $n_1 = 11$, so that $n > n_1$. Hence, this is a sample of insufficient size to satisfy the stated values of α and β. Therefore, more items are needed.

Solution b   The hypothesis and α and β errors are the same as in solution *a*.

$$n_2 = 21 \qquad s_2 = 11 \qquad \bar{x}_2 = 64 \text{ ksi} \qquad \nu = n_2 - 1 = 20$$

$$s_2 = 11 \text{ ksi} \qquad \bar{x} = 64 \text{ ksi} \qquad t_{\alpha;\nu} = t_{0.05;20} = 1.725 = t_{\beta;\nu}$$

Equation (4.17) gives

$$n = \frac{(t_{\alpha;\nu} + t_{\beta;\nu})^2 s_2^2}{(\mu_0 - \mu_1)^2}$$

$$= \frac{4 \times 1.725^2 \times 11^2}{100}$$

$$= \frac{2.98 \times 484}{100} = 14.4$$

$$\therefore n_2 > n.$$

Hence the sample size is statistically significant. Use Eq. (4.18) to find $\bar{x}_c$.

$$\bar{x}_c = 0 + \frac{\mu_0 + \mu_1}{2}$$

$$= 65 \text{ ksi}$$

$$\bar{x}_2 = 64 \text{ ksi}$$

Since $\bar{x}_2 < 65$ ksi, $H_O: (\mu = 60$ ksi$)$ is accepted, and it is concluded that the mean value of the maximum stress is satisfactory.

Case 2—the two-sided alternative   In this case the procedure is almost identical to that presented above.

Again, one must make a decision on the basis of $\bar{x}$ and s. Knowing that $(\bar{x} - \mu)/(s/\sqrt{n})$ has a t distribution and observing Fig. 4.7, one can write the following equalities:

$$-t_{\alpha/2;\,\nu} = \frac{\bar{x}_{c_1} - \mu_0}{s/\sqrt{n}}$$

$$t_{\beta;\,\nu} = \frac{\bar{x}_{c_1} - \mu_1}{s/\sqrt{n}}$$

where $\mu_1 = \mu_0 - d$

It follows by subtraction and rearrangement that

$$\bar{x}_{c_1} = \frac{\mu_0 + \mu_1}{2} + \frac{(t_{\beta;\,\nu} - t_{\alpha/2;\,\nu})s}{2\sqrt{n}}$$

rewritten as

$$\bar{x}_{c_1} = \frac{1}{2}(t_{\beta;\,\nu} - t_{\alpha/2;\,\nu})\sqrt{\frac{s^2}{n}} + \frac{\mu_0 + \mu_1}{2} \tag{4.19}$$

Addition gives

$$n = \frac{(t_{\alpha/2;\,\nu} + t_{\beta;\,\nu})^2}{(\mu_0 - \mu_1)^2}\, s^2 \tag{4.20}$$

also

$$\bar{x}_{c_2} = \mu_0 + (\mu_0 - \bar{x}_{c_1})$$

$$= 2\mu_0 - \bar{x}_{c_1} \tag{4.21}$$

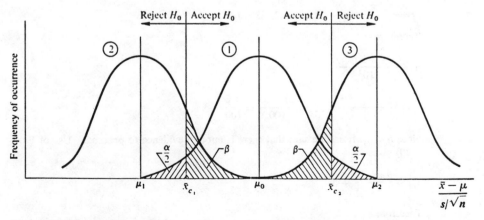

Fig. 4.7 Two-sided alternative.

Here again, in order to determine the sample size n, a trial-and-error procedure is necessary. The following steps are recommended:

1. Decide on $H_O : (\mu = \mu_0)$, $H_D : (|\mu - \mu_0| = d)$, and $H_A : (\mu \neq \mu_0)$.
2. Decide on the magnitude of α and β that can be tolerated.
3. Select a sample of n_1 items. Then find s_1 and $\bar{x}_1$.
4. Use Eq. (4.20) to find n, using s_1 and $v_1 = n_1 - 1$. If $n \leq n_1$, there is sufficient evidence to proceed to step 5; if not, select more items until $n \leq n_1$. Then proceed to step 5.
5. Use Eqs. (4.19) and (4.21) to find $\bar{x}_{c_1}$ and $\bar{x}_{c_2}$. Then compare $\bar{x}$ to determine whether $\bar{x}_{c_1} \leq \bar{x} \leq \bar{x}_{c_2}$. If this relation holds, accept the hypothesis H_O. If it does not hold, reject H_O.

Example 4.8 An engine manufacturer wants to buy automatic torque wrenches. These wrenches are used to tighten the head bolts to a torque of 100 ft·lb. If the torque is less than 90 ft·lb, the connection opens. If the torque is 110 ft·lb, the bolt fails in fatigue. Hence, the values of torque = 90 and 110 ft·lb should be avoided with, say, 97.5 percent certainty. With a 95 percent confidence of detecting that the mean is truly 100 ft·lb, find whether the manufacturer should accept these wrenches. An observation of 11 items indicates that $s_1 = 5$ ft·lb and $\bar{x}_1 = 94$ ft·lb.

Solution

$$H_O : (\mu = 100 \text{ ft·lb}) \qquad H_A : (\mu \neq 100 \text{ ft·lb})$$

$$H_D : (|\mu - \mu_1| = 10 \text{ ft·lb}) \qquad \mu_1 = 90 \text{ ft·lb} \qquad \mu_2 = 110 \text{ ft·lb}$$

$$\alpha = 0.05 \qquad \beta = \frac{100 - 97.5}{100} = 0.025$$

$$n_1 = 11 \qquad s_1 = 5 \text{ ft·lb} \qquad \bar{x}_1 = 94 \text{ ft·lb}$$

$$v_1 = n_1 - 1 = 10 \qquad t_{\alpha/2;\,v} = 2.228 = t_{\beta;\,v}$$

Therefore

$$n = \frac{(t_{\alpha/2;\,v} + t_{\beta;\,v})^2}{10^2} s^2$$

$$= \frac{(2 \times 2.228)^2}{100} \times 25 = \frac{20 \times 25}{100} = \frac{500}{100} = 5$$

Since $n < n_1$, it can be said that there is enough evidence to proceed. Use of Eq. (4.19) gives

$$\bar{x}_{c_1} = \frac{\mu_0 + \mu_1}{2} + 0$$

Therefore

$$\bar{x}_{c_1} = 95 \text{ ft·lb}$$

and $\bar{x}_{c2} = 2\mu_0 - \bar{x}_{c1} = 105$ ft·lb. Since $\bar{x} = 94$ ft·lb, a comparison shows that $\bar{x}_{c1} \not\leq \bar{x} \leq \bar{x}_c$. Therefore, H_0 is rejected and it is concluded that the wrenches do not tighten to the torque of 100 ft·lb, and therefore the manufacturer should not accept these wrenches.

Comparison of the variances   There are many engineering situations where the major concern is the variability of a product rather than the mean value.   In this case one selects a sample of n items and uses s^2, the sample variance, to predict the value of σ^2.   The type I and type II errors still exist, and their meaning is exactly the same as that discussed earlier in this section.   Here again two cases arise, depending on the alternative used.

Case 1—the one-sided alternative   This type of alternative is predominantly used because low variability is of little concern.   If a producer were to manufacture a precision watch part, he might be concerned about having a variance larger than specified, but not if the variance is small.   In most engineering practice the problem is usually to minimize the spread rather than to increase it.

Consider a case in which it is desired to test the null hypothesis H_0: $(\sigma/\sigma_0 = 1)$ against an alternative, such as H_A: $(\sigma > \sigma_0)$ with H_D: $(\sigma/\sigma_0 = \rho_1)$, where $\rho_1 = \sigma_1/\sigma_0$.   Let s_c be the critical value or the value beyond which H_0 will be rejected.   Then, by definition of the type I error, the probability of rejecting H_0: $(\sigma = \sigma_0)$ when it is true is α, and this error occurs when $s > s_c$.   Therefore, the probability of rejecting H_0: $(\sigma = \sigma_0)$ when it is true $= P(s > s_c) = \alpha$ or

$$P\left[\frac{(n-1)s^2}{\sigma_0^2} > \frac{(n-1)s_c^2}{\sigma_0^2}\right] = \alpha$$

But from Fig. 4.8 and from the definition of $\chi^2_{\alpha;\,v}$

$$P\left[\frac{(n-1)s^2}{\sigma_0^2} > \chi^2_{\alpha;\,n-1}\right] = \alpha$$

It follows that

$$\chi^2_{\alpha;\,n-1} = (n-1)\frac{s_c^2}{\sigma_0^2}$$

rewritten as

$$s_c^2 = \frac{(\chi^2_{\alpha;\,n-1})\sigma_0^2}{n-1} \tag{4.22}$$

Similarly, P(accepting H_0 when H_D is true) $= \beta$.   This happens when $s \leq s_c$ and when H_D is true.

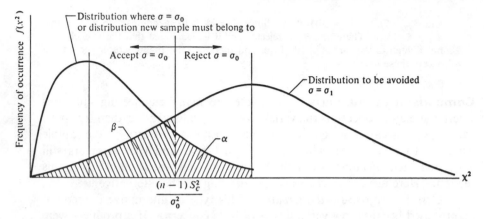

Fig. 4.8 One-sided alternative.

Hence, $\beta = P(s^2 \leq s_c{}^2)$, if H_D is true. Or

$$P(s^2 > s_c{}^2) = 1 - \beta$$

$$P\left[\frac{(n-1)s^2}{\sigma_1{}^2} > \frac{(n-1)s_c{}^2}{\sigma_1{}^2}\right] = 1 - \beta$$

By the definition of $\chi^2_{1-\beta;\,\nu}$ (also see Fig. 4.8)

$$P\left[\frac{(n-1)s^2}{\sigma_1{}^2} > \chi^2_{1-\beta;\,n-1}\right] = 1 - \beta$$

that is,

$$\frac{(n-1)s_c{}^2}{\sigma_1{}^2} = \chi^2_{1-\beta;\,n-1}$$

Therefore

$$\frac{\chi^2_{\alpha;\,n-1}}{\chi^2_{1-\beta;\,n-1}} = \frac{\sigma_1{}^2}{\sigma_0{}^2} = \rho_1{}^2 \tag{4.23}$$

Equation (4.23) gives the sample size, while Eq. (4.22) gives the acceptance region.

The following procedure is suggested:

1. Decide on $H_O : (\sigma = \sigma_0)$, $H_D : (\sigma = \sigma_1)$, and $H_A : (\sigma > \sigma_0)$.
2. Decide on the values of α and β which can be tolerated.
3. Determine the value of the sample size from Eq. (4.23). This is done by entering Table A-2 and by adjusting $\chi^2_{\alpha;\,n-1}$ and $\chi^2_{1-\beta;\,n-1}$ so that the ratio is ρ_1. This is usually achieved in a few trials.

4. Select a sample of n items and find s^2.
5. Find s_c from Eq. (4.22).
6. Compare s_c with s.   If $s \leq s_c$, accept H_0 that $\sigma = \sigma_0$.   If $s > s_c$, reject H_0 that $\sigma = \sigma_0$. The same kind of procedure may be formulated for $H_0 : (\sigma = \sigma_0)$ and $H_A : (\sigma < \sigma_0)$.

Example 4.9   In the field of quality control it is a general practice to determine the tolerance of an assembly (stack-up) through the use of tolerances on each part. Suppliers usually guarantee that the variance on each part will not exceed a given value.   One supplier guaranteed that the σ for a washer which he was selling would be no greater than 0.01225 in with a confidence of 99 percent ($\alpha = 0.01$).   If you were on the receiving end and if you knew that a value of σ larger than 0.0245 in should not occur more than 1 percent of the time, how would you test the supplier's claim?

Solution

$$\sigma_0 = 0.01225 \text{ in} \qquad \sigma_1 = 0.02450 \text{ in} \qquad \rho = \frac{\sigma_1}{\sigma_0} = 2.0 \qquad \rho^2 = 4$$

$$H_0 : \left(\frac{\sigma}{\sigma_0} = 1\right) \qquad H_D : \left(\frac{\sigma_1}{\sigma_0} = \rho = 2.0\right) \qquad H_A : \left(\frac{\sigma}{\sigma_0} > 1\right)$$

$$\alpha = 0.01 \qquad \beta = 0.01$$

The first step is to determine the statistically significant sample size n.   Using Table A-2, determine $\nu = n - 1$ such that

$$\frac{\chi^2_{\alpha;\nu}}{\chi^2_{1-\beta;\nu}} \leq \rho^2$$

From Table A-2, when $\nu = 24$,

$$\frac{\chi^2_{\alpha;\nu}}{\chi^2_{1-\beta;\nu}} = \frac{42.980}{10.856} \approx 4 = \rho_1^2$$

hence

$$n = \nu + 1 = 25$$

A sample of 25 items is drawn; suppose

$$s^2 = 0.0384 \times 10^{-2}$$

$$s_c^2 = \frac{\chi^2_{\alpha;\nu}\sigma_0^2}{n-1}$$

Substituting,

$$s_c^2 = \frac{42.98}{24} \times 0.01225^2 = 0.0270 \times 10^{-2}$$

Comparing, $s_c^2 < s^2$; hence $H_0 : (\sigma_0 = 0.01225)$ is rejected, and it is concluded that the supplier's statement is not correct.

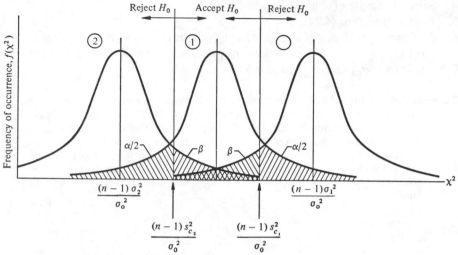

Fig. 4.9 Two-sided alternative.

Case 2—the two-sided alternative This sort of alternative is less frequently encountered in engineering practice than the one-sided alternative.
However, for completeness, it is discussed here.

Here again the population variance σ^2 is determined from the sample
variance s^2 with the aid of a χ^2 distribution. A two-sided test will have
$H_O : (\sigma = \sigma_0)$, $H_A : (\sigma/\sigma_0 \neq 1)$, and $H_D : (\sigma = \sigma_1 \text{ or } = \sigma_2)$.

In Fig. 4.9, curve 1 is the distribution of the null hypothesis. If the
new population comes from distribution 2 or 3, it is desired to detect this
fact with a probability of $1 - \beta$. See Fig. 4.9.

In such a case the probability of making a type I error is equal on both
sides of distribution 1 and is indicated by the equal areas $\alpha/2$.

Knowing that $(n - 1)s^2/\sigma^2$ has a χ^2 distribution with $(n - 1)$ degrees
of freedom, one can conclude the following (as in Case 1):

$$\chi^2_{\alpha/2;\,n-1} = (n - 1)\frac{s_{c_1}^{\;2}}{\sigma_0^{\;2}}$$

rewritten as

$$s_{c_1}^{\;2} = \frac{(\chi^2_{\alpha/2;\,\nu})\sigma_0^{\;2}}{n - 1} \tag{4.24}$$

$$\chi^2_{1-\beta;\,n-1} = (n - 1)\frac{s_{c_1}^{\;2}}{\sigma_1^{\;2}}$$

It follows that

$$\rho_1^{\;2} = \frac{\sigma_1^{\;2}}{\sigma_0^{\;2}} = \frac{\chi^2_{\alpha/2;\,n-1}}{\chi^2_{1-\beta;\,n-1}} \tag{4.25}$$

Also,

$$\chi^2_{1-\alpha/2;n-1} = (n-1)\frac{s_{c_2}^2}{\sigma_0^2}$$

rewritten as

$$s_{c_2}^2 = \frac{(\chi^2_{1-\alpha/2;\,v})\sigma_0^2}{n-1} \tag{4.26}$$

$$\chi^2_{\beta;n-1} = (n-1)\frac{s_{c_2}^2}{\sigma_2^2}$$

Hence

$$\rho_2^2 = \frac{\sigma_2^2}{\sigma_0^2} = \frac{\chi^2_{1-\alpha/2;n-1}}{\chi^2_{\beta;n-1}} \tag{4.27}$$

Steps to be followed:

1. Decide on $H_O : (\sigma = \sigma_0)$, $H_A : (\sigma \neq \sigma_0)$, and $H_D : (\sigma = \sigma_1 \text{ or } \sigma_2)$.
2. Decide on the values of α and β that can be tolerated.
3. Determine the value of the sample size from Eq. (4.25) or (4.27). This is done by referring to Table A-2 and by finding $\chi^2_{\alpha/2;\,n-1}$ and $\chi^2_{1-\beta;\,n-1}$ such that their ratio is ρ_1^2.
4. Select a sample of n items and find s.
5. Find $s_{c_1}^2$ and $s_{c_2}^2$ from Eqs. (4.24) and (4.26).
6. Compare $s_{c_1}^2$, s^2, and $s_{c_2}^2$. If $s_{c_2}^2 \leq s^2 \leq s_{c_1}^2$, accept H_O. If $s_{c_2}^2 > s^2$ or $s^2 > s_{c_1}^2$, reject H_O.

Example 4.10 The wear considerations inside a cylinder bore indicate that the best surface roughness is 12.25 μin (rms). If the surface roughness is half of that, the condition of scuffing wear sets in. This is to be avoided 99 percent of the time. If the surface roughness is twice the optimum, abrasive wear sets in. This is also to be avoided 99 percent of the time. If you were in charge of the machining operations which produce these bores, how would you determine whether a given group of bores meet this requirement at a confidence level of 98 percent?

Solution

$$H_O : (\sigma = 12.25 \ \mu\text{in}) \qquad H_D\left[\sigma = 2(12.25) \text{ or } \sigma = \frac{12.25}{2} \ \mu\text{in}\right]$$

$$H_A : \left(\frac{\sigma}{\sigma_0} \neq 1\right) \qquad \therefore \rho_1 = 2 = \frac{1}{\rho_2}$$

$$\alpha = 0.02 \qquad \beta = 0.01$$

From Table A-2 at $v = 24$,

$$\frac{\chi^2_{\alpha/2;\,v}}{\chi^2_{1-\beta;\,v}} = \frac{\chi^2_{0.01;\,24}}{\chi^2_{0.99;\,24}} \approx 4 \approx \rho_1^2$$

Hence, choose 25 bores, take peak-to-peak roughness readings, and find the $s^2 = 242$.

$$s_{c_1}{}^2 = \frac{\sigma_0{}^2}{n-1} \chi^2_{\alpha/2;n-1}$$

$$= \frac{42.98}{24} \times 12.25^2$$

$$= 270 \ (\mu\text{in})^2$$

$$s_{c_2}{}^2 = \frac{\sigma_0{}^2}{n-1} \chi^2_{1-\alpha/2;n-1}$$

$$= \frac{12.25^2}{24} \times 10.856$$

$$= 68 \ (\mu\text{in})^2$$

Comparing $s_{c_1}{}^2$, $s_{c_2}{}^2$, and s^2, it is found that

$$s_{c_2}{}^2 \leq s^2 \leq s_{c_1}{}^2$$

Hence, H_O is accepted and it is concluded that the bores meet the requirements.

4.2.2 RELATIVE COMPARISON OF TWO PRODUCTS

Several situations arise in this category, as described below:

Comparison of the means when $\sigma_x{}^2$ and $\sigma_y{}^2$ are known

Case 1—*the one-sided alternative* In this case H_0: $[(\mu_x - \mu_y) = 0]$, H_D: $[|(\mu_x - \mu_y)| = |(\mu_x - \mu_y)_1|]$, and H_A: $[(\mu_x - \mu_y) > 0]$. The value $(\bar{x} - \bar{y})_c$ is that value of $\bar{x} - \bar{y}$ beyond which the null hypothesis will be rejected

From Fig. 4.10 it is seen that a type I error is made if H_0 is rejected

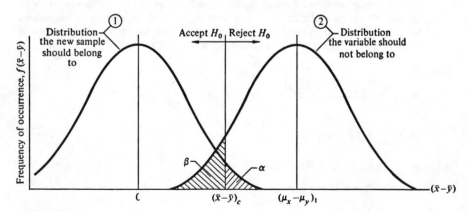

Fig. 4.10 One-sided alternative.

when actually $\mu_x - \mu_y = 0$.   This happens when $(\bar{x} - \bar{y}) > (\bar{x} - \bar{y})_c$.   Hence, by the definition of α,

$$P[(\bar{x} - \bar{y}) > (\bar{x} - \bar{y})_c] = \alpha$$

$$P\left[\frac{(\bar{x} - \bar{y}) - 0}{\sigma_z} > \frac{(\bar{x} - \bar{y})_c - 0}{\sigma_z}\right] = \alpha$$

But since $[(\bar{x} - \bar{y}) - (\mu_x - \mu_y)]/\sigma_z$ has a z distribution and since $\mu_x - \mu_y = 0$ at null hypothesis,

$$P\left(\frac{\bar{x} - \bar{y}}{\sigma_z} > z_\alpha\right) = \alpha$$

It follows that

$$z_\alpha = \frac{(\bar{x} - \bar{y})_c - 0}{\sigma_z}$$

Similarly,

$$-z_\beta = \frac{(\bar{x} - \bar{y})_c - (\mu_x - \mu_y)_1}{\sigma_z}$$

Subtracting,

$$z_\alpha + z_\beta = \frac{(\mu_x - \mu_y)_1}{\sigma_z}$$

Then

$$\frac{1}{\sigma_z{}^2} = \frac{(z_\alpha + z_\beta)^2}{(\mu_x - \mu_y)_1{}^2} \qquad (4.28)$$

but

$$\frac{1}{\sigma_z{}^2} = \frac{1}{\sigma_x{}^2/n_x + \sigma_y{}^2/n_y}$$

If $n_x = n_y = n$, then

$$\frac{1}{\sigma_z{}^2} = \frac{n}{\sigma_x{}^2 + \sigma_y{}^2}$$

It follows that

$$n = \frac{(z_\alpha + z_\beta)^2}{(\mu_x - \mu_y)_1{}^2} (\sigma_x{}^2 + \sigma_y{}^2) \qquad (4.29)$$

This expression gives the sample size. In those situations where n_x is not equal to n_y, the following inequality must be satisfied to ensure an adequate sample size:

$$\frac{(z_\alpha + z_\beta)^2}{(\mu_x - \mu_y)_1^2} \geq \frac{\sigma_x^2}{n_x} + \frac{\sigma_y^2}{n_y}$$

Addition provides

$$z_\alpha - z_\beta = \frac{2(\bar{x} - \bar{y})_c - (\mu_x - \mu_y)_1}{\sigma_z}$$

Therefore

$$(\bar{x} - \bar{y})_c = \frac{(z_\alpha - z_\beta)\sigma_z + (\mu_x - \mu_y)_1}{2}$$

rewritten as

$$(\bar{x} - \bar{y})_c = \frac{1}{2}(z_\alpha - z_\beta)\sqrt{\frac{\sigma_x^2}{n_x} + \frac{\sigma_y^2}{n_y}} + \frac{(\mu_x - \mu_y)_1}{2} \tag{4.30}$$

This expression gives the cutoff point, as shown in Fig. 4.10.

The following procedure is suggested:

1. Decide on $H_O : [(\mu_x - \mu_y) = 0]$, $H_D : [|(\mu_x - \mu_y)| = |(\mu_x - \mu_y)_1|]$, and $H_A : [(\mu_x - \mu_y) > 0]$.
2. Decide on the value of α and β that can be tolerated.
3. Find n, using Eq. (4.29).
4. Pick samples of n items each and find the sample means $\bar{x}$ and $\bar{y}$.
5. Find $(\bar{x} - \bar{y})_c$ from Eq. (4.30).
6. Compare. If $(\bar{x} - \bar{y}) > (\bar{x} - \bar{y})_c$, reject H_O. If $(\bar{x} - \bar{y}) \leq (\bar{x} - \bar{y})_c$, accept H_O.

Example 4.11 It is proposed to substitute grade T black powder in a machine for normal grade C powder to reduce the delay time between the application of the firing current and the explosion. A reduction in the mean time of 0.5 s will make the change desirable. The mean delay period with grade C was $\mu_0 = 1.0$ s, and approximate standard deviation $\sigma_0 = 0.12$ s. If grade T is better, it should be detected with 99 percent confidence ($\beta = 0.01$). The risk of making an incorrect switch to grade T should be 1 percent. Design an experiment to determine whether grade T black powder should be used. Assume that the standard deviation σ_0 is the same in both cases.

Solution   Use subscripts x and y for grades C and T, respectively.

$H_O : [(\mu_x - \mu_y) = 0]$ $H_D : [(\mu_x - \mu_y)_1 = 0.5$ s$]$ $H_A : [(\mu_x - \mu_y) > 0$ s$]$

$\alpha = 0.01$ $\beta = 0.01$

$z_\alpha = 2.33$ $z_\beta = 2.33$

Sample size n is found from Eq. (4.29).

$$n = \frac{(z_\alpha + z_\beta)^2(\sigma_x^2 + \sigma_y^2)}{(\mu_x - \mu_y)_1^2}$$

$$= \frac{(4.66^2)(0.0144 + 0.0144)}{0.5^2}$$

$$= \frac{0.626}{0.25}$$

$$= 2.5$$

Fire three charges each of grade C and grade T black powder. Find the mean delay times $\bar{x}$ and $\bar{y}$. Suppose $\bar{x} - \bar{y} = 0.3$ s.

$$(\bar{x} - \bar{y})_c = \frac{1}{2}(z_\alpha - z_\beta)\sqrt{\frac{\sigma_x^2}{n_x} + \frac{\sigma_y^2}{n_y}} + \frac{(\mu_x - \mu_y)_1}{2}$$

$$= 0 + \frac{0.5}{2}$$

$$= 0.25 \text{ s}$$

Since $(\bar{x} - \bar{y}) > (\bar{x} - \bar{y})_c$, H_0 is rejected, and it is concluded that grade T black powder should be used.

Case 2—the two-sided alternative   Figure 4.11 represents the situation graphically.   In this figure, curve 1 is the distribution in which the two population means are equal.   Hence, this is the distribution representing the null hypothesis; 2 and 3 are the distributions which correspond to the design hypothesis.   The decision theory again will be based on the value of $\bar{x} - \bar{y}$,

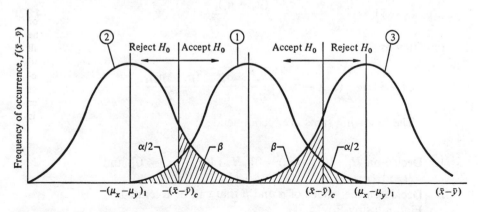

Fig. 4.11   Two-sided alternative.

and rejection will be made when $|(\bar{x} - \bar{y})| > |(\bar{x} - \bar{y})_c|$. In this case the probability of rejecting $H_O : [(\mu_x - \mu_y) = 0]$ when H_O is true will be α. This is indicated by areas $\alpha/2$ on both tails of distribution 1 in Fig. 4.11. The probability of accepting $H_O : [(\mu_x - \mu_y) = 0]$ when $[|(\mu_x - \mu_y)| = |(\mu_x - \mu_y)|_1]$ is true will be given by the areas β of distribution 2 or 3 within the acceptance region. The mathematical arguments here are the same as in the previous section; hence, only the major results are given.

Since $\bar{x} - \bar{y}$ is the variable and variance

$$\sigma_z{}^2 = \frac{\sigma_x{}^2}{n_x} + \frac{\sigma_y{}^2}{n_y}$$

it follows that

$$z_{\alpha/2} = \frac{(\bar{x} - \bar{y})_c - 0}{\sigma_z}$$

$$-z_\beta = \frac{(\bar{x} - \bar{y})_c - (\mu_x - \mu_y)_1}{\sigma_z}$$

Then

$$z_{\alpha/2} + z_\beta = \frac{(\mu_x - \mu_y)_1}{\sigma_z} = \frac{(\mu_x - \mu_y)_1}{(\sigma_x{}^2/n_x + \sigma_y{}^2/n_y)^{1/2}}$$

If $n_x = n_y = n$, then

$$n = \frac{(z_{\alpha/2} + z_\beta)^2 (\sigma_x{}^2 + \sigma_y{}^2)}{(\mu_x - \mu_y)_1{}^2} \tag{4.31}$$

and

$$(\bar{x} - \bar{y})_c = \frac{(z_{\alpha/2} - z_\beta)\sigma_z + (\mu_x - \mu_y)_1}{2} \tag{4.32}$$

rewritten as

$$(\bar{x} - \bar{y})_c = \frac{1}{2}(z_{\alpha/2} - z_\beta)\sqrt{\frac{\sigma_x{}^2}{n_x} + \frac{\sigma_y{}^2}{n_y}} + \frac{(\mu_x - \mu_y)_1}{2}$$

The following steps are recommended:

1. Decide on $H_O : [(\mu_x - \mu_y) = 0]$, $H_A : [(\mu_x - \mu_y) \neq 0]$, and $H_D : [|(\mu_x - \mu_y)| = |(\mu_x - \mu_y)_1|]$.
2. Decide on the values of α and β that can be tolerated.
3. Find n, using Eq. (4.31).

4. Pick samples of n items each and find $\bar{x} - \bar{y}$.
5. Using Eq. (4.32), find $(\bar{x} - \bar{y})_c$.
6. Compare. If $-(\bar{x} - \bar{y})_c \le (\bar{x} - \bar{y}) \le (\bar{x} - \bar{y})_c$, accept H_O. If $(\bar{x} - \bar{y}) < -(\bar{x} - \bar{y})_c$ or $(\bar{x} - \bar{y}) > (\bar{x} - \bar{y})_c$, reject H_O.

Example 4.12   Two types of paints x and y are to be compared for resistance to weather. The paints are rated (on some arbitrary scale) after 6 months of exposure. If the difference in the quality of paints exceeds 10 units, it should be detected 99 percent of the time ($\beta = 0.01$). The risk of making a wrong decision is 0.01. The standard deviation of the two paints is $\sigma_x = 5$, $\sigma_y = 10$, respectively. Design a test for significance and determine which paint is better.

Solution

$$H_O : [(\mu_x - \mu_y) = 0] \qquad H_A : [(\mu_x - \mu_y) \neq 0] \qquad H_D : [|\mu_x - \mu_y|_1 = 10]$$

$$\alpha = 0.01 \qquad \beta = 0.01$$

$$z_{\alpha/2} = 2.575 \qquad z_\beta = 2.33$$

$$n = \frac{(z_{\alpha/2} + z_\beta)^2(\sigma_x^2 + \sigma_y^2)}{(\mu_x - \mu_y)_1^2}$$

$$= \frac{(4.905^2)(125)}{100} = \frac{3{,}000}{100} = 30$$

Suppose two types of 30 cans of paint each were tested, with the result that $\bar{x} = 80$ and $\bar{y} = 72$. Then $\bar{x} - \bar{y} = 8$ units.

$$(\bar{x} - \bar{y})_c = \frac{1}{2}(z_{\alpha/2} - z_\beta)\sqrt{\frac{\sigma_x^2}{n_x} + \frac{\sigma_y^2}{n_y}} + \frac{\mu_x - \mu_y}{2}$$

$$= \frac{1}{2}(0.242)(2.04) + \frac{10}{2}$$

$$= \frac{0.495}{2} + 5$$

$$= 5.247$$

Since $(\bar{x} - \bar{y})_c < (\bar{x} - \bar{y})$, H_O is rejected, and it is concluded that $\mu_x \neq \mu_y$; that is, the paint x is better.

Comparison of the means when σ_x^2 and σ_y^2 are unknown but equal

Case 1—the one-sided alternative   This situation is also represented by Fig. 4.10. In this case $H_O : [(\mu_x - \mu_y) = 0]$, $H_D : [(\mu_x - \mu_y) = (\mu_x - \mu_y)_1]$, and $H_A : [(\mu_x - \mu_y) > 0]$. The value $(\bar{x} - \bar{y})_c$ is that value of $\bar{x} - \bar{y}$ beyond which the null hypothesis will be rejected.

From Eq. (4.3),

$$t_2 = \frac{(\bar{x} - \bar{y}) - (\mu_x - \mu_y)}{\sqrt{\dfrac{(n_x - 1)s_x{}^2 + (n_y - 1)s_y{}^2}{n_x + n_y - 2}} \sqrt{\dfrac{1}{n_x} + \dfrac{1}{n_y}}} \tag{4.3}$$

or

$$t_2 = \frac{(\bar{x} - \bar{y}) - (\mu_x - \mu_y)}{s_z}$$

where

$$s_z = \sqrt{\frac{(n_x - 1)s_x{}^2 + (n_y - 1)s_y{}^2}{n_x + n_y - 2}} \sqrt{\frac{1}{n_x} + \frac{1}{n_y}}$$

By definition of α,

$$P[(\bar{x} - \bar{y}) > (\bar{x} - \bar{y})_c] = \alpha$$

$$P\left[\frac{(\bar{x} - \bar{y}) - 0}{s_z} > \frac{(\bar{x} - \bar{y})_c - 0}{s_z}\right] = \alpha$$

Since $\mu_x - \mu_y = 0$ at null hypothesis, and

$$\frac{(\bar{x} - \bar{y}) - (\mu_x - \mu_y)}{s_z}$$

follows a t distribution,

$$P\left(\frac{\bar{x} - \bar{y}}{s_z} > t_{\alpha;\,\nu}\right) = \alpha$$

Therefore

$$t_{\alpha;\,\nu} = \frac{(\bar{x} - \bar{y})_c - 0}{s_z}$$

Similarly,

$$-t_{\beta;\,\nu} = \frac{(\bar{x} - \bar{y})_c - (\mu_x - \mu_y)_1}{s_z}$$

Subtracting,

$$t_{\alpha;\,\nu} + t_{\beta;\,\nu} = \frac{(\mu_x - \mu_y)_1}{s_z}$$

Therefore

$$s_z{}^2 = \frac{(\mu_x - \mu_y)_1{}^2}{(t_{\alpha;\,\nu} + t_{\beta;\,\nu})^2}$$

Also,

$$s_z^2 = \frac{(n_x - 1)s_x^2 + (n_y - 1)s_y^2}{n_x + n_y - 2} \left(\frac{1}{n_x} + \frac{1}{n_y}\right)$$

If $n_x = n_y = n$, then by equating the above two values of s_z^2, we obtain

$$n = \frac{(t_{\alpha;v} + t_{\beta;v})^2 (s_x^2 + s_y^2)}{(\mu_x - \mu_y)_1^2} \qquad (4.33)$$

This expression gives the sample size. In those situations where n_x is not equal to n_y, the following inequality should be satisfied to ensure an adequate sample size:

$$\frac{(\mu_x - \mu_y)_1^2}{(t_{\alpha;v} + t_{\beta;v})^2} \geq \frac{(n_x - 1)s_x^2 + (n_y - 1)s_y^2}{n_x + n_y - 2} \left(\frac{1}{n_x} + \frac{1}{n_y}\right)$$

where n_x and n_y are the actual number of items tested of x and y, respectively.

Adding $t_{\alpha;v}$ and $-t_{\beta;v}$ gives

$$t_{\alpha;v} - t_{\beta;v} = \frac{(\bar{x} - \bar{y})_c - (\mu_x - \mu_y)_1}{s_z}$$

Therefore

$$(\bar{x} - \bar{y})_c = \frac{1}{2}(t_{u,v} - t_{\beta;v}) \sqrt{\frac{(n_x - 1)s_x^2 + (n_y - 1)s_y^2}{n_x + n_y - 2}} \sqrt{\frac{1}{n_x} + \frac{1}{n_y}}$$

$$+ \frac{(\mu_x - \mu_y)_1}{2} \qquad (4.34)$$

This expression gives the cutoff point as shown in Fig. 4.10.
The following procedure is suggested:

1. Decide on H_O: $[(\mu_x - \mu_y) = 0]$, H_D: $[(\mu_x - \mu_y) = (\mu_x - \mu_y)_1]$, and H_A: $[(\mu_x - \mu_y) > 0]$.
2. Decide on the values of α and β that can be tolerated
3. Pick samples of n_x and n_y items, and find sample means $\bar{x}$ and $\bar{y}$ and variances s_x^2 and s_y^2.
4. From Table A-3, find the values of $t_{\alpha;v}$ and $t_{\beta;v}$, where $v = n_x + n_y - 2$. Find sample size n by using Eq. (4.33). If n_x and n_y are both greater than n, proceed to step 5. Otherwise, go back to step 3 and repeat the steps after testing additional items (i.e., with larger sample size) until this condition is met.
5. Find the cutoff point $(\bar{x} - \bar{y})_c$ by using Eq. (4.34). If $\bar{x} - \bar{y} \leq (\bar{x} - \bar{y})_c$, accept H_O. If $\bar{x} - \bar{y} > (\bar{x} - \bar{y})_c$, reject H_O.

Example 4.13 Octane numbers of two different fuels are to be compared. The following data are available:

Fuel x octane nos. = 81, 84, 79, 76, 82, 85, 88, 84, 80, 79, 82, 81
Fuel y octane nos. = 76, 74, 78, 79, 80, 79, 82, 76, 81, 79, 82, 78

Fuel y is cheaper than x and therefore should be preferred if it has the same octane number as x with 99 percent confidence. However, price difference is not so large as to warrant acceptance of y if $\mu_x - \mu_y \geq 5$. Which fuel should be used? Also find the probability of detecting a difference in means $\mu_x - \mu_y \geq 5$ when it is actually true. Assume $\sigma_x = \sigma_y$.

Solution

$$H_O : [(\mu_x - \mu_y) = 0] \qquad H_D : [(\mu_x - \mu_y) = 5] \qquad H_A : [(\mu_x - \mu_y) > 0]$$

$$n_x = n_y = 12 \qquad \bar{x} = 81.75 \qquad \bar{y} = 78.67$$

$$s_x^2 = 10.20 \qquad s_y^2 = 6.06$$

$\alpha = 0.01$ from Table A-3

$$t_{\alpha;\, n_x + n_y - 2} = t_{0.01;\, 22} = 2.508$$

With a procedure similar to that used for deriving Eq. (4.33), it can be shown that

$$t_{\beta;\, \nu} = -t_{\alpha;\, \nu} + \cfrac{(\mu_x - \mu_y)_1}{\sqrt{\dfrac{1}{n_x} + \dfrac{1}{n_y}} \sqrt{\dfrac{(n_x - 1)s_x^2 + (n_y - 1)s_y^2}{n_x + n_y - 2}}}$$

$$= -2.508 + \cfrac{5}{\sqrt{\dfrac{1}{6}} \sqrt{\dfrac{112.25 + 66.64}{22}}}$$

$$= -2.508 + \frac{5\sqrt{6}}{2.85} = 4.30 - 2.508$$

$$= 1.792$$

From Table A-3, for $\nu = 22$ and $t_{\beta;\, \nu} = 1.792$,

$$\beta \simeq .04$$

Hence, the chance of detecting a difference in means of $\mu_x - \mu_y \geq 5$ when it is true is 96 percent.

The next step would be to determine the required sample size n, using Eq. (4.33):

$$n = \frac{(t_{\alpha;\, \nu} + t_{\beta;\, \nu})^2 (s_x^2 + s_y^2)}{(\mu_x - \mu_y)^2_1}$$

$$= \frac{(2.508 + 1.792)^2 (10.2 + 6.06)}{5^2}$$

$$= 11.9$$

Since $n_x = n_y = 12$ is greater than 11.9, the sample size is adequate to make a statistically significant decision.

Using Eq. (4.34), or referring to Table 4.2,

$$(\bar{x} - \bar{y})_c = \frac{1}{2}(t_{\alpha;\,v} - t_{\beta;\,v})\sqrt{\frac{(n_x - 1)s_x^2 + (n_y - 1)s_y^2}{n_x + n_y - 2}}\sqrt{\frac{1}{n_x} + \frac{1}{n_y}} + \frac{(\mu_x - \mu_y)_1}{2}$$

$$= \frac{1}{2}(2.508 - 1.792)\sqrt{\frac{11 \times 10.2 + 11 \times 6.06}{22}}\sqrt{\frac{1}{12} + \frac{1}{12}} + \frac{5}{2}$$

$$= \frac{1}{2}(0.716)(2.85)\left(\frac{1}{\sqrt{6}}\right) + 2.5$$

$$= 2.92$$

Since $\bar{x} - \bar{y} = 3.08$ and is greater than $(\bar{x} - \bar{y})_c = 2.92$, H_0 is rejected, and it is concluded that the difference $\bar{x} - \bar{y}$ in the octane numbers of fuels x and y is more than 5; therefore the fuel x is accepted since the price difference is not so large (one-sided-alternative case).

Case 2—the two-sided alternative The test procedure for the two-sided alternative still makes use of a t distribution. The derivation of the appropriate equation is identical to the case of the one-sided alternative, except that $t_{\alpha/2;\,v}$ is used instead of $t_{\alpha;\,v}$. This case is represented by Fig. 4.11. The following steps are suggested:

1. Decide on H_0 : $[(\mu_x - \mu_y) = 0]$, H_A : $[(\mu_x - \mu_y) \neq 0]$, and H_D : $[\,|(\mu_x - \mu_y)| = |(\mu_x - \mu_y)_1|\,]$.
2. Decide on α and β errors that can be tolerated.
3. Pick samples of n_x and n_y items arbitrarily. Find the sample means $\bar{x}$ and $\bar{y}$ and the variances s_x^2 and s_y^2.
4. From Table A-3, find the values of $t_{\alpha/2;\,v}$ and $t_{\beta;\,v}$. Find sample size n by using the equation

$$n = \frac{(t_{\alpha/2;\,v} + t_{\beta;\,v})^2(s_x^2 + s_y^2)}{(\mu_x - \mu_y)_1^2} \tag{4.35}$$

If both n_x and n_y are greater than n, proceed to step 5. Otherwise, go back to step 3 and repeat the procedure, using additional items until this requirement is met.
5. Find the cutoff points by means of the equation

$$(\bar{x} - \bar{y})_c = \frac{1}{2}(t_{\alpha/2;\,v} - t_{\beta;\,v})\sqrt{\frac{(n_x - 1)s_x^2 + (n_y - 1)s_y^2}{n_x + n_y - 2}}\sqrt{\frac{1}{n_x} + \frac{1}{n_y}}$$

$$+ \frac{(\mu_x - \mu_y)}{2} \tag{4.36}$$

If $-(\bar{x} - \bar{y})_c \leq -(\bar{x} - \bar{y}) \leq (\bar{x} - \bar{y})_c$, accept H_0. If $(\bar{x} - \bar{y}) < -(\bar{x} - \bar{y})_c$ or $(\bar{x} - \bar{y}) > (\bar{x} - \bar{y})_c$, reject H_0.

Example 4.14 An investigation was conducted to determine whether a certain treatment applied to improve the dyeing properties of yarn produced a change in strength. Strength s was measured of treated (x) and untreated (y) yarns. The following data were obtained:

s_x (lb) = 31, 34, 29, 26, 32, 35, 38, 34, 30, 29, 32, 31
s_y (lb) = 26, 24, 28, 29, 30, 29, 32, 26, 31, 29, 32, 28

It is necessary to determine whether the treatment changes the mean strength of the yarn by more than 5 lb with a confidence of 99.0 percent or more. The variances are assumed equal but unknown.

Solution

$$H_O : [(\mu_x - \mu_y) = 0] \qquad H_D : [|(\mu_x - \mu_y)| = 5 \text{ lb}] \qquad H_A : [(\mu_x - \mu_y) > 0]$$

Since the confidence = 99.0 percent, $\alpha = 0.01$ and $\alpha/2 = 0.005$.

$$t_{\alpha/2; \, n_x + n_y - 2} = t_{0.005; \, 22} = 2.819$$

Sample sizes have already been fixed.

$$s_x{}^2 = 10.2 \qquad \bar{x} = 31.75 \text{ lb}$$
$$s_y{}^2 = 6.06 \qquad \bar{y} = 28.67 \text{ lb}$$

In a manner similar to that of Example 4.13, the following expression can be derived:

$$t_{\beta; \, n_x + n_y - 2} = -t_{\alpha/2; \, n_x + n_y - 2} + \frac{(\mu_x - \mu_y)_1}{\sqrt{\dfrac{1}{n_x} + \dfrac{1}{n_y}} \sqrt{\dfrac{(n_x - 1)s_x{}^2 + (n_y - 1)s_y{}^2}{n_x + n_y - 2}}}$$

$$= -2.819 + \frac{5}{\sqrt{\dfrac{1}{6}} \times \left(\dfrac{112.25 + 66.64}{22}\right)^{1/2}}$$

$$= +4.30 - 2.819$$

$$= 1.481$$

From Table A-3, for $\nu = 22$ and $t_{\beta; \, \nu} = 1.481$,

$$\beta = 0.06$$

or the chance of detecting $|\mu_x - \mu_y| \geq 5$ is 94 percent.

The next step would be to determine the required sample size n, using Eq. (4.35):

$$n = \frac{(t_{\alpha/2; \, \nu} + t_{\beta; \, \nu})^2 (s_x{}^2 + s_y{}^2)}{(\mu_x - \mu_y)_1{}^2}$$

$$= \frac{(2.819 + 1.481)^2 (10.2 + 6.06)}{5^2}$$

$$= 11.9$$

Since $n_x = n_y = 12$ is greater than 11.9, the sample size is adequate to make a statistically significant decision.

Using Eq. (4.36),

$$(\bar{x} - \bar{y})_c = \frac{1}{2}(2.819 - 1.481)\sqrt{\frac{11 \times 10.2 + 11 \times 6.06}{22}}\sqrt{\frac{1}{12} + \frac{1}{12} + \frac{5}{2}}$$

$$= 3.28$$

since $\bar{x} - \bar{y} = 3.08$ and is less than $(\bar{x} - \bar{y})_c = 3.28$, H_O is accepted and it is concluded that treatment does not change the mean strength of the yarn by more than 5 lb, $|\mu_x - \mu_y| > 5$ lb (two-sided-alternative case).

Comparison of the means when $\sigma_x{}^2$ and $\sigma_y{}^2$ are unknown and not equal The procedure for this situation is given below (mathematics is omitted).

Case 1—the one-sided alternative The following steps are suggested:

1. Determine $H_O : (\mu_x - \mu_y) = 0$, $H_A : [(\mu_x - \mu_y) > 0]$.
2. Decide on the value of α that can be tolerated.
3. Select two samples of n_x and n_y items. Find the sample means $\bar{x}$ and $\bar{y}$ and the sample variances $s_x{}^2$ and $s_y{}^2$.
4. Find

$$v = \frac{(s_x{}^2/n_x + s_y{}^2/n_y)^2}{\dfrac{(s_x{}^2/n_x)^2}{n_x + 1} + \dfrac{(s_y{}^2/n_y)^2}{n_y + 1}} - 2 \tag{4.37}$$

and

$$t_1 = \frac{\bar{x} - \bar{y}}{\sqrt{s_x{}^2/n_x + s_y{}^2/n_y}} \tag{4.38}$$

5. Find $t_{\alpha;v}$ and compare. If $t_1 \leq t_{\alpha;v}$, accept H_O. If $t_1 > t_{\alpha;v}$, reject H_O.

Example 4.15 Determine on the basis of the following data with 95 percent confidence whether the resistivity ($\Omega \cdot \text{cm} \times 10^6$) of an experimental resistor y is as high as that of a standard resistor x.

$s_x{}^2 = 10.02 \qquad \bar{x} = 32.25 \; \Omega \cdot \text{cm} \times 10^6 \qquad n_x = 18$

$s_y{}^2 = 6.05 \qquad \bar{y} = 28.5 \; \Omega \cdot \text{cm} \times 10^6 \qquad n_y = 18$

$v = n_x + n_y - 2 = 34$

Solution

$H_O : [(\mu_x - \mu_y) = 0] \qquad H_A : [(\mu_x - \mu_y) > 0]$

$1 - \alpha = 0.95 \qquad \text{or} \qquad \alpha = 0.05$

Find the degrees of freedom ν, using Eq. (4.37):

$$\nu = \frac{(s_x^2/n_x + s_y^2/n_y)^2}{\dfrac{(s_x^2/n_x)^2}{n_x + 1} + \dfrac{(s_y^2/n_y)^2}{n_y + 1}} - 2$$

$$= \frac{(0.556 + 0.333)^2}{\dfrac{0.309}{19} + \dfrac{0.111}{19}} - 2$$

$$= 33.9$$

From Eq. (4.38),

$$t_1 = \frac{\bar{x} - \bar{y}}{\sqrt{s_x^2/n_x + s_y^2/n_y}} = \frac{32.25 - 28.5}{\sqrt{10.02/18 + 6.05/18}}$$

$$= \frac{3.75}{0.945} = 3.97$$

From Table A-3,

$$t_{\alpha;\,\nu} = t_{0.05;34} = 1.692$$

Since $t_1 > t_{\alpha;\,\nu}$, H_O is rejected. Hence it is concluded with 95 percent confidence that the resistivity of the experimental resistor is not as high as that of the standard resistor.

Case 2—the two-sided alternative The following steps are recommended:

1. Determine $H_O : [(\mu_x - \mu_y) = 0]$, $H_A : [(\mu_x - \mu_y) \neq 0]$.
2. Decide on a value for α.
3. Select two samples of n_x and n_y items each and find the sample means $\bar{x}$ and $\bar{y}$ and sample variances s_x^2 and s_y^2.
4. Find ν from Eq. (4.37) and t_1 from Eq. (4.38).
5. Find $t_{\alpha/2;\,\nu}$ from Table A-3 and compare. If $-t_{\alpha/2;\,\nu} \leq t_1 \leq t_{\alpha/2;\,\nu}$, accept H_O; if not, reject it.

Example 4.16 Surface emissivities of aluminum foil manufactured by two different suppliers are to be compared with 95 percent confidence. On the basis of the following data, determine whether there is any significant difference between the two suppliers.

$s_x^2 = 0.0001$ $\quad \bar{x} = 0.070$ $\quad n_x = 15$

$s_y^2 = 0.0002$ $\quad \bar{y} = 0.081$ $\quad n_y = 15$

Solution

$$H_O : [(\mu_x - \mu_y) = 0] \qquad H_A[(\mu_x - \mu_y) \neq 0]$$

$$1 - \alpha = 0.95 \qquad \alpha = 0.05 \qquad \frac{\alpha}{2} = 0.025$$

From Eq. (4.37),

$$\nu = \frac{(s_x{}^2/n_x + s_y{}^2/n_y)^2}{\dfrac{(s_x{}^2/n_x)^2}{n_x + 1} + \dfrac{(s_y{}^2/n_y)^2}{n_y + 1}} - 2$$

$$= \frac{(0.0001/15 + 0.0002/15)^2}{\dfrac{(0.0001/15)^2}{16} + \dfrac{(0.0002/15)^2}{16}} - 2$$

$$= \frac{(9)(16)}{1 + 4} - 2$$

$$= 26.8$$

From Eq. (4.38),

$$t_1 = \frac{\bar{x} - \bar{y}}{\sqrt{s_x{}^2/n_x + s_y{}^2/n_y}}$$

$$= \frac{-0.011}{10^{-2}\sqrt{\dfrac{1}{15} + \dfrac{2}{15}}}$$

$$= \frac{-1.1}{\sqrt{\dfrac{1}{5}}}$$

$$= -2.46$$

From Table A-3, $t_{\alpha/2;\,\nu} = t_{0.025;\,27} = 2.052$. H_O is accepted if $-t_{\alpha/2;\,\nu} \le t_1 \le t_{\alpha/2;\,\nu}$. Since $t_1 < t_{\alpha/2;\,\nu}$, H_O is rejected.

Therefore it is concluded that there is a significant difference in the surface emissivities of aluminum foil manufactured by the two suppliers.

Comparison of the variances

Case 1—the one-sided alternative In this case the null hypothesis is $H_O : (\sigma_x/\sigma_y = 1)$, and the alternative is $H_A : (\sigma_x/\sigma_y > 1)$ when x is the variable with the possible larger variance. If H_A was the alternative, then Fig. 4.12 adequately describes the situation. In addition, the design hypothesis $H_D : (\sigma_x/\sigma_y = \rho_1)$ is selected. In Fig. 4.12 it is seen that a type I error will be made by rejecting H_O when it is actually true. This is done when

$$\frac{s_x{}^2}{s_y{}^2} > \left(\frac{s_x{}^2}{s_y{}^2}\right)_c$$

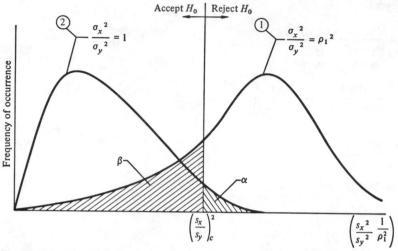

Fig. 4.12   One-sided alternative.

By definition α is the probability of making this type of error.   Hence

$$P\left[\left(\frac{s_x}{s_y}\right)^2 > \left(\frac{s_x}{s_y}\right)^2_c\right] = \alpha$$

when H_O is true.   Since

$$\left(\frac{s_x}{s_y}\right)^2 \left(\frac{\sigma_y}{\sigma_x}\right)^2$$

has an F distribution and by definition of $F_{\alpha;\,\nu_x;\,\nu_y}$

$$P\left[\left(\frac{s_x}{s_y}\right)^2 \left(\frac{\sigma_y}{\sigma_x}\right)^2 > F_{\alpha;\,\nu_x;\,\nu_y}\right] = \alpha$$

When $\sigma_x/\sigma_y = 1$,

$$P\left[\left(\frac{s_x}{s_y}\right)^2 > F_{\alpha;\,\nu_x;\,\nu_y}\right] = \alpha$$

Or at the cutoff point

$$F_{\alpha;\,\nu_x;\,\nu_y} = \left(\frac{s_x}{s_y}\right)^2_c$$

A type II error is made by the acceptance of H_O when H_D is actually true.

$$P\left[\left(\frac{s_x}{s_y}\right)^2 \leq \left(\frac{s_x}{s_y}\right)^2_c\right] = \beta$$

Since $1/\rho_1^2 = \sigma_y^2/\sigma_x^2$,

$$P\left[\left(\frac{s_x}{s_y}\right)^2 \frac{1}{\rho_1^2} \le \left(\frac{s_x}{s_y}\right)_c^2 \frac{1}{\rho_1^2}\right] = \beta$$

From the definition of $F_{1-\beta;\,v_x;\,v_x}$ it follows that

$$P(F \le F_{1-\beta;\,v_x;\,v_y}) = \beta$$

$$P\left[\left(\frac{s_x}{s_y}\right)^2 \frac{1}{\rho_1^2} \le F_{1-\beta;\,v_x;\,v_y}\right] = \beta$$

Or at the cutoff point, as in Sec. 4.1,

$$F_{1-\beta;\,v_x;\,v_y} = \frac{1}{F_{\beta;\,v_y;\,v_x}} = \left(\frac{s_x}{s_y}\right)_c^2 \frac{1}{\rho_1^2} = F_{\alpha;\,v_x;\,v_y} \frac{1}{\rho_1^2}$$

$$(F_{\alpha;\,v_x;\,v_y})(F_{\beta;\,v_y;\,v_x}) = \rho_1^2$$

Moreover, if $n_x = n_y = n$, it follows that $v_x = v_y = n-1$ and that

$$\rho_1^2 = (F_{\alpha;\,n-1;\,n-1})(F_{\beta;\,n-1;\,n-1}) \tag{4.39}$$

and

$$\left(\frac{s_x}{s_y}\right)_c^2 = F_{\alpha;\,n-1;\,n-1} \tag{4.40}$$

Hence, the following procedure is recommended:

1. Determine H_O: $(\sigma_x/\sigma_y = 1)$, H_D: $[(\sigma_x/\sigma_y) = \rho_1]$, and H_A: $(\sigma_x/\sigma_y > 1)$. (x is the symbol for the variable with possibly larger variance.)
2. Decide on the values of α and β which can be tolerated.
3. Use Eq. (4.39) to find $n = n_x = n_y$. This will need a few trials.
4. Select two samples of n_x and n_y items each and find $(s_x/s_y)^2$.
5. Find $(s_x/s_y)_c^2$ from Eq. (4.40) and compare. If $(s_x/s_y)^2 > (s_x/s_y)_c^2$, reject H_O; if not, accept H_O.

Example 4.17 An experiment was conducted to determine whether the water flow (gallons per minute) in an experimental pipe x is as uniform as in a standard pipe y. On the basis of the following data, determine with 99 percent confidence whether the flow uniformity of the pipe x is as high as that of the pipe y. It is necessary to detect with 99 percent confidence whether the variance of x is more than eight times the variance of y.

$s_x^2 = 5.2 \qquad n_x = 25$

$s_y^2 = 3.8 \qquad n_y = 25$

Solution

$$H_O : \left(\frac{\sigma_x}{\sigma_y} = 1 \right) \qquad H_A : \left(\frac{\sigma_x}{\sigma_y} > 1 \right) \qquad H_D : \left[\left(\frac{\sigma_x}{\sigma_y} \right)^2 = 8 \right]$$

so that

$$\rho_1 = \sqrt{8} = 2.82$$

$$\alpha = 0.01 \qquad \beta = 0.01$$

From Eq. (4.39),

$$\rho_1^2 = (F_{\alpha; \, \nu_x; \, \nu_y})(F_{\beta; \, \nu_y; \, \nu_x})$$

From Table A-7, for $\nu_x = \nu_y = 23$,

$$F_{\alpha; \, \nu_x; \, \nu_y} = F_{\beta; \, \nu_y; \, \nu_x} \cong 2.82$$

Hence Eq. (4.39) is satisfied for $\nu_x = \nu_y = 23$. Therefore a sample size of 25 is adequate.

From Eq. (4.40),

$$\left(\frac{s_x}{s_y} \right)_c^2 = F_{\alpha; \, n-1; \, n-1}$$

$$= 2.82$$

$$\left(\frac{s_x}{s_y} \right)^2 = \frac{5.2}{3.8} = 1.37 = F$$

Since

$$\left(\frac{s_x}{s_y} \right)^2 < \left(\frac{s_x}{s_y} \right)_c^2$$

H_O is accepted.

It is therefore concluded with 99 percent confidence that the uniformity of flow in the experimental and the standard pipes is the same.

Case 2—the two-sided alternative   The graphical representation of this test is given in Fig. 4.13. Distribution 1 is the distribution of the null hypothesis. $(s_x/s_y)^2$ comes from the distribution, if $\rho = 1$. Distributions 2 and 3 are distributions of the design hypothesis $H_D : (\sigma_x/\sigma_y = \rho_1$ or $1/\rho_1)$. Distribution 2 has $\rho = \rho_1$. Distribution 3 has $\rho = 1/\rho_1$.

Since $[(s_x/s_y)(1/\rho)]^2$ has an F distribution, the following is true. The mathematical reasoning will be omitted, since it is similar to that in the prior section.

$$F_{\alpha/2; \, n_x-1; \, n_y-1} = \left(\frac{s_x}{s_y} \right)_{c_1}^2$$

$$F_{\beta; \, n_y-1; \, n_x-1} = \frac{\rho_1^2}{(s_x/s_y)_{c_1}^2}$$

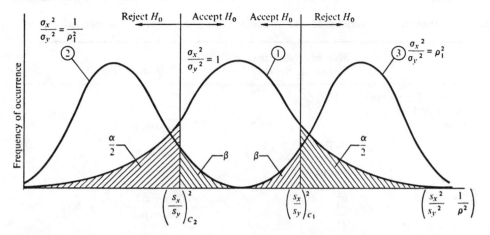

Fig. 4.13 Two-sided alternative.

Therefore

$$\rho_1{}^2 = (F_{\alpha/2;\,n_x-1;\,n_y-1})(F_{\beta;\,n_y-1;\,n_x-1}) \qquad (4.41)$$

and

$$\left(\frac{s_x}{s_y}\right)^2_{c_1} = F_{\alpha/2;\,n_x-1;\,n_y-1} \qquad (4.42)$$

$$\left(\frac{s_x}{s_y}\right)^2_{c_2} = F_{1-\alpha/2;\,n_x-1;\,n_y-1} \qquad (4.43)$$

The following steps are suggested:

1. Determine H_O: $(\sigma_x/\sigma_y = 1)$, H_A: $(\sigma_x/\sigma_y \neq 1)$, and H_D: $(\sigma_x/\sigma_y = \rho_1$ or $1/\rho_1)$.
2. Decide on the values of α and β which can be tolerated.
3. Use Eq. (4.41) to find $n = n_x = n_y$. This will need a few trials.
4. Select two samples of n items from each distribution and find $(s_x/s_y)^2$.
5. Find $(s_x/s_y)^2_{c_1}$ from Eq. (4.42) and $(s_x/s_y)^2_{c_2}$ from Eq. (4.43). Then compare. If $(s_x/s_y)^2_{c_2} \leq (s_x/s_y)^2 \leq (s_x/s_y)^2_{c_1}$, accept H_O; if not, reject H_O.

Example 4.18 Gains of three-stage amplifiers manufactured by two different suppliers x and y are to be tested for their uniformity. Tests were run and the following data were recorded.

$$n_x = n_y = 31 \qquad \nu_1 = \nu_2 = 30$$
$$s_x{}^2 = 41 \qquad s_y{}^2 = 25$$

Determine with 90 percent confidence whether the uniformity of the two amplifiers is the same. Variance ratio greater than 3 should be detected with 90 percent confidence.

Table 4.1 Absolute comparison of two products (designs, materials, processes)

	Must be known	Required test data	D*	Required sample size
I. Comparison of the means				
1. Variance σ^2 known				
a. One-sided				$n = \dfrac{(z_\alpha + z_\beta)^2 \sigma^2}{(\mu_0 - \mu_1)^2}$
	α, β	$n, \bar{x}$	z	
	μ_0, μ_1			
b. Two-sided	σ^2			$n = \dfrac{(z_{\alpha/2} + z_\beta)^2 \sigma^2}{(\mu_0 - \mu_1)^2}$
2. Variance σ^2 unknown				
a. One-sided				$n = \dfrac{(t_{\alpha;\,\nu} + t_{\beta;\,\nu})^2 s^2}{(\mu_0 - \mu_1)^2}$
	α, β	$n, \bar{x}, s^2$	t	
	μ_0, μ_1			
b. Two-sided				$n = \dfrac{(t_{\alpha/2;\,\nu} + t_{\beta;\,\nu})^2 s^2}{(\mu_0 - \mu_1)^2}$
II. Comparison of the variances				
1. One-sided				$\rho^2 = \dfrac{\sigma_1^{\,2}}{\sigma_0^{\,2}} = \dfrac{\chi_{\alpha;\,\nu}^2}{\chi_{(1-\beta);\,\nu}^2}$
	α, β	n, s^2	χ^2	
	$\sigma_0^{\,2}, \sigma_1^{\,2}$			
2. Two-sided				$\rho^2 = \dfrac{\sigma_1^{\,2}}{\sigma_0^{\,2}} = \dfrac{\chi_{\alpha/2;\,\nu}^2}{\chi_{(1-\beta);\,\nu}^2}$

* Distribution.

Cutoff point	Criterion for acceptance
$$\bar{x}_c = \pm \left[\frac{1}{2}(z_\beta - z_\alpha)\sqrt{\frac{\sigma^2}{n}} \right] + \frac{\mu_0 + \mu_1}{2}$$ $+$ For $H_A : (\mu < \mu_0)$ $-$ For $H_A : (\mu > \mu_0)$	$H_O : (\mu = \mu_0)$ For $H_A : (\mu < \mu_0)$ For $H_A : (\mu > \mu_0)$ If $\bar{x} \geq \bar{x}_c$, accept H_O. If $\bar{x} \leq \bar{x}_c$, accept H_O. If $\bar{x} < \bar{x}_c$, reject H_O. If $\bar{x} > x_c$, reject H_O.
$$\bar{x}_{c1} = \left[\frac{1}{2}(z_\beta - z_{\alpha/2})\sqrt{\frac{\sigma^2}{n}} \right] + \frac{\mu_0 + \mu_1}{2}$$ $$\bar{x}_{c2} = 2\mu_0 - \bar{x}_{c1}$$	$H_O : (\mu = \mu_0)$ $H_A : (\mu \neq \mu_0)$ If $\bar{x}_{c1} \leq \bar{x} \leq \bar{x}_{c2}$, accept H_O. If $\bar{x} < \bar{x}_{c1}$ or $\bar{x} > \bar{x}_{c2}$, reject H_O.
$$\bar{x}_c = \pm \left[\frac{1}{2}(t_{\beta;\,v} - t_{\alpha;\,v})\sqrt{\frac{s^2}{n}} \right] + \frac{\mu_0 + \mu_1}{2}$$ $+$ For $H_A : (\mu < \mu_0)$ $-$ For $H_A : (\mu > \mu_0)$	$H_O : (\mu = \mu_0)$ For $H_A : (\mu < \mu_0)$ For $H_A : (\mu > \mu_0)$ If $\bar{x} \geq \bar{x}_c$, accept H_O. If $\bar{x} \leq \bar{x}_c$, accept H_O. If $\bar{x} < \bar{x}_c$, reject H_O. If $\bar{x} > \bar{x}_c$, reject H_O.
$$\bar{x}_{c1} = \left[\frac{1}{2}(t_{\beta;\,v} - t_{\alpha/2;\,v})\sqrt{\frac{s^2}{n}} \right] + \frac{\mu_0 + \mu_1}{2}$$ $$\bar{x}_{c2} = 2\mu_0 - \bar{x}_{c1}$$	$H_O : (\mu = \mu_0)$ $H_A : (\mu \neq \mu_0)$ If $\bar{x}_{c1} \leq \bar{x} \leq \bar{x}_{c2}$, accept H_O. If $\bar{x} < \bar{x}_{c1}$ or $\bar{x} > \bar{x}_{c2}$, reject H_O.
$$s_c^2 = \frac{(\chi^2_{\alpha;\,v})\sigma_0^2}{n-1}$$	$H_O : (\sigma = \sigma_0)$ $H_A : (\sigma > \sigma_0)$ If $s^2 < s_c^2$, accept H_O. If $s^2 > s_c^2$, reject H_O.
$$s_{c1}^2 = \frac{(\chi^2_{\alpha/2;\,v})\sigma_0^2}{n-1}$$ $$s_{c2}^2 = 2\sigma_0^2 - s_{c1}^2$$	$H_O : (\sigma = \sigma_0)$ $H_A : (\sigma \neq \sigma_0)$ If $s_{c2}^2 \leq s^2 \leq s_{c1}^2$, accept H_O. If $s^2 > s_{c1}^2$, or $s^2 < s_{c2}^2$, reject H_O.

Table 4.2 Relative comparison of two products (designs, materials, processes)

	Must be known	Required test data	D*	Required sample size
I. Comparison of the means				
1. Variances σ_1^2 and σ_2^2 known				
a. One-sided				$n = \dfrac{(z_\alpha + z_\beta)^2(\sigma_x^2 + \sigma_y^2)}{(\mu_x - \mu_y)_1^2}$
	α, β $(\mu_x - \mu_y)_1$ σ_x^2, σ_y^2	n_x, n_y $\bar{x}, \bar{y}$	z	
b. Two-sided				$n = \dfrac{(z_{\alpha/2} + z_\beta)^2(\sigma_x^2 + \sigma_y^2)}{(\mu_x - \mu_y)_1^2}$
2. Variances σ_1^2 and σ_2^2 unknown but equal				
a. One-sided				$n = \dfrac{(t_{\alpha:\nu} + t_{\beta:\nu})^2(s_x^2 + s_y^2)}{(\mu_x - \mu_y)_1^2}$
	α, β $(\mu_x - \mu_y)_1$	n_x, n_y s_x^2, s_y^2	t_2	
b. Two-sided				$n = \dfrac{(t_{\alpha/2:\nu} + t_{\beta:\nu})^2(s_x^2 + s_y^2)}{(\mu_x - \mu_y)_1^2}$
3. Variances σ_1^2 and σ_2^2 unknown and not equal				
a. One-sided	α	n_x, n_y $\bar{x}, \bar{y}$	t_1	$\nu = \dfrac{(s_x^2/n_x + s_y^2/n_y)^2}{\dfrac{(s_x^2/n_x)^2}{n_x + 1} + \dfrac{(s_y^2/n_y)^2}{n_y + 1}} - 2$
b. Two-sided		s_x^2, s_y^2		
II. Comparison of the variances				
1. One-sided			F	$\rho_1^2 = \dfrac{\sigma_x^2}{\sigma_y^2} = \dfrac{F_{\alpha:\nu_x:\nu_y}}{F_{(1-\beta):\nu_x:\nu_y}}$ $= (F_{\alpha:\nu_x:\nu_y})(F_{\beta:\nu_y:\nu_x})$
	α, β σ_x^2/σ_y^2	n_x, n_y s_x^2, s_y^2		
2. Two-sided				$\rho_1^2 = \dfrac{\sigma_x^2}{\sigma_y^2} = \dfrac{F_{\alpha/2:\nu_x:\nu_y}}{F_{(1-\beta):\nu_x:\nu_y}}$ $= (F_{\alpha/2:\nu_x:\nu_y})(F_{\beta:\nu_y:\nu_x})$

* Distribution

152

Cutoff point	Criterion for acceptance

$$(\bar{x} - \bar{y})_c = \frac{1}{2}(z_\alpha - z_\beta)\sqrt{\frac{\sigma_x^2}{n_x} + \frac{\sigma_y^2}{n_y}} + \frac{(\mu_x - u_y)_1}{2}$$

$H_0 : [(\mu_x - \mu_y) = 0] \quad H_A : [(\mu_x - \mu_y) > 0]$
If $(\bar{x} - \bar{y}) \le (\bar{x} - \bar{y})_c$, accept H_0.
If $(\bar{x} - \bar{y}) > (\bar{x} - \bar{y})_c$, reject H_0.

$$(\bar{x} - \bar{v})_r = \frac{1}{2}(z_{\alpha/2} - z_\beta)\sqrt{\frac{\sigma_x^2}{n_x} + \frac{\sigma_y^2}{n_y}} + \frac{(\mu_x - \mu_y)_1}{2}$$

If $-(\bar{x} - \bar{y})_c \le (\bar{x} - \bar{y}) \le (\bar{x} - \bar{y})_c$, accept H_0.
If $(\bar{x} - \bar{y}) < -(\bar{x} - \bar{y})_c$ or $(\bar{x} - \bar{y}) > (\bar{x} - \bar{y})_c$, reject H_0.

$$(\bar{x} - \bar{y})_c = \frac{1}{2}(t_{\alpha; v} - t_{\beta; v})\sqrt{\frac{(n_x-1)s_x^2 + (n_y-1)s_y^2}{n_x + n_y - 2}}$$
$$\times \sqrt{\frac{1}{n_x} + \frac{1}{n_y}} + \frac{(u_x - \mu_y)_1}{2}$$

$H_0 : [(\mu_x - \mu_y) = 0] \quad H_A : [(\mu_x - \mu_y) > 0]$
If $(\bar{x} - \bar{y}) \le (\bar{x} - \bar{y})_c$, accept H_0.
If $(\bar{x} - \bar{y}) > (\bar{x} - \bar{y})_c$, reject H_0.

$$(\bar{x} - \bar{y})_c = \frac{1}{2}(t_{\alpha/2; v} - t_{\beta; v})\sqrt{\frac{(n_x-1)s_x^2 + (n_y-1)s_y^2}{n_x + u_y - 2}}$$
$$\times \sqrt{\frac{1}{n_x} + \frac{1}{n_y}} + \frac{(\mu_x - \mu_y)_1}{2}$$

If $-(\bar{x} - \bar{y})_c \le (\bar{x} - \bar{y}) \le (\bar{x} - \bar{y})_c$, accept H_0,
If $-(\bar{x} - \bar{y})_c < -(\bar{x} - \bar{y})_c$ or $(\bar{x} - \bar{y}) > (\bar{x} - \bar{y})_c$. reject H_0.

$$t_1 = \frac{\bar{x} - \bar{y}}{\sqrt{s_x^2/n_x + s_y^2/n_y}}$$

$H_0 : [(\mu_x - \mu_y) = 0] \quad H_A : [(\mu_x - \mu_y) > 0]$
If $t_1 \le t_{\alpha; v}$, accept H_0.
If $t_1 > t_{\alpha; v}$, reject H_0.

$H_0 : [(\mu_x - \mu_y) = 0] \quad H_A : [(\mu_x - \mu_y) \ne 0]$
If $-t_{\alpha/2; v} < t_1 \le t_{\alpha/2; v}$, accept H_0.
If $t_1 \le -t_{\alpha/2; v}$, or $t_1 > t_{\alpha/2; v}$, reject H_0.

$$\left(\frac{s_x^2}{s_y^2}\right)_c = F_{\alpha; v_x; v_y}$$

$H_0 : \left(\frac{\sigma_x^2}{\sigma_y^2} = 1\right) \quad H_A : \left(\frac{\sigma_x^2}{\sigma_y^2} > 1\right)$

If $\frac{s_x^2}{s_y^2} \le \left(\frac{s_x^2}{s_y^2}\right)_c$, accept H_0.

If $\frac{s_x}{s_y^2} > \left(\frac{s_x^2}{s_y^2}\right)_c$ reject H_0.

$$\left(\frac{s_x^2}{s_y^2}\right)_{c_1} = F_{\alpha/2; v_x; v_y}$$
$$\left(\frac{s_x^2}{s_y^2}\right)_{c_2} = F_{(1-\alpha/2); v_x; v_y}$$

$H_0 : \left(\frac{\sigma_x^2}{\sigma_y^2} = 1\right) \quad H_A : \left(\frac{\sigma_x^2}{\sigma_y^2} \ne 1\right)$

If $\left(\frac{s_x^2}{s_y^2}\right)_{c_2} \le \frac{s_x^2}{s_y^2} \le \left(\frac{s_x^2}{s_y^2}\right)_{c_1}$, accept H_0.

If $\frac{s_x^2}{s_y^2} < \left(\frac{s_x^2}{s_y^2}\right)_{c_1}$

or $\frac{\sigma_x^2}{s_y^2} > \left(\frac{s_x^2}{s_y^2}\right)_{c_2}$, reject H_0.

Solution

$$H_O : \left(\frac{\sigma_x}{\sigma_y} = 1\right) \qquad H_A : \left(\frac{\sigma_x}{\sigma_y} \neq 1\right) \qquad H_D : \left(\frac{\sigma_x}{\sigma_y} = \rho_1\right)$$

$$\alpha = 1 - 0.9 = 0.1 \qquad \frac{\alpha}{2} = 0.05 \qquad \beta = 1 - 0.9 = 0.1$$

From Eq. (4.41),

$$\rho_1{}^2 = (F_{\alpha/2;\, n_x-1;\, n_y-1})(F_{\beta;\, n_y-1;\, n_x-1})$$

$$= (1.84)(1.61)$$

$$= 2.96$$

Since this is less than 3, the sample size is adequate.
Find the cutoff points, using Eqs. (4.42) and (4.43). From Table A-5,

$$\left(\frac{s_x}{s_y}\right)^2_{c_1} = F_{0.05;\, 30;\, 30} = 1.84$$

$$\left(\frac{s_x}{s_y}\right)^2_{c_2} = F_{(1-0.05);\, 30;\, 30} = \frac{1}{F_{0.05;\, 30;\, 30}} = 0.544$$

Hence, if $0.544 \leq (s_x/s_y)^2 \leq 1.84$, accept H_O. From the given data $s_x{}^2/s_y{}^2 = 41/25 = 1.64$.

Hence, H_O is accepted and it is concluded with 90 percent confidence that the two amplifiers are equally uniform.

Table 4.3 Summary of the distributions used in Tables 4.1 and 4.2

Table 4.1	Table 4.2
$z = \dfrac{\bar{x} - \mu}{\sqrt{\sigma^2/n}}$	$z = \dfrac{(\bar{x} - \bar{y}) - (\mu_x - \mu_y)}{\sqrt{\sigma_x{}^2/n_x + \sigma_y{}^2/n_y}}$
	$t_1 = \dfrac{\bar{x} - \bar{y}}{\sqrt{s_x{}^2/n_x + s_y{}^2/n_y}}$
$t = \dfrac{\bar{x} - \mu}{\sqrt{s^2/n}}$	
	$t_2 = \dfrac{(\bar{x} - \bar{y}) - (\mu_x - \mu_y)}{\sqrt{\dfrac{(n_x - 1)s_x{}^2 + (n_y - 1)s_y{}^2}{n_x + n_y - 2}} \sqrt{\dfrac{1}{n_x} + \dfrac{1}{n_y}}}$
$\chi^2 = (n - 1)\dfrac{s^2}{\sigma^2}$	$F = \dfrac{s_x{}^2/\sigma_x{}^2}{s_y{}^2/\sigma_y{}^2}$

SUMMARY

In this chapter means for comparing two products (designs, materials, processes) are presented. If it is only desired to determine whether a difference between the products exists, use the preliminary approach (Sec. 4.1). The cases described involve normal and Weibull distributions. To determine which distribution the test data follow, see Sec. 2.8. If the actual differences are to be determined, the detailed approach (Sec. 4.2) is to be used. The cases described involve normal distribution. If the distribution is not known or cannot be established, use nonparametric methods described in Chap. 8.

The detailed approach is summarized in Tables 4.1 and 4.2. Table 4.1 deals with the type of problems in which the characteristics of one product are known and it is desired to determine whether the other product differs from the first by a predetermined amount. Two major categories are involved: (1) when the means are to be compared; (2) when the variances are to be compared. Table 4.2 refers to the type of problems in which the characteristics of neither product are known and it is desired to find whether one product is different from the other by a predetermined amount. As in Table 4.1, means and variances are to be compared.

The comparison between two products is not confined to Chap. 4. Somewhat less directly, methods in Chaps. 5 through 10 can be used, particularly those in Chaps. 7, 8, and 10. In Chap. 7, two products can be compared if the data follow Weibull, normal, or binomial distributions. The methods of Chap. 7 involve less testing time, provided that only the knowledge of the α error is required and not of both the α and β errors (see Glossary), as in Chap. 4. The methods of Sec. 8.3 can also be used for comparing two products, for those cases where the underlying distribution cannot be established. Chapter 10 compares two distributions by forming a third distribution and evaluating it.

PROBLEMS

4.1. In two designs x and y, the random variables x and y are assumed to follow a normal distribution, and their population variances are unknown but equal. From the following data, determine with 90 percent confidence which design has a higher average.

	Design x	Design y
Sample size	21	11
Sample mean	12	10
Sample variance	1	4

4.2. Two groups of specimens x and y, each group with a different type of plating, were tested for resistance to corrosion. If there is a significant difference between their mean resistances, the specimens with the higher resistance will be accepted.   Determine whether there is a significant difference between x and y.   Use $\alpha = 0.01$ and $\beta = 0.01$.   It is known $\sigma_x = 5$ ksi, $\sigma_y = 10$ ksi.   Use the following data as necessary:

Sample size n	Mean $\bar{x}$	Mean $\bar{y}$
10	82	72
20	80	69
30	81	70

4.3. A new and less expensive method has been developed for the fabrication of pistons. The results of diameter measurements of randomly selected samples are:

Diameter, in

Old method x	New method y
1.124	1.127
1.123	1.128
1.127	1.124
1.126	1.122
1.126	1.128
1.125	1.123
1.125	1.123
1.124	$\bar{y} = 1.125$
1.127	$n = 7$
$\bar{x} = 1.125$	$s_y^2 = 6.67 \times 10^{-6}$
$n = 9$	
$s_x^2 = 1.5 \times 10^{-6}$	

Determine with 95 percent confidence whether the precision of the new method is as good as the old.

4.4. The fatigue strengths of two different components x and y, fabricated on the same machines, are to be compared.   Since y is cheaper than x, y will be chosen even though it does not have higher fatigue strength.   Because of the attractiveness of the lower cost, y may be chosen even if it has a lower fatigue strength than x, provided that it is lower by not more than 5 ksi.   Design an experiment which would allow for the decision as to whether y should be chosen, with an error not more than 1 percent ($\alpha = \beta = 0.01$).   Use the following data:

	Component x	Component y
Sample size	8	8
Sample mean, ksi	31	29
Sample standard deviation, ksi	2.0	1.8

4.5. In the manufacture of a certain component, a forming process cannot handle material with a mean hardness greater than 200 BHN, and the stress requirements make it necessary to have a mean hardness of at least 160 BHN. Therefore, 180 BHN was decided on as the required value. Good components can be rejected 5 percent of the time, but bad components can be accepted only 2 percent of the time. From past experience it is known that this material follows a normal distribution with $\sigma = 20$ BHN. Devise and run an experiment which would determine whether the components are acceptable. Use the following data as necessary:

Sample size n	Mean, BHN
10	180.7
15	169.2
20	181.5

4.6. A manufacturer of an electric device uses x brand of thermostats, which operate satisfactorily at 550°F. He receives a competitive bid at a lower price from a supplier, who claims that his thermostats are as good as the x thermostats (that is, they can be set at 550°F). The manufacturer decides that if the difference between the mean settings of x and y is greater than 10.5°, he will conclude that y differs excessively from x and he will not buy the y thermostats. Both suppliers claim that $\sigma = 10$°F. A test is run on 23 x thermostats and 23 y thermostats, with the following results: $\bar{x} = 551.06$; $\bar{y} = 549.93$. On the basis of $\alpha = 0.05$ and $\beta = 0.1$, should y thermostats be accepted?

4.7. Determine, on the basis of the following test data, with 95 percent confidence, whether the means of population x and y are equal:

$$s_x{}^2 = 10.02 \qquad \bar{x} = 32.25 \qquad n_x = 18$$
$$s_y{}^2 = 6.05 \qquad \bar{y} = 28.50 \qquad n_y = 18$$

4.8. Two types of plastic coatings x and y were tested for resistance to wear. The average wear of the population of x and y was compared by assuming that the variances were equal. Test the assumption that the $\sigma_x = \sigma_y$ with 95 percent confidence; it is necessary that $(\sigma_x/\sigma_y)^2 \geq 4$ can be detected with 90 percent probability. The test data are:

$$n_x = 21 \qquad s_x{}^2 = 9.35 \times 10^{-6}$$
$$n_y = 21 \qquad s_y{}^2 = 6.71 \times 10^{-6}$$

4.9. Two metals were tested for their melting points. Before applying a comparison test of the two means, it is necessary to determine whether the variances of the two melting points are equal. Since it is not known which of the two variances is greater, a two-sided test was employed. A 90 percent confidence of detecting equality is desired, and a probability of 90 percent of detecting $(\sigma_x/\sigma_y)^2 = 3$ is necessary. On the basis of the following data, determine whether the two variances are equal: $n_x = n_y = 31$, $s_x{}^2 = 7.57$, $s_y{}^2 = 13.63$.

4.10. The following data represent the lives of turbine rotor blades tested in the laboratory under conditions simulating an actual operation. Design A represents the type presently in use, which experiences some field failures; design B represents a different material thought to be better but involving a cost increase. On the basis of the following data, determine with 90, 95, and 99 percent confidences whether there exists a significant difference between the mean lives of these two designs.

Life of turbine blades, h

Design A	Design B
210	1,010
360	400
320	600
480	225
140	830
520	480
	780

4.11. Compare the standard with the experimental material, for life, by means of the Weibull method. The following data are available:

Standard material life, cycles $\times 10^5$	Experimental material life, cycles $\times 10^5$
0.16	0.25
0.30	0.45
0.40	0.65
0.52	0.75
0.70	0.97
0.85	1.20
1.20	1.40
	1.60
	2.10
	2.50

4.12. A new process was developed for making washers with less variation in thickness than in the standard process. To verify this claim, the thickness of eight standard washers was measured, and the sample standard deviation was found to be 0.006 in. The thicknesses of 10 washers produced by the new process were found to be (inches) 0.123, 0.124, 0.126, 0.129, 0.120, 0.132, 0.123, 0.126, 0.129, and 0.128. Are washers produced by the new process more uniform than those presently being manufactured? Provide the answer with 90 and 95 percent confidence.

4.13. A certain chemical absorbs the oxygen present in the water in a free form. Since a certain amount of oxygen in the water is necessary for fish survival, it is required to reduce the concentration of such chemicals from the waste water before it is dumped in a river. Two water purification processes were recommended; both would reduce the concentration of 300 ppm (parts per million) to an acceptable level. In order to decide which process to use, an experiment was run with the following results (the process which produces the lower concentration level is considered better). Population variances are known to be equal.

	Process x	Process y
Sample mean	80 ppm	72 ppm
Sample standard deviation	2 ppm	3 ppm
Number of tests	20	20

If the difference between the two processes is less than 8 ppm, either is acceptable. Assume $\alpha = \beta = 0.05$ and determine which process is better. Compute the statistically significant required sample size. If it is less than the one used in this experiment, discuss and solve the problem with the necessary trade-off.

4.14. Two water purification processes were available to reduce the concentration of suspended solids (sand, grit, etc.) in the water. Both reduced the concentration of 300 ppm (parts per million) to an average level of 50 ppm. However, one seems to produce more variation in the concentration levels than the other. From the following data, determine with 90 percent confidence level which process is more uniform.

	Process x	Process y
Sample standard deviation	1.5 ppm	3.2 ppm
Number of tests	15	15

4.15. A new engineering process was recommended to reduce the level of air pollution around a given manufacturing plant. At the present time, the concern was to reduce only the level of carbon monoxide (CO) in the air. The new process was tried and the air was sampled several times for a given period of time, and the level of CO was measured. The data on CO level corresponding to the current manufacturing process were also available as shown below. Population variances are known to be equal.

	Current process	New process
Sample size	25	25
Sample mean, ppm	7.3	6.0
Sample standard deviation, ppm	1.8	1.2

With a decision error not more than 1 percent ($\alpha = \beta = 0.01$), determine from the above data whether the new process does produce less pollution.

4.16. Air-pollution levels at two different locations were measured in terms of the amount of suspended particles in micrograms per cubic meter.

	Location 1	Location 2
Sample size	26	26
Sample mean	40	45
Sample standard deviation	2.0	2.5

Determine whether pollution level at location 2 is higher than at location 1 with a decision error not more than 1 percent ($\alpha = 0.01$).

BIBLIOGRAPHY

Bowker, A. H., and G. J. Lieberman: "Engineering Statistics," Prentice-Hall, Inc., Englewood Cliffs, N.J., 1963.

Dixon, W. J., and F. J. Massey, Jr.: "Introduction to Statistical Analysis," 3d ed., McGraw-Hill Book Company, New York, 1969.

Duncan, A. J.: "Quality Control and Industrial Statistics," Richard D. Irwin, Inc., Homewood, Ill., 1959.

Johnson, L. G.: The Statistical Treatment of Fatigue Experiments, *Gen. Motors Res. Rep.* GMR-202, April, 1959.

Moroney, M. J.: "Facts from Figures," Penguin Books, Inc., Baltimore, 1965.

Natrella, M. G.: Experimental Statistics, *Nat. Bur. Stand. Hand.* 91, Aug. 1, 1963.

Wallis, W. A., and H. V. Roberts: "Statistics: A New Approach," The Free Press, New York, 1956.

5
Accelerated Experiments

One of the goals of laboratory tests is to verify the design objectives. If a heavy truck is designed to last for 500,000 m, it is essential for the designer to know as soon as possible whether this objective will be met. To meet the production schedule, it will be unrealistic to test the truck for 500,000 m under normal operating conditions. For this reason, the test is accelerated in the laboratory (and, in the case of a truck, also in the proving grounds) to condense 500,000 m into, say, 50,000 m.

The three factors which govern the degree of acceleration are the environment, the sample size, and the testing time. Which one is employed depends on the circumstances. If the product is complex and expensive (say a large gas turbine), the sample size will of necessity be small and the test will be accelerated by increasing either the test time or the intensity of the environment. On the other hand, if the product is inexpensive and numerous (bearings, resistors, etc.), test time can be considerably reduced by testing a large sample. In other situations, for a given sample size, a reduction in test time can be achieved by increasing the intensity of the environment. It is essential that the mode of failure under accelerated conditions be the same as under the anticipated normal operating conditions for which the part is designed. Various interrelationships between environment, sample size, and testing time are developed in this chapter. Several aspects of this problem are also discussed in Chap. 12.

5.1 RELATIONSHIP BETWEEN TESTING TIME AND THE ENVIRONMENT

By environment is meant here any operating condition to which the part will be subjected in service and which may affect its performance and durability. It can be load, stress, amplitude of vibration, ozone concentration, voltage, temperature, atmospheric corrosion, etc. For the purpose of this discussion these factors will be referred to as "stress."

The relationship between stress and life is generally given by a curve such as shown in Fig. 5.1, where the life decreases as the intensity of stress increases.

In the case of an actual stress (in psi), when plotted on log-log scale, straight-line relationship, as in Fig. 5.2, frequently results. The horizontal portion of the curve corresponds to infinite life, and sometimes it is used as a design criterion (see Chap. 9, Fatigue Experiments). This relationship between stress and life provides an effective means for accelerating the test.

If the design life under normal environment S_1 is N_1, the test time can be reduced to N_2 by increasing the intensity of the environment to S_2. In conducting such a test, it is essential that if the test is run to failure, the mode of failure under accelerated conditions be the same as under the normal operating conditions. If a different mode of failure should result, the

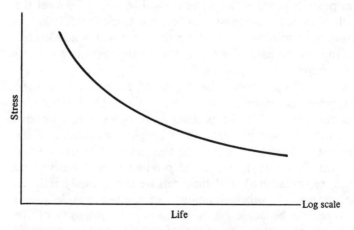

Fig. 5.1 Relationship between stress and life.

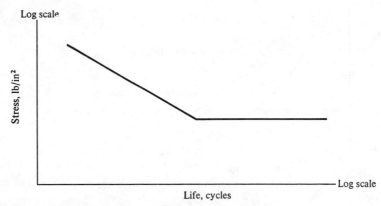

Fig. 5.2 Relationship between actual stress and life.

intensity of the environment S_2 should be reduced (test less accelerated) until a mode of failure identical with S_1 is produced (Fig. 5.3).

At all times the accelerated test should be realistic and meaningful. Consider a test where two products are to be compared. Under conditions shown in Fig. 5.4 the accelerated test is valid because the relative merits of A and B are maintained. However, if the A and B lines are not parallel, an accelerated test would indicate that B has a higher life than A while, in reality, under the operating conditions for which the parts were designed, the opposite is true. To make the test more realistic, stress S_2 should be reduced (Fig. 5.5).

An interesting new technique for accelerated life testing has recently been presented [1]. The theory of this procedure is based on cumulative

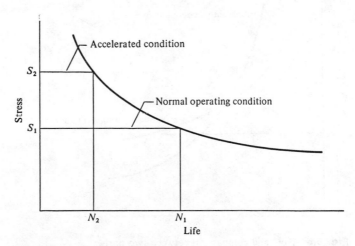

Fig. 5.3 Accelerated test.

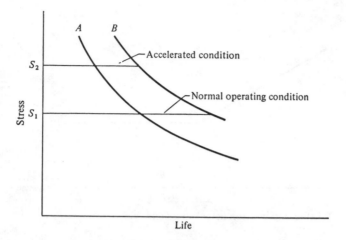

Fig. 5.4 Valid comparison between two parts.

damage (also see Sec. 10.4).   Assume that tests are run (Fig. 5.6) under two modes: stress S_1 and stress S_2, the term "stress" used here to describe an environment.   Let the life of a component be N_1 when run at S_1, and N_2 at S_2.   Then, in some circumstances, if the component is run for a time αN_1 at stress S_1, and the stress is changed to S_2, it will fail after a time βN_2 provided that

$$\alpha + \beta = 1 \qquad\qquad (5.1)$$

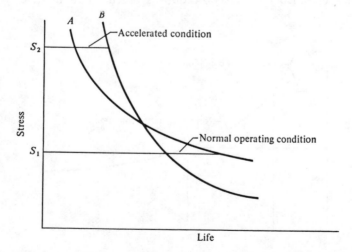

Fig. 5.5 Not-valid comparison between two parts.

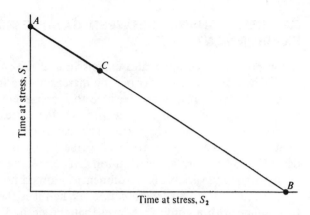

Fig. 5.6 Technique for accelerated testing.

This leads to the following test procedure: One series of tests is run all at the higher stress (point A in Fig. 5.6), while another series is run partly at the lower stress and partly at a higher stress (point C). Extrapolation yields point B, the life at low stress level.   Point B can be interpreted as the design life under normal operating conditions, while A and C are the test times under accelerated conditions.   The above procedure is based on the proposition that the life data in Fig. 5.6 follow a straight line.   Experiments conducted by the authors [1] on four different products seem to confirm this assumption (Fig. 5.7).

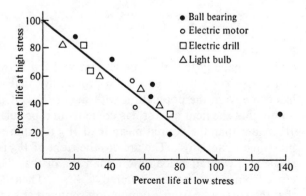

Fig. 5.7 Verification of straight-line relation in Fig. 5.6.

5.2   RELATIONSHIP BETWEEN SAMPLE SIZE AND THE ENVIRONMENT

If the product is complex and expensive and the sample size, therefore, is small, the test can be accelerated by increasing the intensity of the environment. However, if it is not feasible to increase environment intensity and the product is numerous and inexpensive, the same results can be achieved by increasing the sample size. The environment can be load, voltage, temperature, and, in general, any "value" (W) and these can frequently be described by a normal distribution. This section deals with accelerated testing when the variable (environment) follows a normal distribution.

Case 1—when no failure occurs When it is the aim of a test program to evaluate with a confidence level that the desired mean load to failure is exceeded by the actual mean load W to failure, in the case of a normal distribution, the following analysis is used. The probability that a single item would have failed at the test load W_0 or less is

$$F(W_0) = P = \int_{-\infty}^{W_0} \frac{1}{\sigma\sqrt{2\pi}} \exp\left[-\frac{(W - W_D)^2}{2\sigma^2}\right] dW \qquad (5.2)$$

where W_D = desired mean (statistical) load to failure
    σ = population standard deviation
    By defining $(W - W_D)/\sigma = z$ so that $dW = \sigma\, dz$,

$$F(z_0) = P = \int_{-\infty}^{z_0} \frac{1}{\sqrt{2\pi}} e^{-z^2/2} dz \qquad z_0 = \frac{W_0 - W_D}{\sigma} \qquad (5.3)$$

$(W_0 - W_D)/\sigma$ is the number of standard deviations the test load is located from the desired mean load to failure. If this one item on test does not fail, the expression

$$1 - F(z_0) = 1 - P = 1 - \int_{-\infty}^{z_0} \frac{1}{\sqrt{2\pi}} e^{-z^2/2} dz \qquad (5.4)$$

is the probability that this single item will fail between the test load $z = z_0$ and $z = +\infty$. This, then, is the probability that the part on test actually has come from the population with mean W_D. Conversely, P is the probability that the item on test has come from a population with the mean load W greater than the desired mean load W_D to failure. In Fig. 5.8 this fact is presented pictorially. The area to the right of W_0 is the probability that the single part will fail at load more than W_0 or that the single part actually has come from a distribution centered at W_D. Then P must be the probability that the part came from a population centered at some mean value W greater than W_D. This is valid for one part tested to load W_0 without failure.

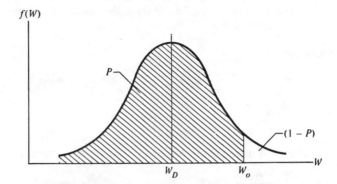

Fig. 5.8 Probability of parts coming from distribution with mean greater than W_D.

If, however, n items are tested in an independent testing operation and there are no failures in any of the specimens, that is, $k = 0$ where k is the number of failures out of the n items tested, the expression for the probability that simultaneous independent events will occur can be used.   That is,

$$P(AB) = P(A) \times P(B) \tag{5.5}$$

Then, in this case, the probability that n items would survive a test load of magnitude W_0 had they come from a population centered at W_D is

$$P(1, 2, \ldots, n) = (1 - P)_1 (1 - P)_2 \cdots (1 - P)_n$$

or, if each comes from the same distribution,

$$P(1, 2, \ldots, n) = (1 - P)^n \tag{5.6}$$

Hence, the probability that these samples actually have come from a population of mean load W greater than W_D is

$$P_n(W > W_D) = 1 - (1 - P)^n \tag{5.7}$$

Tables A-26 to A-32, corresponding to the case when number of failures $k = 0$, provide a quick solution to this confidence problem where σ and W_D are known.   In addition, direct reference to the normal probability table A-1 will result in solutions.

These tables also apply to cases when failures occur (for specific procedure, see the later part of the chapter).   In these applications $N = n + k$, where N is the statistical sample size and n is the physical sample size.

Example 5.1   A given component should be able to sustain a load of 1,000 lb.   At what load should five items be tested without failure in order to have 95 percent confidence that the design goal of 1,000 lb has been achieved? It is known that the loads to failure follow a normal distribution, the standard deviation σ being one-tenth of the average load W_D.

Solution

$$N = n = 5 \qquad \frac{\sigma}{W_D} = 0.1 \qquad W_D = 1,000 \text{ lb} \qquad W_0 = \,?$$

Hence,

$\sigma = 100$ lb

From Table A-27, for $N = 5$, $k = 0$, and 0.95 confidence,

$$\frac{W_0 - W_D}{\sigma} = -0.12$$

or

$$W_0 = W_D - 0.12\sigma = 1,000 - 12 = 988 \text{ lb}$$

Hence, it can be concluded with 95 percent confidence that if all five items can sustain the load of 988 lb without failure, the design goal of 1,000 lb or more has been met.

This result can also be obtained from Table A-1.

$$P_1 = \int_{-\infty}^{z} \frac{1}{\sqrt{2\pi}} e^{-z^2/2} \, dz$$

which is the probability point of $W_0 - W_D/\sigma$ when

$$P_n = 1 - (1 - P_1)^n = 0.95$$

or

$$(1 - P_1)^5 = 0.05$$

or

$$P_1 = 1 - 0.55 = 0.45$$

From Table A-1, $z_z = 0.12$ for $P_1 = 0.45$,

$$\frac{W_0 - W_D}{\sigma} = -0.12$$

or

$$W_0 = 1,000 - 12 = 988 \text{ lb}$$

To demonstrate the reduced-sample-size property of this technique, assume that some design changes are made and only one prototype is available for test. Since only one item is available, it should be tested under more severe conditions in order to meet the design goal at the same confidence level (95 percent). The load at which this item should be tested is computed as follows:

From Table A-26, $k = 0$, $n = N = 1$, and 95 percent confidence,

$$\frac{W_0 - W_D}{\sigma} = +1.63$$

$$W_0 = 1{,}000 + 163 = 1{,}163 \text{ lb}$$

The prototype must be tested at a load of 1,163 lb. This, then, illustrates the trade-off relationship between the sample size and the test severity.

Case 2—when failure occurs The above discussion applies to those situations when the items under test do not fail. When some items do fail during the test, the following analysis applies.

The probability that a single item will fail below the required test load W_0 is

$$p = \frac{1}{\sqrt{2\pi}} \int_{-\infty}^{z_0} e^{-z^2/2} \, dz \qquad (5.8)$$

where $z_0 = (W_0 - W_D)/\sigma$. Then the probability that exactly r_i failures will occur out of n items each tested at load W_0 is found by using the binomial expression given in Eq. (2.18):

$$p(r = r_i) = \binom{n}{r_i} p^{r_i} (1 - p)^{n - r_i}$$

With $r_i = k$,

$$p(r = k) = \binom{n}{k} p^k (1 - p)^{n - k} \qquad (5.9)$$

where

$$\binom{n}{k} = \frac{n!}{(n - k)! \, k!}$$

and p is given by Eq. (5.8). Moreover, the probability that there will be any number of failures r between zero and k is

$$P(r \le k) = \sum_{r_i = 0}^{k} \binom{n}{r_i} p^{r_i} (1 - p)^{n - r_i} \qquad (5.10)$$

With $n = k + 1$,

$$P(r \le n - 1) = \sum_{r_i = 0}^{k} \binom{k + 1}{r_i} p^{r_i} (1 - p)^{k + 1 - r_i} \qquad (5.11)$$

In this case $P(r \le k)$ is the probability that the actual mean load to failure W is less than the desired mean load W_D. It then follows that

$$P_k = 1 - P(r \le k) = 1 - \sum_{r_i = 0}^{k} \binom{r + 1}{r_i} p^{r_i} (1 - p)^{r + 1 - r_i} \qquad (5.12)$$

is the probability that the actual mean load to failure W exceeds the desired mean load to failure where k out of $k + 1$ have failed below the load W_0. By expanding the above expression for various numbers of failures, the following is found:

$$n = 1 \quad k = 0 \qquad P_0 = 1 - (1 - p) = p$$

$$n = 2 \quad k = 1 \qquad P_1 = 1 - \left[\binom{2}{0}(1 - p)^2 + \binom{2}{1}p(1 - p) \right]$$

$$P_1 = p^2$$

$$n = n \quad k = n - 1 \qquad P_k = p^{k+1} = p^n \tag{5.13}$$

Equation (5.13) corresponds to a specific case where the number of failures k is one less than the sample size n.   The solution of the general equation (5.12) is given in Tables A-26 through A-32.

These tables can also be used for the cases where n items are tested, out of which k items failed.   The value N, the statistical sample size, is equal to n.   However, if these k items are repaired and rerun, $N = n + k$.   Other details are discussed in the next section.

Example 5.2   Using the data in Example 5.1, suppose one out of the five items failed below the test load W_0.   How much confidence can be placed in the statement that the actual mean load to failure exceeds the desired mean load to failure?

Solution

$$n = N = 5 \quad k = 1 \qquad \frac{W_0 - W_D}{\sigma} = -0.12$$

From Table A-27, the confidence that the actual mean load exceeds the desired mean load is 0.75, or 75 percent.

5.3   RELATIONSHIP BETWEEN SAMPLE SIZE AND TESTING TIME

One of the situations encountered frequently in testing involves a trade-off between sample size and testing time.   If the item under test is expensive, the test can be accelerated by extending the time of testing on fewer items. For items which are numerous and easily available the opposite will be true: A large sample size is taken, thereby reducing testing time.   Two situations are described below: (1) when the item under test is an assembly or a system; (2) when the item is an individual part.

5.3.1   FOR SYSTEMS

If the variable is testing time and the item is an assembly or a system, the underlying distribution of the variable frequently can be taken as the exponential.

Consider the case of a test of an automotive engine. If the ceramic porcelain in a spark plug cracks and must be replaced, this repair results in a very small change in the system. That is, the system is approximately as old as it was before the failure. However, if during the test a bearing seizes and the crankshaft shatters or a connecting rod breaks, this constitutes more than a small change in the system. This, in fact, affects the system's age so that the system's age can be assumed equal to the accumulated test time. This is an overhauled system. Apparently, the testing of the system after a single repair is approximately equivalent to the case of testing two systems. One system has failed and the other, the repaired system, is still running. It must be noted that this argument is valid for repaired, not overhauled, systems.

Thus it appears that in system testing, testing one prototype with k repairs is approximately equivalent to testing $k + 1$ systems until k systems fail with no repairs. That is, k systems have failed and one system is still running. Here again the sacrifice of test time for the advantage of one sample has been made. This is shown in Fig. 5.9a.

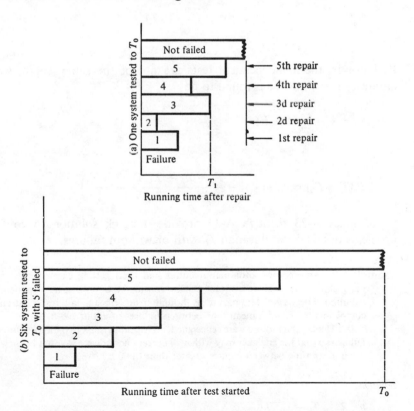

Fig. 5.9 Effective failure pattern for one system with repairs (a) vs. N systems without repairs (b).

Moreover, the analysis described below can also be applied to the case where it is desired to determine that the mean time to failure exceeds the desired mean time to failure for a single part rather than a system. This can be done by running $k + 1$ parts until r parts fail without repair or replacement. This is shown graphically in Fig. 5.9b.

Two cases are considered here: (1) when no failures occur during the test; (2) when some items do fail.

Case 1—when no failure occurs The fact that the exponential distribution describes the normal period of operation of many parts and systems has been discussed before. Here the probability that the actual mean time to failure T exceeds the desired mean time to failure T_D for a single test with no failure is given by

$$P = 1 - e^{-T_0/T_D} \tag{5.14}$$

where T_0 is the test time.

Here, again, if n independent tests of the same length T_0 are conducted without failure, this probability becomes

$$P_n(T > T_D) = 1 - (1 - P)^n = 1 - (1 - 1 + e^{-T_0/T_D})^n$$
$$= 1 - e^{-nT_0/T_D} \tag{5.15}$$

If, however, the n independent tests are not of the same length, with no failures, the confidence is found to be

$$P_n = 1 - \sum_{i=1}^{n} (1 - P)_i \tag{5.16}$$

or

$$P_n(T > T_D) = 1 - \exp\left(-\frac{T_1 + T_2 + T_3 + \cdots + T_n}{T_D}\right) \tag{5.17}$$

Tables A-33 through A-39 provide a quick solution to confidence problems of equal test duration T_0 with or without failures.

Example 5.3 An electric-motor manufacturer had been getting excessive failures of a particular model. Since a motor can be considered a system, the exponential distribution, with its constant hazard, is thought to describe the failures. The customer complaints indicated a mean time between failures of about 200 h.

Hence, the motors were redesigned. How long must two motors run without failures so that the engineer may with a 90 percent confidence say that his new design has a mean time between failures greater than 1,000 h?

Solution

$n = 2$ $T_D = 1,000$ h

From Eq. (5.15)

$$P_n = 1 - e^{-nT_0/T_D}$$

$$P_2 = 1 - e^{-2T_0/T_D} = 0.90$$

$$e^{2T_0/T_D} = 10 = e^{2.3}$$

$$\frac{2T_0}{T_D} \cong 2.3$$

or,

$$T_0 = \frac{2,300}{2} = 1,150 \text{ h}$$

The engineer must run both motors for 1,150 h without failure to be able to conclude with 90 percent confidence that the actual mean time between failures is at least 1,000 h.

This can also be solved by using Table A-33 with $n = 2$ and $k = 0$: For 90 percent confidence:

$$\frac{T_0}{T_D} \cong 1.15$$

$$1.15 \times 1,000 = 1,150 \text{ h}$$

Case 2—when failure occurs As was mentioned above, exponential failures are random failures. If a system is selected randomly from the population, tested, and failed, it may be repaired and retested as though it were new. This is true because the system has failed because of some random mechanism not associated with its age, and a repair makes it new again. This is because the repair makes it vulnerable to the same constant hazard random failure mechanism that has acted upon it from $t = 0$. In this case it appears that the sample size is one more than the number of failures. However, the analysis below also applies where n items are tested. The sample size is n; the total sample size is

$$N = n + \sum_{j=1}^{n} k_j \tag{5.18}$$

where k_j indicates the number of failures on the jth unit of n. The number of failures is

$$k_t = \sum_{j=1}^{n} k_j \tag{5.19}$$

The test time must be the same for all new or repaired units unless failure terminates the test, so that

$$T_0 = T_1 = T_2 = \cdots = T_i = \cdots = T_n \tag{5.20}$$

where T_i is the test time on the ith item of N.

The probability that k failures will occur during the time T_0 is given by

$$P(r = k) = \frac{N!}{k!(N-k)!}(1 - e^{-T_0/T_D})^k(e^{-T_0/T_D})^{N-k} \qquad (5.21)$$

and

$$P(r \le k) = \sum_{r_i=0}^{k} \frac{N!}{r_i!(N-r_i)!}(1 - e^{-T_0/T_D})^{r_i}(e^{-T_0/T_D})^{N-r_i} \qquad (5.22)$$

$P(r \le k)$ is the probability that the actual mean time to failure T is less than the desired mean time to failure T_D. Through the use of mutually exclusive intervals the probability that $T > T_D$ is

$$P_k = 1 - P(r \le k) = 1 - \sum_{r_i=0}^{k} \frac{N!}{r_i!(N-r_i)!}(1 - e^{-T_0/T_D})^{r_i}(e^{-T_0/T_D})^{N-r_i}$$
$$(5.23)$$

Tables A-33 to A-39 give the confidence as calculated from Eq. (5.23) with a physical sample size n, which ultimately will be repaired k times. Therefore, $N = n + k =$ statistical sample size.

Example 5.4 The electric-motor engineer of Example 5.3 tested both motors for 1,150 h, and encountered a failure in one motor before it accumulated 1,150 h. This was repaired and run again. What confidence would he have in the statement $T \ge T_D$ if he continued to run the motors without another failure? The total running time for each motor was 2,200 h.

Solution

$$k = 1 \qquad n = 2 \qquad T_0 = 2,200 \text{ h} \qquad T_D = 1,000 \text{ h} \qquad N = 3$$

$$\frac{T_0}{T_D} = 2.2$$

Using Eq. (5.23),

$$P_k = 1 - \left[\frac{3!}{0!(3-0)!}(1 - e^{-2.2})^0(e^{-2.2})^3\right]$$

$$- \left[\frac{3!}{1!(3-1)!}(1 - e^{-2.2})^1(e^{-2.2})^2\right]$$

$$= 1 - [0.00136 + 3(1 - 0.111)(0.0123)]$$

$$= 1 - (0.00136 + 0.03686)$$

$$= 1 - 0.0382$$

$$= 0.962$$

This means that if one of the two motors fails before 1,150 h, they both (one repaired and one original) should be tested to 2,200 h without failure to be able to conclude with 96.2 percent confidence that the new design has at least 1,000 h mean time to failure.

This can be solved also, with the aid of Table A-33. For $N = 3$, $k = 1$, and $T_r = T_0/T_D = 2.2$, assurance = 96.6 percent that $T_0 \geq T_D$.

5.3.2 FOR COMPONENTS

If the unit under test is a component, in contrast to a system or an assembly discussed above, two situations may arise: (1) the failure of the component is independent of time; that is, the failure is just as likely to occur during the first hour of operation as 1 month later (constant-hazard type of failure). (2) The failure is a function of time. In the first situation the underlying distribution can be taken as exponential, and the methods of Subsec. 5.3.1 will apply. In the second situation Weibull distribution can frequently describe the events, as developed in the following section.

Case 1—when no failure occurs In the Weibull distribution the characteristic life θ is one of the important parameters. In this case one can associate confidence with the statement that the actual characteristic life θ is greater than the desired characteristic life θ_D. The value of the characteristic life θ can be obtained by the expression

$$T_D = \int_0^1 \exp\left[-\left(\frac{T_0}{\theta}\right)^b\right] dT = \theta\Gamma\left(1 + \frac{1}{b}\right) \tag{5.24}$$

where b is the Weibull slope, the lower bound x_0 is assumed to be zero, and θ is the characteristic life as discussed in Sec. 2.3. Here $\Gamma[1 + (1/b)]$ is a gamma function which can be evaluated by means of Table A-10. It follows that

$$\theta = \frac{T_D}{\Gamma(1 + 1/b)} \tag{5.25}$$

The confidence that the mean time to failure T exceeds the desired mean time to failure T_D is

$$P = \int_0^{T_0} f(T)\, dT = \int_0^{T_0} b\left(\frac{T}{\theta}\right)^{b-1} \exp\left[-\left(\frac{T}{\theta}\right)^b\right] dT$$

$$= 1 - \exp\left[-\left(\frac{T_0}{\theta}\right)^b\right] \tag{5.26}$$

when no failure exists. Again, for n independent tests, with no failures,

$$P_n = 1 - (1 - P)^n \tag{5.27}$$

This may be reduced to

$$P_n(T > T_D \text{ or } \theta > \theta_D) = 1 - \exp\left[-n\left(\frac{T_0}{\theta}\right)^b\right] \tag{5.28}$$

Note that Eq. (5.28) applies to both parameters, mean life and characteristic life.

Example 5.5   A bearing engineer developed a new race which he feels will improve the life of the bearing.   The present bearing has a mean life B_{56} of 550 h, and the new bearing must have B_{56} life of 1,000 h.   The company has set 95 percent confidence level for all decisions.   Because these bearings are made on a prototype basis, only two bearings are available for testing.   How long must both bearings run without failure to meet the above condition?   The Weibull slope b is 1.66.

Solution   From Eq. (5.28),

$$P_n = 1 - \exp\left[-n\left(\frac{T_0}{\theta}\right)^b\right]$$

$$= 0.95 = 1 - \exp\left[-2\left(\frac{T_0}{\theta}\right)^{1.66}\right]$$

where

$$\theta = \frac{T_D}{\Gamma(1 + 1/b)} = \frac{1,000}{\Gamma(1.6)}$$

and the gamma function is evaluated from Table A-10.

$$\theta = \frac{1,000}{0.8935} = 1,120 \text{ h}$$

Therefore

$$1 - 0.95 = \exp\left[-2\left(\frac{T_0}{1,120}\right)^{1.66}\right]$$

$$20 = \exp\left[+2\left(\frac{T_0}{1,120}\right)^{1.66}\right]$$

$$\ln 20 = 2\left(\frac{T_0}{1,120}\right)^{1.66}$$

or

$$\left(\frac{\ln 20}{2}\right)^{0.6} = \frac{T_0}{1,120}$$

or

$$\left(\frac{3}{2}\right)^{0.6} = \frac{T_0}{1,120}$$

Therefore

$T_0 = 1,430$ h

This can be obtained from Table A-33; for $N = 2$, $k = 0$, and confidence $= 95$ percent,

$$T_r = 1.5 = \left(\frac{T_0}{\theta}\right)^b$$

or

$$1.5 = \left(\frac{T_0}{1,120}\right)^{1.66}$$

$$(1.5)^{0.6} = \frac{T_0}{1,120}$$

Therefore

$T_0 = 1,430$ h

Case 2—when failure occurs The analysis is valid for components, where testing one item with k repairs is equivalent to testing $k + 1$ items until k components fail with no repairs. The analysis is also valid in determining that the mean time to failure exceeds the desired mean time to failure where N number of parts are tested and out of these k number of parts fail.

The probability that a single item will fail before or at time T_0, or the confidence that the mean time to failure exceeds the desired mean time to failure is [Eq. (5.26)]

$$P = 1 - \exp\left[-\left(\frac{T_0}{\theta}\right)^b\right]$$

The probability that k or fewer failures will occur during time T_0 is, from Eq. (5.10),

$$P(r \le k) = \sum_{r_i=0}^{k} \binom{N}{k} p^{r_i}(1 - p)^{(N-r_i)}$$

This is also the confidence that $T \le T_D$. Thus the confidence that $T > T_D$ or $\theta > \theta_D$ is

$$P(T > T_D \text{ or } \theta > \theta_D) = 1 - P(r \le k)$$

$$= 1 - \sum_{r_i=0}^{k} \frac{N!}{r_i!(N - r_i)!} \left\{1 - \exp\left[-\left(\frac{T_0}{\theta}\right)^b\right]\right\}^{r_i}$$

$$\times \left\{\exp\left[-\left(\frac{T_0}{\theta}\right)^b\right]\right\}^{N-r_i} \tag{5.29}$$

Solutions to Eq. (5.29) have been tabulated and appear in Tables A-33 to A-39. Values of confidence $P(T_0 > T_D$ or $\theta)$ are tabulated for various values of $(T_0/\theta)^b$, N, and k. It can be seen that Eq. (5.29) (for Weibull) reduces to Eq. (5.23) (for exponential) since in the case of exponential $b = 1.0$ and $\theta = T_D$.

Example 5.6   Referring to Example 5.5, suppose that one of the two bearings failed to operate satisfactorily during the testing time. The testing continued after a minor repair. In this case how long must both bearings run without any more failures to meet the 95 percent confidence level?

Solution   Although this example can be solved either by Eq. (5.29) or by Table A-33, the solution using Table A-33 is given below.

$$n = 2 \qquad k = 1 \qquad N = n + k = 3$$

Therefore, corresponding to the confidence of 95 percent,

$$T_r = 2 = \left(\frac{T_0}{\theta}\right)^b \quad \text{or} \quad 2 = \left(\frac{T_0}{1{,}120}\right)^{1.66}$$

$$2^{0.6} = \frac{T_0}{1{,}120}$$

$$T_0 = 1{,}700 \text{ h}$$

Hence, for 95 percent confidence, the two bearings must run to 1,700 h if one failed, as opposed to 1,430 h if none failed.

5.4   RELATIONSHIP BETWEEN SAMPLE SIZE, TESTING TIME, CONFIDENCE, AND RELIABILITY

In this section, a trade-off relationship between sample size, test time, confidence limit, and reliability is developed. The analysis is based on the Weibull distribution and the success-run theorem.

The success-run theorem (Bayes' formula) is a nonparametric equation given by

$$R_C = (1 - C)^{1/n + 1} \tag{5.30}$$

where R_C = reliability at confidence level C
C = confidence level
n = sample size

It is assumed that there are no failures in the sample while running the test for time t. Equation (5.30) can be rewritten

$$C = 1 - (R_C)^{n+1} \tag{5.31}$$

For a fixed reliability R_C, the confidence level increases with increased sample size. Or, for a fixed confidence level, the reliability increases with increased sample size. If sample size is fixed, an increase in reliability causes a decrease in confidence level.

The relationship between reliability and test time for the Weibull failure distribution is [see Eq. (2.9)]

$$F(x) = 1 - \exp\left[-\left(\frac{x}{\theta}\right)^b\right]$$

where b = slope
θ = characteristic life
x = test time

The probability of survival (reliability) is then

$$R(x) = 1 - F(x) = \exp\left[-\left(\frac{x}{\theta}\right)^b\right] \tag{5.32}$$

The following analysis leads to the determination of increased confidence to achieve a desired reliability at time x (durability objective) when test time is extended to $x + y$ without failure. The probability of survival to the durability objective x_2 is

$$R(x_2) = 1 - F(x_2) = \exp\left[-\left(\frac{x_2}{\theta}\right)^b\right] \tag{5.33}$$

Suppose n_2 items are run without failure to x_2, and R_2 is the reliability at x_2 with a confidence C.

$$R_2 = (1 - C)^{1/(n_2+1)} = \exp\left[-\left(\frac{x_2}{\theta}\right)^b\right]$$

or

$$\ln(1 - C) = -\frac{1}{\theta^b}(n_2 + 1)x_2{}^b \tag{5.34}$$

Next, suppose n_1 items are run without failure to a new objective x_1, R_1 is the reliability at x_1 with the same confidence C,

$$R_1 = (1 - C)^{1/(n_1+1)} = \exp\left[-\left(\frac{x_1}{\theta}\right)^b\right]$$

and as before

$$\ln(1 - C) = -\frac{1}{\theta^b}(n_1 + 1)x_1{}^b \tag{5.35}$$

By equating the right-hand sides of Eqs. (5.34) and (5.35), the trade-off relationship between sample size and test time at a given confidence level is obtained.

$$\frac{1}{\theta} b(n_2 + 1)x_2{}^b = \frac{1}{\theta} b(n_1 + 1)x_1{}^b$$

or

$$\frac{n_2 + 1}{n_1 + 1} = \left(\frac{x_1}{x_2}\right)^b \tag{5.36}$$

The Weibull slope b must be either known or assumed.

Example 5.7 A component is known to follow a Weibull failure distribution, with $\theta = 25,000$ h and $b = 2.0$. One part is tested for 12,000 h, the durability objective, without failure. Determine the reliability and confidence level.

Solution From Eq. (5.32),

$$R = \exp\left[-\left(\frac{x}{\theta}\right)^b\right]$$

$$= \exp\left[-\left(\frac{12,000}{25,000}\right)^2\right]$$

$$= 79.5\%$$

From Eq. (5.30),

$$R_C = (1 - C)^{1/(n+1)}$$

$$0.795 = (1 - C)^{1/(1+1)}$$

$$C = 36.8\%$$

Thus, the part is 79.5 percent reliable with 36.8 percent confidence at the durability objective of 12,000 h. It is decided that the reliability of 79.5 percent at 12,000 h is adequate if the confidence level is higher. If one part is tested for 20,000 h without failure, how much will the confidence level increase? In Eq. (5.36), let

$$n_2 = ? \qquad\qquad x_2 = 12,000 \text{ h}$$

$$n_1 = 1 \qquad\qquad x_1 = 20,000 \text{ h}$$

$$\frac{n_2 + 1}{n_1 + 1} = \left(\frac{x_1}{x_2}\right)^b$$

Therefore

$$\frac{n_2 + 1}{1 + 1} = \left(\frac{20,000}{12,000}\right)^2$$

$$n_2 = 4.52$$

$$\simeq 5$$

The interpretation of $n_2 = 5$ is: One item run for 20,000 h without failure is equivalent to five items run for 12,000 h without failure. From the success-run theorem, the trade-off between confidence level and reliability is

$$R = (1 - C)^{1/(n+1)}$$

By knowing that the part is 79.5 percent reliable at 12,000 h, it is possible to calculate the new confidence level demonstrated by testing five parts for 12,000 h without failure:

$$0.795 = (1 - C)^{1/4.52+1}$$

$$C = 72.0\%$$

For 72.0 percent confidence, how reliable is the part at 20,000 h? From Eq. (5.30),

$$R = (1 - 0.72)^{1/(1+1)}$$

$$= 52.8\%$$

Characteristic life θ was assumed to be 25,000 h; the reliability can be calculated from the Weibull function.

$$R = \exp\left[-\left(\frac{x}{\theta}\right)^b\right]$$

$$= \exp\left[-\left(\frac{20,000}{25,000}\right)^2\right]$$

$$= 52.8\%$$

It should also be noted that the characteristic life θ is not really needed. If a reliability or confidence level has been assumed, one could proceed by using the success-run theorem and trade-off relation.

5.5 SUDDEN-DEATH TESTING

To illustrate sudden-death testing, let there be a collection of 50 specimens which are available for a fatigue test. These 50 specimens are randomly divided into 10 sets G of five items C each. Each set of five specimens is then run simultaneously until one item from this group of five fails. After this failure the whole group of five items is removed from test. This is repeated for all 10 groups. In this way 10 numbers are obtained, each representing the least in a random collection of five. According to the theory of ranking, the lowest value, in a set of five, clusters about a median location of the $B_{12.94}$ point of the population. (See table of median ranks, A-11.) These 10 numbers, therefore, represent separate estimates of the $B_{12.94}$ life.

The 10 numbers are plotted on Weibull probability paper, by using the ranks column for a sample size of 10 from Table A-11. A straight line is obtained which describes the distribution of the $B_{12.94}$ life instead of representing the entire population. As a best estimate of the $B_{12.94}$ point the median (not mean), B_{50} life, is taken. This is located at the 50 percent level of the straight line. The failure distribution of the whole population can then be predicted by drawing a line parallel to the sudden-death line passing through a common point A (see Fig. 5.10) of B_{50} life of the sudden-death

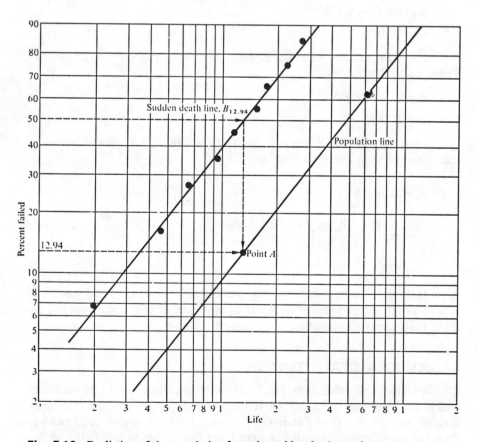

Fig. 5.10 Prediction of the population from the sudden-death test data.

line and $B_{12.94}$ life of the population.   This estimate of the $B_{12.94}$ life is just as reliable as the one obtained if all the 50 specimens had failed.

The time required to arrive at a sudden-death estimate of a life by using a successive testing technique is a certain fraction of the time required to estimate the life through the use of a full failure test.   This fraction is

$$F_T = G\psi \left\{ \frac{(1/c)\log 2}{\log[1/\lambda_n(1)]} \right\}^{1/b} \tag{5.37}$$

where G is the number of groups, c is the number of items in each group, $\lambda_n(1)$ is the mean rank of one failure out of n, and

$$\psi = \frac{(G-1)! \, \Gamma(1+1/b)(G-1+\ln 2)^{1/b}}{\Gamma(G+1/b)} \tag{5.38}$$

in which $\Gamma(\ )$ dėnotes a gamma function which is given in Table A-10.

In some cases it may be possible to run the sudden-death test simultaneously rather than successively.   In this case the median fraction of sudden death to full test time is given by

$$F_T = \left\{ \frac{(1/c)\log[1/\lambda_G(1)]}{\log[1/\lambda_n(1)]} \right\}^{1/b} \tag{5.39}$$

This technique is illustrated by the following example.

Example 5.8   A manufacturer of V-eight automotive engines desires to obtain an estimate of the low life value for exhaust-valve failures.   Ten engines were available for testing.   Each engine was run until the first (not necessarily No. 1) valve failed, and the number of hours to failure was recorded.   This was done for all the engines, and the following failure data were recorded:

Engine No.	Time to first valve failure, h
1	10
2	25
3	18
4	30
5	92
6	42
7	36
8	49
9	72
10	61

Predict the B_{10} life of the valve population.

Solution The data are first arranged in ascending order and median ranks assigned (see Table A-11):

Time to first failure, h	Median rank, %
10	6.70
18	16.32
25	25.94
30	35.57
36	45.19
42	54.81
49	64.43
61	74.06
72	83.68
92	93.30

The data are then plotted on Weibull graph paper and the best-fit straight line is drawn through the points. Since the data represent the first failure in a sample of eight, they are an estimate of the $B_{8.3}$ life (see Table A-11). The straight line through the points is therefore called the $B_{8.3}$ failure line. (See Fig. 5.11.)

Next, the $B_{8.3}$ point is located on the population line. The coordinates of this point are the median $B_{8.3}$ life (50 percent failed) on the abscissa and 8.3 percent failures on the ordinate. The population line is drawn through this point and parallel to the $B_{8.3}$ line.

Read off the population B_{10} life:

$B_{10} = 44.5$ h.

5.5.1 CONFIDENCE LIMITS BY MODIFIED SUDDEN-DEATH APPROACH

When the sudden-death curve in the previous section was located on the Weibull plot, the median-rank tables were used. Therefore, the confidence level on the sudden-death curve and, therefore, on the population, was only 50 percent. However, if confidence bands are constructed about the population curve, it can be stated with increased confidence that the population curve represents the true population within the defined limits. In this section, a method of constructing the confidence band by using the modified sudden-death approach is described.

With the modified sudden-death approach, the sudden-death plot and population curves are located by using the previously described method. The 5 and 95 percent ranks, or any other ranks available in the tables, are obtained from Table A-14 for each failure just as the median ranks were found for locating the sudden-death curve. The points are then plotted and the confidence band drawn about the sudden-death line. The 90 percent confidence band about the population curve is found by projecting downward

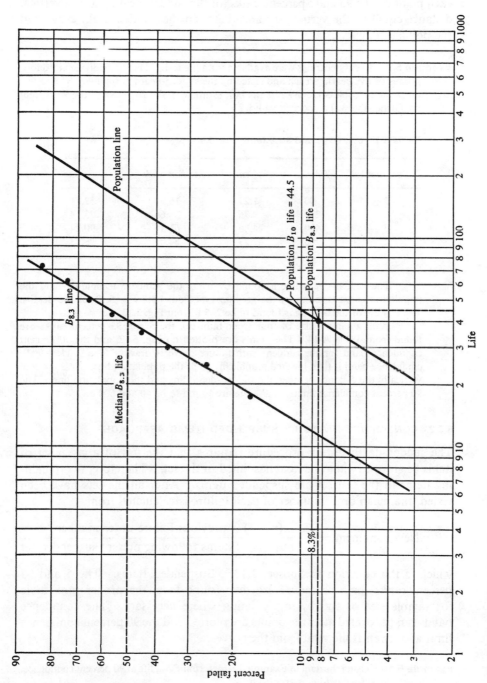

Fig. 5.11 Weibull plot for Example 5.8.

each point on the 95 and 5 percent ranks (of the sudden-death curve) a vertical distance equal to the vertical distance between the sudden-death curve and the population curve.

Example 5.9   Twenty bearings were available for a life test.   They were divided into four groups of five bearings each and tested by the sudden-death technique.   The results are given in Table 5.1.   What is the best estimate of the B_{10} life and the 90 percent confidence interval associated with it?

Table 5.1   5%, 50%, and 95% failure ranks

Failure number	Time, h	5% rank	Median rank %	95% rank
1	90	1.27	15.91	52.71
2	210	9.76	38.64	75.14
3	400	24.86	61.36	90.24
4	700	47.29	84.09	98.73

The sudden-death line is found by using the standard techniques, and the population curve is found by the graphical method.   See Fig. 5.12 for the details. The B_{10} life of the bearing is found to be 235 h from Fig. 5.12.

With a sample size of four (four failures), the 5 and 95 percent ranks are found from Table A-14.   The points are plotted on Fig. 5.12, and the 90 percent confidence band for the sudden-death failure curve is drawn.   The sudden-death confidence band is then lowered graphically to fit the population line.   It is lowered vertically the distance between the sudden-death line and the population line.   The 90 percent confidence limits on B_{10} life are found to be 40 and 600 h.

5.5.2   CONFIDENCE LIMITS BY SUSPENDED-ITEMS APPROACH

The sudden-death testing procedure requires that the test of a given set of items be stopped after one item has failed in the set.   Therefore, the remaining items in the test set are suspended items.   After all the sets have been tested, the mean order number of each failure is calculated from

$$\text{New increment} = \frac{(n+1) - \text{previous failure order number}}{1 + \text{number of items following present suspended set}}$$

which is the equation in Subsec. 1.13.3, Suspended Items.   The 5 and 95 percent ranks for each failure are then found from the proper rank table for sample size n, and by interpolating where necessary.   The confidence bands can be located after the points are plotted and the 90 percent confidence limit at a given B life read from the curves.

Example 5.10   Using the data in Example 5.9, determine B_{10} life at 90 percent confidence, using the suspended-items approach.

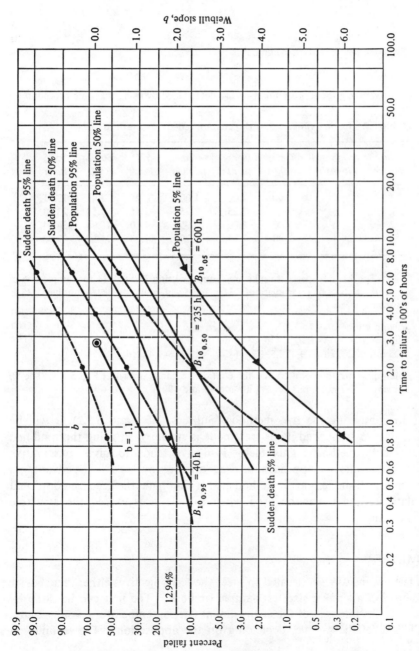

Fig. 5.12 Confidence limits—modified sudden-death approach (Example 5.9).

Solution Data in Table 5.1 are organized as follows: the first item in the table failed, so it has a mean order number of 1. The second failure, however, is preceded by four suspended items. Therefore, to find the increment, as shown above,

$$I = \frac{20 + 1 - 1}{1 + 15}$$
$$= 1.25$$

Thus, the new mean order number is

$$1 + 1.25 = 2.25$$

The remaining mean order numbers are determined by using the same procedure.

Table 5.2   Failure ranks using the suspended-items method

Failure number	Suspended items	Time, h	Incre- ment	Order number	5% Rank	50% Rank	95% Rank
1	4	90	1.00	1.00	0.26	3.41	13.91
2	4	210	1.25	2.25	2.45	9.54	23.52
3	4	400	1.70	3.95	7.10	17.98	34.44
4	4	700	2.84	6.79	17.03	31.80	49.60

The 5 and 95 percent ranks are found by using Table A-14 for sample size of 20, with the aid of interpolation. The 5 percent rank for order number 2.25 is illustrated.

Second out of 20 = 1.83%
Third out of 20 = 4.29%
2.25 out of 20 = 1.83 + (2.25 − 2.0)(4.29 − 1.83) = 2.45%

The points are then plotted and the confidence bands drawn in Fig. 5.13. The 90 percent confidence limits on B_{10} life are 46 and 490 h.

In comparing the two methods (Subsecs. 5.5.1 and 5.5.2), it will be noted that no suspended-items treatment is made when using the modified sudden-death method. This method always tends to give conservative results, but results that approach the exact ones for low lives. The advantages of the modified sudden-death method are: It eliminates the need for additional calculations, and it can be used when sample sizes exceed the available tabular values.

SUMMARY

The test is generally accelerated to meet the test objectives either in a shorter time or with a smaller-size test sample or both. The first can be done by increasing the intensity of the environment (stress, temperature, voltage, etc.) or the size of the sample, depending on the application. The second can

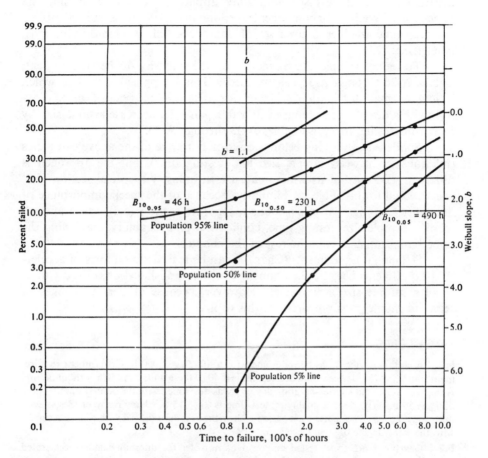

Fig. 5.13 Confidence limits—suspended-items approach (Example 5.10).

be accomplished by increasing the intensity of the environment or the length of the test.   Chapter 5 was organized to provide the interrelationship between these three variables.

To reduce the test time by increasing the environmental intensity, refer to Sec. 5.1, where general rules for accomplishing this are described.   No specific data are provided, as each problem is a special case.   When an increase in the intensity of the environment is not feasible, the sample size must be increased.   Tables A-33 through A-39 relate testing time, sample size, and confidence level, both for the cases when no failures occurred during the test ($k = 0$) and for the cases when some of the items did fail ($k = 1, 2,$

3, ...). The tables can be used if the data follow Weibull or exponential distributions, as these distributions are applicable to data describing life phenomena: Weibull for components, exponential for systems and assemblies. To determine whether a given set of data does follow these distributions, consult Sec. 2.8.

To reduce sample size by increasing the intensity of the environment, Tables A-26 through A-32 can be used. This is applicable to data which follow normal distribution. Means for determining whether given data follow this distribution are given in Table 2.1. To decrease sample size by increasing testing time, use Tables A-33 through A-39.

Sudden-death testing belongs essentially to one of the above categories (relationship between sample size and testing time), but the approach is different. Because of its widespread use in industry a special section (5.5) was devoted to this subject. Similarly, because of the special importance of reliability in industrial applications a separate section (5.4) was prepared, relating sample size, testing time, confidence, and reliability, employing the same basic statistics as in the rest of the chapter.

Sequential Experiments (Chap. 7) can be regarded as a form of accelerated experiments because their aims are to accomplish test objectives with as small a sample size as possible. However, because of their uniqueness a separate chapter was prepared for this topic.

PROBLEMS

5.1. A certain device X has been designed for a life $T_D = 1,000$ h. Government requirement states that this device would be acceptable if 40 of these units run 150 h without failure. A more powerful but shorter-duration device y was developed with the desired running time of $T_D = 50$ h. The desired qualifying test time is $T_0 = 15$ h. How many of these new y units must be tested so that if each passes 15 h without failure the y device will be accepted with the same degree of confidence as x?

5.2. Gearboxes are to be tested on the dynamometer for durability under accelerated conditions. The design goal is 100 h of operation without failure. Past experience indicates that the principal mode of failure is bearing failure due to dirt particles lodging in. If only one gearbox is available for testing, how long must it run without failure so that there is 95 percent assurance that the 100-h design goal will be met? Suppose the gearbox failed during the test, and after a minor repair the test was continued. In this case, how long must this gearbox run so that there is 95 percent assurance that the 100-h design goal will be met? Provide the answers by using the analytical method (equations), and then verify them by using the tables in the Appendix.

5.3. An engine tested on the dynamometer after 30 h of running developed bearing failure because of a sand particle which dislodged from the casting and embedded in the bearing. The bearing was replaced and the pump continued running. Forty-five hours later a chip caught in the water pump and the pump failed. The pump was replaced and the engine continued running with no further failures. The total testing time was $T_0 = 400$ h, and the desired time $T_D = 100$ h. What is the assurance that the engine life will be 100 h?

5.4. Two pumps were tested for durability under accelerated conditions. The design goal under these conditions is 100 h without failure. After 115 h of operation one pump developed failure because of a sand particle which dislodged from the casting and embedded in the bearing. The bearing was replaced and each pump was run for an additional 105 h without failure. What is the assurance that the goal of 100 h has been met?

5.5. A new surface treatment for gears is being considered. If the process will give a desired mean time to failure of 1,100 h, it will be adopted for use. Three test specimens were run for 1,350 h without failure. With what confidence can it be said that the actual mean time to failure exceeds the desired mean time to failure? From experience with this process it is known that the Weibull slope = 2.0.

5.6. A new design of a lighting mechanism is submitted. This design provides better quality, but because of its design complexity there is some question as to its longevity. It is necessary that the mechanism's characteristic life equal the characteristic life of the old mechanism (4,000 h). The tests of the old design showed that the failure distribution is Weibull, with $b = 2$. Five mechanisms were run for 5,000 h each; failure occurred in some cases, requiring a minor repair in order to resume testing. The results are indicated below. Find the assurance associated with the statement that $T > T_D$.

Mechanism, number	1	2	3	4	5
Number of failures	0	0	1	1	0

5.7. Seventy-two electrical resistors are available for life testing. They are divided into six groups of 12 resistors each and tested by the sudden-death technique. The resultant data are given below:

Group 1. Resistor 3 failed at 120 h.
Group 2. Resistor 4 failed at 65 h.
Group 3. Resistor 11 failed at 155 h.
Group 4. Resistor 5 failed at 300 h.
Group 5. Resistor 7 failed at 200 h.
Group 6. Resistor 12 failed at 250 h.

What is the B_{10} life of the resistor population?

5.8. A rash of switch failures are reported from the field. To isolate the cause, it is decided to analyze the failures under simulated field conditions. In order to expedite the investigation, the sudden-death test procedure will be used. Facilities are available to test six switches simultaneously. Ten groups of six switches each were tested with the following results:

Group number	Time to first failure, cycles
1	7.8×10^5
2	1.5×10^6
3	2.0×10^5
4	3.0×10^6
5	1.8×10^6
6	1.2×10^6
7	2.2×10^6
8	9.8×10^5
9	4.2×10^5
10	6.0×10^5

(a) Determine the population B_{10} life.
(b) Determine percent failure below 1.5×10^6 cycles.

5.9. Nine prototype transmissions were installed in trucks to determine their characteristic life. At the end of the 90-day test the following data were obtained:

Transmission number	1	2	3	4	5	6	7	8	9
Condition	F*	S†	F	F	S	F	F	S	F
Life, m	8,800	4,400	5,300	20,000	11,200	9,400	14,600	13,300	2,800

* Failed.
† Did not fail.

    (a) What is the B_{10} life of these transmissions?
    (b) Would you recommend a 5,000-m warranty?

5.10. A certain large computer gave the following record of failures in a highly accelerated operation:

Time to failure, h
12, failed
150, failed
120, failed
32, computer inspected, OK (discontinued for other test purposes)
75, failed
35, failed
76, failed
101, computer inspected, OK (discontinued for other test purposes)
202, failed
79, failed

    How many failures do you anticipate at the end of 1,000 h of operation? The total number of computers originally installed was 28.

5.11. In the design stage of a certain product, it was stated that the reliability must be 90 percent at 5,000 h and 50 percent at 1,000 h, both at 50 percent confidence. The product was designed, and after the parts were produced, a test program was undertaken. Outline three possible test methods for determining whether these goals have been met.

REFERENCE

1. Rabinowicz, E., R. H. McEntire, and B. Shiralkar: A Technique for Accelerated Life Testing, *ASME Paper* 70-Prod.-10, April, 1970.

BIBLIOGRAPHY

Bild, C. F.: Failure from a Material Point of View, *IRE Nat. Symp. Qual. Contr. Rel. Electron.*, 1964.
Bird, C. M., and W. F. Elsbree: Correlation of Laboratory and Field Reliability, *IRE Nat. Symp. Qual. Contr. Rel. Electron.*, 1960.

Bond, M. E.: Accelerated Life Test, *Ind. Qual. Contr.*, October, 1965.

Cary, H., and R. E. Thomas: Accelerated Testing as a Problem of Modeling, *IRE Nat. Symp. Qual. Contr. Rel. Electron.*, 1960.

Johnson, L. G.: *GMR Rel. Manual* GMR-302, General Motors Research Laboratories, August, 1960.

————: "Theory and Technique of Variation Research," Elsevier Publishing Company, Amsterdam, 1964.

Johnson, W. F.: Research on Accelerated Reliability Testing Methods Applicable to Non-electronic Components of Flight Control Systems, *Tech. Rep.* AFFDL-TR-64-181 USAF, 1965.

Jones, H. C.: Reliability Testing, *IRE Nat. Symp. Qual. Contr. Rel. Electron.*, 1965.

Kao, J. H.: Statistical Models in Mechanical Reliability, *IRE Nat. Symp. Qual. Contr. Rel. Electron.*, 1965.

Locati, L., and C. F. Bona: Accelerated Testing at Fiat Laboratories, *SAE Paper* 292 A, 1961.

Rabinowicz, E., R. H. McEntire, and B. Shiralkar: A Technique for Accelerated Life Testing, *ASME Paper* 70-Prod.-10, April, 1970.

Roberts, N. H.: "Mathematical Methods in Reliability Engineering," McGraw-Hill Book Company, New York, 1964.

Thomas, R. E.: When Is a Life Test Truly Accelerated? *Electron. Des.*, Jan. 6, 1964.

Van Alvin, W. H.: "Reliability Engineering," Prentice-Hall, Inc., Englewood Cliffs, N.J., 1964.

Yurkowsky, W.: Nonelectronic Reliability Notebook, *Final Tech. Rep.* RADC-TR-69-458, Rome Air Development Center, Griffiss Air Force Base, New York, March, 1970.

———— and R. E. Schafer: Accelerated Reliability Test Methods for Mechanical and Electromechanical Parts, *Tech. Rep.* RADC-TR-65-46 USAF, 1965.

Zelen, M., and M. C. Dannemiller: Are Life Testing Procedures Robust? *IRE Nat. Symp. Qual. Contr. Rel. Electron.*, 1960.

6
Factorial Experiments

A factorial experiment is an experiment which extracts information on several design factors more efficiently than can be done by the traditional (classical) test. The main objective is to determine the effect of various factors (independent variables) on some characteristic of a product (dependent variable) of interest. For example, this can be applied to the study of wear characteristics of a material, where the effect of load, velocity, lubricant, temperature, etc., may be determined. In addition, this type of experiment enables the engineer to determine whether the major factors affecting a product have been taken into consideration. In this chapter, one-factor, two-factor, three-factor, and fractional factorial experiments are discussed and the analysis is shown.

6.1 THE CLASSICAL EXPERIMENT

A classical experiment is one where, generally, one factor is varied while others are held constant. Suppose the effect of sliding velocity and temperature on the wear rate of a material is to be investigated. The test variables are:

1. The two independent variables (factors): sliding velocity A and temperature B
2. The dependent variable (response): wear rate W

These factors A and B are investigated at two levels (A_1, A_2, and B_1, B_2) where the test is repeated (replicated) at each test condition to obtain a number of observations. These changes in the sliding velocity and the temperature will indicate how they affect the response in wear.

The classical technique of investigating two levels of A while holding temperature B constant is shown in Fig. 6.1. Two items are tested at each level of A to obtain an average effect. The average-effects wear rate W at the B_2 level is 47 mg/d at the high sliding velocity A_2 and 40 mg/d at the low velocity A_1. This indicates a change of 7 mg/d of wear rate when going from the low to a high velocity. This change is called the A effect. Similarly, additional pieces are tested for two levels of temperature B while holding sliding velocity A constant. By holding the sliding velocity A_2 constant, the B effect, the difference between the two average values, is found to be 8.

Since the B effect is larger than the A effect, temperature appears to be more important in affecting wear than the sliding velocity. Although the classical approach will evaluate the effect of A and B on W, it suffers from the following shortcomings:

1. Confidence levels cannot be attached to these estimates of effects A and B.
2. It does not estimate the experimental error (residual error) in test data.
3. It does not estimate the effect of interactions resulting from the two factors. Interaction can be illustrated by the following example (Fig. 6.2): Resistivity of a material to corrosion with 20 percent chromium increases by five units (from 15 to 20) when nickel content increases from 10 to 20 percent. However, when the percent chromium

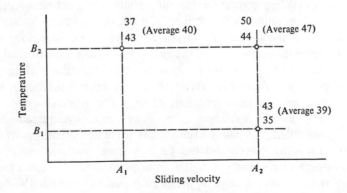

Fig. 6.1 The classical experimental test plan.

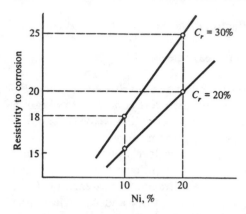

Fig. 6.2 Interactions.

is increased to 30 percent, the change in resistivity is seven units (25 − 18). This indicates the presence of interaction between nickel and chromium. If, at 30 percent chromium, the resistivity increases only by five units, as in the case of 20 percent chromium, no interaction is said to exist.

In contrast, by conducting a factorial experiment, with the aid of the analysis of variance, all the above information can be obtained.

6.2 FACTORIAL EXPERIMENTS

6.2.1 ANALYSIS OF VARIANCE IN FACTORIAL EXPERIMENTS

Analysis of variance is a powerful technique for analyzing experimental data involving quantitative measurements. It is particularly useful in factorial experiments where several independent sources of variation may be present.

When several sources of variation are acting simultaneously on a set of observations, the variance (and not the standard deviation) of the observations is the sum of the variances of the independent sources. This property makes the application of the analysis of variance particularly useful in factorial experiments. By this method, the total variation within an experiment can be broken down into variations due to each main factor, interacting factors, and residual (experimental) error. The significance of each variation is then tested. Variables other than those investigated should be properly controlled. In many experiments there are situations where some variables are either uncontrollable or unknown, such as environmental changes, operator efficiency, and drift in test instruments. Since the variance analysis is based on the laws of probability, the experiment should be conducted so that the influence of these variables is randomly distributed throughout the test. This can be done by randomization of the experiment, where each test combination is selected randomly and then tested.

Suppose K samples each of size n are picked at random from a homogeneous population having a normal distribution. The variance s^2 for each sample is calculated as an estimate of the population variance σ^2 The mean $\bar{x}$ of each sample and the variance of the means s^2 are then calculated (see Table 6.1). In Chap. 3, it was shown that

$$ns_{\bar{x}}^2 = s^2 \tag{6.1}$$

Table 6.1 Means and variances of several samples picked from a homogeneous population

	Sample						
	1	*2*	*3*	*4* $\cdots$ *i*		*K*	
n observations in each sample	x_{11}	x_{21}	x_{31}	x_{41}	x_{i1}	x_{K1}	
	x_{12}	x_{22}	x_{32}	x_{42}	x_{i2}	x_{K2}	
	x_{13}	x_{23}	x_{33}	x_{43}	x_{i3}	x_{K3}	
	x_{1j}	x_{2j}	x_{3j}	x_{4j}	x_{ij}	x_{Kj}	
	x_{1n}	x_{2n}	x_{3n}	x_{4n}	x_{in}	x_{Kn}	
Mean variance s^2 (mean square)	$\bar{x}_1$ s_1^2	$\bar{x}_2$ s_2^2	$\bar{x}_3$ s_3^2	$\bar{x}_4$ s_4^2	$\bar{x}_i$ s_i^2	$\bar{x}_K$ s_K^2	$\bar{X}$ (Grand mean)

The variance s^2 for all K samples, with n observations in each, is

$$s_i^2 = \frac{\sum_{j=1}^{n}(x_{ij} - \bar{x}_i)^2}{n - 1} \tag{6.2}$$

The average of all s_i^2 is an estimate of σ^2:

$$\sigma^2 = \frac{\sum_{i=1}^{K} s_i^2}{K} \tag{6.3}$$

σ^2 can also be determined from the relationship

$$\sigma^2 = n\sigma_{\bar{x}}^2 = \frac{n \sum_{i=1}^{K}(\bar{x}_i - \bar{X})^2}{K - 1} \tag{6.4}$$

Both Eqs. (6.3) and (6.4) provide a good estimate of σ^2; however, σ^2 determined from Eq. (6.3) is more reliable since it is based on more degrees of freedom.

Now suppose that the population from which the samples were picked is not homogeneous, and differences exist between the samples. That is, suppose $A_1, A_2, \ldots, A_K$ are levels of some factor A, and the effect of these levels on the variable x of interest is to be determined. The variation within the samples will be due to the original variation in the population, but the mean values of the samples will vary, additionally, because of the differences between the samples (see Table 6.2).

Table 6.2 Means and variances of several levels of factor A which relates to samples from a nonhomogeneous population

		Factor A					
	A_1	A_2	A_3	A_4	A_K		
n observations in each sample	$x_{11} + W_1$ $x_{12} + W_1$ $x_{13} + W_1$ $x_{1n} + W_1$	$x_{21} + W_2$ $x_{22} + W_2$ $x_{23} + W_2$ $x_{2n} + W_2$	$x_{31} + W_3$ $x_{32} + W_3$ $x_{33} + W_3$ $x_{3n} + W_3$	$x_{41} + W_4$ $x_{42} + W_4$ $x_{43} + W_4$ $x_{4n} + W_4$	$x_{K1} + W_K$ $x_{K2} + W_K$ $x_{K3} + W_K$ $x_{Kn} + W_K$		
Mean variance (mean square)	$\bar{x}_1 + W_1$ s_1^2	$\bar{x}_2 + W_2$ s_2^2	$\bar{x}_3 + W_3$ s_3^2	$\bar{x}_4 + W_4$ s_4^2	$\bar{x}_K + W_K$ s_K^2	$\overline{X} + \overline{W}$ (Grand mean)	

The mean for level A_1 is

$$\frac{\sum_{j=1}^{n} (x_{1j} + W_1)}{n} = \frac{\sum_{j=1}^{n} x_{1j}}{n} + \frac{\sum W_1}{n}$$

$$= \bar{x}_1 + \frac{nW_1}{n}$$

$$= \bar{x}_1 + W_1$$

The variance for level A_1 is

$$\sum_{j=1}^{n} \frac{[(x_{1j} + W_1) - (\bar{x}_1 + W_1)]^2}{n-1} = \frac{\sum_{j=1}^{n} (x_{1j} - \bar{x}_1)^2}{n-1} \tag{6.5}$$

Therefore, the presence of the differences W does not affect the original variance σ^2, where the factor W was not present. Hence, it is a good estimate of variation within the samples (or levels).

The differences W would definitely affect the new mean squares (variance) of the sample means. Let this new mean square be denoted by $(\sigma_{\bar{x}}^2)'$, as contrasted with the original $\sigma_{\bar{x}}^2$.

$$(\sigma_{\bar{x}}^2)' = \frac{\sum_{i=1}^{k} [(\bar{x}_i + W_i) - (\overline{X} + \overline{W})]^2}{K-1} = \frac{\sum_{i=1}^{k} [(\bar{x}_i - \overline{X}) + (W_i - \overline{W})]^2}{K-1}$$

$$= \frac{\sum (\bar{x}_i - \overline{X})^2}{K-1} + \frac{\sum (W_i - \overline{W})^2}{K-1} + \frac{2 \sum (\bar{x}_i - \overline{X})(W_i - \overline{W})}{K-1} \tag{6.6}$$

Since the sum of the deviations of observations from their mean is zero, the last term in Eq. (6.6) drops out.   Therefore

$$(\sigma_{\bar{x}}^2)' = \sigma_{\bar{x}}^2 + \sigma_W^2 \tag{6.7}$$

or

$$n(\sigma_{\bar{x}}^2)' = n\sigma_{\bar{x}}^2 + n\sigma_W^2$$

or

$$n(\sigma_{\bar{x}}^2)' = \sigma^2 + n\sigma_W^2$$

or

Total variance = variance within the samples + variance between the means of the samples

In analyzing the variations in factorial experiments it is sometimes easier to first compute the sum of squares SS and then the mean squares MS, where

$$\text{Mean squares} = \frac{\text{sum of squares}}{\text{degrees of freedom}}$$

The following are the relationships for the determination of the sum of squares SS in an experiment where only one factor (or variable) at several levels is to be investigated:

$$\text{SS}_{\text{total}} = \sum x^2 - \frac{T^2}{N}$$

where $\sum x^2$ = sum of squares of all observations
$\quad\quad T$ = grand total of all observations
$\quad\quad N$ = total number of observations

Let the different levels of the factor to be investigated be represented by columns.

$$\text{SS}_c = \text{SS between column means} = \frac{\sum T_c^2}{n} - \frac{T^2}{N}$$

where T_c = total of each column
$\quad\quad c$ = number of columns
$\quad\quad n$ = number of observations in each column

$\text{SS}_{\text{residual}}$ = SS within the columns, or the experimental error
$\quad\quad\quad\quad\quad$ = $\text{SS}_{\text{total}} - \text{SS}_c$

Thus, the total variation is broken down into two sources of variation; variations within the columns (experimental error) and the variation between the columns.   Each of these variations, which are in terms of the sum of squares, reduces to the mean squares when they are divided by their corresponding degrees of freedom. The ratio of any two of these mean squares provides the basis for the F test of significance. When the F test is applied to the ratio of MS of columns to MS of residual, it will indicate whether a significant difference exists between the columns (or various levels of a factor) or whether the observed difference is due to chance or the experimental error alone.

Computations involved in the analysis of variance are relatively simple for single- or two-factor experiments. They become complicated, however, as the number of factors is increased. The necessary relationships to compute various sums of squares, etc., for single-, two-, and three-factor experiments are summarized in Tables 6.13 to 6.15. The actual derivations of these relationships are not given here, as they can be found in standard statistics books. (See Refs. 1 to 3.) The analyses of single-, two-, and three-factor experiments are illustrated by the following examples.

6.2.2 SINGLE-FACTOR EXPERIMENT

Example 6.1   Four brands of tires, A, B, C, and D, are to be compared for tread loss after 20,000 miles of driving. For this purpose, four tires of each brand were used, and they were installed randomly in four cars. After 20,000 miles of operation, the following tread loss in millimeters was measured in these tires:

A	B	C	D
14	14	12	10
13	14	11	9
17	8	12	13
13	13	9	11

On the basis of the above data, determine, with 90 percent confidence, whether there is any significant difference between the four brands.

Solution

Brands of tires

	A	B	C	D
	14	14	12	10
	13	14	11	9
	17	8	12	13
	13	13	9	11
$T_c \rightarrow$	57	49	44	43

c = number of columns = number of brands of tires = 4
N = 16, total number of observations or tests
n = 4, number of replications (or number of tests) at each test combination
$T = \sum x$, where x is the value of each observation, in this case tread loss
   = 193 (grand total of all observations)
$\sum x^2$ = summation of squares of all observations = 2,409
T_c = total of each column
Sums of squares for the sources of variation are computed with the aid of Table 6.13 as follows:

1. Among columns (tire brands):

$$SS_c = \frac{\sum T_c^2}{n} - \frac{T^2}{N}$$

$$= \frac{57^2 + 49^2 + 44^2 + 43^2}{4} - \frac{193^2}{16}$$

$$= 31$$

2. Total (total variance in the set of observations):

$$SS_{total} = \sum x^2 - \frac{T^2}{N}$$

$$= 2,409 - \frac{193^2}{16}$$

$$= 81$$

3. Residual (measure of experimental error or test repeatability):

$$SS_{residual} = SS_{total} - SS_c$$
$$= 81 - 31$$
$$= 50$$

    As shown in Table 6.3, the mean-square ratio experimentally determined (2.48) is less than the F ratio (2.61) for 90 percent confidence. Hence, on the basis of the above test data it can be concluded with 90 percent confidence that there is no significant difference between the four tire brands.

Table 6.3   Analysis-of-variance table for single-factor experiment

Source of variation	Sum of squares SS	Degrees of freedom DF	Mean square MS (SS/DF)	Mean-square ratio MSR (MS/MS$_{residual}$)	Minimum MSR required for factors to be significant at 90% confidence $F_{0.10;\,3;\,12}$ (Table A-4)
Brands (columns)	31	$c - 1 = 3$	$\frac{31}{3} = 10.33$	$\frac{10.33}{4.17} = 2.48$	2.61
Residual	50	$(N-1)-(c-1)$ $= 12$	$\frac{50}{12} = 4.17$		
Total	81	$N - 1 = 15$			

6.2.3 TWO-FACTOR EXPERIMENT

Example 6.2 The effect of sliding velocity and temperature on the wear-rate characteristic of a material is to be determined. The test results (wear rate, mg/d) are tabulated (see also Fig. 6.3):

		Temperature B	
		B_1	B_2
Sliding velocity A	A_1	34 30	33 41
	A_2	43 37	50 44

Determine with 90 percent confidence whether the sliding velocity and the temperature have any significant effect on the wear rate of the material.

Solution The calculations can be simplified by subtracting 30 from each of the above values. This will not affect the final results and the F statistics. The revised data are:

			Columns		
			Temperature		
			B_1	B_2	T_r
Rows	Sliding velocity	A_1	4 0	3 11	18
		A_2	13 7	20 14	54
		T_c	24	48	

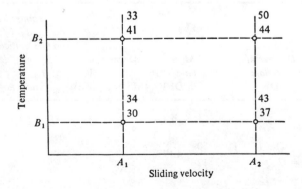

Fig. 6.3 Two-factor two-level experiment.

r = number of rows = number of sliding velocities = 2

c = number of columns = number of temperatures = 2

$N = 8$, total number of observations or tests

$n = 2$, number of replications (or number of tests) at each test combination

$T = \sum x$, where x is the value of each observation, in this case the wear rate
$= 72$

$$\frac{T^2}{N} = \frac{72^2}{8} = 648$$

T_r = total of each row

T_c = total of each column

$\sum x^2 = 4^2 + 0^2 + 3^2 + 11^2 + 13^2 + 7^2 + 20^2 + 14^2 = 960$

$\sum T_c^2 = 24^2 + 48^2 = 2,880$

$\sum T_r^2 = 18^2 + 54^2 = 3,240$

$\sum T_{cr}^2 = 4^2 + 14^2 + 20^2 + 34^2 = 1,768$

From Table 6.14, the following are the sums of squares for the sources of variation:

1. Among columns (temperature):

$$SS_c = \frac{\sum T_c^2}{nr} - \frac{T^2}{N} = \frac{2,880}{2 \times 2} - 648 = 72$$

2. Among rows (sliding velocity):

$$SS_r = \frac{\sum T_r^2}{nc} - \frac{T^2}{N} = \frac{3,240}{2 \times 2} - 648 = 162$$

3. Column-row interaction:

$$SS_{cr} = \frac{\sum T_{cr}^2}{n} - \frac{T^2}{N} - SS_c - SS_r = \frac{1,768}{2} - 648 - 72 - 162 = 2$$

4. Total:

$$SS_{total} = \sum x^2 - \frac{T^2}{N} = 960 - 648 = 312$$

5. Residual:

$$\begin{aligned} SS_{residual} &= SS_{total} - SS_c - SS_r - SS_{cr} \\ &= 312 - 72 - 162 - 2 \\ &= 76 \end{aligned}$$

On the basis of the results in Table 6.4 it can be concluded with 90 percent confidence that sliding velocity significantly affects the wear rate of the material, while the effect of temperature is insignificant when the variation in wear rate due to chance is considered. The interaction effect between velocity and temperature is also insignificant, which means that the effect of velocity is not dependent on temperature.

Table 6.4   **Analysis-of-variance table for two-factor experiment**

Source of variation	Sum of squares SS	Degrees of freedom DF	Mean square MS (SS/DF)	Mean-square ratio MSR ($MS/MS_{residual}$)	Minimum MSR required for factors to be significant at 90% confidence $F_{0.1;1;4}$ (Table A-4)
Among columns (temperature)	72	$c - 1 = 1$	72	3.8	4.54
Among rows (sliding velocity)	162	$r - 1 = 1$	162	8.5	4.54
Column-row interaction	2	$(c - 1)(r - 1) = 1$	2	0.105	4.54
Residual (experimental error)	76	Total $-$ sum of previous $= 4$	19		
Total	312	$N - 1 = 7$			

6.2.4 THREE-FACTOR EXPERIMENT

Example 6.3 Several complaints from the field have been received on an intra-vane pump due to a substantial number of smears and seizures in the cartridge of the pump. It was felt that three factors may have a significant influence on smears and seizures:

1. The type of shaft and rotor combination
2. The type of housing and cover
3. The positioning clearances of pressure and wear plates

The pump was tested, and the matrix layout describing the test variations is shown in Table 6.5. Determine with 90 percent confidence which of the above three factors have significant effect on smears and seizures in the cartridge of the pump.

Solution The three factors in this experiment can be considered as rows, groups, and columns, each at two, two, and six levels, respectively. This gives 24 possible combinations such as $A_1B_1C_1$, $A_1B_1C_2$, ... This experiment was replicated only for 5 out of 24 combinations. In order to apply the techniques of factorial experiment, it is necessary to have the same number of replications for each of these combinations. This was done by keeping the worst response of the two for those five combinations. This is shown in Table 6.6.

With the aid of Table 6.15, the following are the sums of squares for the sources of variation:

1. Among columns:

$$SS_c = \frac{\sum T_c^2}{nrg} - \frac{T^2}{N} = \frac{0^2 + 1^2 + 6^2 + 3^2 + 1^2 + 1^2}{1 \times 2 \times 2} - \frac{144}{24}$$
$$= 6$$

Table 6.5 Three-factor-experiment plan

| | | | Columns (c) | | | | | |
| | | | Positioning clearances | | | | | |
		Groups (g)	c_1	c_2	c_3	c_4	c_5	c_6
Rows (r)	A_1 Square shaft and rotor combination	Standard housing and cover B_1	0	0	0	1 0	0	0
		Doweled housing and cover B_2	0	0	1	1	0	0
	A_2 Production shaft and rotor combination	Standard housing and cover B_1	0	0 0	2	1	0 1	0 0
		Doweled housing and cover B_2	0	1	3 3	0	0	1

Degree of degradation:

0 — no marking
1 = light smear of plate, rotor, or vanes
2 = heavy smear of plate, rotor, or vanes
3 = seizure, large area affected, metal transferred

2. Among rows:

$$SS_r = \frac{\sum T_r^2}{ncg} - \frac{T^2}{N} = \frac{(1+1+1)^2 + (2+1+1+1+3+1)^2}{1 \times 6 \times 2} - \frac{144}{24}$$

$$= \frac{3}{2}$$

3. Among groups:

$$SS_g = \frac{\sum T_g^2}{nrc} - \frac{T^2}{N} = \frac{(1+2+1+1)^2 + (1+1+1+3+1)^2}{1 \times 2 \times 6} - \frac{144}{24}$$

$$= \frac{1}{6}$$

Table 6.6 Single-replication three-factor experiment

			Columns (c)					
	Groups (g)		c_1	c_2	c_3	c_4	c_5	c_6
Rows (r)	A_1	B_1	0	0	0	1	0	0
		B_2	0	0	1	1	0	0
	A_2	B_1	0	0	2	1	1	0
		B_2	0	1	3	0	0	1

r = number of rows = number of shaft and rotor combinations
g = number of groups = number of housing and cover combinations
c = number of columns = number of combinations of positioning clearances (wear plate clearance, ring clearance, pressure plate clearance, etc.) for positioning
$r = 2$; A_1 and A_2
$g = 2$; B_1 and B_2
$c = 6$; $c_1, c_2, c_3, \ldots, c_6$
$N = 24$, number of observations or sample size
$n = 1$, number of replications
$T = \sum x$, where x is the value of each response, in this case the number of smears and seizures in the cartridge of the pump
$\therefore T = 1 + 1 + 1 + 2 + 1 + 1 + 1 + 3 + 1$
    $= 12$
$\therefore T^2 = 12^2 = 144$

4. Column-row interaction:

$$SS_{cr} = \frac{\sum T_{cr}^2}{ng} - \frac{T^2}{N} - SS_c - SS_r$$

$$= \frac{0 + 0 + 1^2 + (1 + 1)^2 + 0 + 0 + 0 + 1^2 + (2 + 3)^2 + 1^2 + 1^2 + 1^2}{1 \times 2}$$

$$- \frac{144}{24} - 6 - \frac{3}{2} = \frac{7}{2}$$

5. Column-group interaction:

$$SS_{cg} = \frac{\sum T_{cg}^2}{nr} - \frac{T^2}{N} - SS_c - SS_g$$

$$= \frac{0 + 0 + 2^2 + (1 + 1)^2 + 1^2 + 0 + 0 + 1^2 + (1 + 3)^2 + 1^2 + 0 + 1^2}{1 \times 2}$$

$$- \frac{144}{24} - 6 - \frac{1}{6} = \frac{11}{6}$$

6. Row-group interaction:

$$SS_{rg} = \frac{\sum T_{rg}^2}{nc} - \frac{T^2}{N} - SS_r - SS_g$$

$$= \frac{1^2 + (1+1)^2 + (2+1+1)^2 + (1+3+1)^2}{1 \times 6} - \frac{144}{24} - \frac{3}{2} - \frac{1}{6}$$

$$= 0$$

7. Column-row-group interaction:

$$SS_{crg} = \frac{\sum T_{crg}^2}{n} - \frac{T^2}{N} - SS_c - SS_r - SS_g - SS_{cr} - SS_{cg} - SS_{rg}$$

$$= \frac{1^2 + 1^2 + 1^2 + 2^2 + 1^2 + 1^2 + 1^2 + 3^2 + 1^2}{1} - \frac{144}{2} - 6 - \frac{3}{2}$$

$$\qquad - \frac{1}{6} - \frac{7}{2} - \frac{11}{6} - 0$$

$$= 1$$

8. Total:

$$SS_{total} = \sum x^2 - \frac{T^2}{N} = (1^2 + 1^2 + 1^2 + 2^2 + 1^2 + 1^2 + 1^2 + 3^2 + 1^2) - \frac{144}{24}$$

$$= 14$$

9. Residual or error:

$$SS_{residual} = SS_{total} - \text{all previous SS}$$

$$= 14 - \left(6 + \frac{3}{2} + \frac{7}{2} + \frac{1}{6} + \frac{11}{6} + 0 + 1\right)$$

$$= 14 - 14$$

$$= 0$$

In the experiments having only *one* replication, the sum of squares of the residual is always equal to zero. This is because the residual error arises only from replicating (repeating) the experiments under the same set of conditions. This does not necessarily mean that the error term (experimental error and error involved due to some other factors unknown to the test designer) does not exist. In single-replication experiments, this error term is generally lost in the interactions (column-row, column-group, row-group, and column-row-group). This is known as confounding, where certain effects cannot be distinguished from others. Hence, the sum of squares SS of the error term (residual) and its degrees of freedom DF may be taken as the total of SS of all the interactions and the sum of their degrees of freedom respectively. This is shown in Table 6.7.

In the single-replication experiment it is therefore not possible to estimate or test the interactions (interacting factors) for the significance. Therefore, only the main effects (among columns, among rows, and among groups) have been investigated here. If the effect of interactions is important, the experiment should be replicated.

Table 6.7 Analysis-of-variance table for three-factor experiment

Source of variation	Sum of squares SS	Degrees of freedom DF	Mean square MS (SS/DF)	Mean-square ratio MSR (MS/MS$_{residual}$)	Minimum MSR required for factor to be significant — Confidence		
					90% $F_{0.10; \nu_1; \nu_2}$	95% $F_{0.05; \nu_1; \nu_2}$	97.5% $F_{0.025; \nu_1; \nu_2}$
Main factors							
1. Among columns (positioning clearances)	6	$c-1=5$	1.2	3.03	2.24	2.85	3.5
2. Among rows (square and production shafts)	$\frac{3}{2}$	$r-1=1$	1.5	3.79	3.05	4.49	6.12
3. Among groups (Standard and doweled housing)	$\frac{1}{6}$	$g-1=1$	0.16	0.405	3.05	4.49	6.12
Interacting factors							
4. Column-row interaction	$\frac{7}{2}$	$(c-1)(r-1)=5$	← Residual or error term →				
5. Column-group interaction	$\frac{11}{6}$	$(c-1)(g-1)=5$					
6. Row-group interaction	0	$(r-1)(g-1)=1$					
7. Column-row-group interaction	1.0	$(c-1)(r-1)(g-1)=5$					
8. Residual or error term nos. (4+5+6+7)	$\frac{7}{2}+\frac{11}{6}+0+1=6\frac{1}{3}$	$5+5+1+5=16$	0.396				

On the basis of values from Table 6.7, it appears that the positioning clearances (columns) and the type of shafts—square and production—(rows) are significant at 90 percent confidence level. That is, these two factors are the cause of the smears and seizures in the cartridge of the pump.

6.3 FRACTIONAL FACTORIAL EXPERIMENTS

When each of the possible combinations of the levels of the variables (factors) is tested, as in the three previous cases, the experiments are called full factorial experiments. In some situations these experiments can be costly and time consuming. For example, the total number of tests required to determine the main effects and interactions for seven factors, each at two levels, is 2^7, or 128. For three levels, the total tests would be 3^7 or 2,187. In such cases, it is not practical to plan an entire experimental program. Instead, a fractional factorial experiment is run. This requires testing only a fraction of the total number of possible test combinations. This "fraction" is the representative test combination carefully selected from the total test combinations. Fractional factorial experiments obviously cannot produce as much information as the full factorial. However, economy is achieved at the expense of assuming that some of the interactions between factors are negligible. Two fractional factorial methods are presented here.

6.3.1 PREDETERMINED TEST COMBINATIONS

In this method, the test combinations are selected from the total of all possible combinations to obtain the desired information about the main factors and their interacting effects. Several fractional factorial design test plans are available [4–6], including up to 10 factors, each at two or three levels. The application of these plans is illustrated by the following example. Yates' procedure [7, 8, 4] was used to analyze the test data.

Example 6.4 Design an experiment to determine the effect of the following four factors on the dynamic load-carrying capacity of a mechanical part: A, hardness; B, heat treatment; C, surface finish; D, temperature. These factors are to be evaluated at two levels each.

Solution The total number of tests required for a full experiment is $2^4 = 16$. Run a fractional experiment of magnitude (replication) = 1/2. Hence, the number of tests to run = $16 \times 1/2 = 8$. These eight tests are run according to predetermined test combinations. These combinations are:

(1), *ad*, *bd*, *ab*, *cd*, *ac*, *bc*, *abcd*

The presence of *a*, *b*, *c*, or *d* in the test combination indicates that factor A, B, C, or D is at its higher level. For example:

$ad = A_2 B_1 C_1 D_2$ test combination
$abcd = A_2 B_2 C_2 D_2$ test combination
$(1) = A_1 B_1 C_1 D_1$ test combination

Each test combination need not be tested in the order shown. It is preferable that the combinations be randomized so as to distribute evenly throughout the total experiment the unknown test variations.

The following are the steps for Yates' procedure.

1. Make a table with $(n + b)$ columns, where

   $n =$ number of factors
   $b = 2$ for full factorial experiment
   $= 1$ for 1/2 fractional factorial experiment
   $= 0$ for 1/4 fractional factorial experiment
   $= -1$ for 1/8 fractional factorial experiment

   In this experiment $n = 4$ and $b = 1$; hence the number of columns is 5. (See Table 6.8.)

2. In the first column, enter the treatment combination in prescribed order, and enter the corresponding load-capacity values in column 2 (Table 6.8).

3. In the upper half of column 3 enter, in order, the sum of consecutive pairs of the load values in column 2; $(30 + 35 = 65, 32 + 45 = 77$, etc.). In the lower half of column 3 enter, in order, the difference of consecutive pairs of the load values, i.e., second value minus first value, fourth value minus third value, etc. $(35 - 30 = 5, 45 - 32 = 13, \ldots)$.

4. In a similar manner obtain columns 4 and 5, using, in each case, the preceding column (Table 6.8).

5. Estimates of the magnitude of the effects are obtained by dividing the values in column 5 by 2^{n-b-1}, which is equal to 4 (Table 6.9).

6. To determine with, say, 90 percent confidence which factors have a significant effect on the load capacity, it would be necessary to have an estimate of the experimental error. However, in the case of fractional replication, the experimental error is confounded in the pure or mixed interactions. In this

Table 6.8   Yates' method of analysis

(1) Test combination	(2) Load capacity, ksi	(3)	(4)	(5)	Estimated effect
(1)	30	65	142	319	T (total)
ad	35	77	177	39	A
bd	32	89	18	11	B
ab	45	88	21	3	$AB + CD$
cd	38	5	12	35	C
ac	51	13	−1	3	$AC + BD$
bc	40	13	8	−13	$BC + AD$
abcd	48	8	−5	−13	D
Total	319				
Sum of squares 13,123				104,984*	

* Sum of squares in column 5 $= 2^{n-b} \times$ sum of squares in column 2.

Table 6.9 Estimated magnitude of various factors affecting the load-carrying capacity

Estimated effect	$\text{Magnitude of effect} = \dfrac{\text{column 5 of Table 6.8}}{4^*}$
T	79.75
A	9.75
B	2.75
$AB + CD$	0.75
C	8.75
$AC + BD$	0.75
$BC + AD$	−3.25
D	−3.25

$* \; 2^{n-b-1} = 2^{4-1-1} = 4$

case, use the sum of squares of interaction as an error s from column 5 of Table 6.8:

$$(3)^2_{AB+CD} + (3)^2_{AC+BD} + (-13)^2_{BC+AD} = 187$$

Divide the sum of squares by $2^{n-b} \nu$, where ν is the number of interactions included.

$$s^2 = \frac{187}{2^3 \times 3} = \frac{187}{24} = 7.8$$

$$s = 2.79$$

7. Compute

$$K = s\sqrt{2^{n-b}}\, t_{\alpha;\nu} \quad \text{where } \alpha = 1 - \text{confidence level} = 1 - 0.9 = 0.1 \text{ and } \nu = 3$$

$$= s\sqrt{8}\, t_{0.1;3} \quad t_{0.1;3} = 1.638 \text{ from Table A-2}$$

$$= 2.79\sqrt{8} \times 1.638$$

$$= 12.9$$

8. For any main factor or the interacting factor to be significant at 90 percent confidence level, the absolute value in column 5 of Table 6.8 should be greater than $K = 12.9$.

   Since $A = 39$, $C = 35$, and $D = -13$, the absolute values of all of which are greater than 12.9, it can be concluded with 90 percent confidence that hardness, surface finish, and temperature have a significant effect on the load-carrying capacity of the part. Heat treatment is not a significant factor.

6.3.2 RANDOMLY SELECTED TEST COMBINATIONS

This method enables one to estimate the effect of a large number of factors, independently, with the amount of testing considerably less than the method of Subsec. 6.3.1. The disadvantage is that confidence levels cannot be established in estimating the influence of the factors since the method involves making a decision from the trend of the results when graphically plotted.

Example 6.5 Investigate which of the following factors have a significant effect on the dynamic load-carrying capacity of a mechanical part: A, hardness; B, heat treatment; C, surface finish; D, temperature; E, time. These factors are to be evaluated at four different levels: K, L, M, and N. The time and test facilities are such that only 16 tests can be run.

Solution First, the test plan is prepared as shown in Table 6.10, where the experiment is designed for five factors, each at four levels, for a total of 16 test runs. A full factorial design would take $4^5 = 1,024$ tests. The recommended number of test combinations necessary for this method should be at least 1 percent of the total combinations under full factorial design.

The following is the procedure to construct the test plan as shown in Table 6.10

To determine the testing sequence for factor A, select a set of 16 consecutive numbers from the random-number table (A-47). The first 16 numbers chosen are:

Random numbers	Test run	Level of A
38001	6	K
37402	5	K
97125	16	K
21826	4	K
73135	13	L
07638	2	L
60528	11	L
83596	14	L
10850	3	M
39820	7	M
40791	8	M
55444	10	M
63315	12	N
03133	1	N
86961	15	N
54525	9	N

Assign a number between 1 and 16 to each random number corresponding to its rank if the random numbers were arranged in ascending order. This creates a set of random numbers between 1 and 16. Let the first four numbers (6, 5, 16, 4) be the test runs where A will be tested at level K. The third four numbers and the fourth four numbers correspond to A at levels M and N.

Fill in the test sequence for A in the test layout constructed in Table 6.10, where K's, L's, M's, and N's are thus in random sequence under A for the 16 tests. Repeat this for the factors B, C, D, and E, each time selecting randomly a set of 16 random numbers from Table A-47. This is shown in Table 6.11.

The test is then run first with factors A, B, C, D, and E at N, M, M, L, and L levels, respectively. The test result is found to be 1,600 lb (Table 6.10). This is repeated until all the 16 tests are run. The results of these tests are shown in Table 6.10.

Table 6.10 Test plan corresponding to five factors, four levels, and sixteen test runs

Test run	A	B	C	D	E	Load-carrying capacity, 100 lb, Test results*
1	N	M	M	L	L	16
2	L	K	N	N	K	9
3	M	M	M	K	K	12
4	K	L	K	K	L	8
5	K	L	L	L	M	7
6	K	M	K	N	N	9
7	M	K	M	N	L	15
8	M	M	L	K	L	14
9	N	K	K	M	M	15
10	M	N	N	K	N	16
11	L	N	N	L	K	11
12	N	N	K	M	M	18
13	L	N	L	M	M	10
14	L	L	M	N	K	12
15	N	K	L	M	N	17
16	K	L	N	L	N	10

* Average $\simeq 12.5$.

Graphical presentation of test results To analyze the test data, plot the test results on the ordinate against the levels of the factor on the abscissa, making a separate plot for each factor (see Fig. 6.4). Draw the response lines through the test points by means of the least-squares fitting method. The response line indicates how the experimental results change by changing the levels of any one factor. The steeper the line the more sensitive, that is, influential, is the factor. The vertical scatter about the response line is a measure of the effect of all other factors. For example, when factor A was at level K four times, each other factor B, C, D, and E was at some random level. Hence this scatter is due to the combined influence of the other four factors.

In Fig. 6.4, the response line of factor A has a higher slope than those of the other factors, with the least scatter around it. Hence, it is concluded that factor A is the most influential. In addition, there seem to exist several interactions because of the nonparallel responses in factors B, C, and E. If two or three plots show this situation, there may be a two- or three-factor interaction, respectively. This can be verified by a small, balanced, full factorial experiment.

Removal of the effect of the most influential factor The predominant factor A considerably influences the plots of the other factors. First, it causes scatter in the data plots of these factors; and, in addition, through a slight imbalance, it disturbs the true slopes of their response lines. After it is found that factor A is the most influential, the next most influential factor, if it exists, can be determined by removing the effect of A from the test results. This is done as follows:

Draw a horizontal line corresponding to the average test result (12.5) on the plot of factor A in Fig. 6.4.

Table 6.11 Determination of the testing sequence for factors B, C, D, and E

Random numbers	Test run	Level of B	Random numbers	Test run	Level of C
10721	2	K	88977	12	K
39755	9	K	15243	4	K
31652	7	K	24335	6	K
87662	15	K	61105	9	K
83651	14	L	19087	5	L
23790	5	L	42678	8	L
18370	4	L	98086	15	L
88318	16	L	94614	13	L
00157	1	M	00582	1	M
30635	6	M	97703	14	M
17340	3	M	32533	7	M
37589	8	M	04805	3	M
70322	11	N	68953	10	N
66492	10	N	02529	2	N
74083	12	N	99970	16	N
80680	13	N	74717	11	N

Random numbers	Test run	Level of D	Random numbers	Test run	Level of E
31347	10	K	11354	2	K
30240	8	K	31312	3	K
23823	4	K	69921	11	K
19051	3	K	79888	14	K
44640	11	L	06256	1	L
00812	1	L	46065	4	L
97207	16	L	52777	7	L
24767	5	L	54563	8	L
48336	13	M	59952	9	M
31224	9	M	50691	5	M
44906	12	M	78430	13	M
96988	15	M	77400	12	M
75172	14	N	80457	15	N
26401	7	N	51878	6	N
10157	2	N	90070	16	N
26017	6	N	66209	10	N

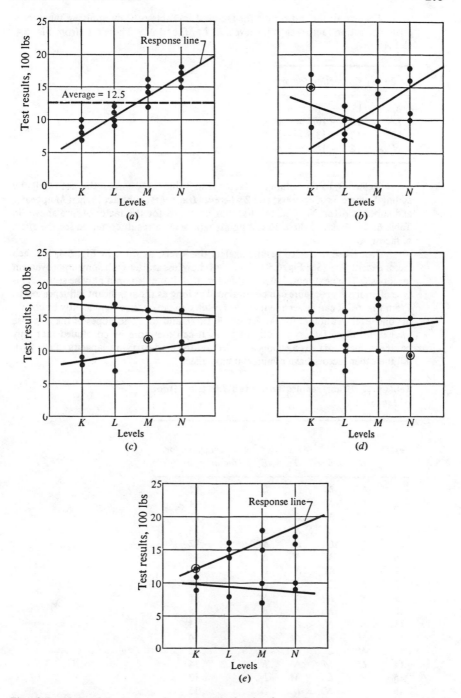

Fig. 6.4 Plots of the test results against the levels of each factor.

Correct all the test results for factor A in Table 6.10 by applying the appropriate correction factors each for levels K, L, M, and N. These are (from Fig. 6.4, plot A):

Levels	Correction factors
K	$12.5 - 7.5 = +5$
L	$12.5 - 10.5 = +2$
M	$12.5 - 13.5 = -1$
N	$12.5 - 16.5 = -4$

Referring to the column of factor A in Table 6.10, correct the test results by adding 5 wherever K appears; add 2 wherever L appears; subtract 1 when M appears; and subtract 4 for N. The results, thus corrected for the factor A, are shown in Table 6.12. Each of these 16 test points represents a result corrected for the effect of factor A.

Plot these corrected results against the levels, as done in Fig. 6.4, for each factor except A. (See Fig. 6.5.) This will show less scatter with some improvement in slopes. The next most influential factor to A, if it exists, can now be seen. This type of analysis procedure can be continued as long as the significant influence exists. From Fig. 6.5 it can be seen that the test results do not change significantly by changing one level to another for the factors B, C, D, and E. The slopes of the response lines of all these factors are almost flat. It can therefore be concluded that only hardness (factor A) has a significant effect on the load-carrying capacity of the part. All the other factors have a minor, if any, effect.

Table 6.12   Test results corrected for the effect of factor A

Test run	B	C	D	E	Test results corrected for factor A, in 100 lb
1	M	M	L	L	12
2	K	N	N	K	11
3	M	M	K	K	11
4	L	K	K	L	13
5	L	L	L	M	12
6	M	K	N	N	14
7	K	M	N	L	14
8	M	L	K	L	13
9	K	K	M	M	11
10	N	N	K	N	15
11	N	N	L	K	13
12	N	K	M	M	14
13	N	L	M	M	12
14	L	M	N	K	14
15	K	L	M	N	13
16	L	N	L	N	15

Table 6.13 Analysis of variance—single factor [9]

Source of variation	Sum of squares SS	Degrees of freedom DF	Mean square MS	Mean-square ratio
Among columns	$\dfrac{\sum T_c^2}{n} - \dfrac{T^2}{N}$	$c - 1$	$\dfrac{SS}{DF}$	$\dfrac{MS_c}{MS_{residual}}$
Residual	By subtraction: Total SS − among-col. SS	By subtraction: $(N-1) - (c-1)$	$\dfrac{SS}{DF}$	
Total	$\sum x^2 - \dfrac{T^2}{N}$	$N - 1$		

T = total of all observations
T_c = total for each column
x = each observation
N = number of total observations
n = number of replications
c = number of columns

Table 6.14 Analysis of variance—two factors [9]

Source of variation	Sum of squares SS	Degrees of freedom DF	Mean square MS	Mean-square ratio
Among columns	$\dfrac{\sum T_c^2}{nr} - \dfrac{T^2}{N}$	$c - 1$	$\dfrac{SS}{DF}$	$\dfrac{MS_c}{MS_{residual}}$
Among rows	$\dfrac{\sum T_r^2}{nc} - \dfrac{T^2}{N}$	$r - 1$	$\dfrac{SS}{DF}$	$\dfrac{MS_r}{MS_{residual}}$
Column-row interaction	$\dfrac{\sum T_{cr}^2}{n} - \dfrac{T^2}{N} - SS_c - SS_r$	$(c-1)(r-1)$	$\dfrac{SS}{DF}$	$\dfrac{MS_{rc}}{MS_{residual}}$
Residual	Total SS − all above	Total DF − all above	$\dfrac{SS}{DF}$	
Total	$\sum x^2 - \dfrac{T^2}{N}$	$N - 1$		

T = total of all observations
T_c = total for each column
T_r = total for each row
T_{cr} = total for each column-row combination
x = each observation
N = number of total observations
n = number of replications
c = number of columns
r = number of rows

Table 6.15   Analysis of variance—three factors [9]

Source of Variation	Sum of squares SS	Degrees of freedom DF	Mean square MS
Among columns	$\dfrac{\sum T_c{}^2}{nrg} - \dfrac{T^2}{N}$	$c-1$	$\dfrac{\text{SS}}{\text{DF}}$
Among rows	$\dfrac{\sum T_r{}^2}{ncg} - \dfrac{T^2}{N}$	$r-1$	$\dfrac{\text{SS}}{\text{DF}}$
Among groups	$\dfrac{\sum T_g{}^2}{ncr} - \dfrac{T^2}{N}$	$g-1$	$\dfrac{\text{SS}}{\text{DF}}$
Column-row interaction	$\dfrac{\sum T_{cr}{}^2}{ng} - \dfrac{T^2}{N} - \text{SS}_c - \text{SS}_r$	$(c-1)(r-1)$	$\dfrac{\text{SS}}{\text{DF}}$
Column-group interaction	$\dfrac{\sum T_{cg}{}^2}{nr} - \dfrac{T^2}{N} - \text{SS}_c - \text{SS}_g$	$(c-1)(g-1)$	$\dfrac{\text{SS}}{\text{DF}}$
Row-group interaction	$\dfrac{\sum T_{rg}{}^2}{nc} - \dfrac{T^2}{N} - \text{SS}_r - \text{SS}_g$	$(r-1)(g-1)$	$\dfrac{\text{SS}}{\text{DF}}$
Column-row interaction	$\dfrac{\sum T_{crg}^2}{n} - \dfrac{T^2}{N}$ minus all six previous sums of squares	$(c-1)(r-1)(g-1)$	$\dfrac{\text{SS}}{\text{DF}}$
Residual	Total SS minus all seven sums of squares	Total DF minus all previous DF	$\dfrac{\text{SS}}{\text{DF}}$
Total	$\sum x^2 - \dfrac{T^2}{N}$	$N-1$	

T = total of all observations
T_c = total for each column
T_r = total for each row
T_g = total for each group
T_{cr} = total for each column-row combination
T_{rg} = total for each row-group combination
T_{cg} = total for each column-group combination
T_{crg} = total for each column-row-group combination
x = each observation
N = number of total observations
n = number of replications
c = number of columns
r = number of rows
g = number of groups

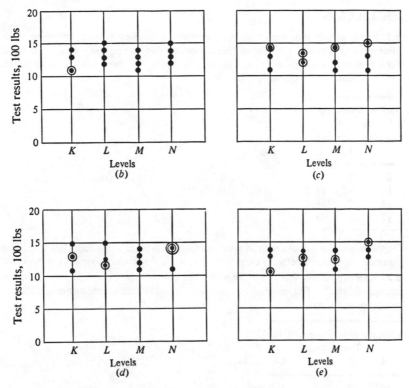

Fig. 6.5 Plots of the test results corrected for the factor A against the levels of each of the remaining factors.

SUMMARY

The principal objective of a factorial experiment is to determine the effect of various factors (temperature, velocity, method of manufacture, etc.) on some characteristic of a product (electrical conductivity, wear, power consumption, etc.). Tables 6.13 to 6.15 are summary tables for the analysis of data obtained from such experiments, in the case of one, two, and three factors, respectively. The values computed from test observations and inserted into these tables are then compared, at a given confidence level, with F values from Tables A-4 through A-9 to determine the significance of each factor on the product characteristic under consideration. The meaning and use of these tables are illustrated by Examples 6.1 to 6.3 for one-, two-, and three-factor experiments. Where the number of factors is large, the above methods become cumbersome, and a recourse may be taken to fractional factorial experiments. Two methods are presented (Subsecs. 6.3.1 and 6.3.2), and the choice depends on the amount of testing involved and the degree of confidence desired.

PROBLEMS

6.1. In a certain industrial process, automatic grinding machines are to be used. The machine that produces the least surface roughness (minimum rms value) is to be selected. Four repeat tests were run, using machines submitted by three different suppliers, and the following data were obtained:

Surface roughness, rms, μin		
Grinding machines		
A	B	C
20	16	12
18	18	14
15	17	16
17	13	15

On the basis of the above data, determine, with 90 percent confidence, whether there is any significant difference between these machines and, if so, which one should be selected.

6.2. An experiment was run to determine the effect of temperature on the ultimate tensile strength of steel. The material was tested at four different temperatures and the data were recorded, as follows:

Ultimate tensile strength, ksi			
Temperatures			
A	B	C	D
66	74	55	52
65	71	56	49
72	60	55	55
69	65	49	53
70	66	53	51

(*a*) Construct an appropriate analysis-of-variance table.

(*b*) Determine with 95 percent confidence whether the temperature has a significant effect on the ultimate strength.

6.3. An interlaboratory calibration check on a testing machine is to be made. Four laboratories are involved in the test, and each laboratory makes five measurements. It can be assumed that the test repeatability is the same for all laboratories. The results are:

Measurements			
Laboratories			
A	B	C	D
73	74	68	71
73	74	69	71
73	74	69	72
75	74	69	72
75	75	70	73

(a) Construct an appropriate analysis-of-variance table.

(b) Determine with 90 percent confidence whether the calibrations among the laboratories are different.

6.4. Four friction lining materials are to be evaluated. Hardness measurements were made on these linings, at two different temperatures, and the following were recorded:

		Hardness, BHN			
Linings		A	B	C	D
Temperature °F	100	35	36	31	38
		33	32	34	36
	200	30	32	28	30
		27	29	33	26

Determine by means of analysis of variance whether there is any significant difference in hardness between the four lining materials at 90 percent confidence. Does the temperature affect the hardness significantly?

6.5. An experiment was run to determine the abrasive wear of three different laminated materials. The samples used were either oven dried or moisture saturated. The machine ran for a fixed amount of time, and the depth of bearing wear was recorded as follows:

		Wear, in $\times 10^{-3}$		
Materials		A	B	C
Heat treatment	Oven dried	38	44	35
	Moisture saturated	32	42	35

Determine whether (a) wear characteristics of the three materials are different; (b) the heat treatments affect wear; and (c) the interaction between materials and the heat treatments are significant. Provide the answers with 90 percent confidence.

6.6. An experiment was run to determine the effect of the type of tool, the bevel angle, and the type of cut on the power consumption for ceramic-tool cutting. Other variables such as cutting speed and depth of cut were kept constant during this experiment. The following data were recorded:

		Power consumption			
Tool type		A		B	
		Bevel angle		Bevel angle	
		15°	30°	15°	30°
Type of cut					
Continuous		32	31	30	32
		27	31	31	38
		35	34	30	32
		28	39	24	30
Interrupted		30	28	23	29
		24	32	24	30
		27	29	30	28
		27	29	26	26

(a) Construct an appropriate analysis-of-variance table.

(b) Determine whether all the main effects and the interactions among them are significant at 90 percent confidence level.

6.7. The following experiment was run to determine the effect of factors A, B, C, and D on the quantity measured. On the basis of the following data, determine with 90 percent confidence which of these four factors are significant.

		C_1		C_2	
		D_1	D_2	D_1	D_2
A_1	B_1	18			26
	B_2		20	28	
A_2	B_1		23	39	
	B_2	33			36

6.8. The following air pollution data for sulfur dioxide were obtained in terms of parts per hundred million (pphm) for 6 months.

Month	Station 1	Station 2
1	1.9	3.0
	2.3	2.8
2	1.6	2.2
	1.5	2.6
3	2.1	2.9
	2.4	2.3
4	2.9	2.1
	2.7	2.5
5	2.6	1.7
	2.5	3.5
6	2.3	2.2
	2.4	2.6

(a) Determine whether the pollution level at station 1 is significantly different from the level at station 2 at 90 percent confidence.

(b) Does the pollution level change significantly from one month to another? (90 percent confidence level.)

6.9. The following data were obtained in an air pollution study in terms of the density of suspended particles, micrograms per cubic meter.

Period	Location 1	Location 2
1	210	150
	200	170
	230	190
2	185	205
	180	200
	170	210

(a) Determine whether the level of pollution is different from one location to another, and from one time period to the next, at 90 percent confidence level.

(b) Determine if the differences in the pollution levels among locations depend on the time period.

REFERENCES

1. Hicks, C. R.: "Fundamental Concepts in the Design of Experiments," Holt, Rinehart and Winston, Inc., New York, 1964.
2. Bowker, A. H., and G. J. Lieberman: "Engineering Statistics," Prentice-Hall, Inc., Englewood Cliffs, N.J., 1963.
3. Guttman, I., and S. S. Wilks: "Introductory Engineering Statistics," John Wiley & Sons, Inc., New York, 1965.
4. Davies, O. L. (ed.): "The Design and Analysis of Industrial Experiments," Hafner Publishing Company, Inc., New York, 1954.
5. National Bureau of Standards: "Fractional Factorial Experiment Designs for Factors at Two Levels," Applied Mathematics Series, No. 48, U.S. Government Printing Office, Washington 25, D.C., 1957.
6. National Bureau of Standards: "Fractional Factorial Experiment Designs for Factors at Three Levels," Applied Mathematics Series, No. 54, U.S. Government Printing Office, Washington 25, D.C., 1959.
7. Yates, F.: The Design and Analysis of Factorial Experiments, *Imp. Bur. Soil Sci.*, *Harpenden, England*, Tech. Commun. 35, 1937.
8. Cochran, W. G., and G. M. Cox: "Experimental Designs," 2d ed., John Wiley & Sons, Inc., New York, 1957.
9. Gionet, Paul A.: Analysis of Variance, "Statistics for the Engineer," *SAE Paper* SP-250, December, 1963.

BIBLIOGRAPHY

Barron, C. L.: Factorial Experiments in Reliability Analysis, *Eighth Nat. Symp. Rel. Qual. Contr., Wash.*; January, 1962.
Bowker, A. H., and G. J. Lieberman: "Engineering Statistics," Prentice-Hall, Inc., Englewood Cliffs, N.J., 1963.
Brownlee, K. A.: "Industrial Experimentation," Chemical Publishing Company, Inc., New York, 1953.

Cochran, W. G., and G. M. Cox: "Experimental Designs," 2d ed., John Wiley & Sons, Inc., New York, 1957.

Davies, O. L. (ed.): "The Design and Analysis of Industrial Experiments," Hafner Publishing Company, Inc., New York, 1954.

Fisher, R. A.: "The Design of Experiments," Oliver & Boyd, Ltd., Edinburgh, 1951.

Gionet, Paul A.: Analysis of Variance, "Statistics for the Engineer," *SAE Paper* SP-250, December, 1963.

Guttman, I., and S. S. Wilks: "Introductory Engineering Statistics," John Wiley & Sons, Inc., New York, 1965.

Hicks, C. R.: "Fundamental Concepts in the Design of Experiments," Holt, Rinehart and Winston, Inc., New York, 1964.

Kempthrone, O. "The Design and Analysis of Experiments," John Wiley & Sons, Inc., New York, 1962.

Moroney, M. J.· "Facts from Figures," Penguin Books, Inc., Baltimore, 1965.

National Bureau of Standards: "Fractional Factorial Experiment Designs for Factors at Two Levels," Applied Mathematics Series, No. 48, U.S. Government Printing Office, Washington 25, D.C., 1957.

National Bureau of Standards: "Fractional Factorial Experiment Designs for Factors at Three Levels," Applied Mathematics Series, No. 54, U.S. Government Printing Office, Washington 25, D.C., 1959.

Scott, J. F.: Design and Analysis of Experiments, *Chem. Ind.*, vol. N29, pp. 1182–1187, July 20, 1963.

Shainin, Dorian: How to Achieve More Productive Experimentation, paper presented to General Session of Conference, "Application of Statistics to Experimentation," Northeastern Section, American Chemical Society, Boston, Dec. 2, 1958.

Statistical Engineering Methods Group of Pratt and Whitney Aircraft, East Hartford: Reliability Considerations in a Development Program, *Ann. Rel. Maintainability*, vol. 4, July, 1965.

Yates, F.: The Design and Analysis of Factorial Experiments, *Imp. Bur. Soil Sci., Harpenden, England, Tech. Commun.* 35, 1937.

7
Sequential Experiments

Experiments are frequently costly and time consuming. Sequential testing represents an attempt to reduce the time of testing. Originally, sequential testing was developed as an inspection tool to determine whether a given lot meets the production requirements. This involved detecting, at a certain level of confidence, the number of defective items in a given lot which might exceed allowed inspection specification. The same statistics as in the inspection problems can be applied to the design and analysis of engineering experiments. Basically, a sequential experiment is a technique by which items are tested in sequence (one after another), the test results are reviewed after each test is completed, and two tests of significance (see Chap. 4) are applied to the data accumulated up to that time. This method has the advantage of being able to locate a product which has come from a distribution that varies widely from that hypothesized.

The technique of sequential experiments is discussed in this chapter under three headings corresponding to the three distributions which test data may follow: Weibull, normal, and binomial.

7.1 THE CONCEPT OF A SEQUENTIAL EXPERIMENT

In analogy with the tests for significance in Chap. 4, consider in Fig. 7.1 the left-hand distribution. Suppose this is the distribution of the life of a compression spring using standard material (call this distribution 1). Now consider the right-hand distribution. This is the distribution below which it is desired that the new material used in the spring has not come (call this distribution 2). One would like to determine from which distribution the new springs actually did come. This is impossible under sequential analysis. The only conclusion that can validly be made is either that the new spring has a distribution of life with $\mu > \mu_1$ or that the new spring has a life distribution with $\mu < \mu_2$. This very indecisiveness is the greatest drawback of the sequential experiment.

Consider the following test for significance:

$$H_{O_1}: (\mu = \mu_1) \quad \text{at } \alpha_I$$
$$H_{D_1}: (\mu = \mu_2) \quad \text{at } \alpha_{II}$$
$$H_{A_1}: (\mu > \mu_1) \quad \text{at } (1 - \alpha_I) \text{ confidence}$$

This is a one-sided test involving an absolute comparison of two tests. Referring to Fig. 7.1, if the sample provides a result such as $\bar{x}_1$, then H_{O_1} is rejected while H_{A_1} is accepted. The hypothesis H_{A_1} states with $(1 - \alpha_I)$ percent confidence or greater that $\mu \neq \mu_1$ but is actually greater than μ_1.

Consider a second test for significance:

$$H_{O_2}: (\mu = \mu_2) \quad \text{at } \alpha_{II}$$
$$H_{D_2}: (\mu = \mu_1) \quad \text{at } \alpha_I$$
$$H_{A_2}: (\mu < \mu_2) \quad \text{at } (1 - \alpha_{II}) \text{ confidence}$$

Here one has a one-sided lower-tailed single-parameter test for significance. In this case, if a sample with mean $\bar{x}_2$ were drawn, one could conclude

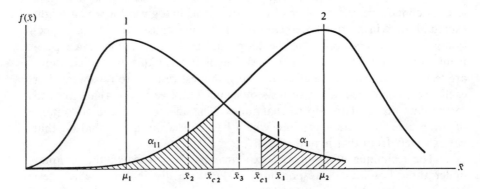

Fig. 7.1 Model for sequential analysis.

with $(1 - \alpha_{II})$ percent or greater assurance that $\mu < \mu_2$ (see Fig. 7.1). That is, $\bar{x}_2$ could not have come from a distribution centered at μ_2 more than α_{II} percent of the time.

There is, of course, the possibility of drawing a sample $\bar{x}_3$ from the lot that is actually less than $\bar{x}_{c_1}$. In the traditional test for significance this would call for an acceptance of H_{O_1}. However, considering the hypothesis based upon μ_2, it is found that the second test for significance has $\bar{x}_3$ greater than $\bar{x}_2$, and so H_{O_2} is accepted. Obviously, this is impossible because the sample cannot belong simultaneously to distributions 1 and 2. This situation is due to the fact that in making the test, the type II error was conveniently eliminated from the discussion. This can justifiably be done if one never accepts H_{O_1} or H_{O_2} in a sequential analysis. Either H_{A_1} or H_{A_2} is accepted. It follows, then, that one is not interested in making a type II error since this error is the probability of accepting H_O when it is not true. This can never be accepted, and so there is no interest in this error.

It was noted in previous chapters that as the sample size increases, the distribution of $\bar{x}$ narrows down ($s_{\bar{x}}$ decreases). If α and β are kept constant, this will result in the case where, in Fig. 7.2, $\bar{x}_{c_1}$ moves to the left and $\bar{x}_{c_2}$ moves to the right as n increases. When $\bar{x}_{c_1}$ and $\bar{x}_{c_2}$ become equal, the sample size can be arrived at through the use of the testing of significance techniques presented in Chap. 4.

In general, in a sequential experiment two basic questions are asked:

1. Can $H_{O_1}: (\mu = \mu_1)$ be rejected in light of the data accumulated to this time and with $H_{D_1}: (\mu = \mu_2)$?

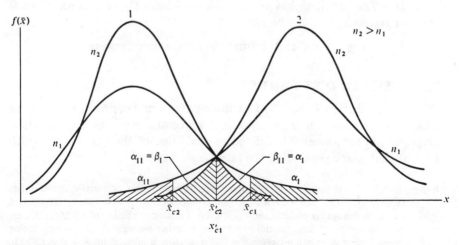

Fig. 7.2 Movement of cutoff point as sample size changes.

2. Can H_{O_2}: $(\mu = \mu_2)$ be rejected in light of the data accumulated to this time and with H_{D_2}: $(\mu = \mu_1)$?

This results in three decisions:

Reject H_{O_1} and accept H_{A_1}. $\hspace{6cm}$ (7.1a)

Reject H_{O_2} and accept H_{A_2}. $\hspace{6cm}$ (7.1b)

Insufficient information to make conclusions 1 or 2.   Continue testing.
$\hspace{13cm}$ (7.1c)

These decisions correspond to:

All $\bar{x}$ is to the right of $\bar{x}_{c_1}$. $\hspace{6.5cm}$ (7.2a)

All $\bar{x}$ is to the left of $\bar{x}_{c_2}$. $\hspace{6.8cm}$ (7.2b)

All $\bar{x}$ is between $\bar{x}_{c_1}$ and $\bar{x}_{c_2}$, respectively, in Fig. 7.1. $\hspace{2cm}$ (7.2c)

If one rejects H_{O_1} and, therefore, accepts H_{A_1}, he does not accept H_{O_2}. Moreover, rejecting H_{O_2} does not lead to accepting H_{O_1}.   The only conclusions that can be made here are:

$\mu > \mu_1$ $\hspace{9cm}$ (7.3a)

$\mu < \mu_2$ $\hspace{9cm}$ (7.3b)

Not enough information $\hspace{6.5cm}$ (7.3c)

This leads to the following decision making:

D_1: The new item has $\mu > \mu_1$ with confidence $(1 - \alpha_{\mathrm{I}})$ percent or greater.

D_2: The new item has $\mu < \mu_2$ with confidence $(1 - \alpha_{\mathrm{II}})$ percent   (7.4) or greater.

D_3: There is insufficient information.   Continue testing.

7.2   WEIBULL DISTRIBUTION

Weibull distribution (Chap. 2) is a three-parameter (x_0, b, θ) function, of which θ is of particular importance in a sequential experiment.   The characteristic life (or value) θ is defined as the value of the variable at which 63.2 percent of the population has failed.

Example 7.1   A compression spring when made from a current production material has a characteristic life θ of 900 h with a Weibull slope of 2.0.   A new material is proposed for use in this spring with the same slope, but a characteristic life of 1,800 h.   Determine by means of a sequential experiment whether the sample of the new spring comes from a population having $\theta \doteq 900$ h or from a population with $\theta = 1,800$ h.

Solution   The first step in setting up a sequential analysis is to set up the decisions D_1, D_2, and D_3.

D_1: The new spring comes from a Weibull population of slope 2.0 and $\theta > 900$ h.
D_2: The new spring comes from a Weibull population of slope 2.0 and $\theta < 1,800$ h.
D_3: There is insufficient evidence.   Continue testing.

It is necessary at this point to stipulate what risks, associated with making an incorrect decision, are allowable.   These risks are:

1. α_I, the probability of rejecting H_{0_1} when it is true
2. α_{II}, the probability of rejecting H_{0_2} when it is true

Suppose $\alpha_I = \alpha_{II} = 0.05$.   This selection means that the engineer wishes to be 95 percent confident that he has not accepted springs with $\theta > 900$ h when θ actually equals 900 h.   Moreover, it means that he wishes to be 95 percent confident that he has not accepted $\theta < 1,800$ h when θ actually is equal to 1,800 h.
In making decisions D_1, D_2, and D_3, the following generalized formulas are useful [1, 2].   Here $\theta_1 = 900$ h and $\theta_2 = 1,800$ h.

1. To accept D_1, the following inequality must be satisfied.   This is analogous to finding a sample to the right of $\bar{x}_{c_1}$ in Fig. 7.1.

$$\sum_{i=1}^{r} x_i^b > \frac{\theta_1^b}{\gamma^b - 1}\left(br \ln \gamma + \ln \frac{1 - \alpha_{II}}{\alpha_I}\right) \tag{7.5}$$

2. To accept D_2, the following inequality must be satisfied.   This is analogous to finding a sample to the left of $\bar{x}_{c_2}$.

$$\sum_{i=1}^{r} x_i^b < \frac{\theta_1^b}{\gamma^b - 1}\left(br \ln \gamma - \ln \frac{1 - \alpha_I}{\alpha_{II}}\right) \tag{7.6}$$

where

r = number of failures
$\gamma = \theta_2/\theta_1$
b = Weibull slope

3. Accept D_3, if neither D_1 nor D_2 is accepted.   Continue testing.   This is analogous to finding a sample between $\bar{x}_{c_1}$ and $\bar{x}_{c_2}$ in Fig. 7.1.
Now, suppose the test was run on the new spring and it failed in 1,000 h.   In order to accept D_1 after the first trial, one must satisfy Eq. (7.5).   Since

$$\gamma = \frac{\theta_2}{\theta_1} = \frac{1,800}{900} = 2.0 \qquad b = 2.0$$

$$\alpha_I = 0.05 \qquad \alpha_{II} = 0.05$$

Therefore

$$\sum_{i=1}^{r} x_i^b = 1,000^2 = 10^6$$

and

$$\frac{\theta_1{}^b}{\gamma^b - 1}\left(br \ln \gamma + \ln \frac{1 - \alpha_{II}}{\alpha_I}\right) = \frac{900^2}{2^2 - 1}(2 \ln 2 + \ln 19) = 1.17 \times 10^6$$

But

$$10^6 \not> 1.17 \times 10^6$$

Hence, the inequality of Eq. (7.5) is not satisfied; therefore, a check is made for decision D_2. To accept D_2 in one trial, it is necessary to satisfy Eq. (7.6). Here

$$10^6 < \frac{900^2}{2^2 - 1}(2 \ln 2 - \ln 19)$$

$$10^6 \not< -0.42 \times 10^6$$

Thus, D_2 is rejected too; therefore, D_3 is accepted. The testing must be continued. The second specimen failed at 2,000 h. Again, for decision D_1,

$$x_1{}^b + x_2{}^b = 10^6 + 4 \times 10^6 > \frac{900^2}{2^2 - 1}(4 \ln 2 + \ln 19)$$

or

$$5 \times 10^6 > 1.543 \times 10^6$$

With this inequality satisfied, the decision D_1 is accepted.

This means that it is more likely that $\theta = 1,800$ h than $\theta = 900$ h. This does not imply that the true characteristic life $\hat{\theta}$ is 1,800 h. However, it can be concluded with 95 percent confidence that the true characteristic life of the new springs does exceed 900 h.

7.3 NORMAL DISTRIBUTION

Since normal distribution is a two-parameter function (μ and σ), sequential tests are classified according to whether the parameter of interest is the mean of the population μ or its standard deviation σ.

7.3.1 TEST TO DETERMINE WHETHER THE SAMPLE BELONGS TO A POPULATION WITH $\mu > \mu_1$ OR $\mu < \mu_2$

Case 1—σ is known If Fig. 7.1 represents two normal distributions with means μ_1 and μ_2 and standard deviation σ, the three possible decisions are:

D_1: If $\bar{x} > \bar{x}_{c_1}$, then $\mu > \mu_1$ with confidence $(1 - \alpha_I)$ percent or greater, where

$$P\left(\frac{\bar{x} - \mu_1}{\sigma/\sqrt{n}} > \frac{\bar{x}_{c_1} - \mu_1}{\sigma/\sqrt{n}}\right) = \alpha_I \qquad \text{(see Chap. 4)}$$

Since $(\bar{x}_{c_1} - \mu_1)/(\sigma/\sqrt{n})$ has a z distribution, and $z_{\alpha I} = (\bar{x}_{c_1} - \mu_1)/(\sigma/\sqrt{n})$,

$$\bar{x}_{c_1} = \mu_1 + z_{\alpha I} \frac{\sigma}{\sqrt{n}} \tag{7.7}$$

D_2: If $\bar{x} < \bar{x}_{c_2}$, then $\mu < \mu_2$ with confidence $(1 - \alpha_{II})$ percent or greater, where

$$P\left(\frac{\bar{x} - \mu_2}{\sigma/\sqrt{n}} < \frac{x_{c_2} - \mu_2}{\sigma/\sqrt{n}}\right) = \alpha_{II}$$

$$-z_{\alpha_{II}} = \frac{\bar{x}_{c_2} - \mu_2}{\sigma/\sqrt{n}}$$

$$\bar{x}_{c_2} = \mu_2 - z_{\alpha_{II}} \frac{\sigma}{\sqrt{n}} \tag{7.8}$$

D_3: If $\bar{x}_{c_2} \leq \bar{x} \leq \bar{x}_{c_1}$, then there is insufficient evidence, and testing should be continued.

Case 2—σ is unknown If σ is unknown, the decisions are the same as in the previous section except that the cutoff points $\bar{x}_{c_1}$ and $\bar{x}_{c_2}$ are determined as follows:

$$P\left(\frac{\bar{x} - \mu_1}{s_1/\sqrt{n}} > \frac{\bar{x}_{c_1} - \mu_1}{s_1/\sqrt{n}}\right) = \alpha_I \qquad \text{(Chap. 4)}$$

By definition of the t distribution

$$t_{\alpha I; \nu} = \frac{\bar{x}_{c_1} - \mu_1}{s_1/\sqrt{n}}$$

$$\bar{x}_{c_1} = \mu_1 + t_{\alpha I; \nu} \frac{s_1}{\sqrt{n}} \tag{7.9}$$

Similarly,

$$-t_{\alpha_{II}; \nu} = \frac{\bar{x}_{c_2} - \mu_2}{s_2/\sqrt{n}}$$

$$\bar{x}_{c_2} = \mu_2 - t_{\alpha_{II}; \nu} \frac{s_2}{\sqrt{n}} \tag{7.10}$$

Summarizing the sequential test procedure:

1. Determine α_I and α_{II}. Find $t_{\alpha I}$ and $t_{\alpha_{II}}$ from Table A-3.
2. Test one item. Calculate $\bar{x}_{c_1}$ and $\bar{x}_{c_2}$. Make a decision: Accept D_1, D_2, or D_3.

3. If D_3 is accepted, increase sample size by 1.   Repeat step 2.
4. The test is terminated when D_1 or D_2 is reached.

7.3.2 TEST TO DETERMINE WHETHER THE SAMPLE BELONGS TO A POPULATION WITH $\sigma > \sigma_1$ or $\sigma < \sigma_2$

The three possible decisions in a sequential test to determine the uniformity of the population are:

D_1: If $s > s_{c_1}$, then $\sigma > \sigma_1$ with $(1 - \alpha_I)$ percent confidence or greater, where s_{c_1} is determined as follows:

$$P\left[\frac{(n-1)s^2}{\sigma_1{}^2} > \frac{(n-1)s_{c_1}{}^2}{\sigma_1{}^2}\right] = \alpha_I$$

By definition of χ^2 distribution

$$\chi^2_{\alpha_I;\, n-1} = \frac{(n-1)s_{c_1}{}^2}{\sigma_1{}^2}$$

$$s_{c_1} = \sigma_1 \sqrt{\frac{\chi^2_{\alpha_I;\, n-1}}{n-1}} \tag{7.11}$$

D_2: If $s < s_{c_2}$, then $\sigma < \sigma_2$ with $(1 - \alpha_{II})$ percent confidence or greater, where

$$P\left[\frac{(n-1)s^2}{\sigma_2{}^2} < \frac{(n-1)s_{c_2}{}^2}{\sigma_2{}^2}\right] = 1 - \alpha_{II}$$

$$\chi^2_{1-\alpha_{II};\, n-1} = \frac{(n-1)s_{c_2}{}^2}{\sigma_2{}^2}$$

$$s_{c_2} = \sigma_2 \sqrt{\frac{\chi^2_{1-\alpha_{II};\, n-1}}{n-1}} \tag{7.12}$$

D_3: If $s_{c_2} \le s \le s_{c_1}$, then there is insufficient evidence, and testing should be continued.

The procedure for performing the sequential test follows the same steps as listed at the end of Sec. 7.1.

Example 7.2 For a particular component the average ultimate strength S_u of steel must be 50 ksi.   By utilizing steel with $S_u = 60$ ksi, the warranty period of this component can be extended.   The supplier claims that the S_u of his steel is 60 ksi or greater. It is desired to verify the supplier's claim with 99 percent confidence.   Standard deviation for this steel is 3 ksi.

Use the following sequential test data:

n	x_i, ksi
1	64.3
2	71.1
3	66.3
4	67.7
5	69.2

Solution

$D_1 : \mu > 60$ ksi.

$D_2 : \mu < 65$ ksi [arbitrarily assumed $H_{o_2} : (\mu = 65$ ksi)].

D_3 : Continue testing.

Determine:

$$\bar{x}_n \qquad \bar{x}_{c1(i)} \qquad \bar{x}_{c2(i)}$$

$$\bar{x}_{c1} = \mu_1 + z_{\alpha 1} \frac{\sigma}{\sqrt{n}} \qquad \bar{x}_{c2} = \mu_2 - z_{\alpha 11} \frac{\sigma}{\sqrt{n}} \qquad \bar{x}_n = \frac{\sum\limits_{i=1}^{n} x_i}{n}$$

For $n = 1$, $\alpha = 0.01$,

$$\bar{x}_{c1} = 60 + 2.322 \left(\frac{3}{1} \right) = 66.96 \text{ ksi} \qquad \bar{x}_{c2} = 65 - 2.322 \left(\frac{3}{1} \right) = 58.03 \text{ ksi}$$

$$\bar{x}_1 = 64.3$$

Since $\bar{x}_{c2} < \bar{x}_1 < \bar{x}_{c1}$, D_3 is accepted and testing is continued.

For $n = 2$, $\alpha = 0.01$

$$\bar{x}_{c1} = 60 + 2.322 \left(\frac{3}{\sqrt{2}} \right) = 64.93 \text{ ksi} \qquad \bar{x}_{c2} = 65 - 2.322 \left(\frac{3}{\sqrt{2}} \right) = 60.07 \text{ ksi}$$

$$\bar{x}_2 = \frac{64.3 + 71.1}{2} = \frac{135.4}{2} = 67.7 \text{ ksi}$$

Since $\bar{x}_2 > \bar{x}_{c1}$ and also $\bar{x}_2 > \bar{x}_{c2}$, D_1 is accepted and $\mu > 60$ ksi with at least 99 percent confidence.

7.4 BINOMIAL DISTRIBUTION

In a system where a variable can have only one of two possible values (a sample is either good or bad, a student either passes or fails, a run is successful or unsuccessful, a coin falls either heads or tails), the probability of one event can be denoted as p and the alternative probability as $1 - p = q$. The problem is to establish the value of p for the population being sampled within some tolerable limits and with an acceptable chance of error.

The possible sequential test decisions are:

D_1: $p > p_1$ with $(1 - \alpha_I)$ percent confidence or greater.
D_2: $p < p_2$ with $(1 - \alpha_{II})$ percent confidence or greater.
D_3: Insufficient evidence; continue testing.
(p is the fraction of defectives or failures.)

To accept D_1, the following inequality must be satisfied [3]: $p > p_{c_1}$, where

$$p_{c_1} = \frac{\ln[(1 - \alpha_{II})/\alpha_I] + n \ln[(1 - p_1)/(1 - p_2)]}{\ln\left(\dfrac{1 - p_1}{1 - p_2} \dfrac{p_2}{p_1}\right)} \tag{7.13}$$

and $n =$ sample size.

To accept D_2, the following inequality must be satisfied: $p < p_{c_2}$, where

$$p_{c_2} = \frac{-\ln[(1 - \alpha_I)/\alpha_{II}] + n \ln[(1 - p_1)/(1 - p_2)]}{\ln\left(\dfrac{1 - p_1}{1 - p_2} \dfrac{p_2}{p_1}\right)} \tag{7.14}$$

and $n =$ sample size.

If neither D_1 nor D_2 is accepted, then D_3 is accepted.

For the purpose of the analysis of test data, one item at a time is tested and the number of failed (defectives) is plotted as ordinate against the total number of items tested as abscissa. p_{c_1} and p_{c_2} can be plotted as parallel lines on the chart by rearranging Eqs. (7.13) and (7.14) into the form [3]

$$p_{c_1} = h_1 + mn \tag{7.15}$$

$$p_{c_2} = h_2 + mn \tag{7.16}$$

where

$$h_1 = \frac{\ln[(1 - \alpha_{II})/\alpha_I]}{\ln\left(\dfrac{1 - p_1}{1 - p_2} \dfrac{p_2}{p_1}\right)}$$

$$h_2 = \frac{-\ln[(1 - \alpha_I)/\alpha_{II}]}{\ln\left(\dfrac{1 - p_1}{1 - p_2} \dfrac{p_2}{p_1}\right)}$$

$$m = \frac{\ln[(1 - p_1)/(1 - p_2)]}{\ln\left(\dfrac{1 - p_1}{1 - p_2} \dfrac{p_2}{p_1}\right)}$$

These quantities represent the intercepts and the slope, respectively (see Fig. 7.3).

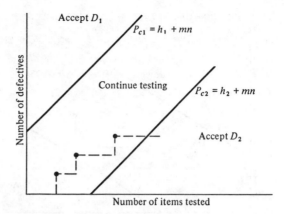

Fig. 7.3   Chart for sequential experiment.

When the plot of number of defectives against sample size crosses either of these two lines, D_1 or D_2 may be accepted at the selected confidence level.   If the plot crosses the lower line, D_2 is accepted; if the plot crosses the upper line, D_1 is accepted.   As long as the plot remains between these two lines, no decision is made.

Example 7.3   In a certain inspection program the desired quality is 1 percent or less defectives.   The poorest quality that will be accepted is 5 percent defectives.   On the basis of the following inspection data, do the parts supplied meet the required quality standard at the 95 percent confidence level?   Out of 190 items inspected, the fiftieth, ninetieth, and hundred and thirty-fifth items were found defective.

Solution

$$p_1 = 0.01 \qquad \alpha_I = 0.05$$
$$p_2 = 0.05 \qquad \alpha_{II} = 0.05$$

Find the equations of the cutoff lines:

$$p_{c1} = h_1 + mn$$
$$p_{c2} = h_2 + mn$$

$$h_1 = \frac{\ln[(1 - \alpha_{II})/\alpha_I]}{\ln\left(\dfrac{1 - p_1}{1 - p_2}\dfrac{p_2}{p_1}\right)} = \frac{\ln(0.95/0.05)}{\ln\left(\dfrac{0.99}{0.95}\dfrac{0.05}{0.01}\right)} = \frac{2.945}{1.651}$$

$$= 1.784$$

$$h_2 = \frac{-\ln[(1 - \alpha_I)/\alpha_{II}]}{\ln\left(\dfrac{1 - p_1}{1 - p_2}\dfrac{p_2}{p_1}\right)} = -1.784$$

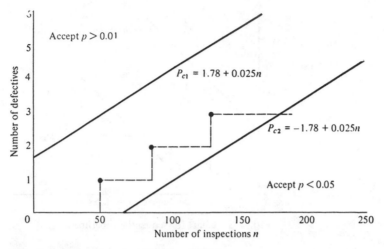

Fig. 7.4　Inspection chart for Example 7.3.

$$m = \frac{\ln[(1 - p_1)/(1 - p_2)]}{\ln\left(\dfrac{1 - p_1}{1 - p_2}\dfrac{p_2}{p_1}\right)} = \frac{\ln(0.99/0.95)}{\ln\left(\dfrac{0.99}{0.95}\dfrac{0.05}{0.01}\right)} = \frac{0.0411}{1.651}$$

$$= 0.025$$

$$p_{c_1} = 1.78 + 0.025n$$

$$p_{c_2} = -1.78 + 0.025n$$

　　　　A plot of these data is shown in Fig. 7.4. At the hundred and ninetieth observation, the data cross the lower control line and the material represented by the sample can be accepted as less than 0.05 defective with a maximum chance of error of 5 percent that there are actually more than 0.05 defectives.

SUMMARY

Sequential experiments are a form of accelerated experiments in that they reduce testing time through a unique statistical method. Because of their importance in industrial applications (engineering and inspection) a separate chapter was devoted to this topic. Sequential experiments are discussed here for the cases when the test data follow Weibull, normal, and binomial distributions. To determine whether a given set of data follows these distributions, consult Sec. 2.8.

　　　　Sequential experiments are also a form of experiments of comparison (see Chap. 4) in that they involve a comparison of two products or a product against a standard. The methods in Chap. 7 involve a smaller sample size (hence less testing time) than those in Chap. 4, provided that only the knowledge of the α error is required and not both the α and β errors (see Glossary).

PROBLEMS

7.1. Because of frequent failures it is desired to increase the characteristic life θ of a certain component, in a major installation, from 2,500 to 4,000 h. A redesigned component, it is claimed, will have a characteristic life of 4,000 h. The new design will be accepted with 99 percent confidence if θ is more than 2,500 h and rejected with 95 percent confidence if θ is less than 4,000 h. It is known that the underlying distribution of the component lives is Weibull with slope $b = 2.5$. The hours to failure of three components of the new design tested are 3,000, 4,000, and 3,800. Determine whether the new design fulfills the life requirement.

7.2. A certain type of velocity pickup is to be tested for its sensitivity. Sensitivity is defined here as the ratio of the voltage amplitude to the velocity amplitude, $10^{-2}V$. The sensitivity, under different velocities, is to be, on an average, at least 10. The average sensitivity of 8 is not acceptable. It is assumed that the sensitivity varies randomly with the velocity. Determine, with 95 percent confidence, whether the velocity pickup is acceptable. The standard deviation of the sensitivity is 1. The experiment was conducted at different velocities, and the following data were found (assume normal distribution):

Test number	Sensitivity, $10^{-2}V/(in/s)$
1	9.5
2	7.5
3	8.0
4	8.5

7.3. Determine by sequential analysis the characteristic life of a component. Two hundred hours characteristic life is satisfactory; one hundred hours is unsatisfactory. The Weibull slope is 1.5, known from previous experience. A confidence of 95 percent is desired. Use the following test data, as necessary:

Test number	Time to failure, h
1	120
2	100
3	140
4	155
5	135

7.4. For a particular application, an average surface hardness of 200 BHN is adequate to prevent wear-out failures in gear teeth, though a higher hardness might be desirable. The supplier claims that an average surface hardness of his gears is 220 BHN, and it could be as high as 240 BHN.   Standard deviation is 10 BHN.   On the basis of the following test data, check on the supplier's claim, with 95 percent confidence.

Test number	Hardness, BHN
1	230
2	210
3	223
4	217

7.5. A batch of parts was fabricated on five identical machines.   Later on it was found that one of these machines was not functioning properly and this produced some defective parts.   On the basis of the following inspection data, can it be concluded, with 90 percent confidence, that the percent defectives in the whole batch is not more than 4 percent, whereas the desired quality is 1 percent or less defectives?   Out of 175 parts inspected, the thirty-sixth and seventy-eighth parts were found defective.

7.6.   What types of engineering problems, situations, or mechanisms are best solved by sequential analysis and what kinds by sudden-death testing?

REFERENCES

1. Johnson, L. G.: "The Statistical Treatment of Fatigue Experiments," Elsevier Publishing Company, Amsterdam, 1964.
2. Wald, A.: "Sequential Analysis," John Wiley & Sons, Inc., New York, 1947.
3. Volk, W.: "Applied Statistics for Engineers," McGraw-Hill Book Company, New York, 1958.

BIBLIOGRAPHY

Brunk, H. D.: "An Introduction to Mathematical Statistics," Ginn and Company, Boston, 1960.
Burr, I. W.: "Engineering Statistics and Quality Control," McGraw-Hill Book Company, New York, 1953.
Columbia University: "Techniques of Statistical Analysis," McGraw-Hill Book Company, New York, 1947.
Davies, O. L.: "Statistical Methods in Research and Production," Hafner Publishing Company, Inc., New York, 1961.
Dixon, W. J., and F. J. Massey, Jr.: "Introduction to Statistical Analysis," 3d ed., McGraw-Hill Book Company, New York, 1969.
Freeman, H. A.: "Industrial Statistics," John Wiley & Sons, Inc., New York, 1942.
Johnson, L. G.: "The Statistical Treatment of Fatigue Experiments," Elsevier Publishing Company, Amsterdam, 1964.
Moses, L. E.: "Elementary Decision Theory," John Wiley & Sons, Inc., New York, 1959.
"Sequential Analysis of Statistical Data," Columbia University Press, New York, 1945.
Shewhart, W. A.: "Economic Control of Quality of Manufactured Products," D. Van Nostrand Company, Inc., Princeton, N.J., 1931.

Simons, L. E.: "An Engineering Manual of Statistical Methods," John Wiley & Sons, Inc., New York, 1941.

Volk, W.: "Applied Statistics for Engineers," 2d ed., McGraw-Hill Book Company, New York, 1969.

Wald, A.: Sequential Tests of Statistical Hypotheses, *Ann. Math. Statist.*, vol. 16, no. 2, 1945.

———: "Sequential Analysis," John Wiley & Sons, Inc., New York, 1947.

———: "Statistical Decision Function," John Wiley & Sons, Inc., New York, 1950.

8
Nonparametric Experiments

In nonparametric experiments "nonparametric" implies running an experiment without necessarily using any statistical parameters, such as mean and standard deviation. In the previous chapters various experiments were discussed with an assumption that the samples were drawn from continuous distributions, such as normal, Weibull, and exponential. There are many engineering situations where very little is known about the type and shape of the underlying distributions, and the data generally available are insufficient to determine the exact nature of the distribution function. In such cases, it may be hazardous to assume any of the above distributions. Nonparametric experiments are designed for situations where the test samples follow a continuous distribution but its type and shape are unknown. In this chapter, several of the simplest and more widely used nonparametric experiments are discussed.

8.1 THE POWER EFFICIENCY OF NONPARAMETRIC EXPERIMENTS [1]

In Chap. 4 the type I, or α, error was defined as the error of rejecting the null hypothesis when it is true, and the type II, or β, error as the error of accepting the null hypothesis when it is false. The power of an experiment is defined as the probability of rejecting the null hypothesis H_O when it is false and accepting the alternate hypothesis H_A when it is true:

$$\text{Power} = 1 - \beta \tag{8.1}$$

This might be called the probability of making the right decision.

Since a nonparametric experiment makes no general assumptions about the nature of the underlying distribution, it does not have as high a power in making a right decision as parametric experiments do. However, since power increases with sample size, the nonparametric experiments can be made as powerful as or more powerful than the parametric experiments by choosing a larger sample size. If the data do not follow normal distribution, the nonparametric experiment may be more powerful than the parametric experiment.

Hence, the power efficiency of a nonparametric experiment is defined as the ratio of the parametric test sample size n_1 to the nonparametric test sample size n_2 for both tests to give the same power when used with a normally distributed population.

$$\text{Power efficiency} = \frac{n_1}{n_2} \qquad n_2 \geq n_1 \tag{8.2}$$

Some nonparametric experiments have power efficiencies of 0.95, which means that a sample size of 20 for the nonparametric experiment is as powerful as a sample size of 19 for a parametric experiment, a normal distribution being used in both cases. However, if the data depart widely from normality, then the situation could change and the nonparametric test could be the more powerful.

8.2 ADVANTAGES AND DISADVANTAGES OF NONPARAMETRIC EXPERIMENTS

8.2.1 ADVANTAGES

1. In nonparametric experiments it is not necessary to know the underlying distribution. Acceptance or rejection of the null hypothesis is generally based on elementary considerations.
2. Nonparametric experiments are usually much easier to calculate than corresponding parametric tests. This ease of calculation has led them to be termed "short-cut statistics" or "quick and easy statistics." This also makes them useful for the preliminary inspection of data.
3. Most parametric experiments require data on an interval scale, where not only the ranking but the magnitude of the difference between values of the random variable must be known. Nonparametric experiments generally require only a knowledge of the ranking order of the data.

4. A given parametric experiment can test only one characteristic of the distribution, such as central tendency or dispersion. Some nonparametric experiments are capable of testing for any difference between the two populations.

8.2.2 DISADVANTAGES

1. If the test samples belong to a normal population, a larger sample size must be used with the nonparametric experiment to have the same power as a parametric experiment.
2. When the exact power of the nonparametric experiment relative to the power of an equivalent parametric experiment is not known, the sample size necessary to make the experiment as powerful as the parametric experiment cannot be exactly determined.

8.3 SIGN AND RUN TESTS

8.3.1 SIGN TEST

The sign test uses plus and minus signs rather than quantitative data for evaluation. It is an extremely simple test for determining whether there is a difference between paired observations. It is useful for the following engineering situations: (1) the underlying distributions are not known, or are too difficult to determine; (2) the pairs of observations are taken on two parts, such that each observation in a given pair is made under similar conditions, and different pairs need not be tested under similar conditions. The two samples being compared must be the same size, and the data are used just as they are taken (i.e., they are not ranked in any way).

The sign test is based on the binomial distribution, which involves analyzing the data based on the sign of the differences: $y_1 - x_1$, $y_2 - x_2$, $y_3 - x_3$, ..., $y_n - x_n$. With a null hypothesis H_O of no difference between pairs of data, it would be expected that the probability of a positive or negative difference is 1/2. Therefore, for no difference in populations there should be very nearly an equal number of positive and negative signs. If one sign appears a number of times so as to be significantly small for the size of the sample, it can be concluded that there is a significant difference between the populations. The equation for calculating the probability that k less frequent signs will occur is

$$P(\alpha) = 2 \sum_{0}^{k} \binom{n}{k} \left(\frac{1}{2}\right)^{n}$$

where k = number of times the less frequent sign occurs

$n = n_1 = n_2$ = number of items in each sample

$1 - \alpha$ = confidence level

$$\binom{n}{k} = \frac{n!}{k!\,(n-k)!}$$

Since the magnitude of the difference between the two values is not used, this test is less powerful than a parametric test. This disadvantage is offset by the advantage of being quick and simple.

The sign test can be applied only where there are no ties in the data. When there is a tie, the practice is to "discard" that piece of data and reduce the sample size by 1. The following procedure is suggested for the application of the sign test:

1. Write down the values for x and y, and set up the null hypothesis H_O that there is no difference between the two pairs of data.
2. Note the sign of the difference between x and y, either $+$, $-$, or 0.
3. Find the sample size excluding 0s.
4. Find the experimental value of k where k denotes the number of times the less frequent sign occurs.
5. If k in step 4 is equal to or less than the k values given in Table 8.1, then reject H_O at the corresponding confidence level and conclude that a difference between the pairs of data does exist.

Table 8.1 Maximum significant number of less frequent signs in n observations when $P(+) = 0.50$ [7]

| n | Probability levels α | | | |
	0.25	0.10	0.05	0.01
8	1	1	0	0
9	2	1	1	0
10	2	1	1	0
12	3	2	2	1
14	4	3	2	1
16	5	4	3	2
18	6	5	4	3
20	6	5	5	3
25	9	7	7	5
30	11	10	9	7
35	13	12	11	9
40	15	14	13	11
45	18	16	15	13
50	20	18	17	15
55	22	20	19	17
60	25	23	21	19
75	32	29	28	25
100	43	41	39	36

Example 8.1   The following are data taken as an index of formability of two different steels.

x	y	Sign of $y - x$
80.6	83.3	+
81.2	80.5	−
81.5	91.0	+
83.2	91.0	+
84.6	84.6	0
82.0	86.1	+
86.1	87.8	+
84.2	84.3	+
82.3	81.0	−
83.0	88.2	+

Is there any difference between the two sets of data?

Solution

H_O : There is no difference between the two sets of data.
$n = 10$; with one tie, $n = 10 - 1 = 9$ is the effective sample size.
$k = 2$

From Table 8.1,

$\alpha = 0.25$	0.10	0.05	0.01
$k = 2$	1	1	0

The experimentally determined k is equal to the tabulated k at the 75 percent confidence level ($\alpha = 0.25$); therefore, reject the null hypothesis H_O of no difference between x and y and conclude with 75 percent confidence that there is a significant difference between the formability of x and y steels.

Example 8.2   Eight different types of fuels were tested for their octane numbers by two operators x and y performing the test using the same test cell.   Each fuel was tested once by each operator, and the following octane numbers were obtained:

Operator x	Operator y	Sign of $y - x$
70	72	+
61	63	+
90	92	+
79	67	−
80	81	+
70	75	+
96	89	−
64	68	+

Can it be concluded with 90 percent confidence that there are differences between the operators x and y performing the test?

Solution

H_O : There is no difference between the operators.

$n = 8$ $k = 2$ $1 - \alpha = 0.9$, or $\alpha = 0.1$

From Table 8.1, for $\alpha = 0.1$ and $n = 8$, $k = 1$. Since the experimental k is not equal to or less than the tabulated k, the null hypothesis H_O cannot be rejected at the 90 percent confidence level. That is, one cannot conclude with 90 percent confidence that there are differences between the operators.

8.3.2 THE WEIGHTED SIGN TEST

The weighted sign test is the same as the sign test of the previous section except that it is used to determine whether one set of data is better than another set by a given amount.

Example 8.3 In Example 8.1 product y was found to be better than x. Determine whether it is 5 percent better.

Solution   Increase the original x data of Example 8.1 by 5 percent and compare it with the original y data.

x' (original $x + 0.05x$)	y	Sign of $y - x$
84.6	83.3	−
85.3	80.5	−
85.7	91.0	+
87.4	91.0	+
88.8	84.6	−
86.0	86.1	+
90.5	87.8	−
88.4	84.3	−
86.4	81.0	−
87.2	88.2	+

Sample size $n = 10$. Experimental value of $k = 4$. From Table 8.1, $k = 2$ for 75 percent confidence and $n = 10$. Experimental k is not less than the tabulated k. Hence it can be concluded that there is no difference between y and x' (original $x + 0.05x$). This means that the product y is 5 percent better than the product x.

8.3.3 THE RUN TEST BETWEEN TWO SAMPLES

The run test for observations between two samples is sometimes referred to as the Wald-Wolfowitz Runs Test. (For mathematical derivation see Ref. 1.) This test is applied to determine whether one group of data has values consistently larger or smaller than the other group. The technique is based on the concept of "run," where a run is defined as a succession of identical symbols followed by different symbols. For example, suppose a series of plus and minus scores occurred in this order:

$$+++ \quad -- \quad + \quad --- \quad ++ \quad -$$

This sample begins with a run of three pluses followed by a run of two minuses, and then a run of one plus, etc. The total number of runs in this sample, therefore, is six. In the sign test (Subsec. 8.3.1), pairs of observations are taken so that a given pair is tested under identical conditions, but not necessarily under the same conditions as another pair from the two samples tested. In the run test all the observations of both samples are made under the same conditions. The following procedure for applying this test is recommended:

1. Arrange the observations of the two samples in one series in ascending order. To differentiate between samples, put the values from one sample in parentheses.
2. Set up the null hypothesis H_0 such that there is no difference between the two samples.
3. Count the total number of runs, referred to here as the experimental runs.
4. Compare the number of experimental runs with the number of runs in Table 8.2. If the number of experimental runs is smaller than the number of runs corresponding to the lower percentage points in Table 8.2, reject H_0 with $(1 - \alpha)$ confidence; if it is larger than the number of runs corresponding to the upper percentage points in Table 8.2, accept H_0 with $(1 - \alpha)$ confidence. If it lies between the two limits, it cannot be concluded with $(1 - 2\alpha)$ confidence that there is a difference between the two samples.

Example 8.4 Two pump designs A and B are to be evaluated for wear and corrosion. Subjective evaluation on five pumps of each design was made, and the degree of degradation established in a numerical score from 0 to 10 (from worst to best condition).

Design A: 4   8   0   6   2
Design B: 10   7   5   3   1

Determine at 95 percent confidence whether there is any significant difference between the two designs.

Solution

H_0 : There is no difference between A and B.
        Arrange the data of both designs in an increasing order with the values of design A in parentheses:

(0)   1   (2)   3   (4)   5   (6)   7   (8)   10

Table 8.2   Distribution of the total number of runs [8]

$n_1 = n_2$	Lower percentage points U'_α		Upper percentage points U''_α	
	$\alpha = 0.05$	$\alpha = 0.01$	$\alpha = 0.05$	$\alpha = 0.01$
5	3	2	9	10
6	3	2	11	12
7	4	3	12	13
8	5	4	13	14
9	6	4	14	16
10	6	5	16	17
11	7	6	17	18
12	8	7	18	19
13	9	7	19	21
14	10	8	20	22
15	11	9	21	23
16	11	10	23	24
17	12	10	24	26
18	13	11	25	27
19	14	12	26	28
20	15	13	27	29
21	16	14	28	30
22	17	14	29	32
23	17	15	31	33
24	18	16	32	34
25	19	17	33	35
26	20	18	34	36
27	21	19	35	37
28	22	19	36	39
29	23	20	37	40
30	24	21	38	41

The total number of experimental runs is $r = 10$.

$n_1 = n_2 = 5$

From Table 8.2, for $n_1 = n_2 = 5$ and $\alpha = 0.05$, $U'_\alpha = 3$, $U''_\alpha = 9$. Since $r = 10$ is larger than $U''_\alpha = 9$, the null hypothesis H_O is accepted; that is, it is concluded with 95 percent confidence that there is no difference between the two pump designs.

8.3.4   THE RUN TEST FOR RANDOMNESS

In engineering experiments it is generally assumed that the test sample was chosen at random. The run test for randomness is performed to determine whether the sample is truly random. This is probably the simplest method of testing for randomness.

The randomness of a sequence is in doubt if there are too few or too many runs. For instance, if a coin is flipped 10 times with outcomes $T, T, T, T, T, H, H, H, H, H$, the randomness of the outcome is immediately in doubt because there are only two runs.   On the other hand, if the outcome of the experiment is $H, T, H, T, H, T, H, T, H, T$, which is a total of 10 runs, the result is also in doubt because there is a certain degree of order to the outcome.   Intuitively, both of these outcomes are questioned because they lack a certain degree of haphazardness.   The conclusion is drawn at a desired level of confidence where the lower and upper confidence limits are obtained from Table 8.2 for 90 and 98 percent confidence limits.   These correspond to $\alpha/2 = 0.05$ and $\alpha/2 = 0.01$.   The following procedure for applying the run test for randomness is suggested:

1.   Write down the values from the experiment as they are obtained.
2.   Find the median of the values, and set up the null hypothesis H_O such that the test data are random.
3.   Differentiate between the values above the median and below the median in the original data by putting parentheses around those below the median.   If an odd number of values is obtained, do not include the median when using the series.
4.   Count the number of experimental runs in the new series.
5.   Compare the number of experimental runs with the number of runs in Table 8.2.   If the number of experimental runs is smaller than the number of runs corresponding to the lower percentage points or larger than the number of runs corresponding to the upper percentage points in Table 8.2, reject H_O with $(1 - \alpha)$ confidence.   If the number of the experimental runs lies between the two limits, H_O cannot be rejected at $(1 - 2\alpha)$ confidence level.

Example 8.5   Do the following data represent a random sample?

77. 87, 81, 83, 79, 84, 80, 86, 78, 85

Solution   Arrange the data in an increasing order.   In this case the median is $(81 + 83)/2 = 82$.   By using step 3, the following are obtained:

(77), 87, (81), 83, (79), 84, (80), 86, (78), 85

The total number of runs is $r = 10$.   Enter Table 8.2 with a sample size of $n_1 = n_2 = 5$ since for the original sample of 10, half the values are above and half the values are below the median.   For $\alpha = 0.05$, $U'_\alpha = 3$ and $U''_\alpha = 9$.   Since $r = 10$ is larger than $U'' = 9$, H_O is rejected.   That is, it is concluded with 95 percent confidence that the sample is not random.

8.4   TEST FOR DETERMINING THE CORRELATION BETWEEN VARIABLES [2]

The test for determining the correlation between variables is sometimes referred to as the Spearman Rank Correlation.   From the viewpoint of the amount of computation required, this is possibly the easiest correlation test available. The power efficiency of this test compared with a parametric test is 0.91.

This correlation test involves ranking both variables in increasing algebraic order. The difference between the ranks is then calculated and used to calculate a statistic r_s, which is related to the correlation [2]. For sample sizes larger than 10 an approximate significance level can be assigned to the correlation.

The degree of correlation between variables is measured in terms of the correlation coefficient r_s (detailed discussion is given in Chap. 13); r_s varies from -1 to $+1$, which correspond to the perfect inverse relationship and the perfect direct relationship, respectively. The larger the absolute value of r_s, the better the relationship. By squaring r_s, an important relationship is obtained. For example, if r_s is .9, then r_s^2 is .81; thus, 81 percent of the variation of either variable is explained by its correlation with the other, while 19 percent is unexplained. The procedure for determining the correlation is (also see Chap. 13):

1. Rank each set of the pair of variables x and y in an increasing algebraic order. The ranks should not be paired (i.e., smallest with smallest, etc.) but should be left in the order in which the data were taken.
2. Take the difference between the ranks $d_i = y_i - x_i$ for each pair of variables.
3. Use this d_i and the sample size n in Spearman's equation

$$r_s = 1 - \frac{1}{n(n^2 - 1)} 6 \sum_{i=1}^{n} d_i^2 \tag{8.3}$$

If $|r_s| \to 1$, it can be concluded that a definite relationship exists between the variables; if $|r_s| \to 0$, the variables are independent, and they do not relate with each other.
4. For sample sizes larger than 10 items, an approximate confidence level for the correlation can be given by

$$t = r_s \sqrt{\frac{n - 2}{1 - r_s^2}} \tag{8.4}$$

which is distributed approximately as the t values with $n - 2$ degrees of freedom. (The discussion of t distribution is given in Chap. 3.)

Example 8.6 Consider the following data relating the abrasion loss in milligrams per 100 test hours, y, and hardness, x, in Brinell hardness numbers. Determine whether correlation exists between abrasion loss and hardness.

Sample number	x	y	Ranks x_i	Ranks y_i	$|d_i|$
1	372	45	11	1	10
2	206	55	9	3	6
3	175	61	8	5	3
4	154	66	5	7	2
5	136	71	4	8.5	4.5
6	112	71	3	8.5	5.5
7	55	81	2	10	8
8	45	86	1	11	10
9	221	53	10	2	8
10	166	60	7	4	3
11	164	64	6	6	0

Solution   Calculating,

$$r_s = 1 - \frac{6 \sum d_i^2}{n(n^2 - 1)}$$

$$= 1 - \frac{1}{11(121 - 1)} 6(10^2 + 6^2 + 3^2 + 2^2 + 4.5^2 + 5.5^2 + 8^2 + 10^2 + 8^2 + 3^2)$$

$$= 1 - 1.984$$

$$= -.984$$

The absolute value of r_s approaches 1; therefore it can be concluded that there is a definite relationship between abrasion loss and hardness. Since r_s is negative, the relationship is inverse (Fig. 8.1). It can further be concluded that $.984^2$, or about 96.7 percent, of the variation of either variable can be explained by its variation with the other. The remaining 3.3 percent is unexplained.

With the value of r_s in Eq. (8.4) to test for significance:

$$t = r_s \sqrt{\frac{n - 2}{1 - r_s^2}}$$

$$= -.984 \sqrt{\frac{11 - 2}{1 - .984^2}}$$

$$= -.984 \sqrt{\frac{9}{.03}}$$

$$= -17.05$$

$$n - 2 = 9$$

From Table A-3, for $\alpha = 0.0005$ and $\nu = 9$, $t = 4.781$. Since the absolute value of t from the test was 17.05, which is much larger than 4.781, a confidence level of 99.95 percent can be attached to this value of the rank correlation coefficient.

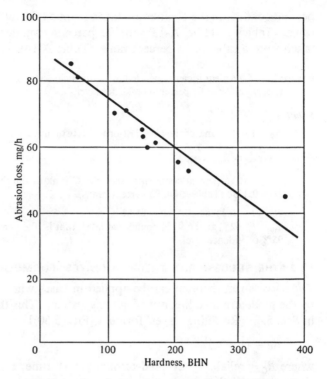

Fig. 8.1 Correlation of abrasion loss and hardness.

8.5 PREDICTION OF PERCENT FAILURES BY SUCCESS TRIALS PROCEDURES

Success trials procedures are particularly well suited to the analysis of test data where the characteristics of the underlying failure distribution are not known, and when the data are based on the successful or unsuccessful (go or no-go type) completion of a specific test program for a specific test objective. If the test runs successfully to the test objective, further testing is discontinued; hence, the time-to-failure data are not available.   Two methods are discussed here.

8.5.1 THE C-RANK THEOREM [3]

The C-rank theorem is used when k failures occur out of n items tested to the predetermined test time (test objective). It predicts the reliability R at this test time (test objective) at a given confidence C, where $R = 100 -$ percent failure.

$$R_C = 1 - [C \text{ rank of the } (k+1)\text{st-order statistic in } (n+1)] \qquad (8.5)$$

The C rank is taken at the desired confidence level.   For example, at 50 percent confidence the C ranks are taken from a table of 50 percent or median ranks (Tables A-11 to A-13); for 95 percent confidence, the C ranks are taken from a table of 95 percent ranks (Table A-14).

Example 8.7 An equipment operated in 15 identical tests with two failures.   Determine the reliability at 95 percent confidence level.

Solution

$R_C = 1 - [C$ rank of the $(k + 1)$st-order statistic in $(n + 1)]$
$k + 1 = 2 + 1 = 3$
$n + 1 = 15 + 1 = 16$

At 95 percent confidence level, the C rank of the three items in a sample of 16 is 0.3438 (Table A-14, 95 percent ranks).

$R_C = 1 - 0.3438 = 0.6562 = 65.62\%$
$R_C = 65.62\%$ at the 95% confidence level, that is, the percent failure is 34.38% at 95% confidence level.

8.5.2 THE SUCCESS-RUN THEOREM (BAYES' FORMULA) [3]

The success-run theorem can be applied in cases where no failures occurred by the predetermined life, out of n items tested.   This theorem was discussed in Sec. 5.4.   Restating Bayes' formula [Eq. (5.30)],

$$R_C = (1 - C)^{1/(n+1)} \qquad\qquad (8.6)$$

where R_C = reliability at predetermined test time x for given confidence
level C
C = confidence level
n = sample size

The reliability and confidence levels determined by using this method are for the time x.   This relationship can be useful in establishing the trade-off among reliability, confidence, and sample size.

For a given confidence level, the reliability increases with increased sample size; and the confidence increases with sample size at any given reliability.   If sample size is fixed, an increase in confidence level causes a decrease in reliability.

Example 8.8 Twenty identical items were tested without failure to a test time objective x.   Determine the reliability R_C of these items in service for time x at 90 percent confidence level.

Solution

$R_C = (1 - C)^{1/(n+1)}$
$= (1 - 0.9)^{1/(20+1)}$
$= (0.1)^{1/21}$
$= 89.7\%$

That is, the percent failure is 10.3% at 90% confidence level.

8.6 POPULATION PREDICTIONS BASED ON THE RANGE OF THE SAMPLE [4–6]

The range is the difference between the largest and smallest values in a group of observations. It can be used for a variety of nonparametric experiments involving the determination, comparison, or prediction of variances and mean values. The tests based on the range are not completely nonparametric, that is, distribution free; their main advantage is that they are quick and relatively simple, and they can be used when the underlying distribution is not known.

8.6.1 PREDICTION OF THE POPULATION STANDARD DEVIATION

Factors for predicting the variance or standard deviation from a given range and sample size are available in quality-control manuals. These factors are generally denoted as $1/d_2$ (see Table 8.3) and are developed from quality-control work where the standard deviation of the population is known and it is desired to know the expected range for a given sample size. In the latter case the conversion constant is generally denoted as d_2. The estimate of σ based on the sample range is

$$\sigma = R \frac{1}{d_2} \qquad (8\ 7)$$

where σ = estimate of standard deviation of population

R = range of sample

$1/d_2$ = conversion constant, given for various values of sample size n in Table 8.3

Table 8.3 Factors for estimating the standard deviation from the range [7]*

n	$1/d_2$	n	$1/d_2$
2	0.886	9	0.337
3	0.591	10	0.325
4	0.486	12	0.307
5	0.430	14	0.294
6	0.395	16	0.283
7	0.370	18	0.275
8	0.351	20	0.268

* $\sigma = R \dfrac{1}{d_2}$

The procedure for estimating the standard deviation σ follows:

1. Calculate the range R of the sample by subtracting the smallest value of the sample from the largest.
2. Find from Table 8.3, for the sample size n, the corresponding value of the conversion constant $1/d_2$.
3. Use these values in Eq. (8.7) to estimate the standard deviation

$$\sigma = R \frac{1}{d_2}$$

Example 8.9   Consider the following data from a test of hardness (BHN) of steel.   The underlying distribution of the hardness is not known.   Estimate the population standard deviation σ.

81.8, 83.3, 84.3, 84.6, 86.1, 86.1, 87.8, 88.2, 91.0, 91.0

Solution

1. $R = 91.0 - 81.8 = 9.2$ BHN
2. From Table 8.3, for $n = 10$, $1/d_2 = 0.325$,

$$\sigma = \frac{1}{d_2} R = (0.325)(9.2)$$

$$= 2.99 \text{ BHN}$$

8.6.2   COMPARISON OF THE SAMPLE MEAN WITH THE POPULATION MEAN

For a normally distributed population, the sample mean $\bar{x}$ can be compared with the population mean μ through a t test (see Chap. 3) where

$$t = \frac{\bar{x} - \mu}{s} \sqrt{n} \tag{8.8}$$

Using Eq. (8.7) as an estimate of s in Eq. (8.8), Lord [4] proposed a new statistic u, which is related to t:

$$u = \frac{|\bar{x} - \mu|}{R} \tag{8.9}$$

The sample mean may then be compared with a standard (population) using a u test in place of a t test.   The procedure follows:

1. Calculate the range R and sample mean $\bar{x}$ from the experimental data.
2. Use the value of the true mean μ (the standard) in Eq. (8.9).

$$u = \frac{|\bar{x} - \mu|}{R}$$

If the value of u calculated from Eq. (8.9) is greater than the values of u from Table 8.4 for the given sample size, the hypothesis that the mean estimated from the sample is equal to the true mean (the standard) may be rejected with the indicated confidence level.

Example 8.10   Consider the following test data from Example 8.9:

81.8, 83.3, 84.3, 84.6, 86.1, 86.1, 87.8, 88.2, 91.0, 91.0

Does the mean of this sample meet the required standard of 84 BHN?

Solution

$$R = 9.2 \text{ BHN} \qquad \bar{x} = 86.4 \text{ BHN} \qquad \mu = 84 \text{ BHN}$$

$$u = \frac{|\bar{x} - \mu|}{R}$$

$$= \frac{|86.4 - 84|}{9.2}$$

$$= \frac{2.4}{9.2}$$

$$= 0.261$$

Table 8.4   Critical values of u for various sample sizes n and probability levels α [7]*

| n | Probability levels α | | | |
	0.10	0.05	0.02	0.01
2	3.157	6.351	15.910	31.828
3	0.885	1.340	2.111	3.008
4	0.529	0.717	1.023	1.316
5	0.388	0.507	0.685	0.843
6	0.312	0.399	0.523	0.628
7	0.263	0.333	0.429	0.507
8	0.230	0.288	0.366	0.429
9	0.205	0.255	0.322	0.374
10	0.186	0.230	0.288	0.333
11	0.170	0.210	0.262	0.302
12	0.158	0.194	0.241	0.277
13	0.147	0.181	0.224	0.256
14	0.138	0.170	0.209	0.239
15	0.131	0.160	0.197	0.224
16	0.124	0.151	0.186	0.212
17	0.118	0.144	0.177	0.201
18	0.113	0.137	0.168	0.191
19	0.108	0.131	0.161	0.182
20	0.104	0.126	0.154	0.175

$$* \; u = \frac{|\bar{x} - \mu|}{R}$$

From Table 8.4, for $n = 10$,

$\alpha =$	0.10	0.05	0.02	0.01
$u =$	0.186	0.230	0.288	0.333

    With 95 percent confidence it can be stated that the mean of this population is significantly different from the required standard. Therefore, the population mean is at least equal to the required standard. That is, the average hardness is at least 84 at 95 percent confidence level.

8.6.3 COMPARISON OF TWO POPULATION MEANS

The range may also be used to establish a difference between two population means in the same way that the t_2 test was used in Chap. 4, where t_2 was given by Eq. (4.3). In this case, the formula is

$$u_R = \frac{|\bar{x} - \bar{y}|}{\bar{R}} \tag{8.10}$$

where $\bar{R} =$ mean range of two samples
    $\bar{x}, \bar{y} =$ two sample means

Table 8.5 Critical values of $u_{\bar{R}}$ for various sample sizes n and probability levels α [7]*

n	\multicolumn{4}{c}{Probability levels α}
	0.10	0.05	0.02	0.01
2	2.322	3.427	5.553	7.916
3	0.974	1.272	1.715	2.093
4	0.644	0.813	1.047	1.237
5	0.493	0.613	0.772	0.896
6	0.405	0.499	0.621	0.714
7	0.347	0.426	0.525	0.600
8	0.306	0.373	0.459	0.521
9	0.275	0.334	0.409	0.464
10	0.250	0.304	0.371	0.419
11	0.233	0.280	0.340	0.384
12	0.214	0.260	0.315	0.355
13	0.201	0.243	0.294	0.331
14	0.189	0.228	0.276	0.311
15	0.179	0.216	0.261	0.293
16	0.170	0.205	0.247	0.278
17	0.162	0.195	0.236	0.264
18	0.155	0.187	0.225	0.252
19	0.149	0.179	0.216	0.242
20	0.143	0.172	0.207	0.232

* $u_{\bar{R}} = \dfrac{|\bar{x} - \bar{y}|}{\bar{R}}$

The null hypothesis is that both samples come from a population with the same mean. Values of u_R are given in Table 8.5. Note that this test requires equal sample sizes $n_x = n_y$.

The procedure follows:

1. Find $\bar{R}$, the mean range of both samples, and $\bar{x}$ and $\bar{y}$, the means of both samples.
2. Calculate

$$u_R = \frac{|\bar{x} - \bar{y}|}{\bar{R}}$$

3. If the calculated value of u_R is greater than or equal to the value from Table 8.5 for the given sample size n, then the null hypothesis that both samples come from a population with the same mean can be rejected at the indicated confidence level.

Example 8.11 Consider the following test data where five repeated flow-rate measurements were made on each of the two hydraulic-pump designs, under identical test conditions. Determine whether the mean flow rates of two designs x and y differ.

$n_x = n_y = 5$
$\bar{x} = 20$ gal/min $\bar{y} = 18.2$ gal/min

$R_x = 2.0$ gal/min $R_y = 1.2$ gal/min $\bar{R} = \dfrac{2.0 + 1.2}{2} = 1.6$ gal/min

Solution

$$u_{\bar{R}} = \frac{|\bar{x} - \bar{y}|}{\bar{R}}$$

$$= \frac{20 - 18.2}{1.6}$$

$$= 1.125$$

From Table 8.5, for $n = 5$,

$\alpha =$	0.10	0.05	0.02	0.01
$u_{\bar{R}} =$	0.493	0.613	0.772	0.896

The calculated value of 1.125 exceeds the tabulated value at $\alpha = 0.01$; therefore, the null hypothesis of equal mean is rejected at the 99 percent confidence level. Hence, the mean flow rates of the two pump designs are significantly different at the 99 percent confidence level.

The statistical techniques presented in this chapter are summarized in Table 8.6.

Table 8.6   Summary of statistical techniques in Chapter 8

		Section number	
Category of experiment	Objective	Nonparametric (underlying distribution unknown)	Parametric (underlying distribution known)
Evaluation of a product	Predict population standard deviation from sample standard deviation	8.6.1	3.1.1
	Predict population mean from sample mean	8.6.2	3.1.2
Comparison between products	Comparison of two population means from the sample range	8.6.3	4.1, 4.2
	Comparison of two populations, the actual difference not sought	8.3.1, 8.3.3	4.1
	Comparison of two populations, actual difference sought	8.3.2	4.2
Prediction of percent failures in a population	When some failures occurred in a sample by a predetermined time	8.5.1	10.3, 10.4
	When no failures occurred in a sample by a predetermined time	8.5.2	10.3, 10.4
Test for randomness	To determine whether the test sample was chosen at random	8.3.4	2.8

SUMMARY

Nonparametric experiments can be used when the underlying distribution of the test data is not known or cannot be established. This is frequently the case with some experiments, and herein lies its main application. The data derived from these tests are also easier to calculate than by the previously described methods (parametric), but the confidence levels are lower. In some situations this limitation is not a serious drawback. The field of application of nonparametric experiments is shown in Table 8.6.

PROBLEMS

8.1. Paper capacitors are required to have 85 percent reliability at 95 percent confidence level under an operating temperature of 175°F. Twenty capacitors are available for test. How many failures can be tolerated to meet the above objective?

8.2. In an experiment to determine the variability in the restriction characteristics of a nozzle, the following readings were obtained from a sample of 15 nozzles. The readings were taken in the order indicated: 52, 62, 61, 65, 49, 60, 59, 62, 61, 59, 54, 58, 63, 57, 65. With 90 percent confidence, is this a random sample?

8.3. Panels fabricated from two materials x and y were tested for resistance to corrosion. The degree of corrosion was recorded in terms of the increasing order of magnitude as normal, excessive, and damaging.

Degree of panel corrosion

Panel number	Material x	Material y
1	Normal	Excessive
2	Excessive	Excessive
3	Normal	Damaging
4	Damaging	Damaging
5	Excessive	Normal
6	Excessive	Damaging
7	Normal	Excessive
8	Normal	Normal
9	Excessive	Damaging
10	Normal	Excessive
11	Damaging	Normal
12	Normal	Excessive
13	Normal	Excessive
14	Excessive	Damaging
15	Normal	Excessive

Determine with 95 percent confidence whether the corrosion characteristics of the materials x and y are different.

8.4. Pick 20 or more random numbers from Table A-47. Run the test of randomness and check whether these numbers are truly random.

8.5. Check whether a pack of cards was thoroughly shuffled by counting the number of runs of black and red cards.

8.6. By means of the sign test, determine with 90 percent confidence whether x values are significantly different from y values.

x	80	72	60	75	82	85	79	50	91	65	87	70	63	68	92	55
y	81	74	58	70	81	84	74	48	90	63	82	69	61	70	84	58

8.7. Estimate the standard deviations σ, from the ranges, of the following data:

Data set number	Sample size in each set, n	Range of the measurements observed	Standard deviation σ
1	5	23	
2	7	26	
3	8	29	
4	4	21	
5	7	27	

8.8. How many items should be tested to objective x without failure to conclude with 95 percent confidence that the product reliability is 90 percent?

REFERENCES

1. Ratkowsky, D. A.: Applications of Nonparametric Statistics, *Brit. Chem. Eng.*, vol. 9, 1964.
2. Tate, M. W., and R. C. Clelland: "Nonparametric and Shortcut Statistics," The Interstate Printers & Publishers, Inc., Danville, Ill., 1957.
3. Johnson, L. G.: *GMR Rel. Manual* GMR-302, General Motors Research Laboratories, August, 1960.
4. Lord, E.: *Biometrika*, vol. 34, p. 41, 1947.
5. Turkey, J. W.: *Trans. N. Y. Acad. Sci.*, vol. 16, p. 8, 1953.
6. Jackson, J. E., and E. L. Ross: *J. Amer. Statist. Ass.*, vol. 50, p. 417, 1955.
7. Volk, W.: "Applied Statistics for Engineers," McGraw-Hill Book Company, New York, 1958.
8. Bennett, C., and N. Franklin: "Statistical Analysis in Chemistry and the Chemical Industry," John Wiley & Sons, Inc., New York, 1954.

BIBLIOGRAPHY

Argentiero, P. D., and R. H. Tolson: Some Nonparametric Tests for Randomness in Sequences, *NASA Tech. Note* TND-3766, 1966.
Bradley, J. V.: "Distribution Free Statistical Tests," Prentice-Hall, Inc., Englewood Cliffs, N.J., 1968.
Dixon, W. J.: Power Functions of the Sign Test and Power Efficiency for Normal Alternatives, *Ann. Math. Statist.*, vol. 24, 1953.
———— and F J. Massey, Jr.: "Introduction to Statistical Analysis," 3d ed., McGraw-Hill Book Company, New York, 1969.
Fisz, M.: "Probability Theory and Mathematical Statistics," 3d ed., John Wiley & Sons, Inc., New York, 1963.
Fraser, D. A. S.: Most Powerful Rank Type Tests, *Ann. Math. Statist.*, vol. 28, 1957.
————: "Nonparametric Methods in Statistics," John Wiley & Sons, Inc., New York, 1957.
Johnson, L. G.: *GMR Rel. Manual* GMR-302, General Motors Research Laboratories, August, 1960.
————: "Theory and Technique of Variation Research," Elsevier Publishing Company, Amsterdam, 1964.
Johnson, N. L., and F. C. Leone: "Statistics and Experimental Design," vol. I, John Wiley & Sons, Inc., New York, 1964.
Lehmann, E. L., and C. Stein: On the Theory of Some Nonparametric Hypothesis, *Ann. Math. Statist.*, vol. 20, 1949.
MacStewart, W.: A Note on the Power of the Sign Test, *Ann. Math. Statist.*, vol. 21, 1941.
Massey, F. J., Jr.: A Note on the Power of a Nonparametric Test, *Ann. Math. Statist.*, vol. 21, 1950.
Mosteller, F.: On Some Useful "Inefficient" Statistics, *Ann. Math. Statist.*, vol. 17, 1946.
Nair, K. R.: The Median in Tests by Randomization, *Sankhya*, vol. 4, 1940.
Noether, G. E., "Elements of Nonparametric Statistics," John Wiley & Sons, Inc., New York, 1967.

Ratkowsky, D. A.: Applications of Nonparametric Statistics, *Brit. Chem. Eng.*, vol. 9, 1964.

Siegel, S., "Nonparametric Statistics for the Behavioral Sciences," McGraw-Hill, New York, 1956.

Tate, M. W., and R. C. Clellan: "Nonparametric and Shortcut Statistics," The Interstate Printers & Publishers, Inc., Danville, Ill., 1957.

Walsh, J. E.: "Handbook of Nonparametric Statistics," D. Van Nostrand Company, Inc., Princeton, N.J., 1962.

9
Fatigue Experiments

In general, the term "fatigue" implies the application of an external dynamic load on a part. Under this load the part undergoes progressive localized permanent change which may result, after a number of load applications, in cracks or a complete fracture.

Some engineering situations involve the application of a constant dynamic load (Fig. 9.1). In those cases the average life and the average strength can be determined. In the past, design of fatigue experiments has been largely restricted to the determination of these quantities. Because of the inherent variability in the material and manufacturing processes the resultant scatter in the fatigue strengths and the fatigue lives is beginning to be recognized. Since scatter can be treated with the aid of statistics, modern concepts in fatigue involve recognition of the probability-stress-life relationship. This relationship is taken up in the first part of this chapter.

The second section of the chapter discusses the relationship between the life and strength of laboratory specimens and actual parts. Most of the fundamental work in fatigue has been done on test specimens, but the interest of the engineer lies in the fatigue characteristics of actual manufactured parts.

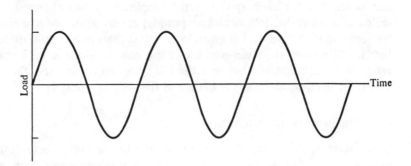

Fig. 9.1 Constant amplitude of loads.

Means of conversion of the fatigue strength of laboratory specimens to fatigue strength of manufactured parts are discussed in this section.

All the above applies to the situations where the fatigue load is of a constant amplitude. There are many situations where parts are subjected to a complicated pattern of randomly varying stress amplitudes and frequencies (Fig. 9.2). This includes the loads produced in automobile suspension components due to random irregularity of road surfaces, the load in aircraft due to atmospheric gusts, etc. These phenomena can best be evaluated in terms of statistical parameters, and the application of these parameters to the prediction of life under random loads is discussed in the last section of the chapter.

9.1 CONSTANT-LOAD EXPERIMENTS

In fatigue experiments the word "constant" is used to describe the situation where the amplitude of dynamic load does not vary with time (Fig. 9.1). Different load amplitudes may be imposed on different specimens in a given

Fig. 9.2 Random-load spectrum.

test series, but for each specimen the amplitude is not changed.   This is called a constant-load (constant-amplitude) experiment.   Mechanical parts subjected to such loads fail at much lower stress than they would under static loads.   Therefore, the main purpose of the constant-load experiments is to estimate the relationship between the life cycles to failure and the imposed stress.   The stress at failure is known as the fatigue strength.

9.1.1 PREDICTION OF AVERAGE LIFE

When seemingly identical parts are subjected to the same fluctuating stress, they generally fail at different lives due to nonhomogeneity in material properties, variation in surface conditions, etc.   Since the best estimate of a sample is generally the average value, the average life is commonly used to describe the fatigue characteristics of specimens and parts.

Conventionally, to predict the average life, a number of test specimens or parts are tested at various stress levels until failure.   The results are then plotted on a log-log paper with the stresses on the ordinate and the corresponding lives on the abscissa, as shown in Fig. 9.3.   A line representing an estimate of average life for a given stress is fitted through the test points. The horizontal line indicates that an infinite life is obtained whenever the part is stressed below the value corresponding to this line.   However, in

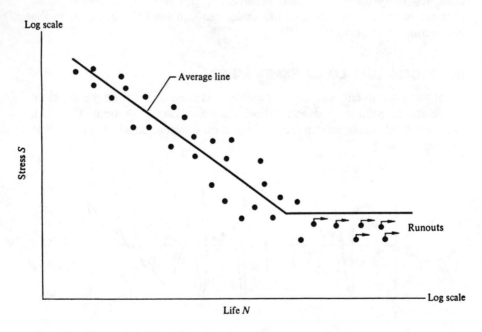

Fig. 9.3 Conventional S-N diagram.

most engineering situations parts are designed for finite life, and the sloping portion of Fig. 9.3 is generally the criterion of design.

The line in Fig. 9.3 drawn through the test points does give a good estimate of the average life of the population from which the test sample was taken, provided that the number of test specimens at each stress level is large. Because the test is generally conducted at several stress levels, the total number of specimens needed in such an experiment is high. This shortcoming can be overcome by restricting the test to three stress levels, which in turn allows a larger number of test specimens at each level. This method is described below, and it involves establishing the relationship between the specimen allocation, the percent error that can be tolerated, the confidence level, and the expected variation in life. It is assumed here that the life at a given stress is a random variable and that it follows the log normal distribution (Fig. 9.4). When the life is expressed in logarithms, the underlying distribution reduces to a normal distribution (discussed in Chap. 2). All the tools and principles applicable to normal distribution would then apply here. Particularly, the t distribution can then be used to estimate the true (population) mean life. When the t distribution equation (3.15) is applied

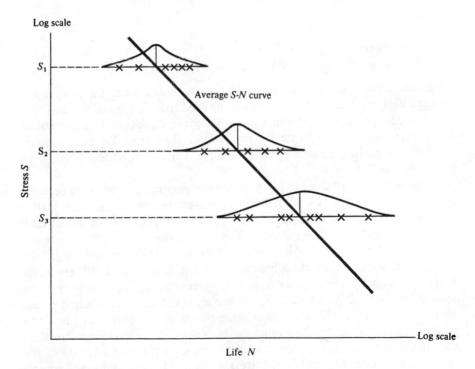

Fig. 9.4 Scatter in fatigue life at a given stress.

to a log normally distributed variable x_l, the following result is obtained:

$$\frac{\bar{x}_l - \mu_l}{s_l/\sqrt{n}} = \pm t_{\alpha/2;\,\nu} \tag{9.1}$$

where $\bar{x}_l$ = sample log average of life, equal to $\sum\limits_{i=1}^{n} \dfrac{(x_l)_i}{n}$

μ_l = population log average of life

s_l = sample log standard deviation of life, equal to $\left[\sum\limits_{i=1}^{n} \dfrac{(x_l - \bar{x}_l)^2}{n-1}\right]^{1/2}$

n = sample size

ν = degrees of freedom $(n - 1)$

$t_{\alpha/2;\,\nu}$ = value of t statistics given in Table A-3

$\alpha/2$ = degree of confidence, where confidence is equal to $(1 - \alpha)$

Rearranging the above equation,

$$\frac{\bar{x}_l - \mu_l}{\bar{x}_l} 100 = \pm \left(\frac{s_l}{\bar{x}_l} \frac{t_{\alpha/2;\,\nu}}{\sqrt{n}}\right) 100 \tag{9.2}$$

or

$$\text{Percent error} = \pm \left(\frac{s_l}{\bar{x}_l} \frac{t_{\alpha/2;\,\nu}}{\sqrt{n}}\right) \tag{9.3}$$

where the percent error is defined as the error made when the sample log average $\bar{x}_l$ is accepted to be the population (the true) log mean μ_l, and $s_l/\bar{x}_l$ is the coefficient of variation or the log standard deviation as a percent of log average life. The percentage error as determined from Eq. (9.3), divided by $s_l/\bar{x}_l$, is plotted against sample size for various confidence levels, as shown in Fig. 9.5.

Generally, the scatter in life at a given stress increases with the decreasing stress level (Figs. 9.4 and 9.6). However, in some cases the scatter remains constant. In the first case the number of specimens to test at each stress level would be different depending on the degree of scatter, whereas in the second case the sample size would be the same. In both cases, the average life is computed at each stress level and it is plotted. Three stress levels, if properly chosen, are generally adequate. These levels may be selected at $0.8S_u$, $0.7S_u$, and $0.6S_u$, where S_u is the ultimate strength of the material under test. This results in three points on the plot. The best line is then fitted through these points by means of the least-squares method (discussed in Chap. 13). This line then represents the best estimate of the true average finite life for different stress levels and for preestablished acceptable error with a given confidence.

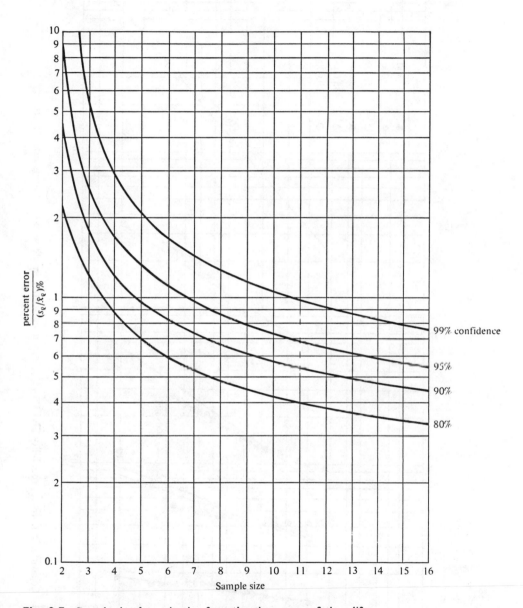

Fig. 9.5 Sample-size determination for estimating average fatigue life.

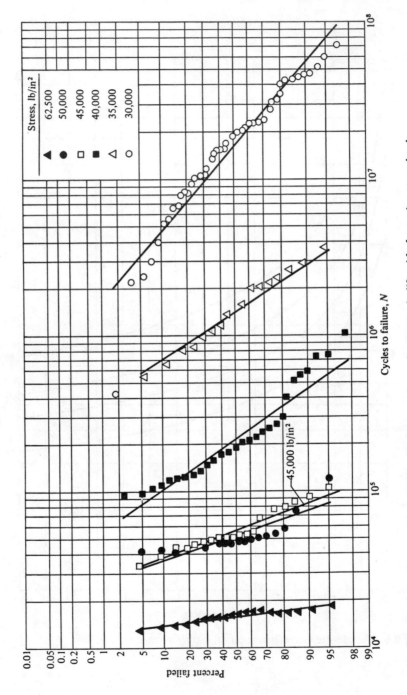

Fig. 9.6 Logarithmic normal probability plot showing increasing scatter in life with decreasing stress levels.

Example 9.1 Design an experiment to determine the relationship between the stress and the average life for a material of 40 ksi ultimate tensile strength. The acceptable error in estimating the true log average life μ_l should be within 5 percent at 90 percent confidence level. (This means that at least 90 percent of the time the error should be less than or equal to 5 percent.) From past experience the expected coefficient of variation $s_l/\bar{x}_l$ is 4 percent at $0.8S_u$ stress level, 5 percent at $0.7S_u$, and 7 percent at $0.6S_u$. (These expected values can be verified after the test is run. In the event the variation is found to be more than the expected at any one or more stress levels, the additional specimens should be run in order to maintain the same percent error and the confidence at all stress levels. The number of necessary additional specimens can be computed by means of Fig. 9.5.)

Solution

$$\text{Percent error} = \left(\frac{\bar{x}_l - \mu_l}{\bar{x}_l}\right) = 5\%$$

Confidence $= 90\%$

The required sample size can be determined for 90 percent confidence from Fig. 9.5. This is shown in the following table:

Stress, ksi	Percent coefficient of variation $s_l/\bar{x}_l$	Percent error/ percent coefficient of variation	Required sample size at a given stress level
32 ($0.8S_u$)	4	1.25	4
28 ($0.7S_u$)	5	1.00	5
24 ($0.6S_u$)	7	0.71	8

Total $= 17$ specimens

The test is then run at the three stress levels, with the proper sample size at each level. Life cycles to failure were determined as follows:

Sample size n	32 ksi stress Life, cycles	Log of life	28 ksi stress Life, cycles	Log of life	24 ksi stress Life, cycles	Log of life
1	2.34×10^3	3.37	1.41×10^4	4.15	1.26×10^5	5.10
2	3.54×10^3	3.55	3.16×10^4	4.50	5.63×10^4	4.75
3	5.00×10^3	3.70	1.17×10^4	4.07	1.78×10^5	5.25
4	3.72×10^3	3.57	1.59×10^4	4.20	3.15×10^5	5.50
5			3.55×10^4	4.55	2.82×10^4	4.45
6					3.16×10^5	5.50
7					7.95×10^4	4.90
8					3.15×10^4	4.50
	$\bar{x}_l = 3.55$		$\bar{x}_l = 4.29$		$\bar{x}_l = 4.99$	
	$s_l = 0.136$		$s_l = 0.216$		$s_l = 0.414$	
	$s_l/\bar{x}_l = 3.9\%$		$s_l/\bar{x}_l = 5.04\%$		$s_l/\bar{x}_l = 8.3\%$	

$$\text{where } x_l = \frac{\sum_{i=1}^{n} \log x_i}{n} \quad \text{and} \quad s_l = \sqrt{\frac{\sum_{i=1}^{n}(\log x_i - \bar{x}_l)^2}{n-1}}$$

Thus, the variation in life at the stress level of 24 ksi was found to be 8.3 percent, which is more than the expected 7 percent. Additional specimens, therefore, should be tested at this level to maintain the same percent error in the life estimate for the same degree of confidence. This is done as follows: The acceptable percent error as stated in the problem is 5 percent, and the coefficient of variation as found for 24 ksi stress is 8.3 percent. The ratio, therefore, is

$$\frac{\% \text{ Error}}{\%(s_l/\bar{x}_l)} = \frac{5}{8.3} = 0.602$$

From Fig. 9.5, with the ordinate of 0.602 at 90 percent confidence, read off 10 as the total sample size required. Since eight specimens were already tested, only two additional specimens should be run. The new average based on the log average $\bar{x}_l$ is 4.95. The revised information is then:

Stress, ksi	Log average life	Average life, cycles
32	3.55	3.55×10^3
28	4.29	1.95×10^4
24	4.95	8.90×10^4

These data are then plotted on a log-log paper as in Fig. 9.7, and a best line is drawn through these points by means of the least-squares method. This line, then, truly represents the average life estimate within 5 percent error at 90 percent confidence.

9.1.2 PREDICTION OF AVERAGE STRENGTH

Several methods have been developed for estimating the average fatigue strength (fatigue limit). The principal ones are the staircase method and increasing amplitude by the Prot method.

The staircase method [2]   The staircase method is primarily used to determine the mean fatigue strength at a given life. The mean value of fatigue strength is estimated for a desired life, and the first test is run at this stress level. If the specimen fails prior to this life, the next specimen is tested at a lower stress level. If the specimen does not fail within this life, the new test is run at a higher level. Thus, each test is dependent on the results from the previous test, and the test continues with stress levels raised or lowered. Since only one specimen can be tested at a time, this kind of test is time consuming.

Selection of proper increments of stress level is important. Generally, this increment d (interval) should be less than about 5 percent of the initial estimate of the mean fatigue strength. The tests should be designed to run at three stress levels, so chosen that about 50 percent of the test specimens survive at the middle stress level and about 30 percent survive at the higher stress [2]. Previous data for the same or similar materials are useful in

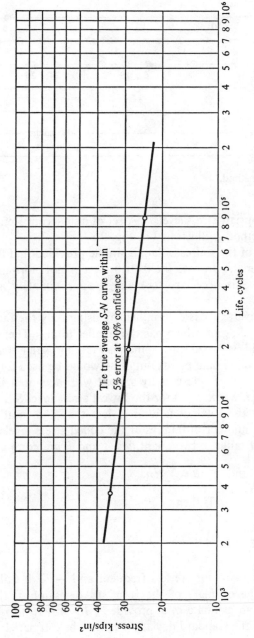

Fig. 9.7 An S-N curve representing the average life within 5 percent error at 90 percent confidence for Example 9.1.

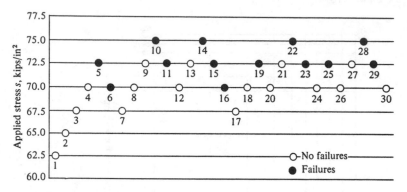

Fig. 9.8   Staircase method.

selecting the appropriate stress levels.   A set of data (from Ref. 2) pertaining to the staircase method is illustrated in Fig. 9.8.

The analysis of the staircase data is simple provided that the underlying distribution is normal.   In case the distribution is not normal, the proper transformation of S is necessary to make it approximately a normal distribution.

The sample average $\bar{x}$ is determined by using only the failures or only the survivals, depending on which has the smaller total.   Therefore, at least 30 specimens should be tested since some may fail and some may not.   As a result, a maximum of 15 data points out of 30 would be used for the analysis. The stress levels S, which are equally spaced with a chosen interval d are given coded scores i, where $i = 0$ for the lowest stress level S_0.

Suppose that the total of failures is less than the total of survivals. Denoting by n_i the number of failures at the coded stress level i and by $\sum n$ the total number of failures, two quantities A and B are computed:

$$A = \sum i n_i \quad \text{and} \quad B = \sum i^2 n_i \tag{9.4}$$

The estimate of the mean is then

$$\bar{x} = S_0 + d\left(\frac{A}{\sum n} \pm \frac{1}{2}\right) \tag{9.5}$$

where $+1/2$ is used if runouts are less frequent and $-1/2$ if failures are less frequent.   As to the estimate of the standard deviation s, the following equation produces an estimate of s, provided $(B \sum n - A^2)/(\sum n)^2$ is larger than 0.3; otherwise the standard deviation cannot be estimated:

$$s = 1.62d\left[\frac{B \sum n - A^2}{(\sum n)^2} + 0.029\right] \tag{9.6}$$

Example 9.2 By means of the staircase method, determine the true mean μ of the fatigue strength at 10^7 cycles for the data given in Fig. 9.8.

Solution For

$i = 0$:	$S_0 = 62.5$ ksi	$n_0 = 0$
$i = 1$:	$S_1 = 65.0$	$n_1 = 0$
$i = 2$:	$S_2 = 67.5$	$n_2 = 0$
$i = 3$:	$S_3 = 70.0$	$n_3 = 2$
$i = 4$:	$S_4 = 72.5$	$n_4 = 7$
$i = 5$:	$S_5 = 75.0$	$n_5 = 4$

$$\sum n = 13$$

d = equal interval of the stress = 2.5 ksi < 5% of 62.5 ksi
From Eq. (9.4):

$$A = \sum_{i=0}^{5} i n_i = 0 + 0 + 0 + (3 \times 2) + (4 \times 7) + (5 \times 4)$$

$$= 54$$

$$B = \sum_{i=0}^{5} i^2 n_i = 0 + 0 + 0 + (9 \times 2) + (16 \times 7) + (25 \times 4)$$

$$= 230$$

From Eq. (9.5):

$$\bar{x} = S_0 + d\left(\frac{A}{\sum n} - \frac{1}{2}\right) \qquad \text{as the failures are less frequent}$$

$$= 62.5 + 2.5\left(\frac{54}{13} - \frac{1}{2}\right)$$

$$= 62.5 + 2.5(4.15 - 0.5)$$

$$= 62.5 + 9.1$$

$$= 71.6 \text{ ksi}$$

Since

$$\frac{B \sum n - A^2}{(\sum n)^2} = 0.438 > 0.3$$

s can be estimated quite accurately as follows:
From Eq. (9.6):

$$s = 1.62d\left[\frac{B \sum n - A^2}{(\sum n)^2} + 0.029\right]$$

$$= 1.62 \times 2.5\left(\frac{230 \times 13 - 54^2}{13^2} + 0.029\right)$$

$$= 1.62 \times 2.5\left(\frac{2{,}990 - 2{,}916}{169} + 0.029\right)$$

$$= 1.62 \times 2.5(0.438 + 0.029)$$

$$\simeq 1.9 \text{ ksi}$$

Hence, the sample average $\bar{x}$ and the sample standard deviation s of the fatigue strength at 10^7 cycles are 71.6 ksi and 1.9 ksi, respectively.

In order to determine the true mean μ, the confidence limits on $\bar{x}$ should be established.   This is done as follows.

The standard deviation of a sample mean $\bar{x}$ is generally computed from

$$s_{\bar{x}} = \frac{s}{\sqrt{n}}$$

In the present case, this must be modified by a factor G [3].   The new expression then becomes

$$s_{\bar{x}} = \frac{Gs}{\sqrt{n}} \tag{9.7}$$

where the value of G depends on the ratio d/s, and n is the total number of failures. The value of G is found from Fig. 9.9.   In this example, the value of $d/s = 2.5/1.9$ $= 1.3$.   Therefore $G \simeq 1.1$ from Fig. 9.9.   The confidence interval for the true mean μ may now be determined by

$$\bar{x} - \frac{Gs}{\sqrt{n}} t_{\alpha/2;\,v} \leq \mu \leq \bar{x} + \frac{Gs}{\sqrt{n}} t_{\alpha/2;\,v}$$

where $t_{\alpha/2;\,v} = 2.18$ for 95 percent confidence and $v = n - 1 = 12$.   Therefore

$$71.6 - \frac{1.1 \times 1.9}{\sqrt{12}} \times 2.18 \leq \mu \leq 71.6 + \frac{1.1 \times 1.9}{\sqrt{12}} \times 2.18$$

or

$$70.2 \text{ ksi} \leq \mu \leq 73.0 \text{ ksi}$$

Increasing amplitude by the Prot method   The fatigue limit may be determined by increasing the stress amplitude continuously at a uniform rate per cycle until failure occurs.   The specimen is tested at an initial stress

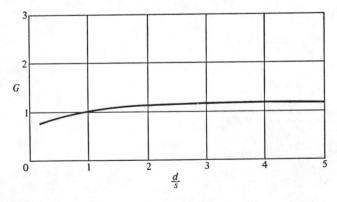

Fig. 9.9 The relation between d/s and G.

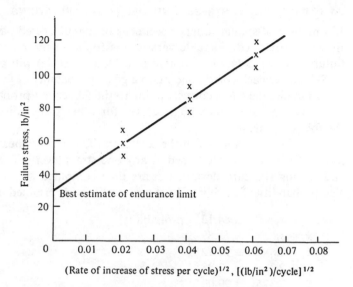

Fig. 9.10 Increasing amplitude by the Prot method.

level estimated to be 60 to 70 percent of the endurance limit, and the stress is raised at a constant rate. This procedure is repeated with a group of specimens. In this method at least three rates of loading are used to establish the relation between the failure stress and the rate of loading (Fig. 9.10). The lowest rate should be as small as possible, and the highest rate should not exceed the rate causing yielding of the specimen. This type of test requires 10, and preferably 20, tests for each rate. The data are plotted in the manner shown in Fig. 9.10.

The failure stress is linearly related to the square root of the rate of increase of stress for ferrous materials [4]:

$$S = S_n + k(d)^{1/2} \tag{9.8}$$

where S = failure stress
S_n = endurance limit
k = constant (slope of the line)
d = rate of increase of stress per cycle

The best estimate of the endurance limit is then the intercept value of this line, or it is the value of S when d is equal to zero in the above equation.

The principal advantage of this method is that the information on each specimen is used in the analysis since each is tested to failure.

This method is more promising for ferrous metals with a well-defined endurance limit than for nonferrous metals where the endurance limit is less well defined.

9.1.3 PROBABILITY, STRESS, AND LIFE (P-S-N) RELATIONS

If a number of similar fatigue specimens or manufactured parts are tested at the same stress, considerable variation is found in the number of cycles to failure. These data when plotted (on a log-log scale) will provide a family of S-N curves each corresponding to a given probability of failure (Fig. 9.11). For example, the curve corresponding to the 50 percent probability of failure is the average S-N curve. The curves for other probabilities are found in the following manner:

In those cases in which the scatter in life values is constant at all stress levels, the scatter is determined at any one stress level. The log average $\bar{x}_l$ and the log standard deviation s_l are then computed as shown in Sec. 2.2. The probability of having at least a given life is determined as

$$\bar{x}_l - s_l \quad \equiv 84.13\% \text{ probability}$$
$$\bar{x}_l + s_l \quad \equiv 15.87\%$$
$$\bar{x}_l - 2s_l \quad \equiv 97.72\%$$
$$\bar{x}_l + 2s_l \quad \equiv 2.28\%$$
$$\bar{x} - 2.325s_l \equiv 99.00\%$$
$$\bar{x} + 2.325s_l \equiv 1.00\%$$

To obtain a curve of, say, 99 percent probability, the value of antilog $(\bar{x} - 2.325s_l)$ against the corresponding stress, say S_a, is plotted. The line passing through this point and parallel to the average line is a 99 percent

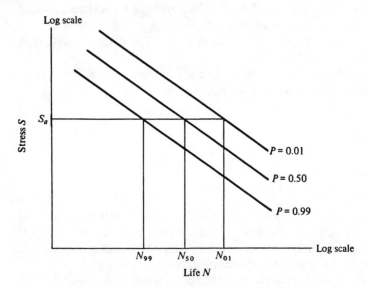

Fig. 9.11 P-S-N curves illustrating the scatter in life at a given stress: constant scatter. (Each curve represents the probability of having at least the corresponding life at any given stress level.)

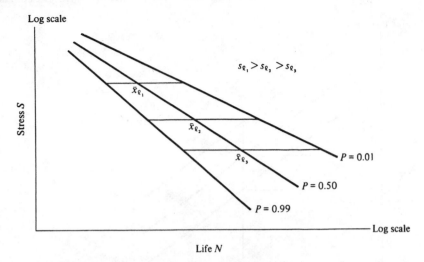

Fig. 9.12 *P-S-N* curves illustrating the scatter in life at a given stress: scatter not constant. (Each curve represents the probability of having at least the corresponding life at any given stress level.)

probability line.   In this manner a family of curves each representing a given probability can be generated (Fig. 9.11).

In those cases where the scatter is not constant at all stress levels, $\bar{x}_l$ and s_l should be computed for more than one stress.   The corresponding values of $\bar{x}_l + zs_l$ should be found at these stress levels and for a chosen value of probability.   A line through these points is then the *S-N* curve of a desired probability (Fig. 9.12).

The curves in Figs. 9.11 and 9.12 indicate that lives of the test specimens would be between the $P = 0.99$ life and $P = 0.01$ life 98 percent of the time with only 1 percent falling on either side of this range.   The scatter in life can be obtained for various values of P (the proportion of specimens having at least the life N) and for various stress levels.

In a similar manner, the scatter in the fatigue strength at any desired life can be obtained for various values of P, where P is the fraction of the total specimens that have the fatigue strengths of at least S_a (Fig. 9.13).

9.1.4 DETERMINATION OF LIFE DISTRIBUTION

Determination of the statistical distribution of life involves handling the scatter for the random variable N (fatigue life) shown in Fig. 9.11.   It is apparent that the magnitudes of the number of cycles N are cumbersome to work with, and since the distribution is more homogeneous in log N than in N, it is reasonable to express this random variable as log N.   It is a common practice to express the scatter in fatigue life by either log normal or Weibull two- or three-parameter distributions.   (The detailed discussions on log normal and Weibull distributions are given in Chap. 2.)

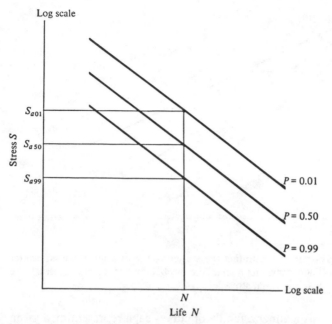

Fig. 9.13 *P-S-N* curves illustrating the scatter in fatigue strength at a given life. (Each curve represents the probability of having at least the corresponding strength at any given life.)

The method of determining the distribution of life is:

1. Arrange the values of life cycles to failure at a given stress (as may be obtained from Fig. 9.11) in an increasing order.
2. Assign the corresponding median rank to each of the corresponding values of life. (This is discussed in Example 2.4.)
3. Plot these data on several probability papers (such as normal, log normal, Weibull, exponential, and extreme value) with the median-rank values on the ordinate and the life values on the abscissa. Draw a best-fitting line through these points; determine the correlation coefficient *r* (as shown in Chap. 13), a measure of the best fit, for each distribution. The distribution with the maximum value of *r* is the best-fitting distribution. In most cases it was found that the life characteristic can be expressed well by either log normal or Weibull distribution.

9.1.5 DETERMINATION OF STRENGTH DISTRIBUTION

Most fatigue testing involves subjecting a number of specimens or parts to the same stress and repeating this process at various stress levels. The data thus obtained, known as life data, are used to construct the conventional *S-N* diagram. In this case, the scatter obtained is the scatter in life at a given stress.   In determining the distribution of fatigue strength, one is concerned with the nature of the scatter in the fatigue strength at a given life.   To obtain such data it is necessary to fatigue-test all the specimens with different stresses imposed on them in such a manner that all would fail at a predetermined life. Practically, this is impossible; therefore, two alternative methods, described below, are considered.

Graphical method   The fatigue-life data are plotted on the conventional *S-N* diagram.   Here, it is assumed that to each specimen of the population can be attributed an individual *S-N* curve, and that there exists for any population of specimens (at fixed test conditions) a family of nonintersecting *S-N* curves, which can be determined with any desired accuracy, each curve corresponding to a given probability [5].

The average *S-N* curve is then fitted to all the test points on the *S-N* diagram by using the least-squares method (see Chap. 13).   Passing through each test point, an *S-N* curve parallel to the average *S-N* curve is shown. These will make a family of *S-N* curves.   (See Fig. 9.14.)   Now if the fatigue-strength distribution at $N = N_1$ life is required, a vertical line is drawn at

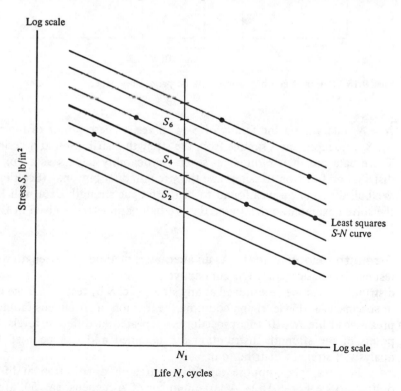

Fig. 9.14   *S-N* diagram for converting life data to strength data.

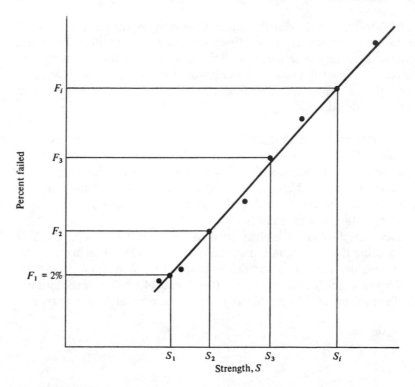

Fig. 9.15   Plot of strength response data on probability paper.

$N = N_1$ intersecting the family of S-N curves.   These points of intersection $S_1, S_2, \ldots$ represent a sample from the strength distribution at a desired life. These data are then plotted on several probability papers as a cumulative distribution function to determine the strength distribution.   (See Fig. 9.15.) Weibull distribution was found to fit the fatigue-strength data well [6], and the solved example to determine the Weibull parameters is given in Example 2.7.

Strength response test   As an alternative method, the strength response test can be considered.   The cumulative percentage point of fatigue-strength distributions can be determined at any stress level S by testing a large number of specimens at this level and counting the fraction of specimens failing at the preassigned life N.   If this procedure is repeated at different levels, several points of the strength distribution are obtained and can be used for the analysis of strength distribution.

For example, suppose the fatigue-strength distribution at life N_1 is desired.   (See Fig. 9.16.)   A large number of specimens, say 50, are tested

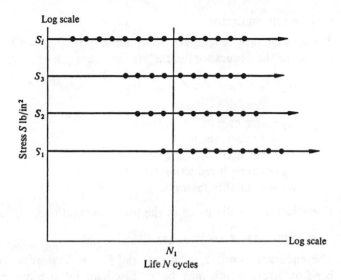

Fig. 9.16 *S-N* diagram for strength response test.

at stress level S_1; and if only one out of these 50 fails before or at the pre-assigned life N_1, then it can be said that on an average only $F_1 = 2$ percent from the lot of specimens have fatigue strength less than or equal to S_1. The same procedure can be repeated for several other stress levels S_2, S_3, ..., S_i, and corresponding percentage points (F_2, F_3, ..., F_i) can be determined. These points represent the cumulative behavior of strength, and can be plotted (as discussed before) on the several probability papers (such as Weibull, normal, logistic, and extreme value) with S as the abscissas and percent F as the ordinates (Fig. 9.15).

The percentage points of the strength distribution measured by this method are independent of each other, and accordingly the method of least squares can be applied.

As this method requires testing of a large number of specimens at any one stress level, very limited data of this type are available, although recently a method was proposed for generating such data [7].

9.2 LABORATORY SPECIMENS VS. ACTUAL PARTS

Most of the fundamental work in fatigue has been done on test specimens. In actual engineering practice principal interest lies in the fatigue characteristics of actual manufactured parts. Methods of conversion of the fatigue strength of the laboratory specimens into the corresponding strength of the fabricated parts are discussed below.

9.2.1 BASIC RELATION

The average fatigue strength of an actual part, S_n, can be approximated by correcting the endurance limit S_n' (fatigue strength of a laboratory specimen):

$$S_n = S_n' \times K_1 \times K_2 \times K_3 \times K_4 \times K_5 \times K_6 \qquad (9.9)$$

where K_1 = type of loading

K_2 = size of part

K_3 = surface finish

K_4 = surface treatment

K_5 = strength-reduction factor

K_6 = reliability factor

These factors are discussed in the following sections.

9.2.2 ENDURANCE LIMIT S_n' [8]

The endurance limit S_n' of a material is the maximum completely reversed bending stress which may be repeated an infinite number of times on a polished standard test specimen without causing failure.   Experimentally, it has been found that the closest correlation with static properties exists between endurance limit and ultimate tensile strength.   For steels having up to 200,000 psi tensile strength, the fatigue ratio of endurance limit to ultimate strength is approximately 0.50.   In equation form,

$$S_n' = 0.50 S_u \qquad (9.10)$$

In gray iron, the endurance limit is between 35 and 50 percent of the tensile strength.   For aluminum and magnesium, the endurance limit is approximately 30 to 40 percent of the tensile strength.   Relationships between tensile strength and endurance limit for several metals are graphically illustrated in Ref. 8.   In the case in which the experimentally determined values are not available, these relations may be used to estimate the endurance limit.   Since nonferrous materials do not possess a true endurance limit, it is customary to use the endurance strength for those cases at 10^8 or 5×10^8 cycles.

9.2.3 TYPE OF LOADING K_1

The three types of fluctuating loading encountered in designing members are axial, bending, and torsion.   Since the endurance limit is commonly determined from specimens loaded in bending, no correction factor is necessary for the bending load.   Hence

K_1 (reversed bending) $= 1.0$

However, it has been found that the endurance strength of polished specimens subjected to reversed axial loading is less than the endurance limit determined by using a bending load.   Although there is a great amount of scatter among the data available, a reasonable and conservative assumption appears to be that the endurance limit for reversed axial loads is 15 percent less than the value for reversed bending; thus

K_1 (reversed axial load) = 0.85

The reasons that the axial endurance limit is not as high as the bending endurance limit probably include:

1. The difficulty of applying axial loads with no eccentricity
2. The zero stress gradient, which causes the minimum cross section of material to be subjected to maximum stress

Fatigue tests in torsion on steel show a torsional endurance limit of a polished specimen to be approximately 58 percent of the endurance limit in bending, thereby indicating good agreement with the distortion energy theory.

K_1 (reversed torsion) = 0.58

An S-N curve provides the information about the fatigue strength at both finite and infinite lives, and it also conforms reasonably well to a straight-line relationship on log-log coordinates between 10^3 and 10^6 cycles. (See Fig. 9.3.) The above-mentioned factors of type of loading correspond to 10^6 cycles, and the influence of type of loading at 10^3 cycles is:

1. For reversed bending, use $0.9S_u$.
2. For reversed axial load, use $0.9S_u$.
3. For reversed torsion, use $0.9S_{us}$.

S_{us} is the ultimate strength in shear. When specific data for S_{us} are not available, it is recommended that S_{us} be taken as $0.82S_u$. For nonferrous metals the data available are not quite so complete but show an average value of 0.75, the range being from 0.66 to 0.88.

9.2.4 SIZE OF PART K_2

For bending and torsional loading, the endurance strength tends to decrease as size increases. Wire has a higher endurance limit than a standard test specimen (nominal 0.3 in diameter). For sizes larger than the standard specimen, strength decreases approximately 15 percent up to 1/2 in diameter, after which and up to about 2 in it varies little. In the case of axial loading, there is no appreciable effect of size on the strength. The reason for this is associated with the zero stress gradient of all size parts. Thus,

K_2 = 0.85 (bending and torsion) for 0.4 in < D ≤ 2.0 in
     = 1.0 (axial)

Large specimens tend to have much lower endurance strengths, but since large-scale testing is costly, there are not enough data for a firm generalization. For very large parts, the endurance strength may be reduced by 25 percent or more below the endurance limit of test specimens.

The influence of size on the 10^3 cycle point on the S-N curve is considerably less than at the 10^6 cycle point and is commonly neglected. Whenever possible the value of S_u or S_{us} should be obtained from test samples of the approximate size range involved.

Although the above-mentioned size correction factors apply specifically to steel, the best available information indicates that they may be applied to other metals too.

9.2.5 SURFACE FINISH K_3

The finish of the surface may considerably affect the fatigue strength, and for this purpose it is classified into five broad categories: polished, ground, machined, hot-rolled, and as-forged.

The reason for this classification is that, at least in the case of steel, the fatigue strength depends very much on the surface finish, being higher for smooth surfaces than for rough surfaces. For example, while the endurance limit of well-polished parts can be taken approximately as 50 percent of the ultimate, in the case of as-forged surfaces it can be less than 10 percent. Intermediate surfaces, such as hot-rolled, turned, and machined, are characterized by endurance-limit values between 10 and 50 percent of the ultimate. Figure 9.17 illustrates the influence of the five surface categories on the endurance limit of steels of various tensile strengths. The strength at 10^3 cycles is not significantly affected by surface finish.

9.2.6 SURFACE TREATMENT K_4

Since most fatigue failures originate at the surface of the part, the condition of the surface is of particular importance. The fatigue strength is substantially affected by subjecting the part to the surface treatment obtained by commonly used processes such as (1) cold working (shot peening, cold rolling, stretching); (2) surface hardening (carburizing, nitriding, cyaniding, flame hardening, induction hardening); and (3) plating (chromium, zinc, cadmium). The effect of these processes on the fatigue strength is reported in Ref. 8.

9.2.7 STRENGTH-REDUCTION FACTOR K_5

A notch, or stress raiser, in a part subjected to fatigue loading can be regarded as a factor causing a local increase in stress or as a factor reducing strength. For example, a notch resulting in a factor of 2 can be thought of as doubling the stress or as halving the strength. In either case the effect is to reduce the magnitude of external load causing failure.

If all parts were made of completely homogeneous materials and had perfectly polished surfaces, the effect of a notch would be to increase the stress by the conventionally determined stress concentration factor K_t.

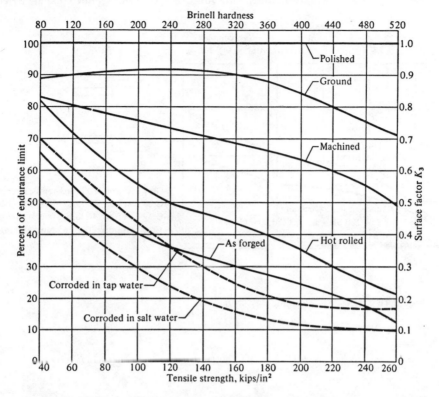

Fig. 9.17 Reduction of endurance strength due to surface finish.

Since actual materials are not perfectly homogeneous and actual surfaces are seldom polished, there exist internal and surface stress raisers in notch-free parts.   For this reason, the addition of a notch to an actual part generally produces a smaller effect than would be predicted from the theoretical stress concentration factor K_t.   The extent to which a notch reduces the endurance limit of a part is referred to as the fatigue stress concentration factor, or the fatigue strength-reduction factor of the notch, and is designated by the symbol K_f. Thus:

$$K_f = \frac{\text{endurance limit of specimen without the notch}}{\text{endurance limit of specimen with the notch}}$$

K_s, the strength-reduction factor, is related to K_f, as shown:

$$K_s = \frac{1}{K_f}$$

The approximate relationship between K_f and K_t is given in Fig. 9.18 for ferrous materials and in Fig. 9.19 for nonferrous.   For a more precise determination of K_f from K_t, see Ref. 8.

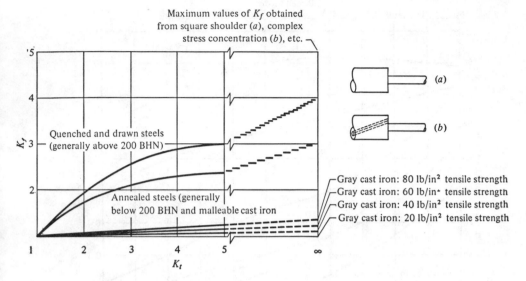

Fig. 9.18 K_f vs. K_t—steel and cast iron.

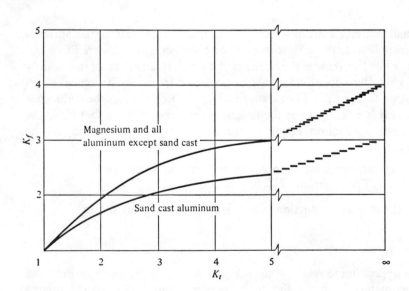

Fig. 9.19 K_f vs. K_t —nonferrous metals.

9.2.8 RELIABILITY FACTOR K_6

Factors K_1 through K_5 discussed in the previous sections referred only to the average fatigue strengths of laboratory specimens as compared with actual manufactured parts. Since from the practical engineering point of view the performance of a part cannot be predicated on the average strength (50 percent reliability), it is necessary to establish a relationship between the fatigue strength at 50 percent reliability and some realistically higher reliability such as 99 percent corresponding to three standard deviations (3σ) from the average. Limited test data suggest that the standard deviation of the fatigue strength is approximately 8 percent of the average [9]. Since 99 percent of the population has strength of at least ($\mu - 2.32\sigma$), $2.32 \times 0.08 = 0.186 \cong 0.2$ is the value to be subtracted from the average strength to arrive at a strength corresponding to 99 percent reliability. Thus, the correction factor K_6 for 99 percent reliability is 0.8. The following example illustrates the procedure for determining the fatigue strength of the actual manufactured parts from the laboratory specimens.

Example 9.3 The shaft shown in Fig. 9.20 is fabricated from steel ($S_u = 70$ ksi) with machined surface and is subjected to completely reversed bending load P of constant amplitude. Determine the fatigue strength of this shaft at 10^5 cycles.

Solution

$$S_n = S'_n K_1 \times K_2 \times K_3 \times K_4 \times K_5 \times K_6$$
$S_u = 70$ ksi
$S'_n = 0.5 \times 70 = 35$ ksi (Subsec. 9.2.2)
$K_1 = 1.0$ for bending load (Subsec. 9.2.3)
$K_2 = 0.85$ for 3/4 in size (Subsec. 9.2.4)
$K_3 = 0.79$ for machined surface with $S_u = 70$ ksi (Fig. 9.17)
$K_4 = 1.0$, no surface treatment (Subsec. 9.2.6)
$K_5 = 1/K_f$ where $K_f = 1.56$ for $K_t = 1.7$ from Fig. 9.18
$\quad = 1/1.56 = 0.64$
$K_6 = 0.8$, the reliability factor (Subsec. 9.2.8)
$\quad$ The fatigue strength of the part is
$S_n = 35 \times 1.0 \times 0.85 \times 0.79 \times 1.0 \times 0.64 \times 0.8$
$\quad = 12.0$ ksi at 10^6 cycles

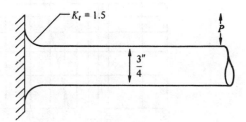

Fig. 9.20 The shaft in Example 9.3.

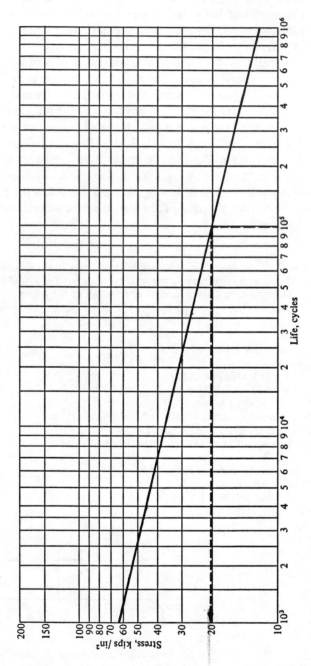

Fig. 9.21 Stress-life relationship for the shaft in Example 9.3.

Fatigue strength at 10^3 cycles $= 0.9S_u$ (Subsec. 9.2.3)
$$= 0.9 \times 70 \text{ ksi}$$
$$= 63.0 \text{ ksi}$$

Plot 12 ksi at 10^6 cycles and 63 ksi at 10^3 cycles on log-log paper (Fig. 9.21). The strength at 10^5 cycles is 21 ksi.

9.2.9 GOODMAN DIAGRAM

Up to this point, the fatigue strength has been considered only with respect to completely reversed load of constant magnitude. The majority of engineering applications involve a combination of alternating stress and static stress, as shown graphically in Fig. 9.22.

A typical example of combined mean and variable loading is provided by the valve springs of an internal-combustion engine. The valve spring is preloaded in assembly to hold the valve on the seat. During operation, as the valve is opened the spring is given greater loading due to the additional deflection. Thus, the loading cycle goes from a low magnitude to a higher magnitude and then back to the low magnitude. That is, the stress resulting from this combination of loads may be broken down into a steady component that raises or lowers the general stress level and an alternating component that tends to fatigue the metal.

A commonly used method of determining the fatigue strength for this type of loading employs the so-called modified Goodman diagram (Fig. 9.23). In this diagram the stress (either maximum or minimum) is plotted on the ordinate and the mean stress on the abscissa, using the same scale on both axes. The strength S_n for a completely alternating load (endurance limit

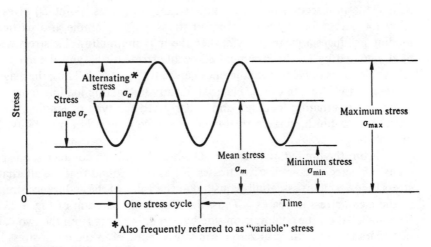

Fig. 9.22 Fatigue terms.

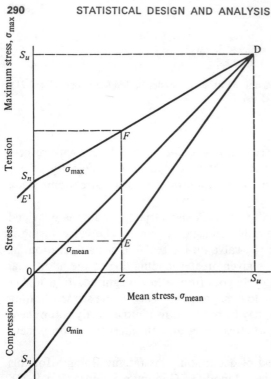

Fig. 9.23 Goodman diagram.

modified for load, surface, size, life, etc.) is laid off on the zero mean-stress ordinate for both tension and compression. A straight line drawn between S_n and the intersection of the ultimate strength S_u lines (point D) gives the allowable stress for any value of mean stress. For example, if a particular loading produces a mean stress Z, then the maximum allowable stress would be designated by F', the minimum allowable stress by E', and the maximum allowable stress range by the difference between F' and E'. Thus, the diagram represents the strength of this material for any fluctuating loading (combined static and alternating loading), and is corrected for type of loading, size of part, surface finish, surface treatment, stress concentration, reliability, and life.

From the limited number of tests which have been conducted on specimens subjected to compressive stresses, it has been found that the alternating stress range remains essentially constant at the value of the endurance strength as the mean stress increases. Thus, the Goodman diagram of Fig. 9.23 can be completed on the compression side by extending lines from the two values of endurance strength S_n at zero mean stress, parallel to the mean-stress line. This, then, gives the constant alternating stress range in compression.

Example 9.4 Suppose that the shaft in Example 9.3 is subjected to a mean stress of 30 ksi in addition to alternating stress fluctuating around the mean stress. (For other situations see Ref. 8.) Determine the fatigue strength (that is, maximum allowable alternating stress) for this shaft at 10^5 cycles.

Solution From Example 9.3, fatigue strength at 10^5 cycles = 21 ksi for zero mean stress, and S_u = 70 ksi.

Plot this value on a Goodman diagram as shown in Fig. 9.24. The fatigue strength at 10^5 cycles for mean stress of 30 ksi is

$$F'B' = (42 - 30) \text{ ksi} = 12 \text{ ksi}$$

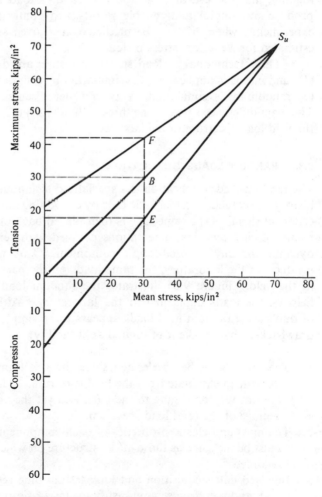

Fig. 9.24 Goodman diagram for Example 9.4.

9.3   RANDOM-LOAD EXPERIMENTS

Random-load experiments are designed to predict the life of components subjected to random loads. Random loads are the loads whose amplitude and frequency vary randomly with time (Fig. 9.2). Random-load experiments are applicable in improving the prediction accuracy of service life of components such as automotive components subjected to random irregularity of road surfaces, aircraft components under random atmospheric gust loads, and marine vehicle components subjected to the random ocean waves. All the tools of constant-load experiments, discussed previously, apply only to those situations where the parts are subjected to the constant-amplitude loading, and the use of these tools to random-load situations would not produce meaningful and reliable results. In contrast with constant-load experiments, where life can be predicted at a given stress, here the life is estimated for a random stress or load.

Three techniques of load spectrum testing are discussed briefly here: (1) random-loads duplication, (2) simulation by programmed step loads, and (3) random-loads simulation. (For a fuller discussion see Refs. 10–12.) The main difference between the three techniques is the method of simulating the field load spectrum in the laboratory.

9.3.1   RANDOM-LOADS DUPLICATION

The random-loads duplication is a specialized technique where the field load history is recorded on a magnetic tape by converting load signal to equivalent electrical signal. The same tape is then played back on the test machine by means of electronic instrumentations coupled with servohydraulic, electrodynamic, or any "closed-loop" equipment. This makes it possible to produce, in the laboratory, the same loads on the part as were encountered in the field. Figure 9.25 illustrates the random loads as recorded in the field vs. the loads duplicated in the laboratory. Although the technique of duplicating random field loads appears to be simple, it has the following drawbacks, which make it of limited applicability:

1. The test cannot be accelerated since the small amplitudes of load are difficult to eliminate from the load spectrum. These small amplitudes are not very damaging to the part, and yet they represent a large percentage of the total load spectrum.
2. The mass and elastic properties of each component in the test structure must be the same as those of the structure on which the field loads were recorded.
3. Involved data acquisition and long testing time result in a high testing cost, and therefore seldom justify testing of simple components.

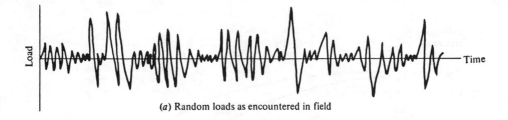

(a) Random loads as encountered in field

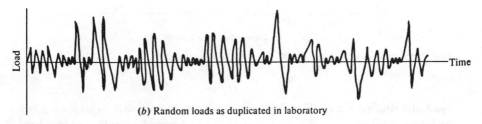

(b) Random loads as duplicated in laboratory

Fig. 9.25 Random-loads duplication.

9.3.2 SIMULATION BY PROGRAMMED STEP LOADS

Another method of simulating random field loads (as shown in Fig. 9.25) in the laboratory is the use of programmed step loads. (For details see Refs. 10, 11, and 13.)

The representative random-load spectrum, as encountered in the field, is recorded in the form shown in Fig. 9.26. This information is then processed to establish the number of times that a given peak load level occurs over the total recording period. This information is plotted as a cumulative

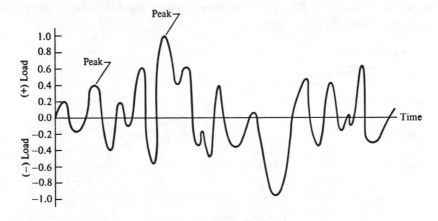

Fig. 9.26 Random field loads.

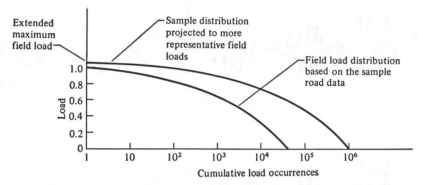

Fig. 9.27 Cumulative load distribution ("load histogram").

load distribution curve (Fig. 9.27) which shows the number of times a given or higher load occurs. The distribution thus obtained from the sample load data is then extended to 10^6 cycles (life corresponding to the endurance limit) to produce a more realistic field-loads history that may occur over the total life of the part.

In order to simulate this extended field-load distribution in the laboratory, this distribution is divided into eight load-level steps [14]. This is known as the step-load histogram (Fig. 9.28). The part is subjected in the laboratory to these step loads with each load imposed for the corresponding number of cycles. The test generally begins with the load level equal to 0.5 of the maximum and follows the ascending-descending sequence (Fig. 9.29). One block consists of 10^6 cycles that cover the complete eight load levels. This block is repeated until failure occurs. The laboratory life in terms of number of blocks to failure is then correlated to the field life. This is discussed in more detail in Chap. 12.

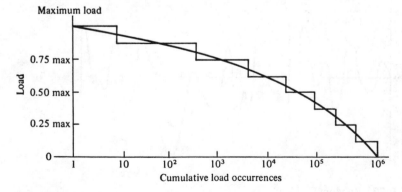

Fig. 9.28 Step-load histogram.

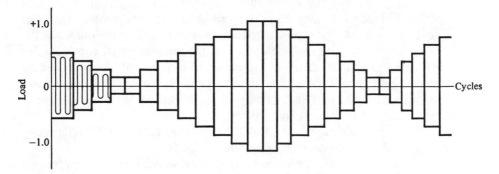

Fig. 9.29 Field loads as simulated in the laboratory.

This method of simulating random field loads in the laboratory assumes that the consecutive load cycle is of the same magnitude until a new load level is attained. In actual field random load this may not be true since "random" implies that the load cycle may be of a different magnitude, and one load cycle followed by another of different magnitude may have a significant influence on the part's life. In the method discussed in Subsec. 9.3.3, Random-loads Simulation, this assumption is not made since the load applied to the part in the laboratory is truly random in nature, having the same statistical properties as the random field loads would have.

9.3.3 RANDOM-LOADS SIMULATION

Since the word "random" itself implies "unpredictable until it has happened," the most appropriate way to treat the random-load spectrum (Fig. 9.30) is to express it in terms of its statistical properties. This section

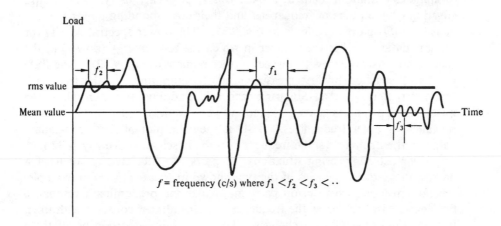

Fig. 9.30 Random-load spectrum.

deals with that technique by which the random-load spectrum in the field can be simulated in the laboratory in terms of its statistical properties. This is done electronically by two methods: (1) properly shaping the output of the random function generator which produces the random signal continuously; and (2) the digital method, where a given length of punched tape representing the varying amplitudes of load is played repeatedly through the appropriate equipment until the part fails. Only the first method is discussed here in detail because of its relative importance.

Random loads can be classified into two broad categories: (1) stationary random loads and (2) nonstationary random loads.

A stationary random process is that process whose statistical parameters do not change with time. These parameters are rms (or intensity) level, mean level, etc. Root mean square (rms) is defined as the square root of the mean (or average) value of the square of the load spectrum. This is obtained by squaring each amplitude of the spectrum, adding them, taking the average, and then computing the square root of this average (Fig. 9.30). This is equal to the standard deviation σ for the case when the overall mean of the load spectrum is zero.

In a nonstationary process the parameter which changes with time is usually the rms level; however, other parameters may change too. An example of a nonstationary random process is an automobile traveling on a highway for a period of time and then turning onto a cobblestone road. This change in road load intensities indicates a nonstationary process where rms levels (load intensity) change from time to time.

Stationary random loads can be simulated by stationary rms testing; the nonstationary random loads, by programmed rms testing.

Stationary rms testing Parts are tested until failure, under a spectrum of stationary random loads at a fixed rms level. Various types of spectra covering a wide range of frequencies and their corresponding power spectral density (PSD) plots are shown in Fig. 9.31. The power spectral density for random data describes the manner in which the total energy (power) of the spectrum is distributed with respect to the frequency (Fig. 9.32). The PSD plot is constructed by first measuring all the amplitudes in the frequency range of f_1 and $f_1 + \Delta f$. Squaring each value, adding, and then taking the mean value produces a mean-square value for the frequency f_1. This is repeated for other frequencies. PSD is then the plot of the mean-square value on the ordinate vs. the frequency on the abscissa. (See Fig. 9.32.)

In most engineering situations the parts are subjected to loads of a narrow range of frequencies of the type shown in Fig. 9.31b. For example, several automobile suspension parts may have one predominant resonance frequency. In such cases the frequency remains almost constant with time, but the amplitude of load changes. The probability distribution of these

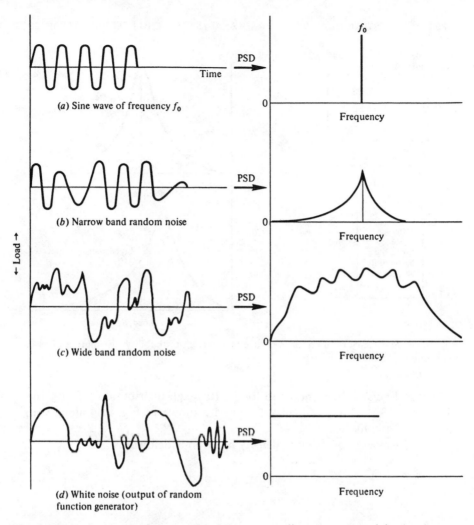

Fig. 9.31 Various types of spectra with their corresponding power spectral density plots.

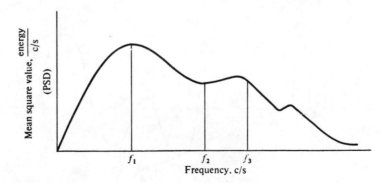

Fig. 9.32 Power spectral density plot.

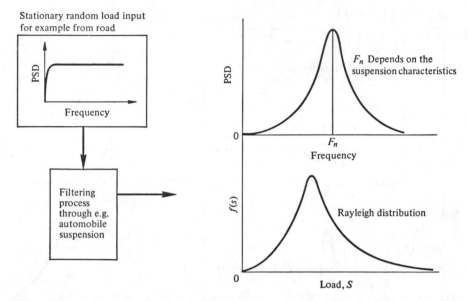

Fig. 9.33 Stationary random loads reduce to Rayleigh probability distribution of peak loads due to the filtering process.

peak load levels is then known to follow Rayleigh distribution [15, 10, 16]. (See Fig. 9.33.)   Peaks of load spectrum are shown in Fig. 9.34 along with valleys and amplitudes.

The Rayleigh probability density function is defined as

$$f(S) = \frac{S}{\sigma^2}\, e^{-1/2(S/\sigma)^2} \qquad \text{for } 0 \leq S < \infty$$

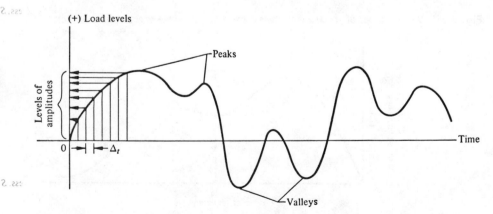

Fig. 9.34 Peaks, valleys, and amplitude levels of a random signal.

and the Rayleigh cumulative distribution function is

$$F(S \leq S_1) = \int_0^{s_1} f(S)\, dS$$

$$= \int_0^{s_1} \frac{S}{\sigma^2} e^{-1/2(S/\sigma)^2}\, dS$$

$$= 1 - e^{-1/2(S/\sigma)^2}$$

or

$$F(S > S_1) = e^{-1/2(S/\sigma)^2} = \text{probability of exceeding a certain value } S = S_1$$

where S = load or stress

σ = standard deviation, or rms level (load intensity)

The family of Rayleigh density and cumulative curves is shown in Fig. 9.35.

In many engineering situations, the part may be subjected to somewhat varying frequencies (and not narrowband frequency). In such cases the underlying peak load distribution is not quite Rayleigh. The problem then

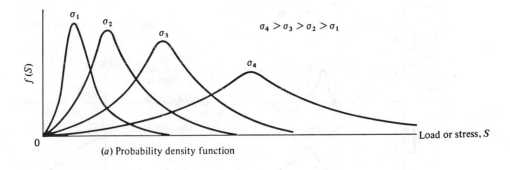

(a) Probability density function

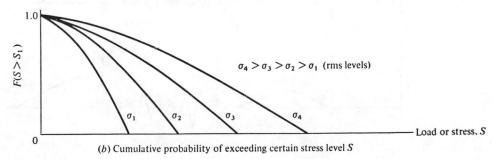

(b) Cumulative probability of exceeding certain stress level S

Fig. 9.35 Family of Rayleigh distributed load curves as found in the field.

becomes more involved. The simpler approach would be to use the Rayleigh distribution since it is somewhat conservative. This is because Rayleigh produces more severe loads than those generated by the non-Rayleigh case. Once the Rayleigh distribution is assumed, the distribution can be simulated in the laboratory electronically, as shown in Fig. 9.36. This should be done in a way such that the shape of the load histogram of peaks produced in the laboratory is matched with the field histogram.

In the laboratory, the part should also be subjected to a spectrum of loads such that the rms level of the load distribution (Fig. 9.35) is the same as the rms level of the field loads, and the number of life cycles to failure is then determined. The rms level can be adjusted to a desired value encountered in the field by using appropriate gain. The appropriate adjustment is made so that the maximum test load applied is not more than the

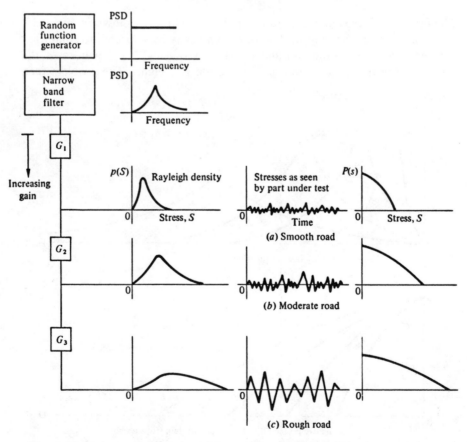

Fig. 9.36 Laboratory simulation of the stationary random field load.

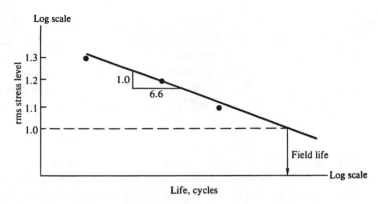

Fig. 9.37 Rms stress-vs.-life relationship.

measured maximum. This is done by establishing what is known as the clipping ratio. This is the ratio of maximum load in the spectrum to rms load of the spectrum. This term is used, for example, in random-function-generator output, where the output is clipped so that the signal from the random function generator does not exceed the desired maximum value. The clipping ratio is also known as the crest factor.

Different specimens are tested each at a different rms level, and the rms stress-vs.-life relationship is established (Fig. 9.37). Some work has been done in this area where the slope of this rms S-N curve has been established for commercial steels. The inverse slope of the curves was found to be -6.6 [17]. As shown in Fig. 9.38, random-load testing produces considerably lower scatter in life than constant-amplitude testing [18]. Because of this low scatter in random-load testing, a relatively smaller sample size is required

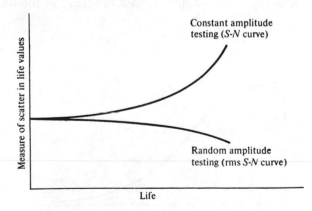

Fig. 9.38 Relative scatter in life values for constant- and random-amplitude testing.

to estimate the true life.   Thus (low scatter and an inverse slope of −6.6) one can accelerate the test by subjecting the parts to, say, 1.2 times the actual field rms load and determine the life to failure.   This point is then plotted on a log-log paper where a line with the inverse slope of −6.6 is then drawn passing through this point.   The field life is then predicted by means of this line; the predicted life is the one which corresponds to the rms level of 1.0.   (See Fig. 9.37.)

Programmed rms testing ⋅ The previous section dealt with the situations where the rms level of the random loads does not change with time.   However, in many situations the rms intensity does change periodically.   For example, in the case of ground vehicles, the rms level changes (1) by changing the vehicle speed and (2) by driving the vehicle on different road surfaces. This results in a continuous step change in the rms levels of loads since each vehicle speed and each road surface constitutes one specific rms intensity. This change in rms levels is random; therefore one rms level may occur more frequently than another level over the total vehicle life.   This results in a distribution of rms levels (Fig. 9.39).   Generally, in actual usage a vehicle is subjected to lower rms levels most of the time and very high levels a few times.   It should also be noted that each rms level represents a Rayleigh distribution of peaks (when the parts are assumed to be subjected to narrow-band load frequency) with the standard deviation σ equal to the rms level. (See Fig. 9.40.)   This leads to the concept of programmed rms testing, where the varying intensities of rms loads as encountered in the field are simulated in the laboratory.   This can be done by controlling the output of the random function generator (with appropriate filters), where any rms level can be set by setting the appropriate gain.   By increasing the gain, the rms level increases, and hence the increase in spread of the Rayleigh distributed peak

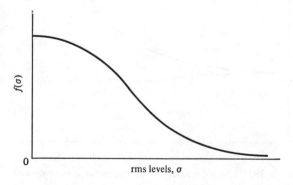

Fig. 9.39   Distribution of rms levels.

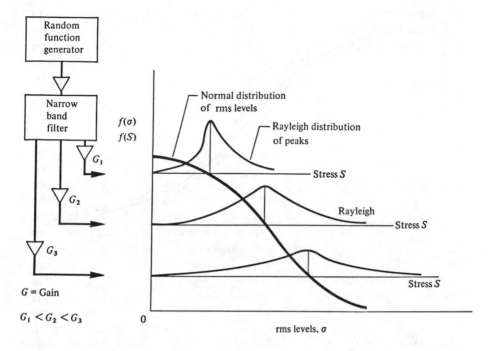

Fig. 9.40 Normally distributed rms levels of Rayleigh distributed loads where the spread in load increases with increase in rms (σ) level.

stress levels. (See Fig. 9.40.) Since the change in rms levels occurs randomly in the field, the only way that this randomness can be expressed is in terms of its statistical distribution parameters. This distribution should be simulated in the laboratory by randomly changing the rms levels so that the test is run at each level for a predetermined period of time. This simulation of a distribution of rms levels is done through a tape which controls the gain on the random function generator. A set of random numbers, each representing a given gain of the rms level, is punched on the tape which, when run through, produces the rms levels whose distribution corresponds to the distribution of rms levels found in the field. The distribution of the rms levels may be assumed to be the positive half of normal (gaussian) distribution [11, 13].

In order to run a programmed rms test where the distribution of rms levels is normal, a control tape is necessary which has a large set of random numbers (in a random sequence on tape) with an underlying normal distribution. (See Fig. 9.41.) This means that the rms level whose intensity is 25 units occurs about 130 times over the period of time, whereas the rms level with intensity of 200 units occurs only about seven times.

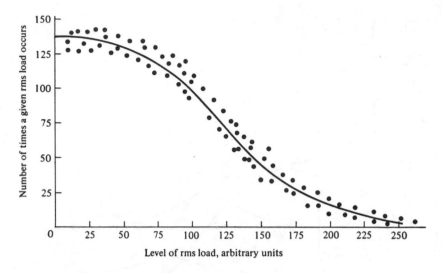

Fig. 9.41   Normally distributed rms levels.

When the rms level is a random variable distributed normally and at each given rms level the peak load is also a random variable with the underlying Rayleigh distribution, the resultant load distribution or the final load spectrum that the part sees is the joint probability distribution of the normal and Rayleigh distributions. This joint probability reduces to log linear cumulative distribution of peak loads [19], where a linear relationship exists between load, when plotted on linear scale, and the number of occurrences, when plotted on log scale. (See Fig. 9.41.)

$$F(S) = \int_{\sigma_{min}}^{\sigma_{max}} \int_0^{S \le 4\sigma_i} G(\sigma_i) e^{-(S^2/2\sigma_i)^2} \, dS \, d\sigma$$

where $G(\sigma)$ is the normal density function. The plot in Fig. 9.42 represents a cumulative distribution where a given percent of the total load cycles in the random spectrum has the corresponding magnitude of load or higher.

The load spectrum thus produced in the laboratory by means of the random function generator, with appropriate programming of the rms levels, would match exactly the field load spectrum if found log linear. (A brief description on how to construct the plot of load vs. the load occurrences from the load spectrum is given in Subsec. 9.3.2.) In the absence of any specific information, random-load spectra in the field may be assumed to have log linear relationship. Figure 9.43 shows the field load spectrum

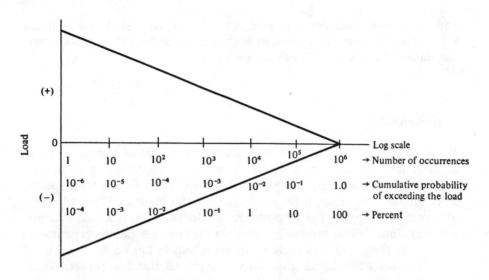

Fig. 9.42 Cumulative load distribution.

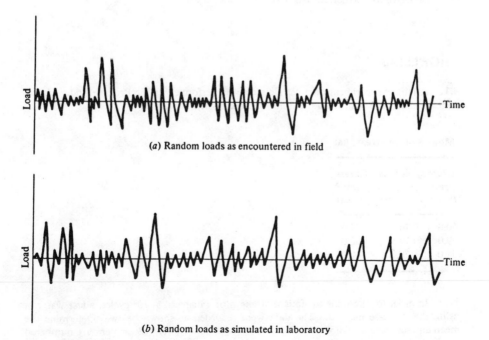

(a) Random loads as encountered in field

(b) Random loads as simulated in laboratory

Fig. 9.43 Random-loads simulation.

and the random spectrum as created in the laboratory. It should be noted that the instantaneous load value of both spectra are not the same; however, the statistical properties of both are the same.

SUMMARY

While Chaps. 3 to 8 deal with methods for the design of experiments and analysis of data which can be applied to a variety of engineering fields, such as mechanical, chemical, and electrical, Chap. 9 is devoted to a specific field: fatigue. The reasons for its inclusion in this book are (1) its considerable industrial application to problems of design, materials, manufacture, reliability, and failure prevention; and (2) the fact that fatigue experiments employ the same statistical tools as the other tests in this book.

The basis for fatigue experiments is the relationship between stress and life, at a given confidence level. Basic relationships are developed first for constant loads, as frequently employed in the laboratory (Sec. 9.1), and then for random loads, characteristic of actual service (Sec. 9.3). Some of these relationships are translated into actual fabricated parts and structures because of their industrial importance (Sec. 9.2).

PROBLEMS

9.1. A fatigue experiment on a certain steel was performed by the Prot method. Determine its endurance limit from the following test data:

Mean failure stress, ksi

(Rate of increase of stress per cycle)$^{1/2}$, (psi/cycle)$^{1/2}$		
0.010	0.025	0.040
60.0	76.5	99.0
59.0	79.5	102.0
58.0	78.0	100.0

9.2. In order to determine the fatigue strength of a material at 10^5 cycles, a test was run using the staircase method. The data were recorded as shown below. Determine the mean and standard deviation of the fatigue strength at this life. (Specimens are numbered in chronological order. The number of cycles for each test is 10^5 unless failure occurred beforehand.)

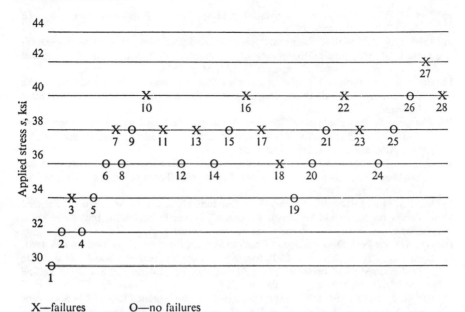

X—failures   O—no failures

9.3. Predict the endurance limit of a bolt with 99 percent reliability from the following data: S_u of the material is 120 ksi. The nominal diameter of the bolt is 1.0 in. The threads were machined and $K_f = 2.0$ at the root of the threads. The bolt is axially loaded, with a prestress of 8 ksi, and the dynamic stress ± 16 ksi.

REFERENCES

1. Sinclair, G. M., and T. J. Dolan: Effect of Stress Amplitude on Statistical Variability in Fatigue Life of 755-T6 Aluminum Alloy, *ASME Paper* 52-A-82, 1952.
2. A Guide for Fatigue Testing and the Statistical Analysis of Fatigue Data, *ASTM Spec. Tech. Publ.* 91-A, 2d ed., 1963.
3. Dixon, W. J., and F. J. Massey, Jr.: "Introduction to Statistical Analysis," 3d ed., McGraw-Hill Book Company, New York, 1969.
4. Corten, H. T., T. Dimoff, and T. J. Dolan: An Appraisal of the Prot Method of Fatigue Testing, *Proc. ASTM*, vol. 54, p. 875, 1954.
5. Weibull, W.: "Fatigue Testing and Analysis of Results," The Macmillan Company, New York, 1961.
6. Lipson, C., N. J. Sheth, and R. Disney: Reliability Prediction—Mechanical Stress/Strength Interference, *Final Tech. Rep.* RADC-TR-66-710, Rome Air Development Center, Research and Technological Division, Air Force Systems Command, Griffiss Air Force Base, New York, March, 1967.
7. Little, R. E.: "Multiple Specimen Testing and the Associated Fatigue Strength Response," The University of Michigan Press, Ann Arbor, 1966.
8. Lipson, C., and R. C. Juvinall: "Handbook of Stress and Strength," The Macmillan Company, New York, 1963.
9. Stulen, F. B., H. N. Cummings, and W. C. Schulte: Preventing Fatigue Failures, part 5, *Mach. Des.*, vol. 33, p. 161, June 22, 1961.

10. Jaeckel, H. R.: Duplication, Simulation, and Synthesis of Random Field Loads, *SAE Paper* 700032, January, 1970.
11. Jaeckel, H. R., and S. R. Swanson: "Random Load Spectrum Test to Determine Durability of Structural Components of Automotive Vehicles," 12th International Automobile Technical Congress (FISITA), May, 1968.
12. Bussa, S. L.: "Fatigue Life of a Low Carbon Steel Notched Specimen under Stochastic Conditions," master's thesis, Department of Engineering Mechanics, Wayne State University, Detroit, 1967.
13. Swanson, S. R.: Evaluating Component Fatigue Performance under Programmed Random, and Programmed Constant Amplitude Loading, *SAE Paper* 690050, SAE Annual Meeting, Detroit, January, 1969.
14. Gassner, E., and W. Schutz: "Evaluating Vital Vehicle Components by Programmed Fatigue Tests," Ninth International Automobile Technical Congress (FISITA), London, May 4, 1962.
15. Rice, S. O.: Mathematical Analysis of Random Noise, in N. Wax (ed.), "Selected Papers on Noise and Stochastic Processes," Dover Publications, Inc., New York, 1954.
16. Borch, J. T.: Peak Distributions of Random Signals, *Bruel and Kjaer Tech. Rev.* 3, 1963.
17. Bussa, S. L., N. J. Sheth, and S. R. Swanson: "Development of a Random Load Life Prediction Model," presented at the Annual meeting of American Society of Testing and Materials, Toronto, Canada, June, 1970.
18. Schijve, J., and F. A. Jacobs: Program-fatigue Tests on Notched Light Alloy Specimens of 2024 and 7075 Material, *N.L.L. (Natherland. National Aeronautical Research Institute) Tech. Rep.* M-2070, July, 1961.
19. Swanson, S. R., and H. Akaike: "Load History Effects on Structural Fatigue," 1969 Annual Meeting Session on Failure Mechanisms and Reliability Prediction in Structural Dynamics, Anaheim, Calif., Apr. 23, 1969.

BIBLIOGRAPHY

Armitage, P. H.: Statistical Aspects of Fatigue, *Met. Rev.*, vol. 6, no, 23, London, Institute of Metals, 1961.
Battelle Memorial Institute: "Fatigue of Metals and Structures," United States Government Printing Office (NAVWEPS 00-25-534), 1960.
Bloomer, N. T.: Time Saving in Statistical Fatigue Experiments, *Engineering*, vol. 184, no. 4783, p. 603, Nov. 8, 1947.
Broch, J. T.: Peak Distributions of Random Signals, *Bruel and Kjaer Tech. Rev.* 3, 1963.
Bussa, S. L.: "Fatigue Life of a Low Carbon Steel Notched Specimen under Stochastic Conditions," master's thesis, Department of Engineering Mechanics, Wayne State University, Detroit, 1967.
———, N. J. Sheth, and S. R. Swanson: "Development of a Random Load Life Prediction Model," *Material Research and Standards* (ASTM), March, 1972.
Conover, J. C., H. R. Jaeckel, and W. J. Kippola: Simulation of Field Loading in Fatigue Testing, *SAE Paper* 660102, January, 1966.
Crandall, S. H., and W. D. Mark: "Random Vibration," Academic Press, Inc., New York, 1963.
Dieter, G. E., and R. F. Mehl: Investigation of the Statistical Nature of the Fatigue of Metals, *NACA Tech. Note* 3019, September, 1953.
Dixon, W. J., and F. J. Massey, Jr.: "Introduction to Statistical Analysis," 3d ed., McGraw-Hill Book Company, New York, 1969.

Dolan, T. J., F. E. Richart, and C. E. Work: Influence of Fluctuations in Stress Amplitude on the Fatigue of Metals, *ASTM Proc.*, vol. 49, 1949.

Epremian, E., and R. G. Mehl: Investigation of Statistical Nature of Fatigue Properties, *NACA Tech. Note* 2719, June, 1952.

Findley, W. N.: Test Procedure and Technique, "Manual on Fatigue Testing," ASTM STP-91, Section V, 1949.

Forrest, P. G.: "Fatigue of Metals," Addison-Wesley Press, Inc., Cambridge, Mass., 1962.

Fuller, J. R.: Research on Techniques of Establishing Random Type Fatigue Curves for Broad Band Sonic Loading, ASD TDR 62-501 USAF, October, 1962; *SAE Paper* 671C, April, 1963.

Gassner, E., and W. Schutz: "Evaluating Vital Vehicle Components by Programmed Fatigue Tests," Ninth International Automobile Technical Congress (FISITA), London, May 4, 1962.

Grover, H. J., S. A. Gordon, and L. R. Jackson: "Fatigue of Metals and Structures," Department of the Navy, 1954.

A Guide for Fatigue Testing and the Statistical Analysis of Fatigue Data, *ASTM Spec. Tech. Publ.* 91-A, 2d ed., 1963.

Harris, J. P., and C. Lipson: Cumulative Damage Due to Spectral Loading, *Aerosp. Rel. Maintainability Conf. Proc.*, July, 1964.

Jaeckel, H. R., and S. R. Swanson: "Random Load Spectrum Test to Determine Durability of Structural Components of Automotive Vehicles," 12th International Automobile Technical Congress (FISITA); also reprinted by MTS Systems Corp., Minneapolis, Minn., Report No. 900, 22-1, May, 1968.

Johnson, L. G.: Fatigue Tests Proved by Three Statistical Checks, *SAE J.*, March, 1958.

———: "The Statistical Treatment of Fatigue Experiments," Elsevier Publishing Company, Amsterdam, 1964.

Juvinall, R. C.: "Engineering Considerations of Stress, Strain, and Strength," McGraw-Hill Book Company, New York, 1967.

Lipson, C., and R. C. Juvinall: "Handbook of Stress and Strength," The Macmillan Company, New York, 1963.

——— and N. J. Sheth: Prediction of Percent Failures from Stress/Strength Interference, *SAE Paper* 680084, January, 1968.

———, ———, and R. Disney: Reliability Prediction—Mechanical Stress/Strength Interference, *Final Tech. Rep.* RADC-TR-66-710, Rome Air Development Center, Research and Technological Division, Air Force Systems Command, Griffiss Air Force Base, New York, March, 1967.

Miner, M. A.: Cumulative Damage in Fatigue, *J. Appl. Mech.*, vol. 12, no. 3, p. A-159, September, 1945.

Rice, J. R., F. P. Beer, and P. C. Paris: On the Prediction of Some Loading Characteristics Relevant to Fatigue, in W. J. Trapp and D. M. Forney (eds.), "Acoustical Fatigue in Aerospace Structures," pp. 121–144, Syracuse University Press, Syracuse, N.Y., 1965.

Rice, S. O.: Mathematical Analysis of Random Noise, in N. Wax (ed.), "Selected Papers on Noise and Stochastic Processes," Dover Publications, Inc., New York, 1954.

Schijve, J.: The Analysis of Random Load-time Histories with Relation to Fatigue Tests and Life Calculations, "Fatigue of Aircraft Structures," Pergamon Press, New York, 1963.

——— and F. A. Jacobs: Program-fatigue Tests on Notched Light Alloy Specimens of 2024 and 7075 Material, *N.L.L. (Natherland. National Aeronautical Research Institute) Tech. Rep.* M-2070, July, 1961.

Schjelderup, H. C., and A. E. Galef: Simulation of the Fatigue Effects of Random Stresses by a Minimum Number of Discrete Stress Levels, *Mater. Res. Stand.*, October, 1962.

Sinclair, G. M., and T. J. Dolan: Effect of Stress Amplitude on Statistical Variability in Fatigue Life of 755-T6 Aluminum Alloy, *ASME Paper* 52-A-82, 1952.

Sines, G., and J. L. Waisman (eds.): "Metal Fatigue," McGraw-Hill Book Company, New York, 1959.

Sjostrom, S.: On Random Load Analysis, *KTH Rep.* 181, *Trans. Roy. Inst. Technol.*, *Stockholm*, 1961.

Stulen, F. B.: On the Statistical Nature of Fatigue, *ASTM Symp. Statist. Nature Fatigue*, STP 121, 1951.

Swanson, S. R.: An Improved Law of Cumulative Damage in Metal Fatigue, *Coll. Aeronaut., Cranfield, Engl., Rep.* **148**, August, 1961.

———: Random Load Fatigue Testing: A State of the Art Survey, *Mater. Res. Stand.*, vol. 8, no. 4, April, 1968.

———: Evaluating Component Fatigue Performance under Programmed Random, and Programmed Constant Amplitude Loading, *SAE Paper* 690050, SAE Annual Meeting, Detroit, January, 1969.

——— and H. Akaike: "Load History Effects on Structural Fatigue," 1969 Annual Meeting Session on Failure Mechanisms and Reliability Prediction in Structural Dynamics, Anaheim, Calif., Apr. 23, 1969.

Weibull, W.: "Fatigue and Fracture of Metals," pp. 182–196, John Wiley & Sons, Inc., New York, 1950.

———: Statistical Aspects of Fatigue Strength, *Engineer's Dig.*, February, 1951.

———: Statistical Design of Fatigue Experiments, *J. Appl. Mech.*, vol. 19, no. 1, March, 1952.

———: "Fatigue Testing and Analysis of Results," The Macmillan Company, New York, 1961.

———: Research on Statistical Evaluation of Data from Small Test Series, *Annu. Rep. ARDC Contract*, 1958.

10
Analysis of Interference Data

Interference is a measure of the amount by which one parameter exceeds another parameter, the two being meaningfully related. This may arise from situations such as the following: when stress imposed on a part exceeds its strength; when the diameter of a shaft exceeds the diameter of the bore. The analysis of interference data results in the determination of the probability of interference, which in engineering situations corresponds essentially to the probability of failure. Although the analysis of the interference data can be applied to any general situation, the following discussion is oriented to the mechanical-design problems.

10.1 THE CONCEPT OF INTERFERENCE

A given part has certain physical properties which, if exceeded, will cause failure. For mechanical design the fatigue strength at a given number of cycles is frequently the limiting property. This property, as all properties of nonhomogeneous materials, varies from part to part, and frequently it can be expressed as a statistical distribution.

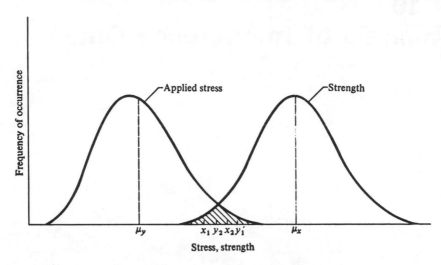

Fig. 10.1 Interference of stress and strength distributions.

The operating stress imposed on the part varies too, from time to time in a particular part, from part to part in a particular design, and from environment to environment. As in the case of strength, stresses can be represented by a statistical distribution.

Thus, two consistent distributions are available. These distributions can be, graphically, placed side by side at their respective means and compared. However, the shaded area of the distribution represents a pseudo area of interference. This area is only representative of the interference and is not a numerical measure of interference. Thus in Fig. 10.1 a stress Y_1 compared with strength X_1 does result in interference (failure) because $Y_1 > X_1$. However, stress Y_2 compared with strength X_2 will not produce failure, even though both are in the shaded area.

10.2 ANALYTICAL TREATMENT OF INTERFERENCE THEORY

Interference theory is concerned with the interplay of two variables x and y. Each variable is considered to arise as a consequence of performing some action and measuring the resulting value of the variable. It is assumed that both x and y are random variables, with known distribution or density functions $F(x)$ and $G(y)$ or $f(x)$ and $g(y)$, respectively. Failure occurs whenever y exceeds x. Hence, the probability of failure is

$$P(\text{failure}) = P(y > x)$$

Thus, to determine the probability of failure one needs to explore the probability that one random variable exceeds another random variable.

In practical application it is to be expected that the random variables are independent of each other in the sense that knowledge of one does not allow one to predict the other any more closely than would the absence of such knowledge.   Thus, random variables x and y are independent if

$$P(x|y) = P(x)$$

This statement says that the probability of x is the same whether one knows the exact value of y or not.

There are three main ways to determine the probability of failure from the above considerations.

1. One can fix attention on some particular value of one of the random variables, say y, and determine the probability that the other random variable does not exceed this fixed value, say Y.   The probability that x does not exceed a fixed, given value of y is

    $$P(x \leq Y | y = Y) \tag{10.1}$$

    In terms of density and distribution functions this is equivalent to

    $$\int_0^Y f(x)\, dx$$

    for those cases where x takes only nonnegative values   If one now multiplies Eq. (10.1) by the probability that y is in the neighborhood of Y, he obtains a joint probability function

    $$P(x \leq Y;\ Y \leq y \leq Y + dY) = \int_0^Y f(x)g(y)\, dx$$

    The probability that $x \leq y$ for any value that the random variable y can take is

    $$P(x \leq y) = \int_0^\infty \int_0^y f(x)g(y)\, dx\, dy \tag{10.2}$$

    in the case in which the random variable y is distributed on the non-negative axis.   Since failure occurs whenever $x \leq y$, Eq. (10.2) gives the probability of failure sought.   It is expressed in terms of the double integral of the known density functions.
2. One can define a new variable

    $$z = x - y$$

    Since x and y are random variables, their difference z is a random variable.   Further, if x and y are distributed on $(0,\infty)$, z is distributed on $(-\infty,\infty)$.   The probability of failure then is equivalent to the probability that z is nonpositive, $P(z \leq 0)$. The problem then is to find $h(z)$, the probability density function for z. From this the desired probability of failure can be obtained in a simple fashion.

Assume that x has a probability density function $f(x) = 1/6$ and y is identically distributed. Both are distributed on the integer $1, 2, \ldots, 6$. Therefore,

$$f(x) = \begin{cases} 1/6 & \text{for } x = 1, 2, 3, 4, 5, 6 \\ 0 & \text{elsewhere} \end{cases}$$

and y is independent and identically distributed.

Now consider a table of x and y values and the difference $x - y = z$.

	x value	1	2	3	4	5	6
	1	0	1	2	3	4	5
	2	−1	0	1	2	3	4
y value	3	−2	−1	0	1	2	3
	4	−3	−2	−1	0	1	2
	5	−4	−3	−2	−1	0	1
	6	−5	−4	−3	−2	−1	0

For $x = 1$ and $y = 1$, $z = 0$. The probability that $x = 1$ and $y = 1$ is $1/36$ since x and y are independent. The probability associated with each cell in the above table is $1/36$. Now if $z = 0$, then $x = 1$, $y = 1$; or $x = 2$, $y = 2$; or $x = 3$, $y = 3$; etc. Hence the probability that $z = 0$ is given by $1/36 + 1/36 + 1/36 + 1/36 + 1/36 + 1/36 = 1/6$. If now, $h(z) =$ the probability that $x - y = z$ for fixed z, then $h(z)$ is the desired probability density function. From the above discussion, it is clear that $f(x)f(y)$ is the probability that both x and y take on desired values. In every case $x = z + y$ for the x, y, z of interest. Hence $f(y + z)f(y)$ is the probability that $x = z + y$ and $y = y$ for any y and fixed z. The above probability is the joint distribution of y, $y + z$, say $g(y,z)$. It is well known that to get the marginal distribution $h(z)$ from $g(y,z)$ one merely "sums over all y." One must remember that both x and y are distributed on some interval $(1, 2, \ldots, 6$ in this example) and hence the sum must be over "permissible values of y."

x is distributed on $1, 2, \ldots, 6$, and $f(z + y)$ is the probability distribution of x. Hence, $z + y$ cannot exceed 6 nor fall below 1. Thus at the upper limit $z + y = 6$, and at the low limit $z + y = 1$. Or $y = 6 - z$ and $y = 1 - z$. Now, in the yz plane (Fig. 10.2), clearly $g(y,z)$ can be summed only over the y values defined in the rectangle.

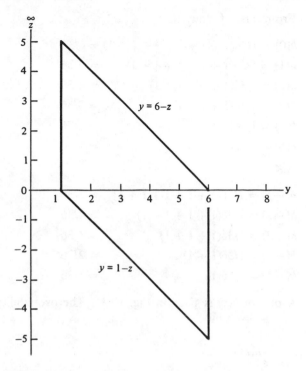

Fig. 10.2   Permissible y values for $z = x - y$.

But below the y axis this means y is summed from $1 - z$ to 6; and above the y axis, y is summed from 1 to $6 - z$.   Hence two parts of the sum must be considered as follows:

$$h(z) = \sum_{y} f(z + y)f(y)$$

$$= \sum_{y=1-z}^{6} f(z + y)f(y) \qquad \text{if } 0 > z \geq -5$$

$$= \sum_{y=1}^{6-z} f(z + y)f(y) \qquad \text{if } 0 \leq z \leq 5$$

(Note that $z = 0$ could be in either sum, but not both; arbitrarily it has been put into the second.)

Now $f(x) = 1/6$ for all $x = 1, 2, \ldots, 6$ and similarly for y.   Hence

$$h(z) = \begin{cases} \displaystyle\sum_{y=1-z}^{6} 1/36 & 0 > z \geq -5 \\[2ex] \displaystyle\sum_{y=1}^{6=z} 1/36 & 0 \leq z \leq 5 \end{cases}$$

From this it follows that

$$h(0) = 1/36(1 + 1 + 1 + 1 + 1 + 1) = 6/36$$
$$h(1) = 1/36(1 + 1 + 1 + 1 + 1) \quad\quad = 5/36$$
$$h(2) = 1/36(1 + 1 + 1 + 1) \quad\quad\quad = 4/36$$
$$h(3) = 1/36(1 + 1 + 1) \quad\quad\quad\quad = 3/36$$
$$h(4) = 1/36(1 + 1) \quad\quad\quad\quad\quad = 2/36$$
$$h(5) = 1/36(1) \quad\quad\quad\quad\quad\quad = 1/36$$

and

$$h(-1) = 1/36(1 + 1 + 1 + 1 + 1) \quad = 5/36$$
$$h(-2) = 1/36(1 + 1 + 1 + 1) \cdot \quad = 4/36$$
$$h(-3) = 1/36(1 + 1 + 1) \quad\quad\quad = 3/36$$
$$h(-4) = 1/36(1 + 1) \quad\quad\quad\quad = 2/36$$
$$h(-5) = 1/36(1) \quad\quad\quad\quad\quad = 1/36$$

A plot of $h(z)$ is given in Fig. 10.3. The probability of failure in this case is given by

$$\sum_{z=-5}^{0} h(z) = 1/2$$

Having solved the foregoing simple problem, one is able to generalize. Because of the special nature of the interference problem it can be assumed that x has the probability density function $f(x)$, y has probability density function $g(y)$, and both are distributed on $(0,\infty)$. Note that f and g are allowed to be different. Hence, the

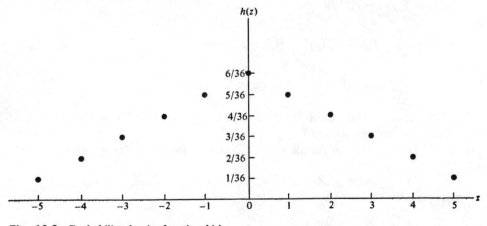

Fig. 10.3 Probability density function $h(z)$.

assumption of identically distributed random variables has been dropped although it is still assumed that they are independent.

As in the previous work it is clear that the only difficulty in finding $h(z)$ is in determining the correct limits on the integrals. The probability arguments are trivial. Since it was assumed that both x and y are distributed on $(0,\infty)$, a complete solution to this problem can be given. For consider the yz plot again. Since $x = 0$ is the minimum value that x can take, it follows that $y + z = 0$ or $y = -z$ is the lower bound of the area to be considered. Since $x = \infty$ is the maximum value that x can take, there is no upper bound on area. Hence all values of permissible y's are included in the area above. These are shown in Fig. 10.4, and one can pursue the probability arguments precisely as in the example to show

$$h(z) = \int_y f(z + y)g(y)\, dy$$

$$= \int_0^\infty f(z + y)g(y)\, dy \qquad z \ge 0 \tag{10.3}$$

$$= \int_0^\infty f(z + y)g(y)\, dy \qquad z \le 0$$

Clearly, this solves the problem in general for nonnegative x, y; extensions to other domains for x, y follow quite readily.

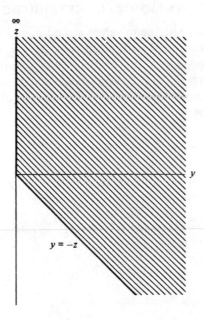

Fig. 10.4 Area of integration for the difference of two non-negative random variables.

It follows, then, that with the above formulation one need not resort to Monte Carlo simulation (see Subsec. 10.4.2). At worst one must evaluate, numerically, the above integrals. In some cases the $h(z)$ can be obtained in closed form. In the often considered case in which x and y are normally distributed, the probability density function of z is known to be normal. Hence, $P(z \leq 0)$ is obtainable from tables of the normal curve. It will be shown to be true in Subsec. 10.4.1.

3. From the definition of a probability distribution function one sees from Eq. (10.2) that the probability of failure can be expressed as

$$P(x \leq y) = \int_0^\infty F(y)g(y)\,dy \tag{10.4}$$

where $F(y)$ is the probability distribution function of x evaluated at the point y. Equation (10.4) is convenient to use when $F(y)$ is easily determined, as in the case of parameters that are Weibull distributed.

An equivalent representation obtainable from Eq. (10.2) is

$$P(x \leq y) = \int_0^\infty [1 - G(x)]f(x)\,dx \tag{10.5}$$

where $G(x)$ is the distribution function of the random variable y. Again this is convenient to work with in some cases, such as when the parameters are Weibull distributed.

10.3 INTERFERENCE OF TWO NORMAL DISTRIBUTIONS

If x and y are normally distributed with mean values μ_x and μ_y and variances σ_x^2 and σ_y^2, then $z = x - y$ is distributed with mean value $\mu_z = \mu_x - \mu_y$ and variance $\sigma_z^2 = \sigma_x^2 + \sigma_y^2$. Consequently, the probability of failure will be given by the area under the normal curve whose mean and variance are μ_z and σ_z^2, respectively. This area is found on the interval $(-\infty, 0)$.

The normal density function is

$$f(x) = \frac{1}{\sqrt{2\pi}\,\sigma_x} \exp\left[-\frac{(x - \mu_x)^2}{2\sigma_x^2}\right] \qquad -\infty < x < \infty$$

The probability density function of the random variable $z = x - y$, say $h(z)$, is given by

$$h(z) = \frac{1}{\sqrt{2\pi}\,\sigma_x \sigma_y} \exp\left[-\frac{(y - \mu_y)^2}{2\sigma_y^2}\right] \exp\left[-\frac{(z + y - \mu_x)^2}{2\sigma_x^2}\right]\,dy$$

Using the relation

$$\int_{-\infty}^\infty e^{-\frac{1}{2}r^2}\,dr = \sqrt{2\pi}$$

it can be shown that

$$h(z) = \frac{1}{\sqrt{2\pi}\,(\sigma_x^2 + \sigma_y^2)^{1/2}}\, \exp\left\{-\frac{[z - (\mu_x - \mu_y)]^2}{2(\sigma_x^2 + \sigma_y^2)}\right\} \qquad -\infty < z < \infty$$

That is, z is normally distributed with mean value $\mu_x - \mu_y$ and variance $\sigma_x^2 + \sigma_y^2$. From this it follows that the probability of failure $P(z \le 0)$ is the integral of the normal curve over $(-\infty, 0)$. (See Fig. 10.5.) The standardized normal variate (see Sec. 2.1) is then given by

$$z = \frac{(x - y) - (\mu_x - \mu_y)}{\sqrt{\sigma_x^2 + \sigma_y^2}} \qquad\qquad (10.6)$$

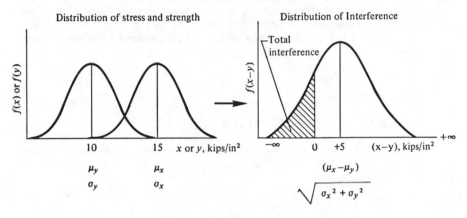

(a) When the mean strength μ_x is *more* than the mean stress μ_y

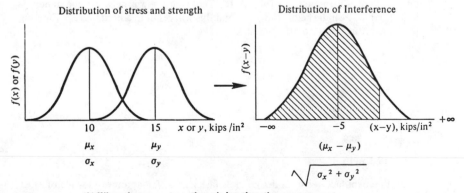

(b) When the mean strength μ_x is *less* than the mean stress μ_y

Fig. 10.5 Total interference (probability of failure) from the distribution of stress x and strength y.

Example 10.1 A leading fastener manufacturer has been approached by a company which desires a custom-made fastener. This fastener is similar to a standard cap screw, but has a tapered unthreaded shank with a specially designed head that will shear off when a torque of 350 in·lb is applied. The purchaser also specifies that the manufacturer is liable to the price of $2 per head that does not shear off during installation. This liquidated damage is to cover the expenses involved in the removal and reinstallation of a new fastener. To ensure that the manufacturer will not have more than 80 heads refuse to shear in the 20,000 to be produced, design the point of shear of the head. The standard deviation of the shear point of the material expressed in terms of the applied torque is 10 in·lb. The standard deviation of the torque wrenches used to tighten these fasteners is 6 in·lb of torque. Normal distributions are assumed.

Solution The probability of interference in this case is the probability of having a nonfailure, since the strength distribution will now be placed below the stress distribution. This probability of interference must be 80/20,000 = 0.004 or less.

The first task is to form a new random variable, where x = strength and y = stress,

$$z = y - x \quad \text{in·lb}$$

and standard deviation of

$$\sigma_z = \sqrt{\sigma_x^2 + \sigma_y^2} = \sqrt{100 + 36} = 11.66 \text{ in·lb}$$

When z is 0 or negative, the head will not shear off, since the strength is equal to or greater than the stress. Using Eq. (10.6) at $y - x = 0$ and the torque wrenches set at $\mu_y = 350$ in·lb,

$$y - x = 0$$

$$z_\alpha = \frac{\mu_x - 350}{11.66}$$

This is schematically shown in Fig. 10.6. The z_z below which 0.4 percent of the distribution lies is found from Table A-1. That is, if $y - x$ is negative, the standardized variable z_α will lie below this point and this will result in a nonfailure. Entering the table at the 0.004 point,

$$z_\alpha = -2.65$$

Then

$$z_\alpha = -2.65 = \frac{\mu_x - 350}{11.66}$$

$$\mu_x = 350 - 30.9$$

$$= 319.1 \text{ in·lb}$$

Thus, the manufacturer must design these fasteners to fail at a mean torque of 319.1 in·lb.

Figure 10.7 represents a convenient plot of Eq. (10.6).

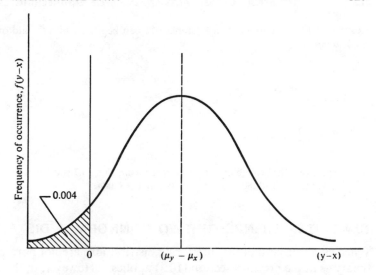

Fig. 10.6 Distribution of $y - x$.

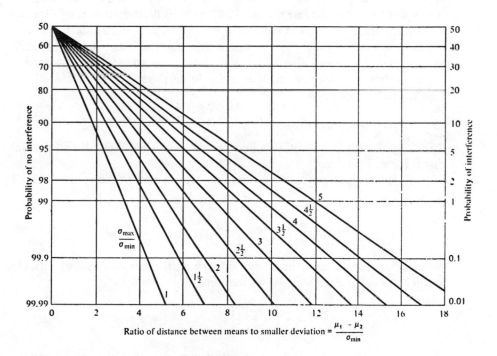

Fig. 10.7 Probability of interference of two normal distributions.

Example 10.2   The problem in Example 10.1 can be solved with the aid of Fig. 10.7 as follows:

Solution

$$\frac{\mu_1 - \mu_2}{\sigma_{min}} = \frac{350 - 319.1}{6} = 5.15$$

$$\frac{\sigma_{max}}{\sigma_{min}} = \frac{10}{6} = 1.667$$

Therefore, the probability of interference from Fig. 10.7 is 0.4 percent, which checks satisfactorily with 80 heads in 20,000, that is 0.004.

10.4   INTERFERENCE OF TWO NONNORMAL DISTRIBUTIONS

When either or both of the interfering distributions are not normal, the same theory as in the previous section (10.3) applies.   However, in the case of two normal distributions the resultant distribution of interference is also normal, and therefore the simple z distribution applies.   When the interfering distributions are not normal, the resultant distribution may not be normal. Hence, the solution can get quite involved.   Exact solution can be obtained by what is known as the integral method.   This is illustrated in the following section for the case when Weibull and normal distributions interfere.   For the exact solution of other cases of nonnormal distributions, see the appropriate references given in Table 10.1.   For those cases where only an approximate solution is adequate, a Monte Carlo method, discussed in Subsec. 10.4.2, can be used.

10.4.1   INTEGRAL METHOD

The integral method is illustrated here for the case when Weibull and normal distributions interfere.

Suppose the strength x is a random variable and it follows the Weibull density function

$$f(x) = \frac{1}{\theta - x_0} \left(\frac{x - x_0}{\theta - x_0}\right)^{b-1} \exp\left[-\left(\frac{x - x_0}{\theta - x_0}\right)^{b-1}\right] \quad \text{for} \quad x_0 \le x < \infty$$

Now, suppose the stress y imposed on these parts is also a random variable and it follows the normal density functions

$$f(y) = \frac{1}{\sqrt{2\pi}\,\sigma_y} \exp\left[-\frac{1}{2}\left(\frac{y - \mu_y}{\sigma_y}\right)^2\right] \qquad -\infty < y < +\infty$$

Then the resulting distribution of interference is an unknown function, which cannot be evaluated in a simple closed form.   Numerical analysis is

Table 10.1   Nonnormal distributions and references for the solutions

Case	Stress distribution	Strength distribution	Reference
1	Normal	Weibull	1 and 2
2	Weibull	Weibull	1
3	Normal	Smallest extreme value	2
4	Normal	Largest extreme value	2
5	Exponential	Exponential	1*
6	Gamma	Gamma	1*

*These cases are discussed only. The solutions are not tabulated.

available [1, 2] which lists the interference (probability of failure) values for various values of the parameters b, $(\theta - x_0)/\sigma$, and $(x_0 - \mu)/\sigma$. These are tabulated in Refs. 1 and 2, and sample tables are given in Tables 10.4 and 10.5.

The following discussion indicates how to determine the statistical distributions of the strength and stress parameters and the resulting interference.

The statistical distribution of strength   Most fatigue testing involves subjecting a number of specimens or parts to the same stress and repeating this process at various stress levels. The data thus obtained, known as life data, are used to construct the conventional S-N diagram. In this case, the scatter obtained is the scatter in life at a given stress. To obtain the scatter data for fatigue strength at a given life, it is necessary to fatigue-test all the specimens with different stresses imposed on them in such a manner that all would fail at a predetermined life. Practically, this is impossible; therefore, an alternative method is suggested.

The fatigue life data are plotted on the conventional S-N diagram. Here, it is assumed that to each specimen or part of the population can be attributed an individual S-N curve, and that there exists for any population (at fixed test conditions) a family of nonintersecting S-N curves, which can be determined with any desired accuracy, each curve corresponding to a given probability. The average S-N curve is then fitted to all the test points on the S-N diagram by using the least-squares method. Passing through each test point, S-N curves parallel to the average S-N curve are drawn. These will make a family of S-N curves (see Fig. 10.8). Now if the fatigue-strength distribution at $N = N_1$ life is required, a vertical line is drawn at $N = N_1$ intersecting the family of S-N curves. These points of intersection S_1, S_2, ... represent a sample from the strength distribution at a desired life [3].

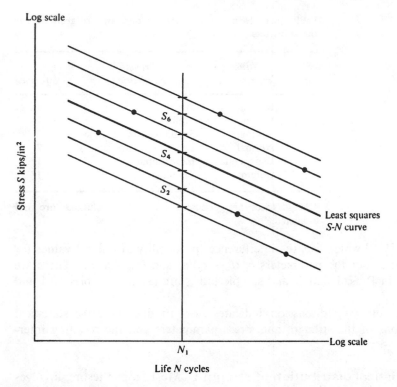

Fig. 10.8   *S-N* diagram for converting life data to strength data.

These data are then plotted on the Weibull probability paper as a cumulative distribution function, and the three Weibull parameters x_0, b, and θ are determined in the manner described in Chap. 2.

The statistical distribution of stress   The problem of stress distribution, in the interference theory, appears to be much more involved than the problem of strength distribution.   Consider, for example, the problem of a connecting rod in a reciprocating engine.   Because of the variation in hardness, surface finish, etc., the fatigue strength will vary from one rod to another. This will result in a distribution curve, in which the strength will be plotted on the abscissa and the number of rods having a given strength (i.e., frequency of occurrence) on the ordinate.

Consider now the stress distribution in the connecting rods.   The stresses in the rod result from the combined effect of gas pressure and inertia loading.   If the attention is now focused on a single rod, then the variation in the two types of loading will produce a distribution of stresses in this particular rod.   The resultant curve will be a plot of the stresses in the rod on the abscissa and the number of times that this stress occurs in this particular rod on the ordinate (Fig. 10.9a).

This, however, is not what is wanted in the application of the interference theory, because this distribution of stresses cannot be matched with the distribution of strength. In the strength distribution, the ordinate gives the number of rods having a given strength. Therefore, in stress distribution the ordinate must read the number of rods having given stress (and not number of times given stress occurs in a single rod). This can be obtained by considering the fact that different engines will be subjected in service to different operating conditions and, therefore, the distribution of gas pressure loading and inertia loading will vary from engine to engine. As pointed out in the following section, a spectrum of stresses must be converted to an

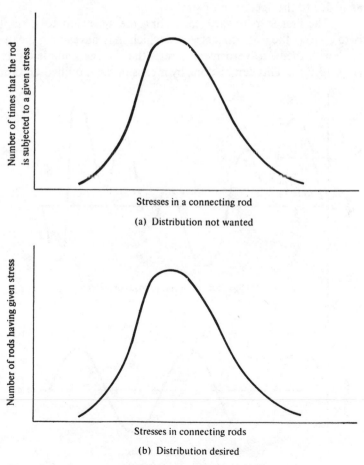

Fig. 10.9 Stress distribution for the interference theory.

equivalent stress for the purpose of interference theory. Therefore, if a spectrum of loading due to different service conditions varies from engine to engine, in a population of connecting rods, the equivalent stress will vary from rod to rod. Thus, the statistical stress distribution desired for the interference theory may be obtained (Fig. 10.9*b*). In this distribution the equivalent stress will be plotted on the abscissa and the number of rods (frequency of occurrence) having that stress on the ordinate. This distribution can be compared with the strength distribution to obtain the probability of interference.

By definition, equivalent stress is a completely reversed stress of constant amplitude which, when imposed on a part, should cause failure at the same life as if the stress spectrum were imposed instead. Thus, the damage accumulated at any given life, due to this equivalent stress, will be the same as if due to the spectrum of stresses.

The first step toward converting the spectrum to a single stress S_{equ} is to convert the operating stresses, which may have some mean stress associated with them, to zero mean stress, that is, the completely reversed stress (Fig. 10.10). This can be done by means of the modified Goodman diagram.

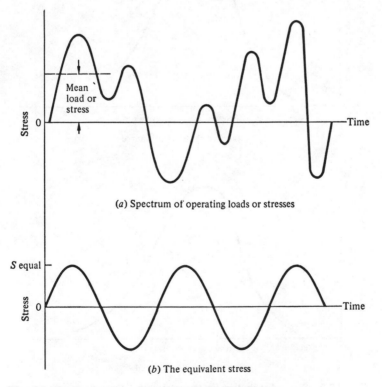

(*a*) Spectrum of operating loads or stresses

(*b*) The equivalent stress

Fig. 10.10 Conversion of stress spectrum to equivalent stress.

Draw the Goodman diagram as shown in Fig. 10.11. From the spectrum of operating stresses, plot each stress cycle on this diagram as shown (for example, line *AB*). Connect *DA* and *DB* and extend to the vertical line where mean stress is equal to zero. Hence, *XY* is the zero mean stress equivalent to *AB*. After reducing all such stress cycles to zero mean stress, the stress spectrum will have all the stress cycles completely reversed. The magnitude *XY* will be different for different stress cycles. Therefore, the original operating stress spectrum (Fig. 10.10*a*), with various mean stress levels, is thus reduced to a stress spectrum with zero mean stress level, that is, to a completely reversed stress (Fig. 10.12). This spectrum can then be reduced to a single equivalent stress of constant amplitude, by means of Miner's or Corten-Dolan's rule.

Miner's rule   Miner's rule [4] assumes that the total life of a component can be estimated by simply adding the fraction of life consumed by each

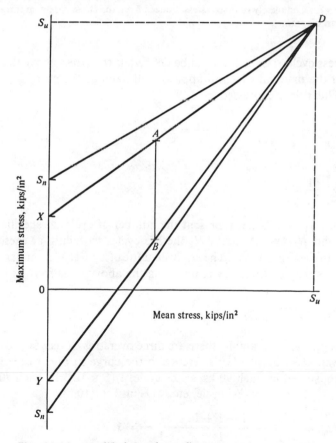

Fig. 10.11   Modified Goodman diagram.

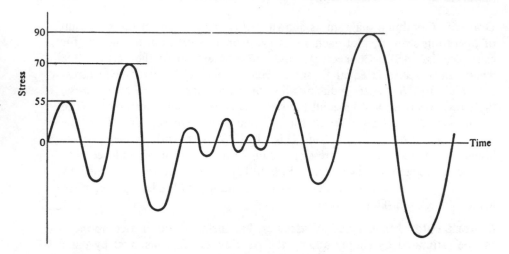

Fig. 10.12 Completely reversed stress reduced from the stress spectrum through modified Goodman diagram.

overstress cycle.   Overstress can be defined as the stress above the endurance limit of the material which, if applied, will damage the part.

This rule is expressed as

$$\frac{n_1}{N_1} + \frac{n_2}{N_2} + \frac{n_3}{N_3} + \cdots + \frac{n_k}{N_k} = 1$$

or

$$\sum_{i=1}^{i=k} \frac{n_i}{N_i} = 1 \tag{10.7}$$

where $n_1, n_2, n_3, \ldots, n_k$ represent the number of cycles at specific overstress levels, and $N_1, N_2, N_3, \ldots, N_k$ the life cycles to failure at these levels, as read from the S-N curve.   The equivalent life of a part (N_{equ}) under a spectrum of stresses may be found by rearranging the above equation:

$$N_{equ} = \frac{\sum_{i=1}^{i=k} n_i}{\sum_{i=1}^{i=k} n_i/N_i} \tag{10.8}$$

Suppose, for example, there are three overstress levels, 90, 70 and 55 ksi, in a given spectrum.   With reference to the curve in Fig. 10.13, $1/(6 \times 10^4)$ life is consumed by each 90-ksi stress cycle, $1/(5 \times 10^5)$ by each 70-ksi cycle, $1/(8 \times 10^5)$ by each 55-ksi cycle, etc.   Using Eq. (10.8),

$$N_{equ} = \frac{1 + 1 + 1}{\dfrac{1}{6 \times 10^4} + \dfrac{1}{5 \times 10^5} + \dfrac{1}{8 \times 10^5}} = 1.5 \times 10^5 \text{ cycles}$$

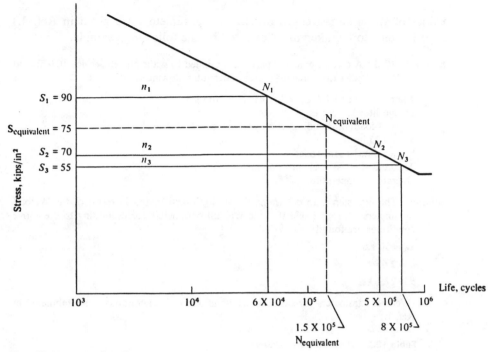

Fig. 10.13 Miner's rule.

Thus, the life of the part under the above spectrum of stresses will be equivalent to a life of 1.5×10^5 cycles. The stress equivalent to this life is (from Fig. 10.13) 75 ksi. Hence, the damage that the part accumulates due to the above spectrum of varying stress amplitude will be the same as if stress cycles of constant amplitude equal to S_{equ} (in this case, 75 ksi) were imposed for N_{equ} (1.5×10^5 cycles). Thus, the spectrum of stresses can be replaced by a single stress.

Miner's rule, as stated in Eq. (10.7), gives 1.0 as the criterion for failure. Miner's original tests showed that the value for the summation in Eq. (10.7) actually varied between 0.61 and 1.45. His more recent data [5] give a range of 0.7 to 2.2. Other sources [6] quote a range as high as 0.18 to 23.0. In view of this scatter it is generally agreed that the value of 1.0, originally proposed by Miner, is probably the best overall estimate that can be made at this time.

Determination of interference Once the parameters of the strength distribution (x_0, b, θ) and stress distribution ($\mu = S_{equ}$ and $\sigma = k\mu$, where k represents a fraction of the average stress) are determined, the percent interference, that is, percent failures, can be computed with the aid of Tables 10.4 and 10.5. (Tables of interference values corresponding to a wide range of

values of stress and strength distribution parameters are given in Ref. 1.)
Specific steps to be taken are illustrated by the following example.

Example 10.3   A certain machine part was designed to withstand in service 10,000 load
cycles.   Predict the percent failures under the following conditions:

Material: Ti-6Al-4V, $S_u = 177$ ksi, $S_y = 166$ ksi
Design life: 10^4 cycles
Type of loading: Bending, completely reversed
Size: 0.25 in
Surface finish: Hot-rolled
Theoretical stress concentration factor: $K_t = 1.0$
Operating temperature: 600°F

Solution   The first step is to determine the strength distribution in terms of the Weibull
parameters.   From Table 10.2 the Weibull parameters corresponding to the above
conditions are found:

$x_0 = 50$ ksi

$b = 2.65$

$\theta = 77.1$ ksi

(Weibull parameters for other materials under various conditions are tabulated in
Ref. 1.)

Table 10.2   Weibull parameters*

T, °F	H.T.[†]	S_m, ksi	x_0, ksi			b			θ, ksi		
Life, cycles:			10^4	10^5	10^6	10^4	10^5	10^6	10^4	10^5	10^6
Effect of temperature											
80	A	0	67.0	57.0	50.0	2.8	2.85	3.35	89.2	78.0	68.0
400	A	0	75.0	56.0	41.0	2.75	2.95	3.1	96.1	72.7	55.3
600	A	0	50.0	40.0	33.0	2.65	2.7	3.0	77.1	60.5	47.0
800	A	0	43.0	34.0	26.0	2.9	3.1	3.25	71.0	55.0	44.0
900	A	0	45.0	34.0	25.0	3.5	4.0	4.1	65.0	49.7	38.0
80	B	0	70.0	55.0	46.0	3.05	3.35	3.48	97.7	80.2	66.1
400	B	0	45.0	35.0	25.0	1.9	2.2	2.7	78.8	57.5	44.0
600	B	0	42.0	32.0	22.0	3.15	3.2	3.6	79.2	59.9	44.5
800	B	0	40.0	30.0	20.0	3.25	3.5	4.1	71.7	50.4	36.6
Effect of heat treatment											
80	A	0	70.0	40.0	26.0	2.9	3.8	4.1	118.6	77.2	50.2
80	B	0	70.0	55.0	46.0	3.05	3.35	3.48	97.7	80.2	66.1

* Ti-6Al-4V; $S_u = 177$ ksi, $S_y = 166$ ksi; rotary bending.
† Heat treatments:
A: Solution treated 1690°F, 12 min, WQ, aged 900°F, 4 h, air cooled.
B: Solution treated 1675°F, 20 min, WQ, aged 900°F, 4 h, air cooled.

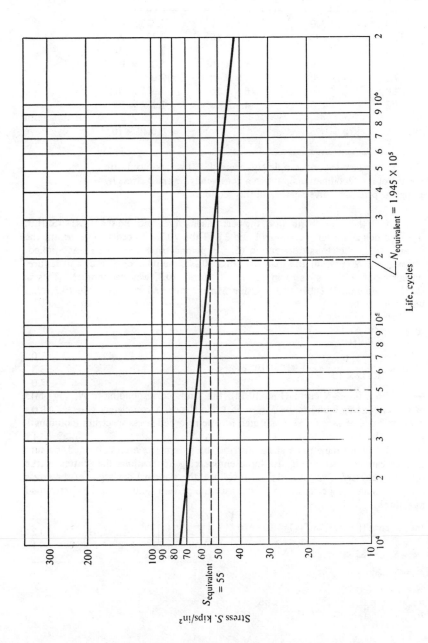

Fig. 10.14 S-N relationship for Example 10.3.

Table 10.3   Stress and life data for Example 10.3

(1) Completely* reversed stress S, ksi	(2) Occurrences n, cycles	(3) Number of cycles to failure, N	(4) $\dfrac{n}{N}$
52.0	200	3.5×10^5	5.710×10^{-4}
54.1	80	2.4×10^5	3.333×10^{-4}
56.5	50	1.6×10^5	3.125×10^{-4}
58.0	60	1.2×10^5	5.000×10^{-4}
59.3	20	1.0×10^5	2.000×10^{-4}
62.0	10	6.6×10^4	1.515×10^{-4}
64.8	5	4.3×10^4	1.162×10^{-4}
	$\sum n_i = 425$		$\sum \dfrac{n_i}{N_i} = 21.845 \times 10^{-4}$

* Actually, stress was not completely reversed.   It was reduced with the aid of the Goodman diagram to a completely reversed stress by using the procedure given earlier in this section.

As to the stress distribution, the part was instrumented and the stress spectrum was recorded as shown in columns 1 and 2 of Table 10.3.   In order to determine the parameters of the stress distribution ($S_{equ} = \mu$, and σ), Miner's rule was applied. From the S-N curve of the material (Fig. 10.14), the number of cycles to failure, N, corresponding to stresses in column 1, Table 10.3, was determined.   This is shown in column 3, Table 10.3.   Using Miner's rule and data in Table 10.3, N_{equ} was found to be

$$N_{equ} = \frac{\sum n_i}{\sum n_i/N_i}$$

$$= \frac{425}{21.845 \times 10^{-4}} = 1.945 \times 10^5 \text{ cycles}$$

From the S-N curve (Fig. 10.14), the stress corresponding to $N_{equ} = 1.945 \times 10^5$ cycles was found to be $S_{equ} = 55$ ksi.   Hence, a completely reversed stress application of 55 ksi can be substituted for the recorded stress spectrum (columns 1 and 2, Table 10.3).

Once the strength and stress distribution parameters are established, percent failures can be determined.   In some engineering applications the scatter in the operating stresses is very small and, therefore, the standard deviation of the stress can be assumed to be zero.   In those cases the percent failures can be determined as follows:

$$\text{Interference (failures)} = F(x) = 1 - \exp\left[-\left(\frac{x - x_0}{\theta - x_0}\right)^b\right] \text{ where}$$

$x = S_{equ} = 55$ ksi

$x_0 = 50$ ksi

$b = 2.65$

$\theta = 77.1$ ksi

$$F(x) = 1 - \exp\left[-\left(\frac{55-50}{77.1-50}\right)^{2.65}\right]$$

$$= 1 - e^{-0.0114}$$

$$= 0.0113$$

Percent failures $= 1.13\%$

Table 10.4   Interference values [1]*

A \ C	10	15	20	25	30	35	40	45	50	55
.8	.0011	.0005	.0003	.0002	.0001	.0001	.0001	.0001	.0000	.0000
.6	.0017	.0008	.0004	.0003	.0002	.0001	.0001	.0001	.0001	.0001
.4	.0025	.0011	.0006	.0004	.0003	.0002	.0002	.0001	.0001	.0001
.2	.0035	.0016	.0009	.0006	.0004	.0003	.0002	.0002	.0001	.0001
.0	.0049	.0022	.0012	.0008	.0006	.0004	.0003	.0002	.0002	.0002
−.2	.0067	.0030	.0017	.0011	.0008	.0006	.0004	.0003	.0003	.0002
−.4	.0089	.0040	.0023	.0014	.0010	.0007	.0006	.0004	.0004	.0003
−.6	.0116	.0052	.0030	.0019	.0013	.0010	.0007	.0006	.0005	.0004
−.8	.0149	.0067	.0038	.0024	.0017	.0012	.0010	.0008	.0006	.0005
−1.0	.0188	.0085	.0048	.0031	.0021	.0016	.0012	.0009	.0008	.0006
−1.4	.0284	.0128	.0073	.0047	.0032	.0024	.0018	.0014	.0012	.0010
−1.8	.0407	.0185	.0105	.0067	.0047	.0034	.0026	.0021	.0017	.0014
−2.2	.0557	.0254	.0144	.0093	.0065	.0047	.0036	.0029	.0023	.0019
−2.6	.0733	.0336	.0191	.0123	.0086	.0063	.0048	.0038	.0031	.0026
−3.0	.0935	.0431	.0246	.0158	.0110	.0081	.0062	.0049	.0040	.0033
−3.4	.1159	.0538	.0308	.0198	.0138	.0102	.0078	.0062	.0050	.0041
−3.8	.1406	.0658	.0377	.0243	.0170	.0125	.0096	.0076	.0062	.0051
−4.2	.1671	.0789	.0453	.0293	.0205	.0151	.0116	.0092	.0074	.0061
−4.6	.1954	.0930	.0536	.0347	.0243	.0179	.0137	.0109	.0088	.0073
−5.0	.2251	.1082	.0626	.0406	.0284	.0210	.0161	.0127	.0103	.0086
−5.5	.2640	.1286	.0748	.0486	.0341	.0251	.0193	.0153	.0124	.0103
−6.0	.3043	.1504	.0879	.0573	.0402	.0297	.0228	.0181	.0147	.0121
−6.5	.3457	.1735	.1020	.0667	.0468	.0346	.0266	.0211	.0171	.0142
−7.0	.3876	.1977	.1170	.0767	.0539	.0399	.0307	.0244	.0198	.0164
−8.0	.4713	.2490	.1493	.0985	.0695	.0516	.0398	.0316	.0256	.0212
−9.0	.5525	.3032	.1845	.1226	.0869	.0646	.0499	.0396	.0322	.0267
−10.0	.6285	.3591	.2222	.1488	.1059	.0790	.0611	.0486	.0396	.0328

* Stress distribution, normal; strength distribution, Weibull.

$$B(x) = 2.00, \quad C = \frac{\theta - x_0}{\sigma}, \quad A = \frac{x_0 - \mu}{\sigma}.$$

In those engineering applications where the scatter of stress is appreciable, percent failures may be found by noting that in engineering practice, standard deviation generally lies in the range

$$0.01 \leq \frac{\sigma}{\mu} \leq 0.10$$

Table 10.5   Interference values [1]*

A \ C	10	15	20	25	30	35	40	45	50	55
.8	.0001	.0000	.0000	.0000	.0000	.0000	.0000	.0000	.0000	.0000
.6	.0002	.0001	.0000	.0000	.0000	.0000	.0000	.0000	.0000	.0000
.4	.0004	.0001	.0000	.0000	.0000	.0000	.0000	.0000	.0000	.0000
.2	.0005	.0002	.0001	.0000	.0000	.0000	.0000	.0000	.0000	.0000
.0	.0008	.0002	.0001	.0001	.0000	.0000	.0000	.0000	.0000	.0000
−.2	.0011	.0003	.0001	.0001	.0000	.0000	.0000	.0000	.0000	.0000
−.4	.0016	.0005	.0002	.0001	.0001	.0000	.0000	.0000	.0000	.0000
−.6	.0022	.0007	.0003	.0001	.0001	.0001	.0000	.0000	.0000	.0000
−.8	.0030	.0009	.0004	.0002	.0001	.0001	.0000	.0000	.0000	.0000
−1.0	.0041	.0012	.0005	.0003	.0002	.0001	.0001	.0000	.0000	.0000
−1.4	.0069	.0021	.0009	.0004	.0003	.0002	.0001	.0001	.0001	.0000
−1.8	.0111	.0033	.0014	.0007	.0004	.0003	.0002	.0001	.0001	.0001
−2.2	.0169	.0051	.0022	.0011	.0006	.0004	.0003	.0002	.0001	.0001
−2.6	.0247	.0075	.0032	.0016	.0009	.0006	.0004	.0003	.0002	.0002
−3.0	.0349	.0106	.0045	.0023	.0013	.0008	.0006	.0004	.0003	.0002
−3.4	.0475	.0145	.0062	.0032	.0018	.0012	.0008	.0005	.0004	.0003
−3.8	.0630	.0193	.0082	.0042	.0024	.0015	.0010	.0007	.0005	.0004
−4.2	.0815	.0252	.0108	.0055	.0032	.0020	.0014	.0010	.0007	.0005
−4.6	.1031	.0322	.0138	.0071	.0041	.0026	.0017	.0012	.0009	.0007
−5.0	.1279	.0404	.0173	.0089	.0052	.0033	.0022	.0015	.0011	.0008
−5.5	.1634	.0524	.0225	.0116	.0067	.0043	.0029	.0020	.0015	.0011
−6.0	.2037	.0665	.0287	.0148	.0086	.0054	.0036	.0026	.0019	.0014
−6.5	.2485	.0828	.0360	.0186	.0108	.0068	.0046	.0032	.0023	.0018
−7.0	.2973	.1013	.0443	.0230	.0134	.0084	.0057	.0040	.0029	.0022
−8.0	.4039	.1454	.0645	.0336	.0196	.0124	.0083	.0059	.0043	.0032
−9.0	.5165	.1985	.0897	.0471	.0276	.0175	.0117	.0083	.0060	.0045
−10.0	.6269	.2600	.1202	.0636	.0374	.0237	.0160	.0112	.0082	.0062

* Stress distribution, normal; strength distribution, Weibull.

$$B(x) = 3.00, \quad C = \frac{\theta - x_0}{\sigma}, \quad A = \frac{x_0 - u}{\sigma}.$$

In the absence of any specific information, an approximate average value of σ/μ = 0.05 can probably be assumed. By using this value, percent failures are determined:

Strength	Stress
$x_0 = 50$ ksi	$\mu = S_{equ} = 55$ ksi
$b = 2.65$	$\sigma = 0.05\mu$
$\theta = 77.1$ ksi	$= 0.05 \times 55$ ksi
	$= 2.75$ ksi

From the above data, parameters C, A, and $B(x)$, to be used in the tables of interference values (Tables 10.4 and 10.5), are computed:

$$C = \frac{\theta - x_0}{\sigma} = \frac{77.1 - 50}{2.75} \simeq 10$$

$$A = \frac{x_0 - \mu}{\sigma} = \frac{50 - 55}{2.75} = -1.82$$

$$B = b = 2.65$$

The interference value corresponding to these parameters can be determined by the interpolation between Table 10.4 [for $B(x) = 2.0$] and Table 10.5 [for $B(x) = 3.0$]:

Interference $\simeq 0.0245$

or

Percent failures $= 2.45\%$

Thus, probabilities of failure to be expected are:

In the event of no scatter in stresses, 1.13 percent failures
For the scatter of the order of 0.05μ, 2.45 percent failures

10.4.2 MONTE CARLO METHOD

Essentially, the Monte Carlo method is a sophisticated means of randomly selecting a sample from one distribution and comparing it with a sample taken from a different distribution. The use of this technique in the interference theory involves a comparison between two distributions (stress vs. strength, shaft dimensions vs. bore dimensions, etc.). However, the Monte Carlo method can be extended to a variety of complex system problems consisting of more than two components. If the statistical distribution of each component in the system is known, the distribution of the system output can be estimated. (In case of stress and strength the output is failure, and for shaft and bore it is the dimensional interference.)

The technique employs random numbers, which are essentially a collection of random digits. If the probability of occurrence is constant

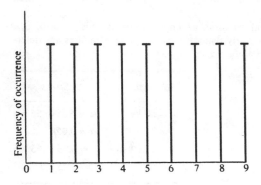

Fig. 10.15 Rectangular density function from which random numbers are selected.

(rectangular distribution), random numbers are generated from the digits 0, 1, ..., 9 (Fig. 10.15).

Physically, a set of equal-probability random numbers can be generated by placing 10 slips of paper in a hat. On each slip one digit, 0, 1, ..., or 9, is written. Then, at random, one slip is selected, the number is recorded in a vertical column, and the slip is returned to the hat. This process is continued until sufficient random numbers are generated. Tests for randomness of sample are available (see Subsec. 8.3.4). It is apparent that this is a laborious task. Therefore, tables are published which provide a convenient source of these numbers [7].

There are many situations where the probability of occurrence of a random variable is not constant, as in the case of data which follow normal, Weibull, exponential, etc., distributions. Tables of random numbers based on each of these distributions can be generated. Here, only random numbers for the normal distribution are given. These numbers are in terms of the random normal deviates (Fig. 10.16). Five hundred random normal deviates are given in Table A-48. An additional 10,000 random normal deviates are available [7].

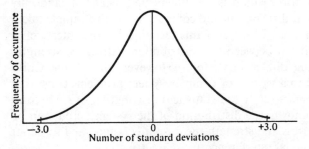

Fig. 10.16 Normal density function from which random numbers (random normal deviates) are selected.

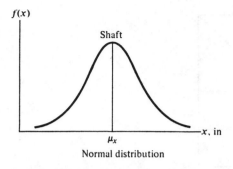 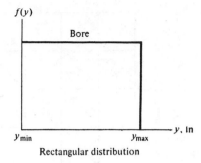

Fig. 10.17 Distributions of shaft and bore dimensions.

Example 10.4 A shaft whose dimension is normally distributed is to be assembled into a bore with a rectangular distribution. The corresponding dimensions are **given** below (also see Fig. 10.17). Dimensions of the shafts and the bores are:

	Shaft	*Bore*
	$\mu_x = 0.999$ in	$y_{min} = 1.000$ in
	$\sigma_x = 0.0005$ in	$y_{max} = 1.002$ in

Find:

1. Mean value of clearance
2. Percent assemblies having clearance above 0.003 in
3. Percent assemblies having interference (zero clearance or less)

Solution The problem requires simulation by the Monte Carlo method since the distribution of a new random variable formed from the normal and the rectangular density function is unknown. This is done by first constructing cumulative density function (CDF) plots for the shaft and the bore.

For the shaft, the CDF plot of normal distribution is a straight line on a cartesian paper when the ordinate is in terms of number of standard deviations and the abscissa in inches of shaft diameter. Hence, plot $\mu_x = 0.999$ in on the abscissa with the corresponding value 0 on the ordinate, and $\mu_x + \sigma_x = 0.9995$ in on the abscissa and $+1$ on the ordinate (Fig. 10.18). A line through these two points represents the CDF plot.

For the bore, the CDF plot of rectangular distribution is a straight line on cartesian paper. Plot $y_{min} = 1.000$ in on the abscissa and the corresponding value zero on the ordinate; $y_{max} = 1.002$ in on the abscissa and 1.0 on the ordinate (Fig. 10.19). A line through these two points represents the CDF plot.

Enter the appropriate random-number table arbitrarily, where these numbers are used to represent the cumulative probability of occurrence of the shaft and the

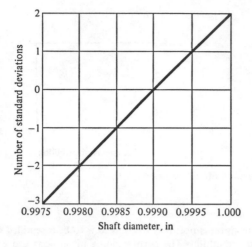

Fig. 10.18 CDF plot for the shaft.

bore values. Random normal deviates from Table A-48 are used for the normal distribution since the probability of occurrence of the shaft diameter is not constant. For the bore dimensions, the rectangular random numbers from Table A-47 are used because the probability of occurrence of the bore dimension is constant. Fifteen random normal deviates are obtained for the shaft from Table A-48, and fifteen random numbers for the bore from Table A-47 (see Table 10.6). The shaft and the bore dimensions corresponding to these numbers are read off from Figs. 10.18 and 10.19, respectively. These values are also given in Table 10.6.

Table 10.6   Monte Carlo simulation of a shaft-bore clearance

| | *Shaft* | | *Bore* | | |
No.	*Random normal deviates from Table A-48*	*Corresponding values of shaft diameter from Fig. 10.18*	*Random numbers from Table A-47*	*Corresponding values of bore diameter from Fig. 10.19*	*Clearance (bore-shaft)*
1	−0.958	0 998525	0.4285	1.00086	0.00233
2	1.742	0.999875	0.6614	1.00132	0.00144
3	−0.646	0.998675	0.5475	1.00110	0.00242
4	−0.370	0.998820	0.7553	1.00151	0.00269
5	0.825	0.999420	0.6041	1.00122	0.00180
6	0.259	0.999125	0.9967	1.00199	0.00287
7	0.859	0.999430	0.2659	1.00054	0.00111
8	−0.428	0.998790	0.6911	1.00138	0.00260
9	0.408	0.999200	0.6400	1.00128	0.00208
10	−0.042	0.998960	0.1471	1.00030	0.00134
11	−0.993	0.998500	0.4147	1.00083	0.00233
12	0.190	0.999090	0.9456	1.00190	0.00281
13	−0.119	0.998940	0.4162	1.00083	0.00189
14	0.142	0.999070	0.5027	1.00110	0.00203
15	0.203	0.999100	0.4140	1.00083	0.00173

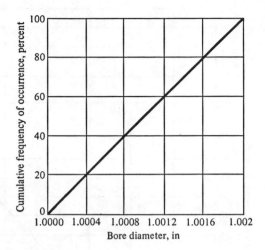

Fig. 10.19 CDF plot for the bore.

Fifteen random numbers are taken here for the purpose of illustrating the procedure; for better simulation and higher accuracy a larger number may be necessary. This can be done with the aid of a digital computer. By subtracting shaft dimensions from bore dimensions, a set of clearances is obtained. Arrange the clearances in an increasing order and assign median ranks:

Clearance, in	Median ranks
0.00111	0.0452
0.00134	0.1101
0.00144	0.1751
0.00173	0.2401
0.00180	0.3051
0.00189	0.3700
0.00203	0.4350
0.00208	0.5000
0.00233	0.5650
0.00233	0.6300
0.00242	0.6949
0.00260	0.7599
0.00269	0.7599
0.00269	0.8249
0.00281	0.8899
0.00287	0.9548

The clearance values and their corresponding median ranks (cumulative frequency of occurrence) should be plotted on various probability papers to determine the best-fitting distribution. Here these data are plotted only on a normal probability paper as a first try, and they give a good linear plot indicating that the distribution of clearance is approximately normal (Fig. 10.20). The clearance values are plotted on the abscissa and the corresponding median ranks on the ordinate.

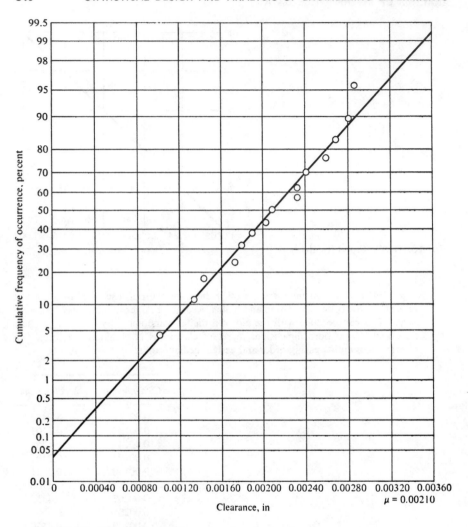

Fig. 10.20 Distribution of clearance values (normal).

Conclusion:

1. The mean value of clearance (50 percent value) is 0.0021 in.
2. Percent clearance above 0.003 in is 7.5.
3. Percent interference (zero clearance) is 0.04.

In order to determine whether different sets of 15 random numbers selected arbitrarily from different parts of the tables would yield essentially the same results, the above procedure was repeated for two more sets of random numbers. The results of all three sets are plotted in Fig. 10.21. These results show closer agreement near the median value of the clearance than at the tail end of the distribution. This indicates that in the present example more than 15 random numbers should have been taken.

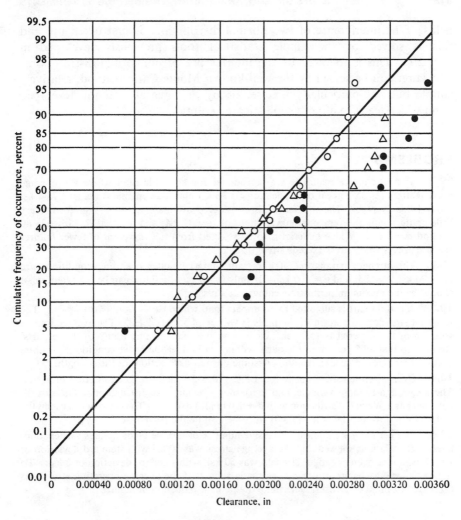

Fig. 10.21 Comparison of results based on three different sets of random numbers selected arbitrarily from the same table.

SUMMARY

Chapters 3 through 9 were devoted to the organization of experiments so that the resultant data would be statistically significant, and also to the analysis of these data. The present chapter deals specifically with the analysis of data, the type of data which are obtained when two distributions interfere with each other. The amount of interference can then be taken as failures, scrap, rejects, etc. These are the principal fields of application described here, although the techniques developed can be applied to a variety of other fields. The interference of two normal distributions is first taken up, and this is solved by the simple statistical tools previously developed in Sec. 2.1. This is followed by the interference of nonnormal distributions. The latter can be solved by the well-known Monte Carlo method, relatively simple but not particularly precise, or by an integral method, developed here, which leads to a more sophisticated solution.

PROBLEMS

10.1. Two alternative materials are considered for an application where the average applied stress for 95 percent of applications is 30 ksi and the dynamic stress is ± 7.84 ksi around this average. The two materials considered are material A, very uniform (very little scatter) with an average fatigue strength of $\mu = 37$ ksi; and material B, with $\mu = 40$ ksi and $\sigma = 5$ ksi. Which material is to be selected from the viewpoint of fewer failures (failure occurs when stress exceeds strength)?

10.2. The strength parameters of a certain material under operating conditions are $x_0 = 70$ ksi, $b = 2.5$, and $\theta = 100$ ksi. If only 0.5 percent failures can be tolerated, what should be the maximum permissible equivalent stress?

10.3. A certain part is subjected to an impact load which varies from part to part. The resultant stress on 95 percent of the parts is within 14.5 to 15.5 ksi. The average impact strength of the material is 16.0 ksi. Determine the required uniformity σ of the parts' strength so that 99.5 percent of the parts will not fail under the above condition. Failure occurs when stress exceeds strength. Assume normal distribution for the strength.

10.4. On a given assembly line a rod 1.995 ± 0.005 in long is to be inserted in a slot. The length is a random variable, and it follows a rectangular distribution. The slot dimension has a Weibull distribution with $\theta = 2.0$ in and $b = 2.5$. The clearance between the rod and the slot must be at least 0.001 in. Determine the percent scrap.

10.5. In a certain gas turbine an excessive number of turbine blade failures was encountered. Calculations showed that the average stress was 30 ksi with standard deviation of 10 ksi, while the average fatigue strength was 50 ksi, with standard deviation of 2 ksi. To reduce failure, a change in design was made such that the average operating stress was decreased from 30 to 20 ksi. Determine the percent reduction in failures.

REFERENCES

1. Lipson, C., N. J. Sheth, and R. Disney: Reliability Prediction—Mechanical Stress/Strength Interference, *Final Tech. Rep.* RADC-TR-66-710, Rome Air Development Center, Research and Technology Division, Air Force Systems Command, Griffiss Air Force Base, New York, March, 1967.
2. Lipson, C., N. J. Sheth, R. L. Disney, and M. Altun: Reliability Prediction—Mechanical Stress/Strength Interference (Nonferrous), *Final Tech. Rep.* RADC-TR-68-403, Rome Air Development Center, Research and Technology Division, Air Force Systems Command, Griffiss Air Force Base, New York, February, 1969.
3. Weibull, W.: "Fatigue Testing and Analysis of Results," The Macmillan Company, New York, 1961.
4. Miner, M. A.: Cumulative Damage in Fatigue, *J. Appl. Mech.*, vol. 12, no. 3, p. A-159, September, 1945.

5. Miner, M. A.: Estimation of Fatigue Life with Particular Emphasis on Cumulative Damage, in G. Sines and J. L. Waisman (eds.), " Metal Fatigue," chap. 12, McGraw-Hill Book Company, New York, 1959.
6. Dolan, T. J., F. E. Richart, and C. E. Work. Influence of Fluctuations in Stress Amplitude on the Fatigue of Metals, *ASTM Proc.*, vol. 49, 1949.
7. Rand Corporation: "A Million Random Digits with 100,000 Normal Deviates," The Free Press of Glencoe, Ill., Chicago, 1955.

BIBLIOGRAPHY

Baur, E. H.: Skewed Load-Strength Distribution in Reliability, *Aero General Corp. Rep.* 9200 6 64, Sacramento, Calif., AD 434-414, Feb. 10, 1964.
Bratt, M. J., G. Reethof, and G. W. Weber: A Model for Time Varying and Interfering Stress-Strength Probability Density Distributions with Consideration for Failure Incidence and Property Degradation, *Aerosp. Rel. Maintainability Conf., Washington, D.C.*, July, 1964.
Bussiere, R.: Method for Critiquing Designs and Predicting Reliability in Advance of Hardware Availability, *SAE Paper* 343A, 1961.
Corten, H. T.: Application of Cumulative Fatigue Damage Theory to Farm and Construction Equipment, *SAE Paper* 735 A, September, 1963.
————: Overstressing and Understressing in Fatigue (Cumulative Fatigue Damage), " Metals Engineering-Design," 2d ed., ASME Handbook, 1965.
———— and T. J. Dolan: Cumulative Fatigue Damage, "The International Conference on Fatigue of Metals," London, Sept. 10–14; New York, Nov. 28–30, 1956.
Dieter, G. E., and R. F. Mehl: Investigation of the Statistical Nature of the Fatigue of Metals, *NACA Tech. Note* 3019, September, 1953.
Dolan, T. J., F. E. Richart, and C. E. Work: Influence of Fluctuations in Stress Amplitude on the Fatigue of Metals, *ASTM Proc.*, vol. 49, 1949.
Eckert, L. A.: Design Reliability Prediction for Low Failure Rate Mechanical Parts, "Engineering Application of Reliability," The University of Michigan, Engineering Summer Conference, 1962.
Freudenthal, A. M., and E. J. Gumbel: Distribution Functions for the Prediction of Fatigue Life and Fatigue Strength, "The International Conference on Fatigue of Metals," London, Sept. 10–14; New York, Nov. 28–30, 1956.
Hanna, R. W., and R. C. Varnum: Interference Risk When Normal Distributions Overlap, *Ind. Qual. Contr. J.*, pp. 26–27, September, 1950.
Harris, J. P., and C. Lipson: Cumulative Damage Due to Spectral Loading, *Aerosp. Rel. Maintainability Conf. Proc.*, July, 1964.
Kaechele, L. E.: Probability and Scatter in Cumulative Fatigue Damage, *RAND Rep.* RM-3688-PR, December, 1963.
Kececioglu, D., and D. Cormier: Designing a Specified Reliability Directly into a Component, *Aerosp. Rel. Maintainability Conf., Washington, D.C.*, pp. 546–564, 1964.
Kullback, S.: The Distribution Laws of the Differences and Quotient of Variables Distributed in Pearson Type III Laws, *Ann. Math. Statist.*, vol. 7, no. 1, pp. 51–53, March, 1936.
Lipson, C., and R. C. Juvinall: "Handbook of Stress and Strength," The Macmillan Company, New York, 1963.
————, N. J. Sheth, and R. Disney: Reliability Prediction—Mechanical Stress/Strength Interference, *Final Tech. Rep.* RADC-TR-66-710, Rome Air Development Center, Research and Technology Division, Air Force Systems Command, Griffiss Air Force Base, New York, March, 1967.

————, ————, ————, and M. Altun: Reliability Prediction—Mechanical Stress/Strength Interference (Nonferrous), *Final Tech. Rep.* RADC-TR-68-403, Rome Air Development Center, Research and Technology Division, Air Force Systems Command, Griffiss Air Force Base, New York, February, 1969.

————, ————, and D. B. Sheldon: Reliability and Maintainability in Industry and the Universities, *Fifth Rel. Maintainability Conf.*, vol. 5, 1966.

Little, R. E.: "Multiple Specimen Testing and the Associated Fatigue Strength Response" The University of Michigan Press, Ann Arbor, 1966.

Miner, M. A.: Cumulative Damage in Fatigue, *J. Appl. Mech.*, vol. 12, no. 3, p. A-159, September, 1945.

————: Estimation of Fatigue Life with Particular Emphasis on Cumulative Damage, in G. Sines and J. L. Waisman (eds.), "Metal Fatigue," chap. 12, McGraw-Hill Book Company, New York, 1959.

Mittenbergs, A. A.: Fundamental Aspects of Mechanical Reliability, "Mechanical Reliability Concepts," ASME Design Engineering Conference, New York, May 17–20, 1965.

Rand Corporation: "A Million Random Digits with 100,000 Normal Deviates," The Free Press of Glencoe, Ill., Chicago, 1955.

Ransom, J. T., and R. F. Mehl: The Statistical Nature of the Endurance Limit, *Trans. AIME*, vol. 185, January, 1949.

Sinclair, G. M., and T. J. Dolan: Effect of Stress Amplitude on Statistical Variability in Fatigue Life of 75S-T6 Aluminum Alloy, *Trans. ASME*, vol. 75, p. 867, July, 1963.

Stulen, F. B.: On the Statistical Nature of Fatigue, *ASTM Spec. Tech. Publ.* 121, 1951.

———— and H. N. Cummings: Statistical Analysis of Fatigue Data, in J. Marin (ed.), "Proceedings for Short Course in Mechanical Properties of Metals," The Pennsylvania State University, Department of Engineering Mechanics, 1958.

Svensson, N. L.: Factor of Safety Based on Probability, *Des. Eng.*, vol. 191, no. 4845, pp. 154–155, Jan. 27, 1961.

Weibull, W.: "Fatigue Testing and Analysis of Results," The Macmillan Company, New York, 1961.

11
Analysis of Systems

The concept of a system is encountered in everyday life. A system may range from the simplest two-component type, such as a jar and its lid, to a four-component unit involving a power plant, leading wire, switch, and a lamp, and finally, to a complex system involved in a space program. Basically, there are two types of arrangements: (1) a system in series and (2) a system in parallel. In the series system the failure of one component will result in an inoperative system. In a parallel system if one component fails, the system will still function.

11.1 CONCEPTS AND DEFINITIONS

In a system in series a law of multiplication applies. For example, for a four-component system the probability of survival (reliability) R is

$$R(ABCD) = R(A) \times R(B) \times R(C) \times R(D) \tag{11.1}$$

or, in general,

$$R(x_1, x_2, \ldots, x_n) = R(x_1) \times R(x_2) \times \cdots \times R(x_n) \tag{11.2}$$

In the case of parallel circuits the probability of failure of one component is independent of the probability of failure of any other. Again, the multiplication rule is used, but in this case the probability of all components' failing is desired since this situation and only this situation results in system failure. If P represents the probability of failure, then

$$P(AB) = P(A) \times P(B) \tag{11.3}$$

$$R(AB) = 1 - P(A) \times P(B) \tag{11.4}$$

or, in general,

$$R(x_1, x_2, \ldots, x_n) = 1 - P(x_1) \times P(x_2) \times \cdots \times P(x_n) \tag{11.5}$$

In the development of a mathematical basis of series and parallel systems the following terms will be used:

$N =$ number of cycles or times a part has been in operation

$N_0 =$ total number of cycles or times that a component is designed to withstand

$i = i$th component in group of n

$n =$ total number of components

$C =$ confidence level

$R_i(N) =$ probability of survival at time N for ith component in system

$R_i(N_0) =$ probability of survival up to and including design time N_0 of ith component in system

$R_{i_c}(N_0) =$ reliability at time N_0 to which there is a confidence C attached that $R_i(N_0)$ actually is greater than or equal to the value $R_{i_c}(N_0)$

$P_{i_c}(N_0) =$ probability of failure at time N_0 of ith component to which there is a confidence C attached that $P_i(N_0)$ actually is less than or equal to $P_{i_c}(N_0)$; here $P_{i_c}(N_0) = 1 - R_{i_c}(N_0)$

11.2   SERIES DESIGN

With these new definitions Eq. (11.2) can be rewritten:

$$R_{123 \cdots n}(N_0) = R_1(N_0) \times R_2(N_0) \times \cdots \times R_n(N_0) \tag{11.6}$$

Applying the confidence level to these reliabilities,

$$R_{(123 \cdots n)c} = R_{1c}(N_0) \times R_{2c}(N_0) \times \cdots \times R_{nc}(N_0) \tag{11.7}$$

It will be noted that the product rule in Eq. (11.7) requires that all components have a common confidence level C, so that the resultant assembly reliability will be known to have the same confidence level. A special case of interest is that of components with equal reliabilities at a particular confidence level,

$$R_{(12 \cdots n)c}(N_0) = [R_{1c}(N_0)]^n \tag{11.8}$$

Example 11.1 A certain city is interested in determining how many street lights will fail to function, due to any cause, during an average period of 3 months.

Solution From the city maintenance files the following data were obtained:

1. In 5 years and 10 months the power supply failed twice, once affecting 1,076 lights and later affecting 274 lights (power transformer).
2. In 5 years and 10 months there were 25 line failures corresponding to a total of affected lights of 285.
3. Only one switching failure was experienced, causing 520 lights to fail to operate.
4. Average burning time in one day is 8 h. The average 3-month period is 91 d long. Total burning time is 728 h.
5. The probability of failure for 728 h was supplied by the bulb manufacturer as being 0.041.
6. The total average number of street lights was 2,072.
7. The total number of street lights at present is 3,052.

Summarizing the above data:

Desired time of operation = 3 months (728 h of burning time)
Number of lights failed out of 2,072 lights during 70-month period:

Power failure   = 1,076 + 274 = 1,350 lights
Line failure   = 285 lights
Switching failure = 520 lights

Number of lights at present = 3,052
Probability of bulb failure for 728 h operation = 0.041

The probabilities of failure are

$$P_{xc}(N_0) = \frac{\text{number of failures} \times 3 \text{ months of desired operation}}{\text{time for above failures} \times \text{number of lights in service}}$$

With the subscripts 1-power, 2-line, 3-switch, 4-light, the following are obtained with 50 percent confidence since the above data relate to sample:

$$P_{(1)50}(728 \text{ h}) = \frac{(1,076 + 274)(3)}{70(2,072)} = \frac{(1,350)(3)}{(70)(2,072)} = 0.028$$

$$P_{(2)50}(728 \text{ h}) = \frac{(285)(3)}{(70)(2,072)} = 0.0059$$

$$P_{(3)50}(728 \text{ h}) = \frac{(520)(3)}{(70)(2,072)} = 0.01075$$

$$P_{(4)50}(728 \text{ h}) = 0.041$$

It follows that

$$R_{(1)50}(728 \text{ h}) = 0.9720$$
$$R_{(2)50}(728 \text{ h}) = 0.9941$$
$$R_{(3)50}(728 \text{ h}) = 0.9893$$
$$R_{(4)50}(728 \text{ h}) = 0.9590$$

The total system probability of survival for 728 h of operation for one bulb, from Eq. (11.7), is

$$R_{(1234)50}(728) = 0.9720 \times 0.9941 \times 0.9893 \times 0.9590$$
$$= 0.915$$

The city should then expect on the average

$$3,052(1 - 0.915) = 260 \text{ failures per 3 months}$$

Hence, 260 street lights should cease to function out of 3,052 lights during any 3-month period started with brand-new bulbs.

11.3 PARALLEL DESIGN

Using the definitions presented in Sec. 11.1, the equation for the reliability of a parallel system, Eq. (11.5), can be written as

$$R_{12\cdots n}(N_0) = 1 - P_1(N_0) \times P_2(N_0) \times \cdots \times P_n(N_0) \tag{11.9}$$

where

$$P_1(N_0) = 1 - R_1(N_0) \tag{11.10}$$

Applying a confidence level to this expression,

$$R_{(123\cdots n)c}(N_0) = 1 - P_{1c}(N_0) \times P_{2c}(N_0) \times \cdots \times P_{nc}(N_0) \tag{11.11}$$

where

$$P_{1c}(N_0) = 1 - R_{1c}(N_0) \tag{11.12}$$

As before, the C associated with each component must be identical so that the system confidence will be equal to C. In the special case where the reliabilities of each component are equal at equal confidence levels,

$$R_{(123\cdots n)c} = 1 - [P_{1c}(N_0)]^n \tag{11.13}$$

11.4 SERIES AND PARALLEL DESIGN

In the case of a series and parallel design, the configuration is solved by a method known as the analysis by sub-assembly, illustrated below.

Example 11.2 Develop an expression for the reliability of the system shown in Fig. 11.1, which is used in a certain control system.

Solution The method of subassemblies suggests that any system can be broken down into small subassemblies. Thus (see Fig. 11.2)

$$R_{1c}(N_0) = R_{1c}(N_0) \times R_{2c}(N_0)$$
$$R_{1vc}(N_0) = R_{11c}(N_0) \times R_{12c}(N_0)$$

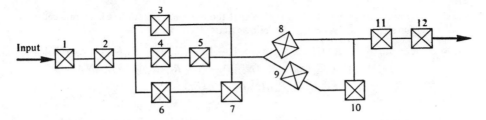

Fig. 11.1 Series and parallel design.

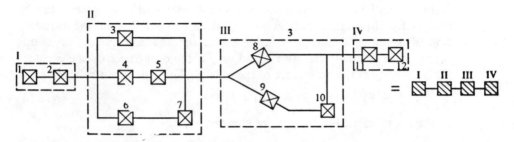

Fig. 11.2 Subassembly technique.

In Fig. 11.3 subsystem II is subdivided further. The reliability is

$$R_{(IIb)C}(N_0) = R_{4C}(N_0) \times R_{5C}(N_0)$$
$$R_{(IIa)C}(N_0) = R_{6C}(N_0) \times R_{7C}(N_0)$$

Then

$$R_{IIC}(N_0) = 1 - [1 - R_{3C}(N_0)][1 - R_{4C}(N_0) \times R_{5C}(N_0)][1 - R_{6C}(N_0) \times R_{7C}(N_0)]$$

Similarly,

$$R_{IIIC}(N_0) = 1 - [1 - R_{8C}(N_0)][1 - R_{9C}(N_0) \times R_{10C}(N_0)]$$

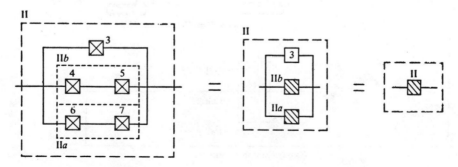

Fig. 11.3 Reduction of a subsystem.

By referring to Fig. 11.2, it is seen that the complex system has been reduced to a four-element series design. It follows that

$$R_{(system)_C}(N_0) = R_{1_C}(N_0) \times R_{2_C}(N_0) \times \{1 - [R_{3_C}(N_0)][1 - R_{4_C}(N_0) \times R_{5_C}(N_0)]$$
$$[1 - R_{6_C}(N_0) \times R_{7_C}(N_0)]\}\{1 - [1 - R_{8_C}(N_0)][1 - R_{9_C}(N_0)$$
$$\times R_{10_C}(N_0)]\}\{R_{11_C}(N_0) \times R_{12_C}(N_0)\}$$

11.5 PARALLEL DESIGN WITH MUTUALLY EXCLUSIVE EVENTS

With the mathematical relations given so far, many different systems can be analyzed. However, one situation cannot be handled without the introduction of a new mathematical relation called the addition rule. Two events E and F are mutually exclusive if and only if the occurrence of one excludes the possibility of the occurrence of the other. This rule is generally stated as

$$P(E \text{ or } F) = P(E + F) = P(E) + P(F) \tag{11.14}$$

In the notation used in the present chapter,

$$R_{(1r2r\cdots rn)_C}(N_0) = R_{1_C}(N_0) + R_{2_C}(N_0) + \cdots + R_{n_C}(N_0) \tag{11.15}$$

where r indicates " or."

Example 11.3 Figure 11.4 shows the power-actuation system in a convertible-top mechanism of an automobile. The electric motor will run on any two battery cells,

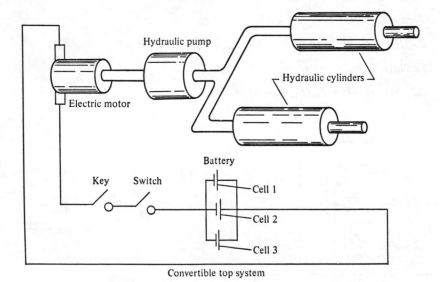

Convertible top system

Fig. 11.4 System with mutually exclusive events.

if the third is dead, and will still operate the top. Develop the reliability of the battery with respect to the top operation in terms of the individual cell reliabilities. Find the expression for the reliability of the whole system.

Solution It will be noted that four events, connected with the battery, will result in the successful operation of the convertible top: $E1$, all cells operative; $E2$, cells I and II operative, III failed; $E3$, cells II and III operative, I failed; $E4$, cells I and III operative, II failed. In this case each event is mutually exclusive. That is, if event $E1$ occurs at time N_0, events $E2$, $E3$, and $E4$ are excluded from possible occurrence. Similarly, if event $E2$ occurs, then events $E1$, $E3$, and $E4$ are impossible. This means that the addition rule of Eqs. (11.14) and (11.15) can be used to solve this problem.

First determine the probability of occurrence (reliability) of events $E1$ through $E4$. Event $E1$ represents a case where if any one cell fails, the whole event changes to $E2$, $E3$, or $E4$. This parallel arrangement is then similar to a series design with respect to a reliability analysis. It follows that

$$R_{E1_C}(N_0) = R_{I_C}(N_0) \times R_{II_C}(N_0) \times R_{III_C}(N_0) \tag{11.16}$$

Then the probability of all three cells' surviving to time N_0 is $R_{E1_C}(N_0)$. Considering event 2, the probability of occurrence of this event is the product of the probabilities of the individual occurrences. In this case the probability of occurrence of cell III's failure is $[1 - R_{III_C}(N_0)]$, while the probability of both cells I and II's survival is

$$R_{III_C}(N_0) = R_{I_C}(N_0) \times R_{II_C}(N_0) \tag{11.17}$$

It follows that

$$R_{E2_C}(N_0) = R_{I_C}(N_0) \times R_{II_C}(N_0)[1 - R_{III_C}(N_0)] \tag{11.18}$$

Similarly,

$$R_{E3_C}(N_0) = R_{II_C}(N_0) \times R_{III_C}(N_0)[1 - R_{I_C}(N_0)] \tag{11.19}$$

Moreover,

$$R_{E4_C}(N_0) = R_{I_C}(N_0) \times R_{III_C}(N_0)[1 - R_{II_C}(N_0)] \tag{11.20}$$

Consequently, by Eq. (11.15),

$$R_{(battery)_C}(N_0) = R_{(E1)_C}(N_0) + R_{(E2)_C}(N_0) + R_{(E3)_C}(N_0) + R_{(E4)_C}(N_0) \tag{11.21}$$

Substituting the above equations into Eq. (11.21) and collecting terms,

$$R_{(battery)_C}(N_0) = R_{I_C}(N_0) \times R_{II_C}(N_0) \times R_{III_C}(N_0)\left[-2 + \frac{1}{R_{I_C}(N_0)} + \frac{1}{R_{II_C}(N_0)} + \frac{1}{R_{III_C}(N_0)}\right]$$

The complete system reliability can be seen from Fig. 11.4 to be essentially a series system up to the hydraulic cylinders. That is, if the key, switch, etc., fail, the whole system fails. Moreover, even though one hydraulic cylinder appears mechanically to be in parallel with another, they are actually in series. The fact that the cylinders are essentially in series is due to the fact that a failure of one cylinder causes the top linkage to twist sufficiently to make the system inoperative. Therefore, the failure of one cylinder results in the failure of the whole system, the system, therefore, being in series. Hence,

$$R_{(system)_C}(N_0) = R_{(key)_C}(N_0) \times R_{(switch)_C}(N_0) \times R_{(battery)_C}(N_0) \times R_{(electric\ motor)_C}(N_0)$$
$$\times R_{(hydraulic\ pump)_C}(N_0) \times R_{(hydraulic\ cylinder)_C}(N_0) \times R_{(hydraulic\ cylinder)_C}(N_0)$$

11.6 REDUNDANCY

The application of the principle of redundancy is probably the oldest known method of reducing the probability of failure. Redundancy is simply the application of dual subsystems (sometimes more) at locations where failure might have disastrous results. The dual subsystems are not always identical. For example, the inclusion of hydraulic and mechanical actuators on aircraft landing gear is an example of redundancy with different subsystems. Some redundant systems are the type where if one fails, the other is activated, like the landing-gear actuator.

Obviously, redundancy increases the complexity, weight, and cost of the total system. Thus, in the design of the system, these factors must be traded off with the required reliability. In some systems, such as missiles, the redundant systems requiring manual changeover are not readily applicable, as no one is available to make the change. When the chances of failure cannot be decreased by any other way, redundancy is a method worth trying.

Basically, redundancy provides extra components which remain nonfunctional until the original components cease to function properly. This type of configuration is shown in Fig. 11.5. In this case 2' is the redundant component. Suppose hydraulic fluid were to flow from component 1 to component 3. Under normal operation the fluid flows from 1 through the booster pump 2 and then to 3. The redundancy 2' is provided so that in case of the failure of the booster pump 2 the switching device will operate and

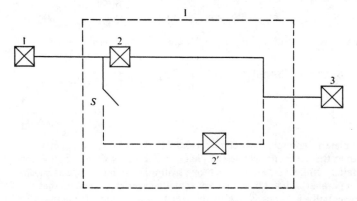

Fig. 11.5 Redundancy in design.

cause the hydraulic circuit to alter to the path 1 to 2' to 3. This system then has two alternate paths:

1. If 2 does not fail: 1 to 2 to 3
2. If 2 fails: 1 to S to 2' to 3, assuming the failure of 2 does not affect the system.

Reducing 2, S, and 2' to a parallel subsystem I,

$$P_{2c}(N_0) = 1 - R_{2c}(N_0) \tag{11.22}$$

$$P_{(S2')c}(N_0) = 1 - R_{Sc}(N_0 - \delta) \times R_{2'c}(N_0 - \delta) \tag{11.23}$$

where $P_{S_2'}(N_0)$ is the probability that either S or 2' will fail before N_0, having started at time δ. It follows that

$$P_{1c}(N_0) = [1 - R_{2c}(N_0)][1 - R_{Sc}(N_0 - \delta) \times R_{2'c}(N_0 - \delta)] \tag{11.24}$$

where $P_{1c}(N_0)$ is the probability that neither branch of subsystem I will survive to time N_0,

$$R_{1c}(N_0) = 1 - [1 - R_{2c}(N_0)][1 - R_{Sc}(N_0 - \delta) \times R_{2'c}(N_0 - \delta)] \tag{11.25}$$

where $RI_c(N_0)$ is the probability that subsystem I will have at least one branch service to time N_0. Hence,

$$R_{(system)c}(N_0) = R_{1c}(N_0)$$
$$\times \{1 - [1 - R_{2c}(N_0)][1 - R_{Sc}(N_0 - \delta) \times R_{2'c}(N_0 - \delta)]\} \times R_{3c}(N_0) \tag{11.26}$$

To minimize the probability of survival (reliability), it can be assumed that component 2 fails at $N = 0$ so that $\delta = 0$. Rewriting Eq. (11.26),

$$R_{(system)c}(N_0) = R_{1c}(N_0)$$
$$\times \{1 - [1 - R_{2c}(N_0)][1 - R_{Sc}(N_0) \times R_{2'c}(N_0)]\} \times R_{3c}(N_0) \tag{11.27}$$

Equation (11.27) is then the lower bound on the system reliability of a redundant circuit with confidence C attached to each component. However, Eq. (11.26) should be evaluated at the time δ corresponding to a confidence level C. That is, δ should be chosen such that δ would not be expected to be less than δ_C more than $(1 - C)$ percent of the time. Equation (11.26) then reads

$$R_{(system)c}(N_0) = R_{1c}(N_0)$$
$$\times \{1 - [1 - R_2(N_0)][1 - R_{Sc}(N_0 - \delta) \times R_{2'c}(N_0 - \delta)]\} \times R_{3c}(N_0) \tag{11.28}$$

As an example, assume that all the reliabilities are equal to 0.07. With this it is found that a simple series design (1-2-3) has a reliability of 34.30 percent, that a redundant circuit has a reliability of 41.403 percent, and that a simple parallel circuit (no switch) has a reliability of 44.59 percent at time N_0.

SUMMARY

Chapters 3 through 9 were devoted to the organization of experiments and the analysis of the resultant data. Chapter 10 dealt specifically with the analysis of data, the type of data which are obtained when two distributions interfere with each other. The present chapter also deals with the analysis of data, specifically with reliability data derived from testing components, and predicting the reliability of systems consisting of these components without being able to test the whole system. The test data are expressed in terms of either reliability or the probability of failure. The term "failure" is used here in a broad sense as any variation beyond the allowable limits, which may mean structural failure, malfunction, deterioration in performance, etc. The order of presentation follows traditional lines, first with the discussion of the series design, then parallel design, series and parallel design, parallel design with mutually exclusive events, and redundancy.

PROBLEMS

11.1. Two different systems made of several components can be employed in a given application. These two systems are exactly similar for functional requirements.

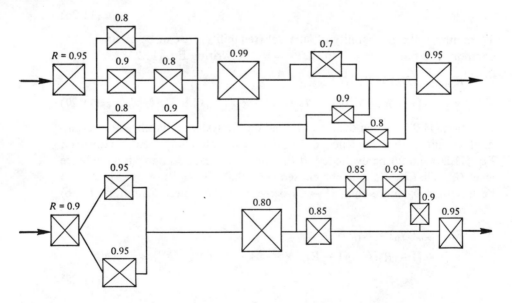

The reliability of each component is as shown. Strictly from the viewpoint of the overall system reliability, which system should be chosen?

11.2. A certain system consists of the components shown in the following sketch. The life distribution of each component is Weibull with the following parameters:

Component number	θ, hours	b
1	1,000	3.0
2	1,200	3.5
3, 4	900	2.5
5	1,100	3.1
6, 7	850	2.0
8	1,400	4.0

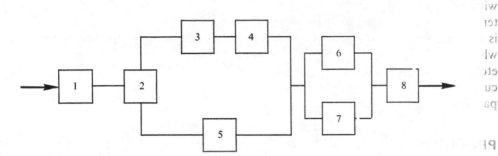

Compute the overall system reliability for 400, 500, and 600 h of operation.

BIBLIOGRAPHY

Bazovsky, I.: "Reliability Theory and Practice," Prentice-Hall, Inc., Englewood Cliffs, N.J., 1961.

Sandler, G.: "System Reliability Engineering," Prentice-Hall, Inc., Englewood Cliffs, N.J., 1963.

Shooman, M. L.: "Probabilistic Reliability: An Engineering Approach," McGraw-Hill Book Company, New York, 1968.

12
Analysis of Laboratory and Field Data

This chapter deals with situations common in the consumer industry (automobiles, refrigerators, television sets, etc.), in which the product is generally guaranteed against defect within a certain warranty period. The customer expects the product to be good and reliable. If the product is found defective, the cost of making corrections can be very high; and the further along the distribution system the trouble is found, the more costly the solution to the problem. The most expensive place to find the defective part is in customers' hands. The least expensive place is in the laboratory.

For this reason, manufacturers of industrial products frequently spend much time and money conducting tests on their products before they are released for production. These tests are of considerable importance, because any error which slips through the testing program will be built into every item produced. This is the most expensive type of error, calling for corrections in the field and changes in design; and these, once production has started, are difficult and expensive to handle. In contrast, the manufacturing error is usually easier to find.

Such tests are conducted with a variety of purposes: to establish design concepts; to verify design objectives; to determine the significance of changes; to satisfy specifications; to determine the relation between variables; to establish the optimum condition; etc. These tests are generally run under controlled conditions which simulate the operating conditions, since in the final analysis the worth of the product can be judged only in terms of its behavior in the hands of the customer. Because of time and cost limitations, and to meet production schedules, laboratory tests are frequently accelerated (see Chap. 5) to condense, say, 10 years of refrigerator life into 10 weeks of laboratory testing. This chapter discusses, in the form of illustrated examples, several aspects of the analysis, simulation, and correlation of laboratory tests and field (customer) usage. It is understood that if the laboratory test is run to failure, the mode of failure must be the same as the one experienced or anticipated in the field. Table 12.1 [1] shows various environmental factors which, depending on application, should be considered for laboratory simulation. This table also gives principal effects of these factors and typical failure modes that may be expected.

In spite of care exercised in the planning and execution of the laboratory tests, field failures will and do occur. It is very important for the manufacturer to know, as soon as possible, how the product is performing in the field, because if it is found to be defective the cost of making corrections can be very high. An example is presented here to illustrate how a prediction can

Table 12.1 Various environmental factors, their effects, and expected failure modes [1]

Environment	Principal effects	Typical failures induced

High temperature	Thermal aging:
	  Oxidation
	  Structural change
	  Chemical reaction
		Insulation failure
		Alteration of electrical properties

	Softening, melting, and sublimation	Structural failure
	Viscosity reduction and evaporation	Loss of lubrication properties
	Physical expansion	Structural failure Increased mechanical stress Increased wear on moving parts
Low temperature	Increased viscosity and solidification	Loss of lubrication properties
	Ice formation Embrittlement	Alteration of electrical properties Loss of mechanical strength Cracking, fracture
	Physical contraction	Structural failure Increased wear on moving parts

Table 12.1 *(continued)*

Environment	Principal effects	Typical failures induced
High relative humidity	Moisture absorption	Swelling, rupture of container Physical breakdown Loss of electrical strength
	Chemical reactions: Corrosion Electrolysis	Loss of mechanical strength Interference with function Loss of electrical properties Increased conductivity of insulators
Low relative humidity	Desiccation: Embrittlement Granulation	Loss of mechanical strength Structural collapse Alteration of electrical properties, "dusting"
High pressure	Compression	Structural collapse Penetration of sealing Interference with function
Low pressure	Expansion	Fracture of container Explosive expansion
	Outgassing	Alteration of electrical properties Loss of mechanical strength
	Reduced dielectric strength of air	Insulation breakdown and arcover Corona and ozone formation
Solar radiation	Actinic and physicochemical reactions: Embrittlement	Surface deterioration Alteration of electrical properties Discoloration of materials Ozone formation
Sand and dust	Abrasion	Increased wear Interference with function
	Clogging	Alteration of electrical properties
Salt spray	Chemical reactions:	Increased wear Loss of mechanical strength
	Corrosion	Alteration of electrical properties Interference with function
	Electrolysis	Surface deterioration Structural weakening Increased conductivity
Wind	Force application	Structural collapse Interference with function Loss of mechanical strength
	Deposition of materials	Mechanical interference and clogging Abrasion accelerated
	Heat loss (low velocity) Heat gain (high velocity)	Accelerates low-temperature effects Accelerates high-temperature effects

Environment	Principal effects	Typical failures induced
Rain	Physical stress Water absorption and immersion Erosion Corrosion	Structural collapse Increase in weight Aids in heat removal Electrical failure Structural weakening Removes protective coatings Structural weakening Surface deterioration Enhances chemical reactions
Temperature shock	Mechanical stress	Structural collapse or weakening Seal damage
High-speed particles (nuclear irradiation)	Heating Transmutation and ionization	Thermal aging Oxidation Alteration of chemical, physical, and electrical properties Production of gases and secondary particles
Zero gravity	Mechanical stress Absence of convection cooling	Interruption of gravity-dependent functions Aggravation of high-temperature effects

Ozone	Chemical reactions:
	  Crazing, cracking
	  Embrittlement
	  Granulation
	Reduced dielectric strength of air
		Rapid oxidation
		Alteration of electrical properties
		Loss of mechanical strength
		Interference with function
		Insulation breakdown and arcover

Explosive decompression	Severe mechanical stress	Rupture and cracking Structural collapse
Dissociated gases	Chemical reactions Contamination Reduced dielectric strength	Alteration of physical and electrical properties Insulation breakdown and arcover
Acceleration	Mechanical stress	Structural collapse
Vibration	Mechanical stress Electrical Fatigue	Loss of mechanical strength Interference with function Increased wear Structural collapse
Magnetic fields	Induced magnetization	Interference with function Alteration of electrical properties Induced heating

be made of the total number of failures, within a warranty period, on the basis of the number of failures to date.

Example 12.1 A manufacturer puts out 10,000 television sets, with a 1-year warranty on the picture tube.   On the basis of the field data accumulated during the first 4 months, predict the total number of failed tubes during the warranty period of 1 year.

Table 12.2 Field failure information for television picture tubes collected by the end of 4 months

Time in service, months (difference between delivery month and failure reporting month)	Number of units having corresponding time in service	Number of failures	Number of suspensions
0 to 1	3,894	30	3,864
1 to 2	2,340	20	2,320
2 to 3	1,255	14	1,241
3 to 4	1,108	11	1,097
	1,403		1,403*
Total	10,000	75	9,925

* Sets operating at end of 4 months in service

Solution From Table 12.2 it can be seen that 3,864 sets were operating satisfactorily with 1 month or less in service; therefore they will be considered as suspended.   Thirty tubes failed in the first month.   The above data are arranged in the following manner:

Months in service	Field data for picture tubes
0 to 1	3,864 suspended (OK) 30 failed
1 to 2	2,320 suspended 20 failed
2 to 3	1,241 suspended 14 failed
3 to 4	1,097 suspended 11 failed
	1,403 sets operating at the end of fourth month in service, therefore suspended

By means of the method of suspended items discussed in Subsec. 1.13.3, a mean order number for the last failure in each failed set is computed, and then the corresponding median rank is found for constructing a Weibull plot.

For the first failed set (30 failures), compute (see Subsec. 1.13.3):

$$\text{New increment} = \frac{n + 1 - \text{previous failure order number}}{1 + \text{number of items following present suspended set}}$$

$$= \frac{10,000 + 1 - 0}{1 + 6,136} = \frac{10,001}{6,137} = 1.63$$

The mean order number for the thirtieth failure in the first month in service is

$$0 + 30(1.63) = 48.9$$

The median rank of this thirtieth failure is (Subsec. 1.13.1)

$$\frac{48.9 - 0.3}{10,000 + 0.4} = \frac{48.6}{10,000.4} = 0.00486 = 0.486\%$$

Thus, this gives one point on the Weibull plot (Fig. 12.1) with 1 month on the abscissa and 0.486 percent on the ordinate.

The information for the set of 20 failures which occurred during the second month in service is found in a similar manner:

$$\text{New increment} = \frac{10,001 - 48.9}{1 + 3,786} = \frac{9,952}{3,787} = 2.625$$

The mean number for the twentieth failure in the second month is

$$48.9 + 20(2.625) = 101.4$$

The median rank for the twentieth failure in the second month is

$$\frac{101.4 - 0.3}{10,000 + 0.4} = 0.0101 = 1.01\%$$

This is then plotted with 2 months on the abscissa and 1.01 percent on the ordinate (Fig. 12.1).

For the third failed set (14 failures):

$$\text{New increment} = \frac{10,001 - 101.4}{1 + 2,525} = 3.94$$

The mean order number for the fourteenth failure is

$$101.4 + 14(3.94) = 156.6$$

The median rank for the fourteenth failure in the third month is

$$\frac{156.6 - 0.3}{10,000 + 0.4} = 0.0156 = 1.56\%$$

For the fourth failed set (11 failures):

$$\text{New increment} = \frac{10,001 - 156.6}{1 + 1,414} = 6.95$$

The mean order number for the eleventh failure in the fourth month is

$$156.6 + 11(6.95) = 233.1$$

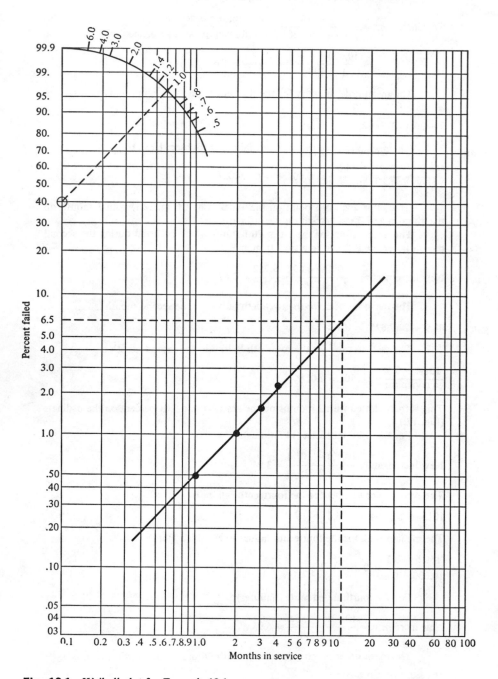

Fig. 12.1   Weibull plot for Example 12.1.

The median rank for this failure is

$$\frac{233.1 - 0.3}{10,000 + 0.4} = 0.0233 = 2.33\%$$

The above information is plotted on the Weibull probability paper (Fig. 12.1), and a best-fitting line is drawn through these points. Read 6.5 percent (expected failures) on the ordinate corresponding to 12 months on the abscissa. Thus, at the end of a 1-year (12 months) warranty period the manufacturer should expect 6.5 percent (that is, 650) failed picture tubes.

Example 12.2 In a certain application the life of the vacuum-system rubber hoses is significantly affected by the ozone concentration in the environment, the assembly load on the hoses, and the operating temperature. The operating conditions in the field are:

Ozone concentration: 50 pphm (parts per hundred million)
Load: 20% elongation
Temperature: 70°F

The required life in the field, under these conditions, is 5 years. Design an accelerated laboratory test to fulfill these requirements.

Solution A laboratory experiment was run, as shown in Table 12.3. Four sets of rubber-hose specimens each at a given percent elongation were tested simultaneously in an ozone chamber with 100 pphm ozone concentration and 150°F temperature. Each set consisted of three specimens, and life to failure for each specimen was recorded. This was repeated for other ozone and temperature combinations, and the time to failure was recorded as shown in Table 12.3.

The average effect of each of the main factors, at its different levels, is

$$\text{Average effect} = \frac{\text{total of all observations at one level}}{\text{number of observations at that level}}$$

For example, from Table 12.3:

$$\text{Average effect for 100 pphm level of ozone} = \frac{5,040}{36}\,\text{h} = 140\,\text{h}$$

The average effects of all factors at different levels are thus found to be:

Ozone factor, pphm	Life, h	Temp. factor, °F	Life, h	Load factor, % elongation	Life, h
100	140	100	126	100	152
175	58	150	66	150	82
250	29	200	36	200	46
				250	27

Table 12.3 Test results, in terms of hours to failure, for Example 12.2

Ozone concentration, pphm	Temperature, °F	Load, percent elongation 100	150	200	250
100	100	560 380 470	285 250 215	120 150 135	80 70 90
	150	242 250 248	130 150 110	70 60 80	33 47 40
	200	126 114 120	65 80 50	52 38 45	20 30 25
175	100	183 197 190	120 80 100	68 52 60	31 39 35
	150	90 97 83	56 50 44	36 30 24	20 24 26
	200	56 50 44	35 29 41	18 22 26	13 16 10
250	100	90 80 100	45 40 50	20 25 30	18 12 15
	150	69 51 60	26 34 30	15 17 13	10 9 8
	200	40 30 35	12 18 15	8 10 6	4.4 3.1 5.0

In order to extrapolate these results to the field conditions listed in the statement of the problem, each of these factors is plotted on a log-log scale with their levels on the ordinate and the corresponding life on the abscissa (Fig. 12.2). An average line is fitted through the data points by means of the least-squares method of fit. (See Chap. 13.) In addition to the log-log fit the higher-order functions should be fitted and their corresponding correlation coefficients (a measure of fit) should be computed. The function with the highest correlation coefficient should be chosen. A method for determining the functional relationship between the factor (say ozone) and its effect (say life) is given in Chap. 13.

From Fig. 12.2,

$$\text{Ozone correlation} = \frac{\text{life at 50 pphm}}{\text{life at 250 pphm}} = \frac{507}{29} = 17.5$$

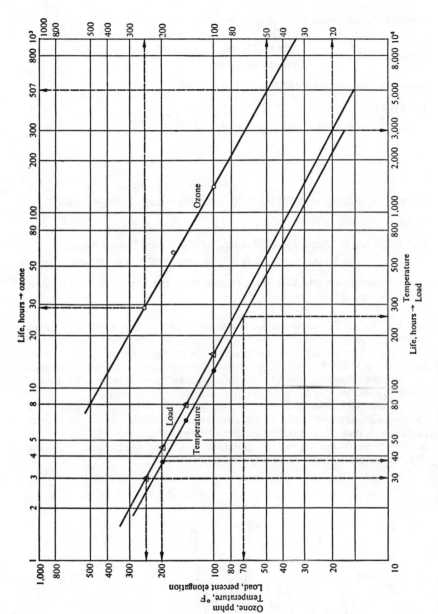

Fig. 12.2 Plot of average effect when the interactions are not significant.

$$\text{Temperature correlation} = \frac{\text{life at } 70°\text{F}}{\text{life at } 200°\text{F}} = \frac{240}{38} = 6.3$$

$$\text{Load correlation} = \frac{\text{life at } 20\% \text{ elong.}}{\text{life at } 250\% \text{ elong.}} = \frac{3,000}{30} = 100$$

Total correlation factor $= 17.5 \times 6.3 \times 100 = 11,000$

Laboratory:

Average life at 250 pphm, 200°F, and 250% elongation $= \dfrac{4.4 + 3.1 + 5.0}{3} = 4.2$ h

(from Table 12.3)

Field:

Average life at 50 pphm, 70°F, and 20% elongation $= 4.2 \times 11,000 = 46,200$ h

Hence,

4.2 h of life test in laboratory at 250 pphm, 200°F, and 250% elongation

$= 5.3$ years of life in field at 50 pphm, 70°F, and 20% elongation

    Approximately 4 h of accelerated testing under the above conditions is equivalent to 5 years of field operation.

Example 12.3 Mechanical components and assembly, in service, are generally subjected to dynamic loads of constant or varying magnitude resulting from the operation of the units. The purpose of this example is to illustrate how the life objective of the part in the field can be predicted or verified through a laboratory test. Although the following discussion relates to structural testing of ground vehicles, the general approach indicated may be applied to other engineering situations.

    Loads were measured in the field over 200 mi of representative terrain, and a spectrum such as shown in Fig. 12.3 was obtained. The life of a critical component of the vehicle under these conditions is estimated to be 300,000 mi. Design an accelerated laboratory test for this component to determine whether the 300,000-mi objective has been met.

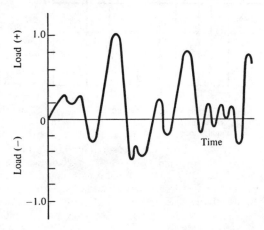

Fig. 12.3 Load data from field.

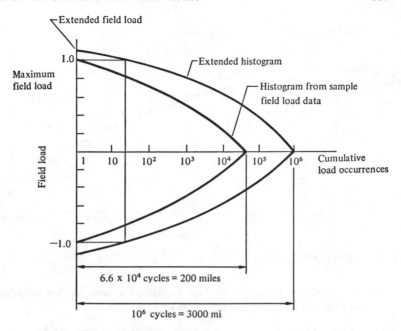

Fig. 12.4 Cumulative load histogram.

Solution The loads in Fig. 12.3 obtained for a 200-mi span are reduced to a cumulative histogram form, as shown in Fig. 12.4, where the ordinate represents load magnitude and the abscissa the number of times a given load occurred.

The histogram made from the sample data in the field consists of 6.6×10^4 load cycles which represent 200 mi of travel on a mixture of road terrains. This is extended to 10^6 cycles (Fig. 12.4), as this gives adequate representation of the whole field load spectrum, including some very severe loads which may occur only once in 10^6 cycles. In Fig. 12.4, 10^6 cycles represent 3,000 mi of travel.

Also shown in this histogram is the number of low-amplitude loads that are considered not damaging; hence they are omitted to reduce the test time. Loads with amplitudes less than 12.5 percent of the maximum are generally below the endurance limit of the part tested; therefore, they can be omitted. This reduces the test time by a factor of $2 - [10^6/(5 \times 10^5) = 2.0]$. Thus, 3,000 mi of field operation can now be represented by 5×10^5 cycles instead of 10^6 cycles.

In order to simulate these loads in the laboratory, the histogram in Fig. 12.4 is reduced to a step-load histogram (Fig. 12.5). The step-load histogram is used as a load program for the laboratory test where the total damage done to the part under test would be the same as the damage in the field from which the data were acquired. (For the discussion of damage see Chap. 10.) Generally, the histogram is divided into eight steps, as this appears to be sufficient to duplicate the field history [2, 3].

One histogram, consisting of 5×10^5 cycles, thus represents 3,000 mi of field operation. Since the objective is 300,000 mi, this would necessitate, in the laboratory, a repetition of 100 histograms for a total of $100 \times 5 \times 10^5$ cycles, that is, 5×10^7 cycles. In testing large structures, the speed of load application is relatively low; therefore, the time for completion of the test is frequently prohibitively high. Therefore, the test must be accelerated. This can be done in one of two ways, depending

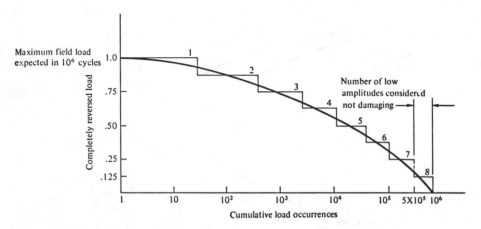

Fig. 12.5 Step-load histogram.

on circumstances: (1) increase the load, retaining the same number of cycles; (2) increase the number of cycles, keeping the load the same.

1. *Increase the load.* In the conventional *S-N* test the time can be reduced by increasing the load. In the same manner, the programmed load histogram can be intensified, as shown in Fig. 12.6. If the relation between this load intensity and life is known, the number of cycles to failure under the increased load can be predicted. It is found that the slope of the *S-N* curve for the step-load histogram does not change significantly, both for notched and for un-notched specimens [2, 4–7]. The value of the inverse slope *K* of the *S-N* curve is approximately 6.5 for steel.

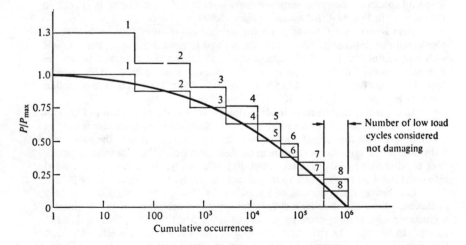

Fig. 12.6 Step-load histogram with increased load.

The equation relating the load intensity factor with cycles to failure is

$$N_2 = N_1 \left(\frac{\sigma_1}{\sigma_2}\right)^x$$

or when the load intensity is increased by 30%,

$$\frac{N_2}{N_1} = (1.3)^{6.5} \simeq 5.5$$

which represents the test acceleration factor of 5.5. That is, if the magnitude of the load in the load histogram is increased 1.3 times, the test time to failure will decrease by a factor of 5.5, that is, from 5×10^5 cycles in one histogram to approximately 9.1×10^4 cycles, or from 100 histogram applications to 100/5.5 $\cong$ 18 applications.

2. *Increase the number of cycles.* In some situations it is not feasible to increase the load. In this case the magnitude of the load is kept the same, but the duration of this load is increased so that the total cumulative effect remains constant. This is shown in Fig. 12.7.

Suppose the damage done from one load histogram (5×10^5 cycles) application is 5 percent (when 100 percent damage is accumulated, the failure occurs). In order to reduce the test time, the loads with low magnitudes are eliminated from the test and they are replaced by fewer high-magnitude load cycles so that the damage done by the eliminated low-load cycles is the same as that by the additional few high-load cycles. This number of additional high-load cycles is computed by the cumulative damage analysis as shown in Chap. 10.

For example, referring to Fig. 12.7, the dotted-line histogram with 10^5 cycles is equivalent to the histogram with 5×10^5 cycles, thereby reducing the test time by a factor of 5.

It will be noted that the principle of cumulative damage applied to the case when the number of cycles was increased (item 2 above) also can be applied to the case when the load was increased (item 1 above).

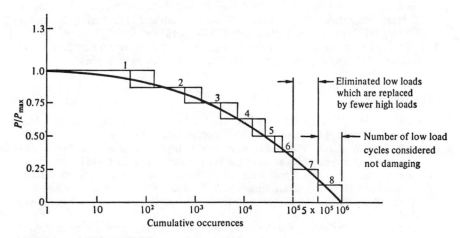

Fig. 12.7 Step-load histogram with increased load cycles.

SUMMARY

Chapters 3 through 9 were devoted to the organization of experiments and the analysis of the resultant data. Chapters 10 through 14 are specifically concerned with the analysis of data, of which Chap. 10 dealt with interference and Chap. 11 with systems. The present chapter is concerned with the application of the disciplines previously developed to the analysis of data derived from laboratory tests and field operations. The disciplines employed are Weibull plots, suspended items, accelerated experiments, and stress-life relations. These are employed to the prediction of the total number of field failures on the basis of preliminary failures, prediction of life objective in the field through laboratory tests, and design of an experiment which would satisfy certain field requirements.

REFERENCES

1. Jones, H. C.: Design Testing, paper presented at the Engineering Summer Conference, "Designing Engineering Experiments," University of Michigan, 1968.
2. Gassner, E., and W. Schutz: "Evaluating Vital Vehicle Components by Programmed Fatigue Tests," Ninth International Automobile Technical Congress (FISITA), London, May 4, 1962.
3. Gassner, E., and W. Schutz: "The Significance of Constant Load Amplitude Tests for the Fatigue Evaluation of Aircraft Structures," Proceedings ICAF—Agard Symposium on Full Scale Fatigue Testing of Aircraft Structures, Pergamon Press, New York, 1960.
4. Freudenthal, A. M.: "Fatigue Sensitivity and Reliability of Mechanical Systems," SAE Congress, Detroit, Michigan, 1962; SAE Reprint No. 459A.
5. Schutz, W.: "Sichere und wirtschaftliche Konstructionen durch Betriebsfestigkeits-versuche," Landtechnische Forschung, vol. 1, 1959.
6. Haibach, E.: "Fatigue Strength under Constant Amplitude and Program Loading of Titanium Base Materials Compared to Those of Steels and Light Metals," Material Prufung Bd. 7 (1965), No. 4 VDI, Verlag Düsseldorf.
7. Gassner, E., and W. Lipp: "Wirklich—Keitsgetreue Lebensdauerfunktion fuer Fahr-zeugbauteile," Special publication of the Fraunhofer Gesellschaft, 1963.

BIBLIOGRAPHY

Freudenthal, A. M.: "Fatigue Sensitivity and Reliability of Mechanical Systems," SAE Congress, Detroit, Michigan, 1962; SAE Reprint No. 459A.
Gassner, E., and W. Lipp: "Wirklich—Keitsgetreue Lebensdauerfunktion fuer Fahr-zeugbauteile," Special publication of the Fraunhofer Gesellschaft, 1963.
—— and W. Schutz: "The Significance of Constant Load Amplitude Tests for the Fatigue Evaluation of Aircraft Structures," Proceedings ICAF—Agard Symposium on Full Scale Fatigue Testing of Aircraft Structures, Pergamon Press, New York, 1960.
—— and ——: "Evaluating Vital Vehicle Components by Programmed Fatigue Tests," Ninth International Automobile Technical Congress (FISITA), London, May 4, 1962.

Haibach, E.: "Fatigue Strength under Constant Amplitude and Program Loading of Titanium Base Materials Compared to Those of Steels and Light Metals," Material Prufung Bd. 7 (1965), No. 4 VDI, Verlag Düsseldorf.

Jaeckel, H. R.: "Correlating Laboratory Test Results to Proving Ground Miles," Reliability Working Paper No. 8, Ford Motor Company, 1965.

Johnson, G.: "The Statistical Treatment of Fatigue Experiments," Elsevier Publishing Company, Amsterdam, 1964.

Ponta, Peter H.: Process and Product Control, *Annu. Conv. Trans., Amer. Soc. Qual. Contr.*, Milwaukee, Wis., pp. 193–204, 1964.

Simpson, B. H.: The Ford Engineering Reliability Program, *Ann. Rel. Maintainability*, vol. 4, pp. 171–181, 1965.

————: Reliability Prediction from Warranty Data, *SAE Paper* 660060, January, 1966.

13
Correlation, Regression, and Variation Analysis

In the analysis of engineering data, it is frequently desirable to determine whether variables are associated with each other, and if so, how one variable changes with respect to another. For example, what is the association between the vertical and the fore-and-aft load inputs to vehicles on the road? Does one depend on the other, or are they unrelated? Is there any relation between the school grades and the IQ of students? These situations are covered by the correlation analysis where variables are observed as they occur, and neither is fixed at any predetermined level.

In other situations one may be interested in establishing the functional relationship between the variables. In this case regression analysis applies. Here, observations are made on one variable (dependent variable) while the other (the independent variable) is held at a known fixed level. For example, how does the material hardness change with the heat-treat time? What is the functional relationship between the electrical resistance and temperature? In this chapter correlation and regression analyses are developed, with the regression analysis presented for several different cases.

In some engineering situations it is important to determine the true variation in the product characteristics.　The variation in the measurements made on the product consists of the product variations and the test variations. In the section on variation analysis a method is presented for isolating these errors.

13.1 CORRELATION AND REGRESSION ANALYSIS

13.1.1 REGRESSION VS. CORRELATION

Regression analysis is used to determine the functional relationship between variables, while correlation analysis is used to determine the degree of association between variables.　For example, the effect of temperature on tensile strength and the effect of sliding velocity on wear fall under the category of regression analysis.　However, it will be meaningless to inquire about the association between temperature and strength since the association implies correlation, which considers the variation in both measurements simultaneously.　Similarly, one cannot determine the effect of laboratory life on field life, since these lives can only be associated.　A situation which can be treated only by correlation analysis is the following: How does the vibration isolation of different types of rubber when measured on one machine compare with the vibration isolation measured on another machine (correlation among machines)?

In regression analysis, once a functional relationship between the dependent and the independent variable is established, it can be used only to determine how the dependent variable changes with the independent variable, and it does not produce any information about why it changes. In the case of correlation, a significant correlation coefficient (discussed here in detail) may indicate a presence of strong association between variables, and it does not imply a causation.

The regression can be applied to all distributions (normal, Weibull, exponential, etc.); however, the regression method assumes that the error of the dependent variable is normally distributed, and it requires no assumption about the distribution of the independent variable.　In the case of correlation analysis, errors of both variables must be normally distributed. This means that correlation method may not be used to establish the functional relationship where the values of the independent variables are fixed.

Since the computational procedures are basically the same for regression and for correlation problems, only regression analysis is discussed in detail in this chapter.

13.1.2 CORRELATION ANALYSIS

Correlation analysis is a method of determining a degree of association between variables.　These variables may be continuous or discrete.　In most engineering situations, the continuous variables are encountered frequently. The major portion of this chapter is devoted to these variables.

Correlation coefficient The correlation coefficient r is defined as the quantitative measure of association between the variables. When r is 1.0, this indicates the perfect correlation between the variables; and where r is zero, there is no correlation. (See Fig. 13.1.) The correlation coefficient can also be taken as the reliability of the association between variables. The equation for computing the value of r is derived below.

Suppose x and y are two random variables; the correlation coefficient is then defined as

$$r = \frac{\text{covariance of } x \text{ and } y}{\text{square root of product of variances of } x \text{ and } y}$$

$$= \frac{\text{covariance}(x, y)}{(\sigma_x^2 \sigma_y^2)^{1/2}}$$

$$= \frac{[1/(n-1)] \sum_{i=1}^{n}(x_i - \bar{x})(y_i - \bar{y})}{\left[\dfrac{\sum(x_i - \bar{x})^2}{n-1} \dfrac{\sum(y_i - \bar{y})^2}{n-1}\right]^{1/2}}$$

or

$$r = \frac{\sum_{i=1}^{n}(x_i - \bar{x})(y_i - \bar{y})}{[\sum(x_i - \bar{x})^2 \sum(y_i - \bar{y})^2]^{1/2}} \tag{13.1}$$

This expression can be simplified to obtain an alternative form:

$$r = \frac{n \sum xy - \sum x \sum y}{\{[n \sum x^2 - (\sum x)^2][n \sum y^2 - (\sum y)^2]\}^{1/2}} \tag{13.2}$$

where

$$\sum xy = \sum_{i=1}^{n} x_i y_i$$

$$\sum x = \sum_{i=1}^{n} x_i$$

$$\sum x^2 = \sum_{i=1}^{n} x_i^2$$

. . . .

The range of r can be from -1 to $+1$, depending on the degree of association. Table A-49 can be used to determine the significance of the

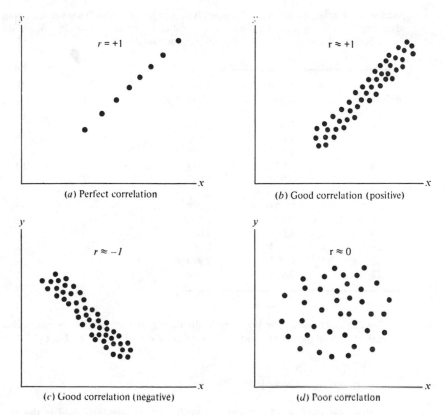

Fig. 13.1 Illustration of various correlations between variables.

correlation coefficient computed from a sample at a certain confidence level. This table provides the maximum values of r which can be expected by chance alone when actually no correlation exists. The 95 percent confidence indicates there is only a 5 percent chance of having r as large as those in the table when no correlation exists. In order to conclude at a given confidence level that the correlation does exist, the calculated r should exceed the tabulated value of r.

Example 13.1 A certain manufacturing process produces a considerable variation in the dimensions (nominal diameter and length) of a type of bolt. These dimensions must be controlled to meet the tolerance specifications. It would be desirable to know whether independent controls on both dimensions are necessary, or whether controlling one dimension would also reduce the variation in the other dimension. To help make such a decision, an experiment was designed to find out whether variation in one dimension is due to the variation in the other. If such association (that is, correlation) between the two dimensions exists, control on one would be sufficient. Hence, the purpose of this analysis was to determine from the following data whether both dimensions should be controlled or the control of any one would be sufficient. Make the decision with 99 percent confidence.

Bolt number	Nominal diameter, in (x)	Length, in (y)
1	0.990	2.988
2	1.002	3.005
3	1.010	3.012
4	0.995	2.993
5	0.998	2.997
6	1.005	3.008
7	1.003	2.999
8	0.997	2.994
9	1.001	3.001
10	1.003	3.004
11	0.997	2.999
12	0.999	3.000

Solution The degree of association between the diameter x and the length y is established through the correlation coefficient r. The value of r is computed by Eq. (13.1):

$$r = \frac{\sum_{i=1}^{n}(x_i - \bar{x})(y_i - \bar{y})}{[\sum (x_i - \bar{x})^2 \sum (y_i - \bar{y})^2]^{1/2}}$$

From the given data, $\bar{x} = 1.000$ and $\bar{y} = 3.000$. Other quantities used in the calculation of r are tabulated below:

Bolt number	$x_i - \bar{x}$	$y_i - \bar{y}$	$(x_i - \bar{x})(y_i - \bar{y})$ $\times 10^4$ in^2	$(x_i - \bar{x})^2$ $\times 10^4$ in^2	$(y_i - \bar{y})^2$ $\times 10^4$ in^2
1	−0.010	−0.012	1.20	1.00	1.44
2	0.002	0.005	0.10	0.04	0.25
3	0.010	0.012	1.20	1.00	1.44
4	−0.005	−0.007	0.35	0.25	0.49
5	−0.002	−0.003	0.06	0.04	0.09
6	0.005	0.008	0.40	0.25	0.64
7	0.003	−0.001	−0.03	0.09	0.01
8	−0.003	−0.006	0.18	0.09	0.36
9	0.001	0.001	0.01	0.01	0.01
10	0.003	0.004	0.12	0.09	0.16
11	−0.003	−0.001	0.03	0.09	0.01
12	−0.001	0.000	0.00	0.01	0.00
Total			3.62	2.96	4.90

Substituting these values in the expression for r above:

$$r = \frac{3.62}{\sqrt{2.96 \times 4.90}}$$

$$= \frac{3.62}{3.81} = .945$$

$$= 94.5\%$$

$$r^2 = .893 \text{ or } 89.3\%$$

This value of r is compared with the value from Table A-49 to test its significance. From Table A-49, with degrees of freedom $\nu = n - 2 = 10$ ($\nu = n - 2$, as discussed in Subsec. 13.1.3), number of variables 2 (x and y), and 99 percent confidence, the value of r is .708.

Since $.945 > .708$, it can be concluded with 99 percent confidence that the variations in both dimensions are interdependent; and 89.3 percent of the total variation in one dimension can be accounted for by the variation on the other. Therefore, the control on any one of the two dimensions should be sufficient.

Rank correlation All the above relates to the situations where the variables follow a continuous distribution, and when they can be measured in terms of quantitative numbers. However, in some cases (e.g., scoring, vehicle ride comfort, etc.), one cannot express variables in terms of numbers. Hence, a discrete scale is assigned to such measurements. This is usually done (as discussed in Chap. 8) by ranking the variables in some order such as by numbers 1 to n. These data are then analyzed to establish the correlation between the ranks of the variables by the method known as rank correlation.

The correlation coefficient r for continuous data is determined either by Eq. (13.1) or by Eq. (13.2); whereas for the discrete case the rank correlation coefficient r is computed from [1] [restated Eq. (8.3)]

$$r = 1 - \frac{6 \sum D^2}{n(n^2 - 1)} \tag{13.3}$$

where n is the number of pairs of data (x, y), and D is the difference between the ranks of corresponding values of x and y (for derivation see Ref. 8 in Chap. 8). This is illustrated by the following example.

Example 13.2 Correlation between two different proving grounds that represent the typical customer routes is to be determined in terms of vehicle ride performance. Eleven passenger vehicles were selected for subjective ride evaluation, and each was tested on both proving grounds. Determine whether the ride correlation between the two proving grounds exists. The ranks were assigned from 1 to 11 to the vehicles for the ride performance.

		Vehicle numbers											
		1	*2*	*3*	*4*	*5*	*6*	*7*	*8*	*9*	*10*	*11*
Ranks	Proving ground 1	7	2	9	3	11	5	4	1	8	6	10
	Proving ground 2	8	2	10	1	9	7	4	3	6	5	11

Solution

D	1	0	1	−2	−2	2	0	2	−2	−1	1	
D^2	1	0	1	4	4	4	0	4	4	1	1	24 Total

where D is the difference between the two ranks of each vehicle evaluated on two proving grounds.

$$\sum D^2 = 24 \qquad n = 11$$

By Eq. (13.3),

$$r = 1 - \frac{6 \sum D^2}{n(n^2 - 1)}$$

$$= 1 - \frac{6 \times 24}{11(121 - 1)}$$

$$= 1 - \frac{144}{1,320}$$

or

$$r = .891$$
$$= 89.1\%$$

This indicates that the vehicle ride evaluated on one proving ground does correlate with the ride evaluated on the other, since $.891 > .735$, where $.735$ is the value of r from Table A-49 for 99 percent confidence.

13.1.3   LINEAR REGRESSION ANALYSIS

Regression analysis is a method for establishing the functional relationship between two variables where the variation in one measurement is considered while the other is held fixed. In many engineering situations a straight-line relationship can express the dependence of one variable on another. This is referred to as the linear regression analysis. The relationship is

$$y = a_0 + a_1 x \qquad \text{first-order equation}$$

where y = dependent variable, which is measured at given level of x
  x = independent variable, which is held fixed
  a_0, a_1 = regression parameters to be determined from sample data

This equation can be extended to the higher order, such as

$$y = a_0 + a_1 x + a_2 x^2 \qquad \text{second-order equation}$$
$$y = a_0 + a_1 x + a_2 x^2 + a_3 x^3 \qquad \text{third-order equation}$$

In each of these equations, one is interested in determining the values of the regression parameters a_0, a_1, a_2, a_3, ... from the experimental data. The methods for determining these parameters, and the confidence associated with them, are discussed in the following sections.

Random variation vs. physical variation   In all the preceding chapters (except Chap. 6, Factorial Experiments) variation in the data has been assumed to be due to random causes, and the test conditions were assumed to be constant during the experiment. This type of variation is called random variation. Consider a problem of wear loss in metals sliding against each other. At any sliding velocity, a scatter in wear loss may be found when more than one observation is taken at that velocity. (See Fig. 13.2.) This

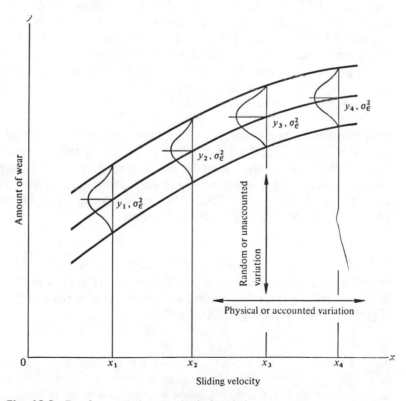

Fig. 13.2 Random variation vs. physical variation.

scatter is denoted by σ_ε^2 (random variation), and it is not caused by the functional relationship between the variables. This variation σ_ε^2 is a result of three factors or their combination:

1. Some variables which have a significant effect on the outcome of the experiment may not have been considered in the analysis, even though they might have been present during the experiment
2. Other uncontrollable variables may be present. This can be accounted for by the experimental error.
3. The mathematical equation model used to establish the functional relationship may not have been appropriately chosen to describe the underlying physical mechanisms.

Therefore, the true (mean) value of wear $\bar{y}$ can only be predicted with certain error due to the random causes.

In this type of experiment it is of primary importance to determine how wear $\bar{y}$ varies when the sliding velocity is varied. This intentional variation made on the independent variable x is called the physical variation. (See Fig. 13.2.) The important thing is to establish the true functional relationship between y and x in the presence of the random variation. Figure 13.2 indicates that the overall upward trend is due to the physical changes made on x. In regression analysis, the ratio of the random variation to the physical variation, when subtracted from 1.0, is an index of how well the function (straight line) fits the experimental data. This is expressed by r^2, where r is the correlation coefficient.

The method of least squares   The principal use of the method of least squares is to determine the "best fit" of an assumed implicit function to the experimental data. This is done by minimizing the sum of the squares of the deviations in the y direction of the data points (x_i, y_i) from the most probable curve. (See Fig. 13.3.) This curve can be linear or nonlinear. This section deals only with the linear function

$$y = a_0 + a_1 x \tag{13.4}$$

When the sum of squares of the deviations is minimized in the y direction, it is called the regression of y on x. It applies only when the random variation exists in y values, and when x is held reasonably constant (Fig. 13.4). The resulting equation is

$$y = a_0 + a_1 x$$

and the values of y can be estimated from the knowledge of x. An erroneous conclusion would be made if this equation were used to estimate x from y, since the error is minimized only in the y direction and not in the x direction.

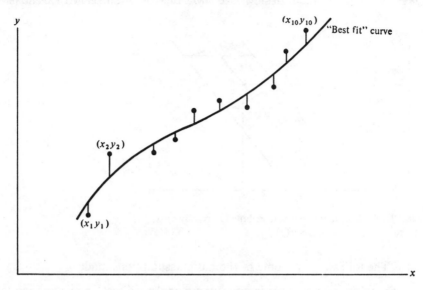

Fig. 13.3 Sum of squares of deviations in y direction for least-squares method.

When the sum of squares of the deviations is minimized in the x direction, the process is the regression of x on y (Fig. 13.5). The equation for this is

$$x = a_0' + a_1'y \tag{13.5}$$

where the variation in the x direction actually exists and the values of y are held constant; x can then be estimated from y. Note that the parameters of Eqs. (13.4) and (13.5) are not the same.

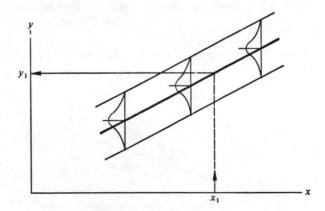

Fig. 13.4 Constant variance distributed normally on y. Regression of y on x, $y = f(x)$.

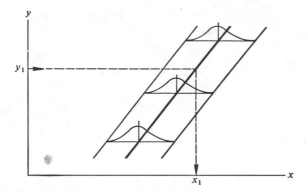

Fig. 13.5 Constant variance distributed normally on x. Regression of x on y, $x = f(y)$.

The following are some of the basic assumptions made:

1. The variation measured in either the x or the y direction must be random and normally distributed, the mean of the random error being zero and the variance equal to σ^2 (unknown). (See Figs. 13.4 and 13.5.)
2. The variance σ^2 should be constant when measured, say, in the y direction for any value of x. This means the random error is additive to the true physical value.
   True y = physical y + random y
3. In the event that σ^2 is not constant (Fig. 13.6), and it either increases or decreases with x, the appropriate transformation should be made, so that the error is additive.

The next step is to derive the necessary equations for determining the "best fit" to the data. Suppose the line that fits the data the best is

$$y = a_0 + a_1 x + \varepsilon$$

where ε is the measure of the deviation of the data points in the y direction.

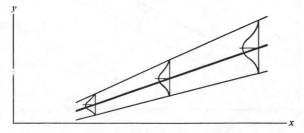

Fig. 13.6 Variance increases as x increases, and is distributed normally on y.

According to the method of least squares, the total sum of squares of deviation E is given by

$$E = \sum_{i=1}^{n} \varepsilon_i^2 = \sum_{i=1}^{n} (y_i - a_0 - a_1 x_i)^2 \tag{13.6}$$

where (x_i, y_i) are the data points. The objective here is to determine the regression parameters a_0 and a_1 (also called the least-squares parameters) so that their substitution in Eq. (13.6) results in the least possible value of E. This is done by partial differentiation of Eq. (13.6) with respect to a_0 and a_1, and setting the results equal to zero.

$$\frac{\partial E}{\partial a_0} = -2 \sum_{i=1}^{n} (y_i - a_0 - a_1 x_i) \tag{13.7}$$

$$\frac{\partial E}{\partial a_1} = -2 \sum_{i=1}^{n} (y_i - a_0 - a_1 x_i) x_i \tag{13.8}$$

The estimates of a_0 and a_1 are

$$\sum_{i=1}^{n} (y_i - a_0 - a_1 x_i) = 0$$

$$\sum (y_i - a_0 - a_1 x_i) x_i = 0$$

By writing $\sum y$ for $\sum_{i=1}^{n} y_i$, $\sum xy$ for $\sum_{i=1}^{n} x_i y_i$, etc., and simplifying, one obtains

$$\sum y = a_0 n + a_1 \sum x \tag{13.9}$$

$$\sum xy = a_0 \sum x + a_1 \sum x^2 \tag{13.10}$$

Equations (13.9) and (13.10) are called the "normal equations."

By solving these equations simultaneously, the parameters a_0 and a_1 become

$$a_0 = \frac{\sum y \sum (x^2) - \sum x \sum (xy)}{n \sum (x^2) - (\sum x)^2} \tag{13.11}$$

$$a_1 = \frac{n \sum (xy) - (\sum x)(\sum y)}{n \sum (x^2) - (\sum x)^2} \tag{13.12}$$

The "normal equation" (13.9) can also be obtained by first multiplying Eq. (13.4) by 1.0 and then summing on both sides of Eq. (13.4), that is, $\sum y = \sum (a_0 + a_1 x)$ or $\sum y = a_0 n + a_1 \sum x$; while the second equation (13.10) is obtained by first multiplying both sides of Eq. (13.4) by x and then summing up, that is, $\sum (xy) = \sum [(a_0 + a_1 x)x]$ or

$$\sum (xy) = a_0 \sum x + a_1 \sum (x^2)$$

It is recommended not to extrapolate the linear regression equation unless a sound theoretical basis exists for doing that. Whenever possible, the range of the independent variables which occurred in the data should be given.

After the least-squares line and its parameters are established, one frequently wishes to know how well this line fits the data. The measure of the "best fit" is the correlation coefficient r, and it can be determined by Eqs. (13.1) and (13.2). Several alternative forms can be found in standard books on statistics. These equations are

$$r = \sqrt{1 - \frac{s_{y \cdot x}^2}{s_y^2}} \tag{13.13}$$

where

$$s_y = \sqrt{\frac{\sum_{i=1}^{n} (y_i - \bar{y})^2}{n - 1}}$$

s_y is the sample standard deviation of y, and

$$s_{y \cdot x} = \sqrt{\frac{\sum_{i=1}^{n} (y_i - y_{i_c})^2}{n - 2}} \tag{13.14}$$

where y_i = actual values of y
   y_{i_c} = values of y computed from the equation of line
   n = number of points

$s_{y \cdot x}$ is called the standard error of estimate, which is the measure of scatter of points around the line; or, as brought up in the previous section, it is the measure of the random variation. This is also an estimate of the standard deviation of the y population for a given x.

It should be noted that $(n - 2)$ is the degrees of freedom used as a divisor in Eq. (13.14), since the number of parameters to be estimated is two, a_0 and a_1. In previous chapters, $(n - 1)$ was used since only one parameter was required to be estimated.

An alternative equation for $s_{y \cdot x}$ is (the proof is available in Ref. 2)

$$s_{y \cdot x} = \sqrt{\frac{n - 1}{n - 2} (s_y^2 - a_1^2 s_x^2)} \tag{13.15}$$

where s_y and s_x are the sample standard deviations of y and x, respectively; a_1 is the slope of the line, which can be computed by Eq. (13.12).

Another equation for determining $s_{y \cdot x}$ is [3]

$$s_{y \cdot x} = \sqrt{\frac{\sum y^2 - a_0 \sum y - a_1 \sum xy}{n - 2}} \tag{13.16}$$

When the number of points, n, is relatively large the correlation coefficient r can also be defined as

$$r = a_1 \frac{s_x}{s_y} \tag{13.17}$$

The value of r may range between -1 and $+1$; when $r = \pm 1$, the function (in this case the straight line) fits the data perfectly. The positive or negative sign indicates that y increases or decreases respectively with x. After r is computed from the sample data, it can be tested for its significance by comparing it with the values of r given in Table A-49. This is discussed in Subsec. 13.1.2.

Example 13.3 In a certain application, the excessive amount of wear in a journal bearing was due to the excessive operating temperature, controlled by an oil bath. It is necessary that the amount of wear not exceed 8 mg/100 h of operation. What should be the bearing operating temperature?

An experiment was designed where 10 bearings were tested for wear, and each bearing was run under a different operating temperature controlled by the oil bath. The following test data were obtained:

Operating temperature, °F (x)	200	250	300	400	450	500	550	600	650	700
Amount of wear, mg/100 h (y)	3	4	5	5.5	6	7.5	8.8	10	11.1	12

Solution Tabulate the data for applying Eqs. (13.11) and (13.12).

x	y	x^2 in 10^4	xy in 10^2
200	3	4	6
250	4	6.25	10
300	5	9	15
400	5.5	16	22
450	6	20.2	27
500	7.5	25	37.5
550	8.8	30.3	48.2
600	10	36	60
650	11.1	42.2	72.1
700	12	49	84
4600	72.9	237.95	381.8

From the above, $\sum x = 4600$, $\sum y = 72.9$, $\sum x^2 = 237.95 \times 10^4$, and $\sum xy = 381.8 \times 10^2$. By Eq. (13.11),

$$a_0 = \frac{\sum y \sum (x^2) - \sum x \sum (xy)}{n \sum (x^2) - (\sum x)^2}$$

$$= \frac{(72.9)(237.95 \times 10^4) - (4600)(381.8 \times 10^2)}{10(237.95 \times 10^4) - (4600)^2}$$

$$= \frac{17,320 - 17,550}{2379.5 - 2120} = \frac{-230}{267.5} \simeq -0.85 \text{ mg/100 h}$$

By Eq. (13.12),

$$a_1 = \frac{n \sum (xy) - (\sum x)(\sum y)}{n \sum (x^2) - (\sum x)^2}$$

$$= \frac{10(381.8 \times 10^2) - (4600)(72.9)}{267.5 \times 10^4}$$

$$= \frac{38.18 - 33.5}{267.5} = \frac{4.68}{267.5} = 0.0175 \text{ mg/(100 h) (°F)}$$

Substituting a_0 and a_1 in Eq. (13.4),

$$y = a_0 + a_1 x$$

or

$$y = -0.85 + 0.0175x$$

A plot of these results is shown in Fig. 13.7.

    In order to know how well this line fits the data, the correlation coefficient r is computed. r can be found through various equations such as (13.1), (13.13), or (13.17). Equation (13.17) would be used here:

$$r = a_1 \frac{s_x}{s_y}$$

where, using Eq. (1.14):

$$s_x = \sqrt{\frac{\sum (x_i - \bar{x})^2}{n - 1}}$$

$$= 171.2$$

$$s_y = \sqrt{\frac{\sum (y_i - \bar{y})^2}{n - 1}}$$

$$= 3.086$$

and

$$a_1 = 0.0175 \text{ (as computed above)}$$

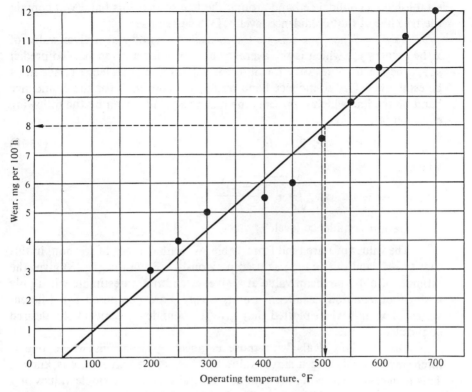

Fig. 13.7 Plot for Example 13.3.

Hence

$$r = \frac{.0175 \times 171.2}{3.086} = .97$$

or

$$r^2 = .94$$

Obviously, the value of r is significant; and this indicates that 94 percent (r^2) of the total variation (s_y^2) in wear can be accounted for by this line, and 6 percent is un-accountable. Therefore, physical variation is 94 percent and random variation ($s_{y \cdot x}^2$) is only 6 percent. The high value of r (.97) indicates that the above linear relationship can reasonably be used to represent the underlying physical phenomenon (the effect of temperature on wear). From Fig. 13.7 it can, therefore, be seen that the operating temperature should not exceed 505°F for the amount of wear to be less than 8 mg.

Confidence limits for the least-squares line After the straight-line relationship is established, one frequently wishes to determine the confidence level at which this line can be accepted as the true line. This can be done by establishing a confidence band around the line so that this band will contain the true line at that confidence level. Two cases arise:

Case 1   When one repeats the measurements of y at a given value of x, he obtains $\bar{y}_x$, which is the average value of y for a given x. To predict $\mu_{y \cdot x}$, the mean values of y for different values of x confidence bands must be established. A confidence band on mean values $\mu_{y \cdot x}$ (or the confidence band on the line which represents the mean values) is given by the following equation [4]:

$$\bar{y}_x - A_1 \le \mu_{y \cdot x} \le \bar{y}_x + A_1 \qquad (13.18)$$

where

$$A_1 = t_{\alpha/2; \, (n-2)} s_{y \cdot x} \sqrt{\frac{1}{n} + \frac{(x - \bar{x})^2}{(n-1)s_x^2}}$$

$1 - \alpha$ = confidence level

The values of t are read from Table A-3 with degrees of freedom in this case $(n - 2)$ and not $(n - 1)$ as used previously. Equation (13.18) gives the estimate of only the mean value at a given x and not an estimate when only a single observation is taken. Limits on $\mu_{y \cdot x}$ are computed for different values of x, and when plotted they give the confidence band at the desired confidence.

Case 2   This deals with those engineering situations where one is interested in predicting a single value of y when the value of x is known. This requires constructing the confidence band around the single values of y and not around the mean of y. The appropriate equation is

$$\bar{y}_x - A_2 \le y \le \bar{y}_x + A_2 \qquad (13.19)$$

where

$$A_2 = t_{\alpha/2; \, (n-2)} s_{y \cdot x} \sqrt{1 + \frac{1}{n} + \frac{(x - \bar{x})^2}{(n-1)s_x^2}}$$

Example 13.4   In Example 13.3 it was concluded that the amount of wear will not exceed 8 mg/100 h when the operating temperature is kept below 505°F. However, only 50 percent confidence can be associated with this conclusion. Hence, for higher confidence, a band around this line as shown in Fig. 13.7 must be constructed. Establish this band for 90 percent confidence for Case 1, when the true average wear $\mu_{y \cdot x}$ is predicted for any temperature x; and for Case 2, when the individual value of wear is expected to occur at any temperature x.

Solution
     Case 1   From Eq. (13.18),

$$\bar{y}_x - A_1 \le \mu_{y \cdot x} \le \bar{y}_x + A_1$$

where

$$A_1 = t_{\alpha/2; \, (n-2)} s_{y \cdot x} \sqrt{\frac{1}{n} + \frac{(x - \bar{x})^2}{(n-1)s_x^2}}$$

$\bar{y}_x$ is the value found from Fig. 13.8 corresponding to x, and μ_x is the true average wear y at temperature x.

From Eq. (13.15),

$$s_{y \cdot x} = \sqrt{\frac{n-1}{n-2}(s_y^2 - a_1^2 s_x^2)}$$

From the given data, $s_x = 171.2$, $s_y = 3.086$, $\bar{x} = 460$, and from Example 13.3, $a_1 = 0.0175$. Hence

$$s_{y \cdot x} = \sqrt{\frac{10-1}{10-2}[3.086^2 - (0.0175^2)(171.2^2)]}$$

$$= 0.619 \text{ mg/100 h}$$

From Table A-3,

$$t_{\alpha/2;\,(n-2)} = t_{0.05;8} = 1.86$$

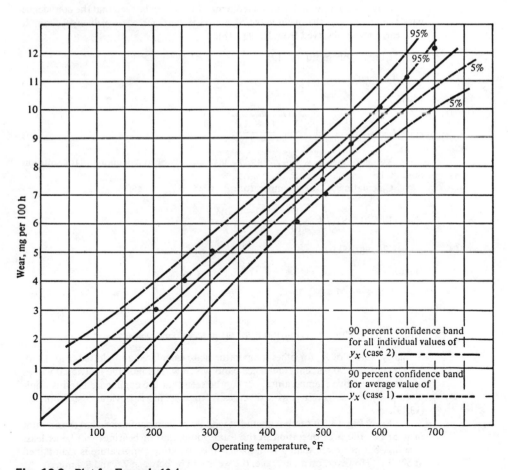

Fig. 13.8 Plot for Example 13.4.

A_1 and $\bar{y}_x$ are computed for various values of x (temperature), and they are plotted on Fig. 13.8 to obtain the desired confidence band. A sample calculation is given for 300°F temperature.

$$A_1 = (1.86)(0.619)\sqrt{\frac{1}{10} + \frac{(300-460)^2}{9(171.2^2)}}$$

$$= 0.51 \text{ mg/100 h}$$

and $\bar{y}_x = 4.4$ mg/100 h for $x = 300$°F from Fig. 13.8. Hence,

$$4.4 - 0.51 \leq \mu_{y \cdot x} \leq 4.4 + 0.51$$

or

3.89 mg/100 h $\leq \mu_{y \cdot x} \leq 4.91$ mg/100 h for 90 percent confidence

Similarly, values of A_1 for other temperatures are calculated, and plotted in Fig. 13.8. This results in the desired 90 percent confidence band which will include the true average value of wear at a given temperature. It can be seen that the confidence band is minimum at the mean value of x, and it increases when x diverges from $\bar{x}$. This can also be observed from Eq. (13.18).

Case 2   From Eq. (13.19),

$$\bar{y}_x - A_2 \leq y \leq \bar{y}_x + A_2$$

where

$$A_2 = t_{\alpha/2;\ (n-2)} s_{y \cdot x} \sqrt{1 + \frac{1}{n} + \frac{(x - \bar{x})^2}{(n-1)s_x^2}}$$

Values of $t_{\alpha/2;\ (n-2)} s_{y \cdot x}$, $\bar{x}$, and s_x have already been calculated in the solution of Case 1.

Calculating A_2 for a temperature x of 300°F,

$$A_2 = (1.86)(0.619)\sqrt{1 + \frac{1}{10} + \frac{(300-460)^2}{9(171.2^2)}}$$

$$= 1.26 \text{ mg/100 h}$$

Hence,

$$4.4 - 1.26 \leq y \leq 4.4 + 1.26$$

or

3.14 mg/100 h $\leq y \leq 5.66$ mg/100 h for 90 percent confidence

Similarly, values of A_2 for other temperatures are calculated and plotted (Fig. 13.8). This results in a 90 percent confidence band which will include the individual values of wear at any given temperature. It can be seen that the confidence band is minimum at $\bar{x}$, and it increases as x diverges from $\bar{x}$. This is obvious also from Eq. (13.19).

From Fig. 13.8 it can be concluded with 90 percent confidence that the amount of wear per 100 h of operation in the total population of bearings will be at least 7.05 mg, but not more than 9.1 mg, when the operating temperature is maintained at 505°F. However, on an average, the wear will be between 7.6 and 8.4 mg.

If it is of interest to have wear not more than 8 mg in 100 h of operation in any one bearing in the total population, then the operating temperature should be maintained at 440°F.   This means that 95 percent of the bearings will have wear less than 8 mg at this temperature.   (See Fig. 13.8.) It can also be concluded with 95 percent confidence that the average wear of the population will not exceed 8 mg if the operating temperature is maintained at 485°F.

Confidence limits for the least-squares parameters   Least-squares parameters a_0, a_1, ... are the same as the regression parameters or the regression coefficient.   In the previous section it was shown how to determine these parameters from experimental data.   However, no mention was made of the confidence associated with them.   In many situations, the true value of one or more of the least-squares parameters a_0, a_1, a_2, ... is of interest rather than the confidence band on the entire curve.   The following is the equation for the confidence intervals on the parameters where the interval contains the true values of the parameters at a given confidence level.   The confidence interval on a_0 is [2, 5]

$$a_0 - A_3 \leq \hat{a}_0 \leq a_0 + A_3 \tag{13.20}$$

where

$\hat{a}_0$ = true intercept (population) of line
a_0 = intercept as found from sample

$$A_3 = t_{\alpha/2;(n-2)} s_{y \cdot x} \sqrt{\frac{1}{n} + \frac{\bar{x}^2}{(n-1)s_x^2}}$$

$1 - \alpha$ = confidence level

Note that Eq. (13.20) is the special case of Eq. (13.18) when $x = 0$.   Similarly, the confidence interval on a_1 (the slope of the line) is [2, 5]

$$a_1 - A_4 \leq \hat{a}_1 \leq a_1 + A_4 \tag{13.21}$$

where

$\hat{a}_1$ = true slope (population) of line
a_1 = slope as found from sample

$$A_4 = t_{\alpha/2;(n-2)} s_{y \cdot x} \sqrt{\frac{1}{\sum(x_i - \bar{x})^2}}$$

$1 - \alpha$ = confidence level

Example 13.5   Using the data from Example 13.3, establish the 90 percent confidence limits for the true value of the parameters a_0 and a_1.

Solution   From Eq. (13.20),

$$a_0 - A_3 \leq \hat{a}_0 \leq a_0 + A_3$$

where $\hat{a}_0$ is the true intercept, and

$$A_3 = t_{\alpha/2;\,(n-2)}\, s_{y\cdot x} \sqrt{\frac{1}{n} + \frac{\bar{x}^2}{(n-1)s_x^2}}$$

Substituting the appropriate values from Example 13.4,

$$A_3 = (1.86)(0.619) \sqrt{\frac{1}{10} + \frac{460^2}{9 \times (171.2^2)}} = 1.09 \text{ mg/100 h}$$

From Example 13.3, $a_0 = -0.85$.   Hence, with 90 percent confidence, limits for the true value of $\hat{a}_0$ are -1.94 mg/100 h $\leq \hat{a}_0 \leq 0.24$ mg/100 h.

From Eq. (13.21),

$$a_1 - A_4 \leq \hat{a}_1 \leq a_1 + A_4$$

where

$$A_4 = t_{\alpha/2;\,(n-2)}\, s_{y\cdot x} \sqrt{\frac{1}{\sum (x_i - \bar{x})^2}}$$

or

$$A_4 = (1.86)(0.619) \sqrt{\frac{1}{9(171.2^2)}} = 0.00224 \text{ mg/(100 h)(°F)}$$

From Example 13.3, $a_1 = 0.0175$.   Hence, with 90 percent confidence, limits for the true value of a_1 are 0.01526 mg/(100 h) (°F) $\leq \hat{a}_1 \leq 0.01974$ mg/(100 h) (°F).

13.1.4   LINEAR MULTIPLE REGRESSION ANALYSIS

Linear multiple regression analysis is a method for establishing a linear functional relationship between one dependent and two or more independent variables.   This is similar to the method discussed in Subsec. 13.1.3, Linear Regression Analysis, in which only two variables (one dependent and another independent) were involved.   Many engineering experiments involve more than one independent variable which affects the dependent variable.   It may then become necessary to isolate the effect of one independent variable from the effects of the others and to determine the contribution of each variable to the total effect.   The general form of the linear function with multiple variables is

$$y = a_0 + a_1 x + a_2 z + \cdots \tag{13.22}$$

where y = dependent variable which varies partially due to variation in x
and partially due to variation in z
x, z = independent variables
a_0, a_1, a_2 = multiple regression parameters

Equation (13.22) is called the linear multiple regression equation of y on x, z, etc. The method of determining the parameters a_0, a_1, and a_2 from the experimental data is given below.

The method of least squares All the principles of the least squares analysis presented in the linear regression analysis apply to the linear multiple regression analysis. Just as a set of data points (x, y) can be represented by a least-squares line on a two-dimensional plot, so a set of points (x, y, z) can be given by a least-squares plane on a three-dimensional plot.

The sum of squares E of the deviations of the data points from the most probable plane [Eq. (13.22)] is

$$E = \sum_{i=1}^{n} (y - a_0 - a_1 x_i - a_2 z_i)^2 \tag{13.23}$$

The best estimate of the parameters a_0, a_1, and a_2 is the one which minimizes the sum of squares E. As before, this is done by first partially differentiating Eq. (13.23) with respect to a_0, a_1, and a_2 and then setting the results equal to zero. With further simplification a set of three simultaneous equations is obtained:

$$\sum y = na_0 + a_1 \sum x + a_2 \sum z$$
$$\sum yx = a_0 \sum x + a_1 \sum x^2 + a_2 \sum xz \tag{13.24}$$
$$\sum yz = a_0 \sum z + a_1 \sum xz + a_2 \sum z^2$$

These are called the "normal" equations. "Normal" equations can also be obtained by multiplying both sides of Eq. (13.22) by 1, x, and z in succession and then summing up. By simultaneously solving these equations, parameters a_0, a_1, and a_2 are computed. The number of computations can be reduced if each variable (x, y, z) is referred to its mean $(\bar{x}, \bar{y}, \bar{z})$. The constant term a_0 becomes zero, and the three "normal" equations reduce to two equations:

$$y = a_0 + a_1 x + a_2 z$$

Therefore,

$$\bar{y} = a_0 + a_1 \bar{x} + a_2 \bar{z}$$

or

$$y - \bar{y} = a_1 (x - \bar{x}) + a_2 (z - \bar{z})$$

This can be written as

$$Y = a_1 X + a_2 Z + \cdots \tag{13.25}$$

where $Y = y - \bar{y}$
$X = x - \bar{x}$
$Z = z - \bar{z}$

The "normal" equations for Eq. (13.25) can then be given as

$$\sum YX = a_1 \sum X^2 + a_2 \sum XZ$$
$$\sum YZ = a_1 \sum XZ + a_2 \sum Z^2$$

By solving these two simultaneous equations, the values of a_1 and a_2 become

$$a_1 = \frac{(\sum YX)(\sum Z^2) - (\sum XZ)(\sum YZ)}{\sum X^2 \sum Z^2 - (\sum XZ)^2} \tag{13.26}$$

$$a_2 = \frac{(\sum X^2)(\sum YZ) - (\sum YX)(\sum XZ)}{\sum X^2 \sum Z^2 - (\sum XZ)^2} \tag{13.27}$$

Once the parameters a_1 and a_2 are determined, the third parameter a_0 is calculated from the equation

$$a_0 = \bar{y} - a_1 \bar{x} - a_2 \bar{z} \tag{13.28}$$

After the multiple regression parameters are determined, the equation that represents the data best is known. The next step is then to establish how well this equation of plane fits the data. This is done by determining the value of the multiple correlation coefficient as developed in the next section. The above concept is illustrated in the following example.

Example 13.6 Humidity and the atmospheric pressure may affect the level of nitrogen oxides (NO) exhaust emission of internal-combustion engines. Design an experiment to determine whether humidity and atmospheric pressure do affect the NO emission level; and if so, establish the relationship.

The level of nitrogen emission of a typical engine was measured at various humidity levels and at the atmospheric pressures prevalent during the experiment. The following data were obtained:

Nitrogen oxide emission level, ppm* (y)	Humidity, gr/lb dry air (x)	Atmospheric pressure, inHg (z)
1,500	20	29.30
1,420	30	29.50
1,430	40	29.40
1,270	50	29.35
1,200	60	29.70
1,100	70	29.60
1,120	80	29.55
1,015	90	29.80
1,040	100	29.78
990	110	29.80

* Parts per million.

Solution The relationship given by Eq. (13.22) is

$$y = a_0 + a_1 x + a_2 z$$

a_1 and a_2 are found by using Eqs. (13.26) and (13.27).

The quantities used in these equations are tabulated below.

$$\bar{x} = 65 \text{ gr} \qquad \bar{y} = 1{,}209 \text{ ppm} \qquad \bar{z} = 29.58 \text{ inHg}$$

$Y = y - \bar{y}$	$X = x - \bar{x}$	$Z = z - \bar{z}$	YX	XZ	YZ	X^2	Z^2
291	−45	−0.28	−13,095	12.6	−81.48	2,025	0.0784
211	−35	−0.08	−7,385	2.8	−16.88	1,225	0.0064
221	−25	−0.18	−5,525	4.5	−39.78	625	0.0324
61	−15	−0.23	−915	3.45	−14.03	225	0.0529
−9	−5	0.12	+45	−0.6	−1.08	25	0.0144
−109	5	0.02	−545	0.1	−2.18	25	0.0004
−89	15	−0.03	−1,335	−0.45	+2.67	225	0.0009
−194	25	0.22	−4,850	5.5	−42.68	625	0.0484
−169	35	0.20	−5,915	7.0	−33.8	1,225	0.04
−219	45	0.22	−9,855	9.9	−48.18	2,025	0.0484
			$\sum YX =$ −49,375	$\sum XZ =$ 44.8	$\sum YZ =$ −277.42	$\sum X^2 =$ 8,250	$\sum Z^2 =$ 0.3226

$$a_1 = \frac{(\sum YX)(\sum Z^2) - (\sum XZ)(\sum YZ)}{\sum X^2 \sum Z^2 - (\sum XZ)^2}$$

Substituting appropriate values,

$$a_1 = -5.35 \text{ ppm/gr}$$

$$a_2 = \frac{(\sum X^2)(\sum YZ) - (\sum YX)(\sum XZ)}{\sum X^2 \sum Z^2 - (\sum XZ)^2}$$

The value of a_2 is calculated to be −116.81 ppm/inHg.
a_0 is found by Eq. (13.28).

$$a_0 = \bar{y} - a_1 \bar{x} - a_2 \bar{z}$$
$$= 1{,}209 - (-5.35)(65) - (-116.81)(29.58)$$
$$= 5{,}011.29 \text{ ppm}$$

Hence, the effect of humidity and atmospheric pressure on NO emission level is given by the following relationship:

$$y = 5{,}011.29 - 5.35x - 116.8z$$

In order to determine whether humidity and pressure affect the NO emission level, it would be necessary to establish how well this equation represents the data. This is done by computing the multiple correlation coefficient as shown in the next example.

Multiple correlation coefficient In the case when only two variables (one dependent and the other independent) are involved, the measure of the "best fit" of a least-squares line is known as the correlation coefficient r. When more than one independent variable is present in the experiment, the measure of the best-fitting equation is referred to as the multiple correlation coefficient. This is defined as

$$r_{y \cdot xz} = \sqrt{1 - \frac{s_{y \cdot xz}^2}{s_y^2}} \tag{13.29}$$

where $r_{y \cdot xz}$ = multiple correlation coefficient of y on x and z
$\qquad s_{y \cdot xz}$ = standard error of estimate of y on x and z
$\qquad s_y$ = sample standard deviation of y

The standard error of estimate $s_{y \cdot xz}$ can be computed from [3]

$$s_{y \cdot xz} = \frac{\sum y^2 - a_0 \sum y - a_1 \sum xy - a_2 \sum zy}{n - 3} \tag{13.30}$$

where n = number of data points
$\qquad n - 3$ = number of degrees of freedom

Note that the number of degrees of freedom in Eq. (13.30) is $n - 3$ since the number of parameters to be estimated is three: a_0, a_1, and a_2.

Example 13.7 Determine the correlation coefficient for the data in Example 13.6 to establish how well the equation (as found in Example 13.6) represents the data.

Solution The correlation coefficient from Eq. (13.29) is

$$r = \sqrt{1 - \frac{s_{y \cdot xz}^2}{s_y^2}}$$

From the data of Example 13.6,

$s_y^2 = 35,089.16$ ppm^2

$s_{y \cdot xz}$ is given by Eq. (13.30). The values used in this expression are tabulated below.

y	y^2 in 10^4	xy in 10^2	zy in 10^2
1,500	225	300	439.5
1,420	201.64	426	492.65
1,430	204.49	572	420.42
1,270	161.29	635	436.25
1,200	144	720	356.4
1,100	121	770	325.6
1,120	125.44	896	330.96
1,015	102.01	913.5	302.47
1,040	108.16	1,040	309.71
990	98.01	1,089	295.02
$\sum y = 12,085$	$\sum y^2 = 1,491.04 \times 10^4$	$\sum xy = 7,361.5 \times 10^2$	$\sum zy = 3,708.98 \times 10^2$

$$s_{y \cdot xz} = \frac{\sum y^2 - a_0 \sum y - a_1 \sum xy - a_2 \sum zy}{n-3}$$

$$= 52.38 \text{ ppm}$$

hence

$$r = \sqrt{1 - \frac{52.38^2}{35,089.16}}$$

$$= \sqrt{.922} = .96$$

or

$$r^2 = .922$$

Therefore the correlation is good; i.e., the equation represents the data very well; 92.2 percent of the total variation $s_y{}^2$ in NO emission level can be related to variation in atmospheric pressure and humidity.

13.1.5 NONLINEAR REGRESSION ANALYSIS

So far, the discussion has been restricted to those situations where the underlying relationships between variables were linear. In some engineering problems, nonlinearity is encountered, in which case the straight-line fit (first-order equation) is not applicable. Two types of such problems are treated in this section: (1) problems where second- and higher-order relationships exist and (2) problems where some sort of transformation of variables is necessary. This is treated by regression analysis involving linearizing transformation. Both types involve establishing a functional relationship between one dependent and one or more independent variables. The case of one independent variable is presented here; however, this concept is the same for two or more independent variables.

Higher-order regression analysis In Subsec. 13.1.3 a method was shown of fitting a straight line (first-order equation) to a set of data. The correlation coefficient was computed to indicate how well the straight line represents (or fits) the data. In the event the straight-line relationship does not exist, the fit will be poor; therefore, higher-order equations should be tried. Some of these equations are:

1. $y = a_0 + a_1 x + a_2 x^2$ second-order (or parabolic) curve

2. $y = a_0 + a_1 x + a_2 x^2 + a_3 x^3$ third-order (or cubic) curve

3. $y = a_0 + a_1 x + a_2 x^2 + a_3 x^3 + a_4 x^4$ fourth-order (or quartic) curve

where a_0, a_1, a_2, a_3, etc., are the regression parameters. With the assumption made in Subsec. 13.1.3, the method of least squares can be applied here to estimate these parameters. A set of normal equations can be derived in the described manner in Subsec. 13.1.3.

The normal equations for second-order fit ($y = a_0 + a_1 x + a_2 x^2$) are

$$\sum y\ \ = na_0 + a_1 \sum x + a_2 \sum x^2$$

$$\sum yx = a_0 \sum x + a_1 \sum x^2 + a_2 \sum x^3 \qquad (13.31)$$

$$\sum yx^2 = a_0 \sum x^2 + a_1 \sum x^3 + a_2 \sum x^4$$

The parameters a_0, a_1, and a_2 are computed by solving these equations simultaneously. The parameters for the third-order equation can be obtained in the same manner.

Example 13.8 In a certain fluidic amplifier design, there are several passages through which the fluid flows. The variation in the dimensions of these passages affects the output flow of the amplifier. It was found that the flow rate is very sensitive to changes in dimension of one particular passage. It is desired to establish the relationship between the passage dimension and the flow rate of the amplifier. Fourteen fluidic amplifiers having different passage dimensions were tested, and the flow rates were measured:

Dimensions, 10^{-3} in (x)	10	15	20	25	30	35	40	45	50	55	60	65	70	75
Flow rate, gal/h (y)	0.26	0.38	0.55	0.70	1.05	1.36	1.75	2.20	2.70	3.20	3.75	4.40	5.0	6.0

Establish the relationship between the dimension of the passage and its effect on the flow rate. Specify the dimension so that the flow rate of 3.2 gal/h is obtained.

Solution In Example 13.3, it was shown how to fit a first-order equation ($y = a_0 + a_1 x$) to a set of data. In this example, a similar procedure would apply for the case when the second- or higher-order equations are considered. Generally, the first-order (linear) equation is tried first. If linear fit is not satisfactory, the higher-order (nonlinear) equation is tried next. In order to illustrate the procedure of establishing nonlinear relationships, the second-order relationship is tried first:

$$y = a_0 + a_1 x + a_2 x^2$$

Coefficients a_0, a_1, and a_2 are found from Eq. (13.31). Tabulate the values used in Eq. (13.31):

x, 10^{-3} in	y, gal/h	yx	x^2	yx^2	x^3	x^4
10	0.26	2.6	100	26.0	1000	10,000
15	0.38	5.7	225	85.5	3375	50,625
20	0.55	11.0	400	220.0	8000	160,000
25	0.70	17.5	625	437.5	15,625	390,625
30	1.05	31.5	900	945.0	27,000	810,000
35	1.36	47.6	1,225	1,666.0	42,875	1,500,000
40	1.75	70.4	1,600	2,800.0	64,000	2,560,000
45	2.20	99.0	2,025	4,455.0	91,125	4,100,625
50	2.70	135.0	2,500	6,750.0	125,000	6,250,000
55	3.20	176.0	3,025	9,680.0	274,625	17,850,625
60	3.75	225.0	3,600	13,500.0	343,000	24,016,000
65	4.40	286.0	4,225	18,590.0	421,875	31,640,625
70	5.0	350.0	4,900	24,500.0		
75	6.0	450.0	5,625	33,750.0		
595	33.30	1,907.3	30,975	117,405		
$\sum x$	$\sum y$	$\sum yx$	$\sum x^2$	$\sum yx^2$	$\sum x^3$ = 1,799,875	$\sum x^4$ = 111,444,375

By substituting these values in Eq. (13.31) and solving for a_0, a_1, and a_2, the following results are obtained:

$a_0 = 0.1959$ gal/h $a_1 = -4.090 \times 10^{-3}$ gal/(h)/10^{-3} in

$a_2 = 1.065 \times 10^{-3}$ gal/(h)/$(10^{-3}$ in$)^2$

Hence, the equation becomes

$$y = 0.1959 - 4.09 \times 10^{-3}x + 1.065 \times 10^{-3}x^2$$

where units of x are in 10^{-3} in, and units of y are in gallons per hour. This equation, therefore, represents the relationship between the dimension of the passage and its effect on the flow rate. The plot is shown in Fig. 13.9, from which it can be concluded that the passage dimension of 0.055 in should be specified to obtain 3.2 gal/h flow rate. To obtain the coefficient of correlation, Eq. (13.13) is used:

$$r = \sqrt{1 - \frac{s_{y \cdot x}^2}{s_y^2}}$$

s_y is found from Eq. (1.14):

$$s_y = \sqrt{\frac{\sum_{i=1}^{n}(y_i - \bar{y})^2}{n-1}}$$

From the data, $\bar{y} = 2.378$, and the standard deviation is $s_y = 1.85$. Also,

$$s_{y \cdot x} = \sqrt{\frac{\sum_{i=1}^{n}(y_i - y_{i_c})^2}{n-3}}$$

which is identical to Eq. (13.14), except that the denominator is $n-3$, where 3 represents the number of parameters estimated. Actual values of y and values of y

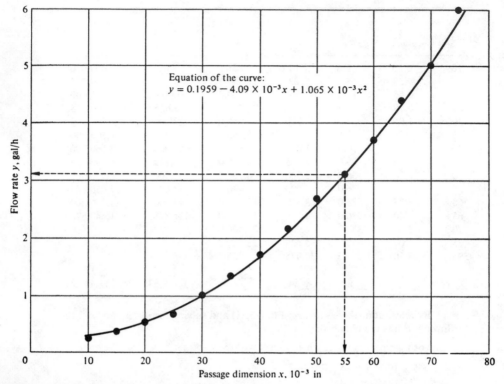

Fig. 13.9   Plot for Example 13.8.

as computed (y_c) from the equation of the curve above corresponding to each actual value of x are tabulated below for calculations of $s_{y \cdot x}$.

i	x actual	y actual (y_i)	y calculated (y_{ic})
1	10	0.26	0.26
2	15	0.38	0.37
3	20	0.55	0.54
4	25	0.70	0.76
5	30	1.05	1.03
6	35	1.36	1.36
7	40	1.75	1.74
8	45	2.20	2.17
9	50	2.70	2.65
10	55	3.20	3.19
11	60	3.75	3.78
12	65	4.40	4.43
13	70	5.00	5.13
14	75	6.00	5.88

Substituting these values in the expression for $s_{y \cdot x}$ gives $s_{y \cdot x} = 0.06$. Hence,

$$r = \sqrt{1 - \frac{(0.06)^2}{(1.85)^2}} = 0.99$$

$r^2 = .978$

Therefore, the value of r is significant, and 97.8 percent of the variation in flow rate can be accounted for by this line.

Regression analysis by linearizing transformation In many engineering situations the relationship between two variables may not be a straight-line or higher-order type. Instead, the underlying functional relationships can be the type shown below:

$$y = \begin{cases} 1 - e^{-a_0 x} & \text{exponential} \\ 1 - \exp\left[-\left(\frac{x}{\theta}\right)^b\right] & \text{Weibull} \\ a_0 + a_1\sqrt{x} & \text{square root} \\ ax^b & \text{logarithmic} \\ a + \dfrac{b}{x} & \text{inverse} \end{cases}$$

These types of equations generally fall under the category having nonlinear parameters. In order to apply the principles of least squares, the equations should be reduced to a linear form. The process of transforming the nonlinear equations into a linear form is called the linearizing transformation. Several types of such transformable functions and their linearized forms are shown in Table 13.1. Suppose, for example, one wishes to fit the data with the assumption that a logarithmic relationship exists.

$$y = ax^b$$

This can be transformed to the following form by taking logarithms of both sides:

$$\ln y = \ln a + b \ln x$$

or

$$Y = a_0 + bX$$

where $Y = \ln y$
$a_0 = \ln a$
$X = \ln x$

Table 13.1   Linearized forms of some nonlinear equations

Nonlinear equation	Linearized equation	Linearized variables Y	Linearized variables X	Linearized variables Z
1. $y = 1 - e^{-a_0 x}$ (Exponential)	$\ln \dfrac{1}{1-y} = a_0 x$	$\ln \dfrac{1}{1-y}$	x	
2. $y = a_0 + a_1 \sqrt{x}$ (Square root)	$y = a_0 + a_1 X$	y	$\sqrt{x}$	
3. $y = 1 - \exp\left[-\left(\dfrac{x-x_0}{\theta-x_0}\right)^b\right]$ (Weibull)	$\ln \ln \dfrac{1}{1-y}$ $= -b \ln(\theta - x_0)$ $+ b \ln(x - x_0)$	$\ln \ln \dfrac{1}{1-y}$	$\ln(x - x_0)$	
4. $y = ax^b$ (Logarithmic)	$\ln y = \ln a + b \ln x$	$\ln y$	$\ln x$	
5. $y = a + \dfrac{b}{x}$ (Inverse)	$y = a + bx$	y	$\dfrac{1}{x}$	
6. $y = e^{(a+bx)}$	$\ln y = a + bx$	$\ln y$	x	
7. $e^y = a_0 x^{a_1}$	$y = \ln a_0 + a_1 \ln x$	y	$\ln x$	
8. $y = a_0(x)^{a_1}(z)^{a_2}$	$\ln y = \ln a_0 + a_1 \ln x$ $+ a_2 \ln z$	$\ln y$	$\ln x$	$\ln z$
9. $y = a_0 + a_1 x + a_2 \sqrt{z}$	$y = a_0 + a_1 x + a_2 Z$	y	x	$\sqrt{z}$
10. $y = a_0 e^{(a_1 x + a_2 z)}$	$\ln y = \ln a_0 + a_1 x + a_2 z$	$\ln y$	x	z

This equation has a linear form; therefore all the equations of the method of least squares discussed in Subsec. 13.1.3 apply here to find a_0 and b. Suppose $a_0 = a_0'$ and $b = b'$. By taking antilogs of both sides, the following results:

$$y = \text{antilog}(a_0')(x)^{\text{antilog}(b')}$$

As an example of nonlinear curve fitting, the case of fitting a line to Weibull data will be considered. The Weibull cumulative equation is

$$y = 1 - \exp\left[-\left(\frac{x - x_0}{\theta - x_0}\right)^b\right] \tag{13.32}$$

Applying the linear transformation,

$$Y = aX + C$$

$$= \ln \ln \frac{1}{1-y}$$

$$X = \ln(x - x_0)$$

$$a = b \text{ (the Weibull slope)} = \frac{n \sum_{i=1}^{n} X_i Y_i - \sum_{i=1}^{n} X_i \sum_{i=1}^{n} Y_i}{n \sum_{i=1}^{n} (X_i^2) - \left(\sum_{i=1}^{n} X_i\right)^2}$$

$$C = -b \ln(\theta - x_0) = \frac{\sum_{i=1}^{n} (X_i^2) \sum_{i=1}^{n} Y_i - \sum_{i=1}^{n} X_i \sum_{i=1}^{n} X_i Y_i}{n \sum_{i=1}^{n} (X_i)^2 - \left(\sum_{i=1}^{n} X_i\right)^2}$$

$$\theta = e^{-C/b} + x_0$$

Thus, with a set of data known to follow a Weibull distribution, $Y = aX + C$ will be the equation of the straight-line fit to the data plotted on the three-cycle Weibull paper. If $x_0 = 0$, the equation of the least-squares line can be calculated directly. If x_0 is unknown, various values must be assumed until the line with the best correlation is obtained. Correlation coefficients are calculated by using Eq. (13.2) and the transformed variables X and Y.

Example 13.9 Test data on the failure of 30 bearings are given below. The data are known to follow a Weibull distribution. Determine the Weibull parameters θ and b (assume $x_0 = 0$) and the equation of the curve that best fits the data given below·

Hours to failure

79	175	240	390	650
98	180	268	420	690
124	190	282	450	750
128	198	300	520	840
150	210	320	550	890
160	215	360	590	930

Solution The data are given in Table 13.2 in the first column. The second column, $y = F(x)$, is taken from a table of median ranks, and successive columns are the transformations as indicated in column headings.
Solving for the Weibull slope b:

$$a = b = \frac{n(\sum XY) - (\sum X)(\sum Y)}{n(\sum X^2) - (\sum X)^2}$$

$$= \frac{(30)(-71.43) - (171.45)(-16.56)}{(30)(993.68) - 171.45^2}$$

$$b = 1.68$$

Solving for the Y intercept C:

$$C = \frac{(\sum X^2)(\sum Y) - (\sum X)(\sum XY)}{n(\sum X^2) - (\sum X)^2}$$

$$= \frac{(993.68)(-16.56) - (171.45)(-71.43)}{(30)(993.68) - 171.45^2}$$

$$= -10.13$$

Thus,

$$Y = 1.68X - 10.13 \qquad (13.33)$$

is the equation of the best-fit curve in transformed coordinates.

Table 13.2   Values for Example 13.9

x	$y = F(x)$	$X = \ln x$	$Y = \ln\ln\dfrac{1}{1-F(x)}$	$X^2 = (\ln x)^2$	$XY = \ln x \ln\ln\dfrac{1}{1-F(x)}$
79	0.0231	4.37	−3.73	19.10	−16.30
98	0.0559	4.58	−2.86	20.98	−13.10
124	0.0888	4.82	−2.36	23.23	−11.38
128	0.1217	4.85	−2.04	23.52	−9.89
150	0.1546	5.01	−1.78	25.10	−8.92
160	0.1875	5.08	−1.57	25.81	−7.99
175	0.2204	5.16	−1.39	26.63	−7.17
180	0.2533	5.19	−1.23	26.96	−6.38
190	0.2862	5.25	−1.09	27.56	−5.72
198	0.3191	5.29	−0.95	27.98	−5.03
210	0.3519	5.35	−0.83	28.62	−4.45
215	0.3848	5.37	−0.72	28.84	−3.87
240	0.4177	5.48	−0.61	30.03	−3.34
268	0.4506	5.59	−0.51	31.25	−2.85
282	0.4835	5.64	−0.42	31.81	−2.37
300	0.5164	5.70	−0.32	32.49	−1.82
320	0.5403	5.77	−0.23	33.29	−1.33
360	0.5822	5.89	−0.14	34.69	−0.82
390	0.6151	5.97	−0.05	35.64	−0.30
420	0.6480	6.04	0.04	36.48	0.24
450	0.6808	6.11	0.13	37.33	0.79
520	0.7137	6.25	0.22	39.06	1.38
550	0.7466	6.31	0.31	39.82	1.96
590	0.7795	6.38	0.41	40.70	2.62
650	0.8124	6.48	0.51	41.99	3.30
690	0.8453	6.54	0.63	42.77	4.12
750	0.8782	6.62	0.75	43.82	4.97
840	0.9111	6.73	0.88	45.29	5.92
890	0.9440	6.79	1.06	46.10	7.20
930	0.9768	6.84	1.33	46.79	9.10
		171.45	−16.56	993.68	−71.43

Solving for θ:

$$\theta = e^{-C/b}$$
$$= e^{10.13/1.68}$$
$$= 423 \text{ h}$$

With b, θ, and x_0 known, the transformation of Eq. (13.33) back to cartesian co-ordinates can be done by substituting directly into Eq. (13.32):

$$y = 1 - \exp\left[-\left(\frac{x - x_0}{\theta - x_0}\right)^b\right]$$

$$= 1 - \exp\left[-\left(\frac{x}{423}\right)^{1.68}\right]$$

Plot the results on Weibull paper (Fig. 13.10).

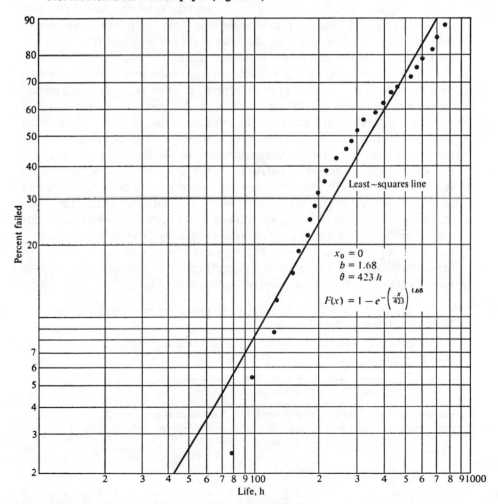

Fig. 13.10 Analysis of bearing-failure data for Example 13.9.

The least-squares method for fitting curves to linear and nonlinear data has been shown to be a very useful tool in analyzing experimental data. While the necessary calculations are quite cumbersome if done by hand, they are easily adapted to the digital computer.

13.2   VARIATION ANALYSIS

When a characteristic of a group of items is measured, the variation in the measurements is encountered.   This variation consists of the sum of two types of variations (1) the variation in the true characteristic from item to item and (2) the variation in measurement.   The first is generally known as the product variation, while the second is called the test variation.   In some engineering situations it is important to control the total variations; therefore it is necessary to isolate the testing variations from the product variation in order to establish how much of the total variation was contributed by the instruments and how much by the product itself.

In the event that repeat measurements on the same item are possible, then the test variation can readily be estimated.   For example, when several repeat measurements for the diameter of a shaft are made, the standard deviation of these measurements is the estimate of the test (or measurement) variation.   However, if a measurement should result in destruction of the item being measured, only one measurement is possible for each item.   In this case isolating the test variation from the product variation is difficult. It is not possible to isolate these variations when several items are measured once with only a single test instrument where the instrument drifts randomly with every item measured.   However, in cases when two or more instruments are available, the following method, as proposed by Grubbs [6], is available for designing an experiment to separate the above variations.   Although the method discussed here uses three instruments, the same procedure applies when the number of instruments is two or more than three.

13.2.1   SEPARATING PRODUCT VARIATION FROM TEST VARIATION BY THREE TEST INSTRUMENTS

When the same n items are measured with three instruments, the following may be found:

Item numbers	Observed values		
	Instrument I	Instrument II	Instrument III
1	$x_1 + e_{11}$	$x_1 + e_{12}$	$x_1 + e_{13}$
2	$x_2 + e_{21}$	$x_2 + e_{22}$	$x_2 + e_{23}$
3	$x_3 + e_{31}$	$x_3 + e_{32}$	$x_3 + e_{33}$
4	$x_4 + e_{41}$	$x_4 + e_{42}$	$x_4 + e_{43}$
$\vdots$	$\vdots$	$\vdots$	$\vdots$
n	$x_n + e_{n1}$	$x_n + e_{n2}$	$x_n + e_{n3}$

where x_i is the true value of the ith item in the sample, and e_{n_1}, e_{n_2}, and e_{n_3} are the measurement errors for the nth item by instruments I, II, and III, respectively.

Variation in the true values x_i is called the product variation and is denoted by σ_x^2; variation in the instrument errors (that is, measurement errors) is called the test variation and denoted by $\sigma_{e_1}^2$, $\sigma_{e_2}^2$, $\sigma_{e_3}^2$, for instruments I, II, and III. The estimates of the above variations are s_x^2, $s_{e_1}^2$, $s_{e_2}^2$, and $s_{e_3}^2$.

When the differences among the above values are taken, the true values of the items measured are eliminated:

Differences between the two observed values on two instruments at a time

$I - II$	$I - III$	$II - III$
$e_{11} - e_{12}$	$e_{11} - e_{13}$	$e_{12} - e_{13}$
$e_{21} - e_{22}$	$e_{21} - e_{23}$	$e_{22} - e_{23}$
$e_{31} - e_{32}$	$e_{31} - e_{33}$	$e_{32} - e_{33}$
$e_{41} - e_{42}$	$e_{41} - e_{43}$	$e_{42} - e_{43}$
$\vdots$	$\vdots$	$\vdots$
$e_{n1} - e_{n2}$	$e_{n1} - e_{n3}$	$e_{n2} - e_{n3}$

The variances for these values (in terms of the differences) within each of these columns can be represented as follows.

The estimates of the variances for each of the individual instruments' errors (test variations) can be computed by the following relationships derived by Grubbs [6]:

$$s_{e_1}^2 = \text{est } \sigma_{e_1}^2 = \tfrac{1}{2}(s_{e_1-e_2}^2 + s_{e_1-e_3}^2 - s_{e_2-e_3}^2) \tag{13.34}$$

$$s_{e_2}^2 = \text{est } \sigma_{e_2}^2 = \tfrac{1}{2}(s_{e_2-e_3}^2 + s_{e_1-e_2}^2 - s_{e_1-e_3}^2) \tag{13.35}$$

$$s_{e_3}^2 = \text{est } \sigma_{e_3}^2 = \tfrac{1}{2}(s_{e_1-e_3}^2 + s_{e_2-e_3}^2 - s_{e_1-e_2}^2) \tag{13.36}$$

The test variations of three instruments, $s_{e_1}^2$, $s_{e_2}^2$, and $s_{e_3}^2$, obtained by the above relationships, are free of the product variation σ_x^2 (the variability in characteristics of the items measured).

The relationship to estimate the product variation σ_x^2 is

$$s_x^2 = \text{est } \sigma_x^2 = \frac{s_{(x+e_1)+(x+e_2)+(x+e_3)}^2}{3} - \frac{1}{18}(s_{e_1-e_2}^2 + s_{e_1-e_3}^2 + s_{e_2-e_3}^2) \tag{13.37}$$

where $s_{(x+e_1)+(x+e_2)+(x+e_3)}^2/3$ is the variance of the averages of the observed values of n items on three instruments.

The above concept is illustrated by the following example:

Example 13.10   In measuring the unburned hydrocarbons (HC) in the exhaust emissions of an engine, the exhaust gas is sampled and the level of HC is measured with an instrument console.   However, the measurement error by the instrument console, as well as the HC level in the exhaust gas, varies randomly from test to test.   Hence, the variation of the measurements is the total variation, consisting of two components: variation due to the changes in the true HC level of the engine and variation due to the changes in the test console.   To determine the adequacy of the test console, it is necessary to establish the variation of the test console alone.   Design a suitable experiment to determine these variations.

Three similar test consoles were used to measure the HC level simultaneously. Ten tests were run, and the following data were obtained:

Test No.	HC level, parts per million (ppm) Test console I	Test console II	Test console III	Average of three test consoles
1	147	151	146	148.00
2	155	155	153	154.33
3	148	150	147	148.33
4	161	164	162	162.33
5	143	146	145	144.67
6	165	168	167	166.67
7	151	150	148	149.67
8	141	146	139	142.00
9	153	151	150	151.33
10	146	149	143	146.00
Average, $\bar{x}$	151	153	150	151.33

Solution   The first step is to tabulate the differences between the HC levels as obtained by the three test consoles for each test:

Test No.	Differences between test consoles values, ppm I − II	I − III	II − III
1	−4	1	5
2	0	2	2
3	−2	1	3
4	−3	−1	2
5	−3	−2	1
6	−3	−2	1
7	1	3	2
8	−5	2	7
9	2	3	1
10	−3	3	6
Variance s^2	$s^2_{e1-e2} = 5.1$	$s^2_{e1-e3} = 4.0$	$s^2_{e2-e3} = 4.9$

The above differences represent the actual differences in measurement errors of the two test consoles involved and are not influenced by the HC level of any one test. That is, the data in the table contain only the measurement errors (test variation) and not any variation in HC levels from test to test. The variance of each of the columns $(I - II, I - III, and II - III)$ represents the variability of the difference in errors of any two of the consoles. The variances were computed by means of Eq. (1.14):

$$s^2 = \frac{\sum_{i=1}^{n}(x_i - \bar{x})^2}{n-1}$$

By substituting $s^2_{e_1-e_2} = 5.1$ ppm^2, $s^2_{e_1-e_3} = 4.0$ ppm^2, and $s^2_{e_2-e_3} = 4.9$ ppm^2 into Eqs. (13.34) to (13.36), test variations for each console are computed, where e_1, e_2, and e_3 represent the errors of measurements by the three consoles:

$$s_{e_1}^2 = \text{est } \sigma_{e_1}^2 = \tfrac{1}{2}(s^2_{e_1-e_2} + s^2_{e_1-e_3} - S^2_{e_2-e_3})$$
$$= \tfrac{1}{2}(5.1 + 4.0 - 4.9)$$
$$= 2.1 \text{ ppm}^2$$

$$s_{e_2}^2 = \text{est } \sigma_{e_2}^2 = \tfrac{1}{2}(s^2_{e_2-e_3} + s^2_{e_1-e_2} - s^2_{e_1-e_3})$$
$$= \tfrac{1}{2}(4.9 + 5.1 - 4.0)$$
$$= 3.0 \text{ ppm}^2$$

$$s_{e_3}^2 = \text{est } \sigma_{e_3}^2 = \tfrac{1}{2}(s^2_{e_1-e_3} + s^2_{e_2-e_3} - s^2_{e_1-e_2})$$
$$= \tfrac{1}{2}(4.0 + 4.9 - 5.1)$$
$$= 1.9 \text{ ppm}^2$$

The next step is to estimate the variation in the characteristic measured (HC level from test to test), which was defined as the product variation. Using Eq. (13.37),

$$s_x^2 = \text{est } \sigma_x^2 = s^2_{1/3(x+e1+x+e2+x+e3)} - \frac{1}{18}(s^2_{e_1-e_2} + s^2_{e_1-e_3} + s^2_{e_2-e_3})$$

The first term is the variance of the averages of the three observations (three consoles average), as shown in the last column of the above table. The variance of this column was computed by Eq. (1.14) and is equal to 60.9. Therefore,

$$s_x^2 = 60.9 - \tfrac{1}{18}(5.1 + 4.0 + 4.9)$$
$$\simeq 60.1 \text{ ppm}^2$$

Hence, the following are the results:

	Product variation (change in HC level from test to test) (s_x^2)	Test variation (consoles measurements error)		
		Console I ($s_{e_1}^2$)	Console II ($s_{e_2}^2$)	Console III ($s_{e_3}^2$)
Variance s^2, ppm^2	60.1	2.1	3.0	1.9
Standard deviation s, ppm	7.76	1.45	1.73	1.38

Thus, the product variation is significantly higher than the test variation, by a factor of 4.5 for the worst console (console II), $7.76/1.73 = 4.5$. The ratio of the product variability to the average test variability of the three consoles is $7.76/[(2.1 + 3.0 + 1.9)/3]^{1/2} = 5.1$. It can therefore be concluded that the test consoles are performing adequately.

13.2.2 SEPARATING PRODUCT VARIATION FROM TEST VARIATION BY N TEST INSTRUMENTS

The above procedure to separate product and test variations can be extended to any number N of test instruments. The following relationships as derived by Grubbs [6] are applicable when $N \geq 3$. The test variation for instrument I can be computed from

$$s_{e_1}^2 = \text{est } \sigma_{e_1}^2 = \frac{1}{N-1} \left(\sum_{r=2}^{N} s_{e_1-e_r}^2 - \frac{1}{N-2} \sum_{2 \leq j < k}^{k=N} s_{e_j-e_k}^2 \right)$$

The formulas for $s_{e_2}^2$, $s_{e_3}^2$, etc., may be found by rotating the subscripts. The product variation can be found by using the relationship

$$s_x^2 = \text{est } \sigma_x^2 = \frac{s_{(x+e_1)+(x+e_2)+\cdots+(x+e_N)}^2}{N} - \frac{1}{N^2(N-1)} \sum_{1 \leq r < s}^{N} s_{e_r-e_s}^2$$

For the case when only two instruments are available, the following relationships may be used to obtain the best estimate of the variations.

For test variations:

$$s_{e_1}^2 = \text{est } \sigma_{e_1}^2 = \tfrac{1}{2}(s_{x+e_1}^2 - s_{x+e_2}^2 + s_{e_1-e_2}^2)$$

and

$$s_{e_2}^2 = \text{est } \sigma_{e_2}^2 = \tfrac{1}{2}(s_{x+e_2}^2 - s_{x+e_1}^2 + s_{e_1-e_2}^2)$$

For product variations:

$$s_x^2 = \text{est } \sigma_x^2 = \frac{s_{(x+e_1)+(x+e_2)}^2}{2} - \tfrac{1}{4}(s_{e_1}^2 - s_{e_2}^2)$$

SUMMARY

Chapters 3 through 12 were concerned with specific experiments such as those of evaluation or comparison, accelerated experiments, and factorial experiments or with the analysis of data for specific applications, such as interference and systems. The present chapter has much broader implications in that it can be applied to all the previous situations. The chapter is concerned with the interrelationship between data, that is, whether one variable depends on another variable or a set of variables, and how one variable changes with the change in other variables. This is covered in sections on correlation analysis, linear and multiple regression analysis, and nonlinear regression analysis. A method is also presented to separate and estimate the sources of variation in a set of experimental observations, in terms of product variations and test variations. This is covered under the section on variation analysis.

PROBLEMS

13.1. The following data relates to a chemical reaction where a precipitate of a particular substance y was obtained by adding different amounts of a reagent x to the solution.

x	y
8.4	10.0
6.0	7.0
6.2	7.9
6.0	8.0
6.6	9.5
7.8	12.8
8.0	14.0
9.5	15.1
7.2	10.0
8.2	11.0
8.7	12.1
5.5	6.7

Plot the above data and, using the method of least squares, estimate the best-fitting line that would represent the above data; determine the correlation coefficient r.

13.2. Using the data of Prob. 13.1, determine the expected amount of precipitate when $x = 8.0$. Determine the 95 percent confidence limits for this value of the precipitate and for the least-squares parameters a_0 and a_1.

13.3. Fatigue-life characteristics of a certain component are to be evaluated, where lives to failure under given dynamic stresses were found to be as follows:

Dynamic stress x, ksi	Fatigue life y, cycles
50	3.5×10^3
40	6.0×10^3
30	2.8×10^4
20	1.0×10^5
15	3.0×10^5
12	8.0×10^5

Assuming that the $y = a_0 e^{-a_1 x}$ relationship exists, determine the best estimates of the parameters a_0 and a_1. What will be the expected life of this component if it is subjected to a 25-ksi stress?

13.4. A calibration curve for a load cell is to be established, where the voltage output of the cell is measured for various values of the load. The data are:

Load applied x, lb	Load-cell output y, mV
10	1.6
50	7.8
100	16.2
150	23.8
200	32.0
300	48.5
500	79.0
700	112.9
900	142.3
1,000	162.0

(a) Determine the best-fitting calibration curve and the correlation coefficient.

(b) Is the output of the load cell linear with the load applied?

(c) What would be the precision of the cell, at 90 percent confidence, at a load of 500 lb? (This would be the 90 percent confidence limits on the voltage output corresponding to 500 lb.)

13.5. The following measurements were recorded simultaneously by two test instruments on each of 15 identical items.

Item number	Characteristic of the item measured Instrument I	Instrument II
1	25	27
2	35	33
3	28	29
4	40	41
5	22	20
6	18	19
7	26	28
8	48	49
9	52	50
10	33	32
11	38	39
12	30	28
13	27	26
14	24	24
15	44	45

(a) Determine the product variation (variation from item to item) and the test variation (instrument measuring error).

(b) How good is the precision of each instrument compared with the variation in the items?

13.6. The characteristic curve of a certain type of triode is to be established for a given plate voltage. With the plate voltage held constant, the grid voltage was varied and the corresponding plate current was measured.

Grid voltage x, volts	Plate current y, mA
−4.5	0.5
−4.0	2.1
−3.5	4.0
−3.0	5.8
−2.5	7.5
−2.0	9.3
−1.5	10.8
−1.0	12.1
−0.5	13.2
0.0	14.2

(a) Determine the relationship between the plate current and the grid voltage.
(b) By means of the method of least squares, check whether this relationship is linear (first order) or nonlinear (second or higher orders).
(c) Determine the coefficient of correlation in each case.

13.7. In this chapter, the equations for estimating the least-squares parameters a_0 and a_1, for the linear model $y = a_0 + a_1 x$, were developed. Develop similarly the equations to estimate the parameters a_0, a_1, and a_2 for the second-order model $y = a_0 + a_1 x + a_2 x^2$.

13.8. It is known from past experience that the coefficient of friction between the brake lining and the drum is significantly dependent on the drum speed and temperature. It is desired to establish the quantitative effects of speed and temperature on friction, and also to determine which factor is more sensitive. The experiment was run, and the following data were obtained:

Drum speed x_1, ft/s	Drum temperature x_2, °F	Coefficient of friction y
50	200	0.30
40	175	0.41
30	200	0.30
20	150	0.50
10	100	0.72
5	125	0.80
40	200	0.35
50	225	0.25
30	170	0.47
20	100	0.58
10	130	0.70
5	80	0.83

Using the multiple regression analysis, determine
(a) the relationship $y = a_0 + a_1 x_1 + a_2 x_2$;
(b) the multiple correlation coefficient, and test its significance at 95 percent confidence level;
(c) the expected value of the coefficient when the drum speed is 25 ft/s and the temperature 140°F.

13.9. On the basis of the following data:

(a) Determine whether there is any definite tendency to rate old employees higher than the more recent ones.

(b) Determine whether there is any correlation between the ratings and the years of service, and check its significance.

(c) Estimate the rating that an employee with 3 years of service may expect to receive.

Employee's number	Rating	Service, years
1	6	1
2	9	5
3	6	3
4	5	4
5	4	5
6	5	1
7	4	4
8	3	3
9	3	1
10	8	3
11	4	7
12	8	3
13	5	6
14	2	1
15	7	2

13.10. Pick 20 random numbers (first two digits) from Table A-47, starting from the top of the first column. Denote these numbers by $x_1, x_2, x_3, \ldots, x_{20}$. Similarly select another 20 numbers, this time by starting from the second column. Denote these numbers by $y_1, y_2, y_3, \ldots, y_{20}$. Are the x's correlated with the y's?

13.11. A subjective evaluation of a given component was made in the laboratory under 10 different test conditions. The result of each test was ranked from 1 to 10. A similar test was run in the field under the same conditions as in the laboratory. From the following data, establish with 95 percent confidence whether the laboratory ranks correlate with the ranks in the field.

Test conditions	Laboratory ranking	Field ranking
1	3	3
2	5	4
3	6	6
4	10	10
5	9	8
6	1	2
7	4	5
8	7	7
9	8	9
10	2	1

REFERENCES

1. Ostle, B.: "Statistics in Research," 2d ed., The Iowa State University Press, Ames, 1963.
2. Guttman, I., and S. S. Wilks: "Introductory Engineering Statistics," John Wiley & Sons, Inc., New York, 1965.
3. Duncan, A. J.: "Quality Control and Industrial Statistics," Richard D. Irwin, Inc., Homewood, Ill., 1959.
4. Dixon, W. J., and F. J. Massey, Jr.: "Introduction to Statistical Analysis," 3d ed., McGraw-Hill Book Company, New York, 1969.
5. Draper, N. R., and H. Smith: "Applied Regression Analysis," John Wiley & Sons, Inc., New York, 1966.
6. Grubbs, F. E.: On Estimating Precision of Measuring Instruments and Product Variability, *J. Amer. Statist. Ass.*, vol. 43, p. 246, 1948.

BIBLIOGRAPHY

Acton, F. S.: "Analysis of Straight-line Data," John Wiley & Sons, Inc., New York, 1959.
Bowker, A. H., and G. J. Lieberman: "Engineering Statistics," Prentice-Hall, Inc., Englewood Cliffs, N.J., 1959.
Brownlee, K. A.: "Statistical Theory and Methodology in Science and Engineering," John Wiley & Sons, Inc., New York, 1960.
Dixon, W. J., and F. J. Massey, Jr.: "Introduction to Statistical Analysis," 3d ed., McGraw-Hill Book Company, New York, 1969.
Draper, N. R., and H. Smith: "Applied Regression Analysis," John Wiley & Sons, Inc., New York, 1966.
Duncan, A. J.: "Quality Control and Industrial Statistics," Richard D. Irwin, Inc., Homewood, Ill., 1959.
Guttman, I., and S. S. Wilks: "Introductory Engineering Statistics," John Wiley & Sons, Inc., New York, 1965.
Johnson, L. G.: "Theory and Technique of Variation Research," Elsevier Publishing Company, Amsterdam, 1964.
Johnson, N. L., and F. C. Leone: "Statistics and Experimental Design," vols. I and II, John Wiley & Sons, Inc., New York, 1964.
Mace, A. E.: "Sample-size Determination," Reinhold Publishing Corporation, New York, 1964.
Miller, I., and J. E. Freund: "Probability and Statistics for Engineers," Prentice-Hall, Inc., Englewood Cliffs, N.J., 1965.
Mosteller, F., R. E. K. Rourke, and G. B. Thomas, Jr.: "Probability with Statistical Applications," Addison-Wesley Publishing Company, Inc., Reading, Mass., 1961.
Natrella, M. G.: Experimental Statistics, *Nat. Bur. Stand. Handb.* 91, Aug. 1, 1963.
Neville, A. M., and J. B. Kennedy: "Basic Statistical Methods for Engineers and Scientists," International Textbook Company, Scranton, Pa., 1964.
Ostle, B.: "Statistics in Research," 2d ed., The Iowa State University Press, Ames, 1963.
Schenck, H.: "Theories of Engineering Experimentation," 2d ed., McGraw-Hill Book Company, New York, 1968.
Spiegel, M. R.: "Theory and Problems of Statistics," Schaum Publishing Co., New York, 1961.
Worthing, A. G., and J. Geffner: "Treatment of Experimental Data," John Wiley & Sons, Inc., New York, 1943.

14
Renewal Analysis

Renewal analysis has been developed primarily for estimating the maintenance and replacement requirements of production components. However, it can be a valuable engineering tool for establishing initial design requirements, testing procedures, etc.

14.1 RENEWAL THEORY [1]

The following are the definitions of the terms used:

1. For one system or component:

   $n(t)$ = number of failures or replacements by time t

   $m(t)$ = average number of failures or replacements by time t

   $s^2(t)$ = variance of number of failures or replacements by time t

2. For more than one system or component:

   $N(t)$ = number of failures or replacements by time t

   $\mu(t)$ = average number of failures or replacements by time t

   $\sigma^2(t)$ = variance of number of failures or replacements by time t

In the simplest case, a one-component system is tested continuously by replacing the component at the instant it fails with an identical new component. The test is discontinued at a preassigned time t_1. The number of failures (or replacements) occurring in this system up to this time t_1 is, say, $n_1(t_1)$. Now the same test is run on another identical system and discontinued at the same time t_1. The number of failures occurring in this second system during the time t_1 is, say, $n_2(t_1)$. By repeating this test a number of times, one obtains $n_1(t_1)$, $n_2(t_1)$, $n_3(t_1)$, etc. Thus, $n(t)$ can be considered a random variable which is defined as the number of failures or replacements needed in a given system by the time t. This random variable $n(t)$ follows a two-parameter distribution function with mean $m(t)$ and variance $s^2(t)$.

In most engineering problems, however, one is not interested in the number of replacements of a component in a single system, but in the total number of replacements of a component in all the systems functioning in the field simultaneously. For example, if 10,000 new lathes are put in service this year, the manufacturer might want to know the replacement demands for a particular transmission gear over the next 10 years. Thus, if $n_i(t)$ is the number of transmission-gear replacements on lathe i, the total number of replacements for R systems is equal to $N(t)$, where

$$N(t) = \sum_{i=1}^{R} n_i(t)$$

and R is the number of lathes in the field, which in this case is 10,000; this assumes that there is only one transmission gear of the given type per lathe.

In cases where R (the number of components in the field) is large, the central limit theorem might be applied, where $N(t)$ is taken as normally distributed with mean and variance (see Fig. 14.1).

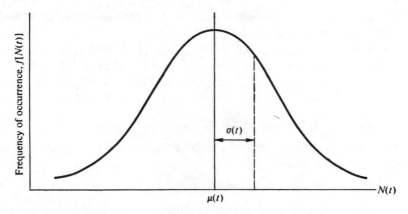

Fig. 14.1 Normal distribution of renewal population.

$$E(N(t)) = RE(n_i(t)) = Rm(t) = \mu(t) \tag{14.1}$$

$$\text{var } [N(t)] = R \text{ var } [n_i(i)] = Rs^2(t) = \sigma^2(t) \tag{14.2}$$

Since the distribution of $n(t)$ depends on the failure distribution of the particular component in question, the parameters $m(t)$ and $s^2(t)$ are determined from the parameters of the failure distribution of the particular component.

14.2   RENEWAL ANALYSIS BASED ON THE WEIBULL DISTRIBUTION [1]

Since the life characteristics of many types of components can be expressed by a Weibull distribution, the renewal-analysis techniques based on the Weibull distribution will be discussed here.

The renewal analysis based on the two-parameter (θ, b) Weibull distribution has been extensively developed by White [1].   Numerical techniques were employed to derive expressions for $m(t)$ and $s^2(t)$ in terms of θ and b. Summarizing White's results:

$$m(t) = E(n(t)) = m_1(t)$$
$$s^2(t) = \text{var } [n(t)] = 2m_2(t) + m_1(t) - m_1{}^2(t)$$

where

$$m_k(t) = \sum_{n=k}^{\infty} (-1)^{n+k} c_k(n) \frac{t^{bn}}{n!}$$

and

b = Weibull slope of component's failure distribution

$c_k(n)$ = function of θ, characteristic life of component

$m(t)$ and $s(t)$ have been calculated for various values of b and t/θ, and they appear in Tables 14.3 and 14.4.

If t, θ, b, and R are known and Eqs. (14.1) and (14.2) are used, the expected number of replacements $\mu(t)$ and the standard deviation of the replacement population $\sigma(t)$ can be calculated.   Furthermore, since the $(1 - \alpha)$ confidence interval for a random variable x is (see Chap. 3)

$$\mu_x - z_{\alpha/2}\,\sigma_x \le x \le \mu_x + z_{\alpha/2}\,\sigma_x$$

the $(1 - \alpha)$ confidence interval can be determined for $N(t)$, the number of replacements in time t.

$$P[\mu(t) - z_{\alpha/2}^{\cdot}\,\sigma(t) \le N(t) \le \mu(t) + z_{\alpha/2}\,\sigma(t)] = 1 - \alpha \tag{14.3}$$

The following four examples illustrate the application of renewal analysis.

Example 14.1   A transmission gear on a lathe follows a Weibull failure distribution with $\theta = 60$ months and $b = 5$.   In order to supply replacement parts, the production department would like to determine customer needs before the orders are received. Assume that all the parts go into the customers' hands at once and that 10,000 units are sold.   What might be the expected replacement-part production required for the next 10 years?   Provide an answer with 99.9 percent confidence.

Solution   After the first year (12 months) in service:

$$\frac{t}{\theta} = \frac{12 \text{ months}}{60 \text{ months}} = 0.2$$

$R = 10,000$ units in service

From Tables 14.3 and 14.4, for $b = 5.0$ and $t/\theta = 0.2$:

$m(t) = 0.000320$ average replacements per number of units in service

$s(t) = 0.017884$ standard deviation around $m(t)$

With 10,000 units in the field,

$\mu(t) = Rm(t) = (10^4)(0.000320) = 3.2$ average number of replacements in 12 months

$\sigma(t) = \sqrt{R}\,s(t) = (10^2)(0.017884) = 1.784$ standard deviation around $\mu(t)$

For 99.9 percent confidence level,

$$\frac{\alpha}{2} = 0.0005$$

$z_{\alpha/2} - 3.2905$

$P[\mu(t) - z_{\alpha/2}\,\sigma(t) \le N(t) \le \mu(t) + z_{\alpha/2}\,\sigma(t)] = 0.999$

$P[3.2 - (3.29)(1.784) \le N(t) \le 3.2 + (3.29)(1.784)] = 0.999$

$P[-2.68 \le N(t) \le 9.08] = 0.999$

or

$P[0 \text{ replacements} \le N(t) \le 9.08 \text{ replacements}] = 0.999$

This means that between 0 and 10 replacements will be required in the first year, with 99.9 percent confidence.   Repeating the above calculations for each successive year up to year 10 (that is, for $t = 24, 36, 48, \ldots$ up to 120 months) yields the results given in Table 14.1.   The last row of Table 14.1 shows the estimate for the maximum production requirement for the year.

Example 14.2   A manufacturer has a module which is a component of a new computer about to be put on the market.   This module has the following Weibull characteristics: $\theta = 12$ months; $b = 7.0$.   The cost of the module is \$100.   Machine rent rate is \$600 per hour.   The customer renting the machine wishes to purchase the manufacturer's maintenance policy.   This policy guarantees machine operation at all times.   No hourly charge will be made for any machine which is not operating properly.   Repair time for the module is constant, and is negligible for a change-over when the machine has not broken down.   It is, however, 1 h when the machine

Table 14.1 Summary of results for Example 14.1

| | \multicolumn{10}{c}{Time since introduction to market, year} | | | | | | | | | |
	1	2	3	4	5	6	7	8	9	10
$m(t)$	0.00032	0.01019	0.07484	0.27982	0.63578	0.93715	1.07243	1.21355	1.44514	1.72166
$s(t)$	0.01788	0.100425	0.26322	0.44983	0.48876	0.31518	0.277099	0.41280	0.508834	0.49647
$\mu(t)$	3.2	101.9	748.4	2,798.2	6,357.8	9,371.5	10,724.3	12,135.5	14,451.4	17,216.6
$\sigma(t)$	1.79	10.04	26.32	44.98	48.88	31.52	27.71	41.28	50.88	49.65
Maximum replacement	9.08	134.9	835.0	2,946.2	6,518.6	9,475.2	10,815.5	12,271	14,618.8	17,380
Minimum replacement	0	68.8	661.8	2,650.2	6,196.9	9,267.8	10,633.1	11,999	14,284.0	17,053
Uncertainty	10	66	173	296	322	208	182	272	334	327
Maximum production requirement for the year	10	125	700	2,112	3,572	2,957	1,340	1,456	2347	2,762

breaks down and a man must be called.   Assume that labor cost is negligible, 24-h operation is required, and 100 machines are expected to be sold.   What is the optimum module block replacement schedule?   Provide an answer with 99.9 percent confidence.

Solution   At the end of 1.2 months of service:

$$\frac{t}{\theta} = \frac{1.2 \text{ months}}{12 \text{ months}} = 0.1$$

$R = 100$ machines

Referring to Tables 14.3 and 14.4, for $b = 7.0$ and $t/\theta = 0.1$:

$m(t) = 0 \qquad\qquad \mu(t) = Rm(t) = (10^2)(0) = 0$

$s(t) = 0.000316 \qquad \sigma(t) = \sqrt{R}\,s(t) = (10^1)(0.000316) = 0.00316$

For a confidence level of 99.9 percent Table A-1 gives $z = 3.2905$; therefore,

$P[\mu(t) - z_{\alpha/2}\,\sigma(t) \le N(t) \le \mu(t) + z_{\alpha/2}\,\sigma(t)] = 1 - \alpha$

$P[0 - (3.2905)(0.00316) \le N(t) \le 0 + (3.2905)(0.00316)] = 0.999$

$P[0 \le N(t) \le 0.0104] = 0.999$

Repeating the above procedure for 2.4 months,

$$\frac{t}{\theta} = \frac{2.4 \text{ months}}{12 \text{ months}} = 0.2$$

Referring to Tables 14.3 and 14.4, for $b = 7.0$ and $t/\theta = 0.2$:

$m(t) = 0.000013 \qquad \mu(t) = Rm(t) = (10^2)(0.000013) = 0.0013$

$s(t) = 0.003578 \qquad \sigma(t) = \sqrt{R}\,s(t) = (10^1)(0.003578) = 0.03578$

Again for a confidence level of 99.9 percent,

$P[0 \le N(t) \le 0.122] = 0.999$

Continuing this process for $t = 3.6$ months to $t = 8.4$ months results in the information shown in Table 14.2.
Now, to determine the minimum expense for a particular changeover period:

1.2-Month Changeover:

Part cost	$= N(t)\max \times 100 \simeq \$$	1.00
Lost computer income	$= 0.0104 \times 600 \quad\simeq$	6.00
Replacement of all modules	$=$	10,000.00

Maximum manufacturer expense for the 1.2-month interval	$= \$10,007.00$ per 1.2 months
	$= \$100,070.00$ per year

2.4-Month Changeover:

Part cost $= 0.122 \times 100 = \$$ 12.20
Lost computer income $= 0.122 \times 600 =$ 73.20
Replacement of all modules $=$ 10,000.00

Maximum manufacturer expense for
the 2.4-month interval $= \$10,085.40$ per 2.4 months
$= \$50,427.00$ per year

Table 14.2   Summary of results for Example 14.2

	Module replacement interval, months						
	1.2	*2.4*	*3.6*	*4.8*	*6.0*	*7.2*	*8.4*
$m(t)$	0.000000	0.000013	0.000219	0.001637	0.007782	0.027606	0.079056
$s(t)$	0.000316	0.003578	0.014786	0.040427	0.087872	0.163841	0.269834
$\mu(t)$	0.00000	0.0013	0.0219	0.1637	0.7782	2.7606	7.9056
$\sigma(t)$	0.00316	0.03578	0.14786	0.40427	0.87872	1.63841	2.69834
Maximum replacement	0.0104	0.122	0.508	1.49	3.58	7.29	16.78
Minimum replacement	0.000000	0.000000	0.000000	0.00000	0.000000	0.000000	0.000000
Maximum expense per year, \$	100,070	50,427	34,500	27,800	25,000	25,200	31,100

The maximum expenses for 3.6-, 4.8-, 6.0-, 7.2-, and 8.4-month changeovers are shown in Table 14.2. A plot of this information is given in Fig. 14.2. For a minimum lost income, scheduled replacement should occur at 6.6-month intervals.

Example 14.3 A manufacturer is planning to produce a part which would cost him $50. The market for this product has been estimated at approximately 10,000 units. This unit is a replacement component for a larger machine. It is estimated that in 10 years of operation the machine will be outdated and scrapped. Management does not wish to have more than $3,000 invested in spare parts due to uncertainty in predicting the replacement market. Determine the possible Weibull failure distributions of the parts such that the above conditions will be satisfied. That is, what are acceptable values of b and t/θ for the product?

Solution

$$\text{Number of uncertain parts} = \frac{\$3,000}{\$50} = 60 \text{ parts}$$

$$2z_{\alpha/2}\,\sigma(t) = 60$$

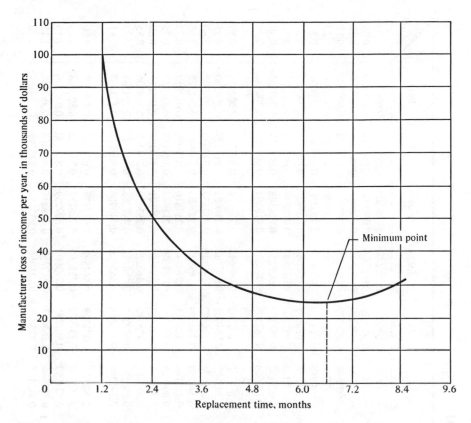

Fig. 14.2 Manufacturer's loss of income as a function of module replacement interval (Example 14.2).

Choosing a confidence level of 99.9 percent, from Table A-1, $z_{\alpha/2} = 3.2905$. Therefore

$$\sigma(t) = \frac{60}{(2)(3.2905)} = 9.12$$

$$s(t) = \frac{\sigma(t)}{\sqrt{R}} = \frac{9.12}{10^2} = 0.0912$$

From Table 14.4 the locus of $s(t)$ can be plotted, as shown in Fig. 14.3. That is, for $s(t) = 0.0912$ the plot involves the acceptable values of b and t/θ. Thus, the product must be designed so that its Weibull parameters θ and b are in the acceptable region of Fig. 14.3. The value of t is arbitrary and depends on how often the manufacturer plans to have production runs of the spare parts over the 10-year interval.

Table 14.3 Weibull renewal, mean $= E(n(t)) = m(t)$ [1]

t/θ	0.5	1.0	1.5	2.0	2.5	3.0	4.0	5.0	7.0	10.0	b \ t/θ
0.00	0.000000	0.000000	0.000000	0.000000	0.000000	0.000000	0.000000	0.000000	0.000000	0.000000	0.00
0.05	0.238138	0.050000	0.011155	0.002498	0.000559	0.000125	0.000006	0.000000	0.000000	0.000000	0.05
0.10	0.345504	0.100000	0.031419	0.009967	0.003158	0.001000	0.000100	0.000010	0.000000	0.000000	0.10
0.15	0.431457	0.150000	0.057413	0.022333	0.008683	0.003370	0.000506	0.000076	0.000002	0.000000	0.15
0.20	0.506365	0.200000	0.087840	0.039474	0.017759	0.007971	0.001599	0.000320	0.000013	0.000000	0.20
0.25	0.574238	0.250000	0.121901	0.061225	0.030856	0.015516	0.003899	0.000976	0.000061	0.000001	0.25
0.30	0.637131	0.300000	0.159021	0.087379	0.048320	0.026675	0.008068	0.002427	0.000219	0.000006	0.30
0.35	0.696263	0.350000	0.198752	0.117696	0.070382	0.042060	0.014897	0.005239	0.000643	0.000028	0.35
0.40	0.752427	0.400000	0.240733	0.151903	0.097158	0.062197	0.025284	0.010188	0.001637	0.000105	0.40
0.45	0.806173	0.450000	0.284658	0.189707	0.128654	0.087503	0.040201	0.018285	0.003730	0.000340	0.45
0.50	0.857901	0.500000	0.330270	0.230794	0.164771	0.118263	0.060642	0.030771	0.007782	0.000976	0.50
0.55	0.907913	0.550000	0.377345	0.274843	0.205309	0.154605	0.087562	0.049093	0.015109	0.002530	0.55
0.60	0.956444	0.600000	0.425690	0.321526	0.249975	0.196487	0.121789	0.074837	0.027606	0.006028	0.60
0.65	1.003679	0.650000	0.475134	0.370520	0.298399	0.243689	0.163925	0.109604	0.047841	0.013373	0.65
0.70	1.049768	0.700000	0.525530	0.421508	0.350144	0.295811	0.214247	0.154816	0.079056	0.027852	0.70
0.75	1.094837	0.750000	0.576746	0.474186	0.404728	0.352285	0.272606	0.211468	0.124964	0.054757	0.75
0.80	1.138988	0.800000	0.628668	0.528267	0.461640	0.412404	0.338350	0.279821	0.189198	0.101811	0.80
0.85	1.182308	0.850000	0.681195	0.583488	0.520365	0.475349	0.410295	0.359099	0.274300	0.178706	0.85
0.90	1.224873	0.900000	0.734239	0.639605	0.580394	0.540242	0.486751	0.447262	0.380228	0.294381	0.90
0.95	1.266746	0.950000	0.787722	0.696402	0.641251	0.606186	0.565626	0.540964	0.502728	0.450497	0.95
1.00	1.307984	1.000000	0.841578	0.753691	0.702507	0.672329	0.644599	0.635778	0.632404	0.632126	1.00
1.05	1.348636	1.050000	0.895747	0.811308	0.763789	0.737907	0.721365	0.726757	0.755702	0.803868	1.05
1.10	1.388744	1.100000	0.950179	0.869117	0.824792	0.802291	0.793900	0.809259	0.858590	0.925296	1.10
1.15	1.428346	1.150000	1.004829	0.927006	0.885284	0.865018	0.860718	0.879875	0.931966	0.982587	1.15
1.20	1.467478	1.200000	1.059660	0.984886	0.945105	0.925810	0.921056	0.937155	0.975599	0.998154	1.20

1.25	1.506169	1.250000	1.114639	1.042693	1.004165	0.984578	0.974957	0.981884	0.997323	1.000355	1.25
1.30	1.544446	1.300000	1.169738	1.100378	1.062439	1.041403	1.023220	1.016762	1.007819	1.000952	1.30
1.35	1.582336	1.350000	1.224933	1.157912	1.119955	1.096515	1.067237	1.045613	1.015443	1.001965	1.35
1.40	1.619859	1.400000	1.280205	1.215276	1.176782	1.150253	1.108735	1.072436	1.024760	1.003904	1.40
1.45	1.657037	1.450000	1.335537	1.272468	1.233025	1.203018	1.149509	1.100642	1.038007	1.007482	1.45
1.50	1.693888	1.500000	1.390915	1.329491	1.288804	1.255235	1.191169	1.132689	1.056723		1.50
1.55	1.730429	1.550000	1.446327	1.386355	1.344248	1.307309	1.234971	1.170080	1.082405		1.55
1.60	1.766677	1.600000	1.501765	1.443077	1.399484	1.359591	1.281729	1.213547	1.116552		1.60
1.65	1.802645	1.650000	1.557219	1.499677	1.454629	1.412358	1.331817	1.263252	1.160498		1.65
1.70	1.838348	1.700000	1.612684	1.556173	1.509783	1.465797	1.385219	1.318914	1.215170		1.70
1.75	1.873797	1.750000	1.668156	1.612586	1.565027	1.520007	1.441614	1.379875			1.75
1.80	1.909003	1.800000	1.723630	1.668937	1.620418	1.575003	1.500464	1.445145			1.80
1.85	1.943979	1.850000	1.779103	1.725242	1.675993	1.630729	1.561104	1.513450			1.85
1.90	1.978733	1.900000	1.834574	1.781519	1.731769	1.687080	1.622808	1.583326			1.90
1.95	2.013275	1.950000	1.890040	1.837781	1.787746	1.743913	1.684861	1.653226			1.95
2.00	2.047614	2.000000	1.945501	1.894039	1.843908	1.801075	1.746606	1.721664			2.00
2.05	2.081757	2.050000	2.000956	1.950302	1.900230	1.858411	1.807489	1.787384			2.05
2.10	2.115713	1.100000	2.056404	2.006577	1.956684	1.915779	1.867097				2.10
2.15	2.149489	2.150000	2.111847	2.062868	2.013233	1.973063	1.925173				2.15
2.20	2.183091	2.200000	2.167283	2.119177	2.069846	2.030172	1.981627				2.20
2.25	2.216525	2.250000	2.222713	2.175506	2.126491	2.087047	2.036524				2.25
2.30	2.249799	2.300000	2.278137	2.231854	2.183141	2.143660	2.090068				2.30
2.35	2.282917	2.350000	2.333556	2.288221	2.239257	2.200005	2.142569				2.35
2.40	2.315885	2.400000	2.388970	2.344605	2.296375	2.256102	2.194432				2.40
2.45	2.348707	2.450000	2.444380	2.401003	2.352934	2.311986	2.246548				2.45

Table 14.4 Weibull renewal, standard deviation $= s(t)$ [1]

t/θ	0.5	1.0	1.5	2.0	2.5	3.0	4.0	5.0	7.0	10.0
0.00	0.000000	0.000000	0.000000	0.000000	0.000000	0.000000	0.000000	0.000000	0.000000	0.000000
0.05	0.518483	0.223607	0.105374	0.049938	0.023635	0.011179	0.002500	0.000559	0.000028	0.000000
0.10	0.639277	0.316228	0.176117	0.099502	0.056125	0.031601	0.009999	0.003162	0.000316	0.000010
0.15	0.726839	0.387298	0.236813	0.148329	0.092854	0.057963	0.022492	0.008714	0.001307	0.000076
0.20	0.798641	0.447214	0.291181	0.196069	0.132295	0.088961	0.039953	0.017884	0.003578	0.000320
0.25	0.860905	0.500000	0.340597	0.242398	0.173442	0.123690	0.062322	0.031227	0.007812	0.000977
0.30	0.916652	0.547723	0.386136	9.287020	0.215470	0.161356	0.089471	0.049206	0.014786	0.002430
0.35	0.967607	0.591608	0.428301	0.329676	0.257639	0.201179	0.121169	0.072189	0.025353	0.005252
0.40	1.014858	0.632456	0.467501	0.370149	0.299259	0.242348	0.157047	0.100425	0.040427	0.010239
0.45	1.059143	0.670820	0.504056	0.408264	0.339684	0.284008	0.196551	0.133990	0.060958	0.018448
0.50	1.100985	0.707107	0.538223	0.443896	0.378320	0.325266	0.238905	0.172718	0.087872	0.031227
0.55	1.140774	0.741620	0.570224	0.476967	0.414631	0.365204	0.283075	0.216108	0.121988	0.050233
0.60	1.178807	0.774597	0.600251	0.507445	0.448158	0.402910	0.327762	0.263216	0.163841	0.077408
0.65	1.215315	0.806226	0.628476	0.535349	0.478531	0.437517	0.371406	0.312564	0.213433	0.114864
0.70	1.250486	0.836660	0.655057	0.560736	0.505484	0.468242	0.412237	0.362035	0.269834	0.164550
0.75	1.284471	0.866025	0.680137	0.583709	0.528862	0.494436	0.448366	0.408886	0.330693	0.227506
0.80	1.317394	0.894427	0.703847	0.604403	0.548631	0.515632	0.477916	0.449834	0.391698	0.302399
0.85	1.349362	0.921954	0.726310	0.622983	0.564880	0.531578	0.499208	0.481304	0.446227	0.383107
0.90	1.380463	0.948683	0.747639	0.639639	0.577814	0.542279	0.510964	0.499855	0.485578	0.455765
0.95	1.410772	0.974679	0.767938	0.654578	0.587749	0.548014	0.512533	0.502775	0.500271	0.497547
1.00	1.440356	1.000000	0.787304	0.668016	0.595094	0.549336	0.504086	0.488756	0.482737	0.482238
1.05	1.469272	1.024695	0.805825	0.680175	0.600334	0.547062	0.486769	0.458526	0.430958	0.397105
1.10	1.497569	1.048809	0.823585	0.691273	0.604001	0.542225	0.462777	0.415282	0.351435	0.263049
1.15	1.525293	1.072381	0.840656	0.701518	0.606644	0.536012	0.435307	0.364800	0.259290	0.131466
1.20	1.552481	1.095445	0.857109	0.711107	0.608802	0.529670	0.408348	0.315187	0.174868	0.047360

1.25	1.579169	1.118034	0.873005	0.720218	0.610966	0.524386	0.386188	0.276002	0.119601	0.023134	1.25
1.30	1.605388	1.140175	0.888400	0.729005	0.613558	0.521167	0.372581	0.255766	0.107354	0.030869	1.30
1.35	1.631166	1.161895	0.903346	0.737602	0.616902	0.520726	0.369701	0.257532	0.125586	0.044280	1.35
1.40	1.656529	1.183216	0.917888	0.746118	0.621216	0.523407	0.377478	0.277099	0.155580	0.062364	1.40
1.45	1.681500	1.204159	0.932067	0.754635	0.626606	0.529179	0.393384	0.307253	0.191250	0.086172	1.45
1.50	1.706099	1.224745	0.945920	0.763215	0.633075	0.537679	0.415933	0.342145	0.231363		1.50
1.55	1.730346	1.244990	0.959478	0.771897	0.640537	0.548312	0.440641	0.378099	0.275064		1.55
1.60	1.754259	1.264911	0.972770	0.780702	0.648842	0.560362	0.465537	0.412808	0.321021		1.60
1.65	1.777854	1.284523	0.985822	0.789637	0.657795	0.573095	0.488775	0.444576	0.367289		1.65
1.70	1.801146	1.303840	0.998655	0.798692	0.667183	0.585840	0.509071	0.471969	0.411300		1.70
1.75	1.824147	1.322876	1.011289	0.807852	0.676793	0.598041	0.525591	0.493728			1.75
1.80	1.846872	1.341641	1.023739	0.817094	0.686430	0.609283	0.537879	0.508834			1.80
1.85	1.869332	1.360147	1.036021	0.826390	0.695929	0.619299	0.545794	0.516603			1.85
1.90	1.891538	1.378405	1.048147	0.835711	0.705158	0.627964	0.549486	0.516812			1.90
1.95	1.913500	1.396424	1.060127	0.845031	0.714026	0.635274	0.549376	0.509786			1.95
2.00	1.935228	1.414214	1.071972	0.854321	0.722479	0.641322	0.546130	0.496472			2.00
2.05	1.956731	1.431782	1.083688	0.863558	0.730494	0.646315	0.540623	0.478434			2.05
2.10	1.978016	1.449138	1.095282	0.872723	0.738084	0.650453	0.533886				2.10
2.15	1.999093	1.466288	1.106761	0.881800	0.745278	0.654004	0.527023				2.15
2.20	2.019968	1.483240	1.118129	0.890775	0.752128	0.657228	0.521108				2.20
2.25	2.040649	1.500000	1.129391	0.899642	0.758693	0.660368	0.517072				2.25
2.30	2.061142	1.516575	1.140549	0.908394	0.765036	0.663632	0.515594				2.30
2.35	2.081453	1.532971	1.151609	0.917030	0.771221	0.667185	0.517017				2.35
2.40	2.101587	1.549193	1.162571	0.925551	0.777306	0.671138	0.521283				2.40
2.45	2.121552	1.565248	1.173440	0.933959	0.783342	0.675549	0.527343				2.45

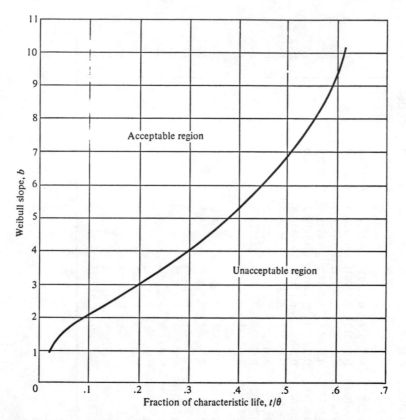

Fig. 14.3 Test and design acceptance regions as a result of Weibull renewal analysis for Example 14.3.

Example 14.4 A Navy vessel is usually at sea for an extended time. In the power train it has one bearing unlike any other bearing on the vessel. The bearing has the following Weibull characteristics: $\theta = 6$ months; $b = 3.0$. With a confidence of 99.9 percent, predict the number of spare bearings the vessel should have on board to maintain itself. Assume that the vessel is at sea for 6-month intervals.

Solution

$$\frac{t}{\theta} = \frac{6 \text{ months}}{6 \text{ months}} = 1.0$$

$$b = 3.0$$

$$R = 1$$

From Tables 14.3 and 14.4, for $b = 3.0$ and $t/\theta = 1.0$:

$$m(t) = 0.672329 \qquad \mu(t) = Rm(t) = 0.672329 \text{ average replacements}$$

$$s(t) = 0.549336 \qquad \sigma(t) = \sqrt{R}\, s(t) = 0.549336$$

$$P[\mu(t) - z_{\alpha/2}\, \sigma(t) \le N(t) \le \mu(t) + z_{\alpha/2}\, \sigma(t)] = 1 - \alpha$$

$$P[0.672329 - (3.2905)(0.549336) \le N(t) \le 0.672329 + (3.2905)(0.549336)] = 0.999$$

$$P[0 \le N(t) \le 2.48] = 0.999$$

Thus, if the ship carries three extra bearings, the crew will be more than 99.9 percent sure of maintaining themselves at sea.

SUMMARY

Renewal analysis, as presented in this chapter, is based on the Weibull distribution because of its widespread application to life, durability, and warranty phenomena. There exists, however, one basic difference between Weibull analysis and renewal analysis which merits a separate chapter. In the Weibull analysis (for the distribution and application see Secs. 2.3, 3.2, 4.1.2, 5.3.2, 5.5, 7.2, and 10.4.1 and Example 12.1) a sample is taken from a lot and tested, and some or all of the items in the sample fail. The time to failure is recorded and plotted, and from this plot certain conclusions pertaining to the sample or the population are drawn. It will be noted that in all the above, the items which have failed are not replaced; that is, in the continuum of time the lot diminishes in size. In actual life situations this is seldom the case, as failed tubes in a TV set are replaced, and so are mufflers in an automobile, or a motor in a heating furnace. A simple Weibull analysis cannot describe these situations, and the renewal analysis must be used instead. Renewal analysis may be termed a Weibull analysis with replacements. The present chapter gives the basic principles of renewal analysis and offers four examples to illustrate its application.

PROBLEMS

14.1. A certain part has a failure pattern which can be described by the equation $F(x) = 1 - \exp[-(x/10)^3]$, where x is in years and $F(x)$ is the cumulative number of failures; 100,000 of these parts are to be produced per year.

(a) Recommend a maximum warranty period in months for this part with 99 percent confidence, assuming that the manufacturer is willing to bear the expense of replacing only 1,700 of these parts.

(b) The manufacturer is required to have replacement parts available for models up to 7 years old. To anticipate the replacement needs of this part over the next 7 years and to take advantage of the cost saving by producing all the parts at once and storing them until needed, how many parts have to be produced? Provide the answer with 99 percent confidence.

14.2. The life of a part in customer usage follows Weibull distribution with $\theta = 4$ years and $b = 3$; 25,000 units were placed in service at the same time. Estimate, with 90 percent confidence, the expected replacement-parts production over the next 7 years.

REFERENCE

1. White, J. S.: Weibull Renewal Analysis, *Third Annu. Aerosp. Rel. Maintainability Conf.*, *Washington, D.C.*, June 29 to July 1, 1964.

BIBLIOGRAPHY

Barlow, R. E., L. C. Hunter, and F. Proschan: Probabilistic Models in Reliability Theory, John Wiley & Sons, Inc., New York, 1965.

Cox, D. R.: "Renewal Theory," John Wiley & Sons, Inc., New York, 1962.

Leadbetter, M. R.: On Series Expansion for Renewal Moments, *Biometrika*, vol. 50, 1963.

Smith, W. L.: Asymptotic Renewal Theorems, *Proc. Roy. Soc. Edinburgh*, ser. A, vol. 64, 1954.

———: Renewal Theory and Its Ramifications, *J. Roy. Statist. Soc.*, ser. B, vol. 20, 1958.

———: On the Cumulants of Renewal Processes, *Biometrika*, vol. 46, 1959.

——— and M. R. Leadbetter: On the Renewal Function for the Weibull Distribution, *Technometrics*, vol. 5, 1963.

Weibull, W.: A Statistical Distribution Function of Wide Applicability, *J. Appl. Mech.*, vol. 18, 1951.

White, J. S.: Estimation for the Weibull Distribution, *Trans. Seventeenth Annu. Conv. Amer. Soc. Qual. Contr.*, *Cincinnati*, 1963.

———: Weibull Renewal Analysis, *Third Annu. Aerosp. Rel. Maintainability Conf.*, *Washington, D.C.*, June 29 to July 1, 1964.

APPENDIX

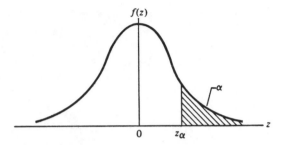

Table A-1   Normal distribution [1]*

z_a	.00	.01	.02	.03	.04	.05	.06	.07	.08	.09
0.0	.5000	.4960	.4920	.4880	.4840	.4801	.4761	.4721	.4681	.4641
0.1	.4602	.4562	.4522	.4483	.4443	.4404	.4364	.4325	.4286	.4247
0.2	.4207	.4168	.4129	.4090	.4052	.4013	.3974	.3936	.3897	.3859
0.3	.3821	.3783	.3745	.3707	.3669	.3632	.3594	.3557	.3520	.3483
0.4	.3446	.3409	.3372	.3336	.3300	.3264	.3228	.3192	.3156	.3121
0.5	.3085	.3050	.3015	.2981	.2946	.2912	.2877	.2843	.2810	.2776
0.6	.2743	.2709	.2676	.2643	.2611	.2578	.2546	.2514	.2483	.2451
0.7	.2420	.2389	.2358	.2327	.2296	.2266	.2236	.2206	.2177	.2148
0.8	.2119	.2090	.2061	.2033	.2005	.1977	.1949	.1922	.1894	.1867
0.9	.1841	.1814	.1788	.1762	.1736	.1711	.1685	.1660	.1635	.1611
1.0	.1587	.1562	.1539	.1515	.1492	.1469	.1446	.1243	.1401	.1379
1.1	.1357	.1335	.1314	.1292	.1271	.1251	.1230	.1210	.1190	.1170
1.2	.1151	.1131	.1112	.1093	.1075	.1056	.1038	.1020	.1003	.0985
1.3	.0968	.0951	.0934	.0918	.0901	.0885	.0869	.0853	.0838	.0823
1.4	.0808	.0793	.0778	.0764	.0749	.0735	.0721	.0708	.0694	.0681

* Tabulation of the values of α versus z_a for the standardized normal curve.

$$\alpha = P(z > z_a) = \int_{z_a}^{\infty} \frac{1}{\sqrt{2\pi}} e^{-z^2/2} \, dz$$

= area under the standardized normal curve from $z = z_a$ to $z = \infty$.

Table A-1 (*cont.*)

z_α	.00	.01	.02	.03	.04	.05	.06	.07	.08	.09
1.5	.0668	.0655	.0643	.0630	.0618	.0606	.0594	.0582	.0571	.0559
1.6	.0548	.0537	.0526	.0516	.0505	.0495	.0485	.0475	.0465	.0455
1.7	.0446	.0436	.0427	.0418	.0409	.0401	.0392	.0384	.0375	.0367
1.8	.0359	.0351	.0344	.0336	.0329	.0322	.0314	.0307	.0301	.0294
1.9	.0287	.0281	.0274	.0268	.0262	.0256	.0250	.0244	.0239	.0233
2.0	.0228	.0222	.0217	.0212	.0207	.0202	.0197	.0192	.0188	.0183
2.1	.0179	.0174	.0170	.0166	.0162	.0158	.0154	.0150	.0146	.0143
2.2	.0139	.0136	.0132	.0129	.0125	.0122	.0119	.0116	.0113	.0110
2.3	.0107	.0104	.0102	.00990	.00964	.00939	.00914	.00889	.00866	.00842
2.4	.00820	.00798	.00776	.00755	.00734	.00714	.00695	.00676	.00657	.00639
2.5	.00621	.00604	.00587	.00570	.00554	.00539	.00523	.00508	.00494	.00480
2.6	.00466	.00453	.00440	.00427	.00415	.00402	.00391	.00379	.00368	.00357
2.7	.00347	.00336	.00326	.00317	.00307	.00298	.00289	.00280	.00272	.00264
2.8	.00256	.00248	.00240	.00233	.00226	.00219	.00212	.00205	.00199	.00193
2.9	.00187	.00181	.00175	.00169	.00164	.00159	.00154	.00149	.00144	.00139

z_α	.0	.1	.2	.3	.4	.5	.6	.7	.8	.9
3	.00135	$.0^3988$	$.0^3687$	$.0^3483$	$.0^3337$	$.0^3233$	$.0^3159$	$.0^3108$	$.0^4723$	$.0^4481$
4	$.0^4317$	$.0^4207$	$.0^4133$	$.0^5854$	$.0^5541*$	$.0^5340$	$.0^5211$	$.0^5130$	$.0^6793$	$.0^6479$
5	$.0^6287$	$.0^6170$	$.0^7996$	$.0^7579$	$.0^7333$	$.0^7190$	$.0^7107$	$.0^8599$	$.0^8332$	$.0^8182$
6	$.0^9987$	$.0^9530$	$.0^9282$	$.0^9149$	$.0^{10}777$	$.0^{10}402$	$.0^{10}206$	$.0^{10}104$	$.0^{11}523$	$.0^{11}260$

$*.0^5541$ means .00000541.

Table A-2 Chi square distribution [2, 3]*

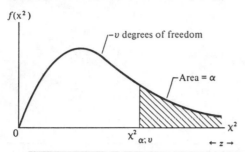

$f(x^2)$

v degrees of freedom

Area = α

$\chi^2_{\alpha;v}$ χ^2

$\leftarrow z \rightarrow$

0

α v	.995	.99	.98	.975	.95	.90	.80	.75	.70	.50
1	.0⁴393	.0³157	.0²628	.0²982	.00393	.0158	.0642	.102	.148	.455
2	.0100	.0201	.0404	.0506	.103	.211	.446	.575	.713	1.386
3	.0717	.115	.185	.216	.352	.584	1.005	1.213	1.424	2.366
4	.207	.297	.429	.484	.711	1.064	1.649	1.923	2.195	3.357
5	.412	.554	.752	.831	1.145	1.610	2.343	2.675	3.000	4.351
6	.676	.872	1.134	1.237	1.635	2.204	3.070	3.455	3.828	5.348
7	.989	1.239	1.564	1.690	2.167	2.833	3.822	4.255	4.671	6.346
8	1.344	1.646	2.032	2.180	2.733	3.490	4.594	5.071	5.527	7.344
9	1.735	2.088	2.532	2.700	3.325	4.168	5.380	5.899	6.393	8.343
10	2.156	2.558	3.059	3.247	3.940	4.865	6.179	6.737	7.267	9.342
11	2.603	3.053	3.609	3.816	4.575	5.578	6.989	7.584	8.148	10.341
12	3.074	3.571	4.178	4.404	5.226	6.304	7.807	8.438	9.034	11.340
13	3.565	4.107	4.765	5.009	5.892	7.042	8.634	9.299	9.926	12.340
14	4.075	4.660	5.368	5.629	6.571	7.790	9.467	10.165	10.821	13.339
15	4.601	5.229	5.985	6.262	7.261	8.547	10.307	11.036	11.721	14.339
16	5.142	5.812	6.614	6.908	7.962	9.312	11.152	11.912	12.624	15.338
17	5.697	6.408	7.255	7.564	8.672	10.085	12.002	12.792	13.531	16.338
18	6.265	7.015	7.906	8.231	9.390	10.865	12.857	13.675	14.440	17.338
19	6.844	7.633	8.567	8.907	10.117	11.651	13.716	14.562	15.352	18.338
20	7.434	8.260	9.237	9.591	10.851	12.443	14.578	15.452	16.266	19.337
21	8.034	8.897	9.915	10.283	11.591	13.240	15.445	16.344	17.182	20.337
22	8.643	9.542	10.600	10.982	12.338	14.041	16.314	17.240	18.101	21.337
23	9.260	10.196	11.293	11.688	13.091	14.848	17.187	18.137	19.021	22.337
24	9.886	10.856	11.992	12.401	13.848	15.659	18.062	19.037	19.943	23.337
25	10.520	11.524	12.697	13.120	14.611	16.473	18.940	19.939	20.867	24.337
26	11.160	12.198	13.409	13.844	15.379	17.292	19.820	20.843	21.792	25.336
27	11.808	12.879	14.125	14.573	16.151	18.114	20.703	21.749	22.719	26.336
28	12.461	13.565	14.847	15.308	16.928	18.939	21.588	22.647	23.647	27.336
29	13.121	14.256	15.574	16.047	17.708	19.768	22.475	23.567	24.577	28.336
30	13.787	14.953	16.306	16.791	18.493	20.599	23.364	24.478	25.508	29.336

* Tabulation of the values of α versus $\chi^2_{\alpha;\,v}$ for different values of v.

$$\alpha = P(\chi^2 > \chi^2_{\alpha;\,v}) = \int_{\chi^2_{\alpha;\,v}}^{\infty} f(\chi^2)\, d\chi^2$$

.30	.25	.20	.10	.05	.025	.02	.01	.005	.001	α ν
1.074	1.323	1.642	2.706	3.841	5.024	5.412	6.635	7.879	10.827	1
2.408	2.773	3.219	4.605	5.991	7.378	7.824	9.210	10.597	13.815	2
3.665	4.108	4.642	6.251	7.815	9.348	9.837	11.345	12.838	16.268	3
4.878	5.385	5.989	7.779	9.488	11.143	11.668	13.277	14.860	18.465	4
6.064	6.626	7.289	9.236	11.070	12.832	13.388	15.086	16.750	20.517	5
7.231	7.841	8.558	10.645	12.592	14.449	15.033	16.812	18.548	22.457	6
8.383	9.037	9.803	12.017	14.067	16.013	16.622	18.475	20.278	24.322	7
9.524	10.219	11.030	13.362	15.507	17.535	18.168	20.090	21.955	26.125	8
10.656	11.389	12.242	14.684	16.919	19.023	19.679	21.666	23.589	27.877	9
11.781	12.549	13.442	15.987	18.307	20.483	21.161	23.209	25.188	29.588	10
12.899	13.701	14.631	17.275	19.675	21.920	22.618	24.725	26.757	31.264	11
14.011	14.845	15.812	18.549	21.026	23.337	24.054	26.217	28.300	32.909	12
15.119	15.984	16.985	19.812	22.362	24.736	25.472	27.688	29.819	34.528	13
16.222	17.117	18.151	21.064	23.685	26.119	26.873	29.141	31.319	36.123	14
17.322	18.245	19.311	22.307	24.996	27.488	28.259	30.578	32.801	37.697	15
18.418	19.369	20.465	23.542	26.296	28.845	29.633	32.000	34.267	39.252	16
19.511	20.489	21.615	24.769	27.587	30.191	30.995	33.409	35.718	40.790	17
20.601	21.605	22.760	25.989	28.869	31.526	32.346	34.805	37.156	42.312	18
21.689	22.718	23.900	27.204	30.144	32.852	33.687	36.191	38.582	43.820	19
22.775	23.828	25.038	28.412	31.410	34.170	35.020	37.566	39.997	45.315	20
23.858	24.935	26.171	29.615	32.671	35.479	36.343	38.932	41.401	46.797	21
24.939	26.039	27.301	30.813	33.924	36.781	37.659	40.289	42.796	48.268	22
26.018	27.141	28.429	32.007	35.172	38.076	38.968	41.638	44.181	49.728	23
27.096	28.241	29.553	33.196	36.415	39.364	40.720	42.980	45.558	51.179	24
28.172	29.339	30.675	34.382	37.652	40.646	41.566	44.314	46.928	52.620	25
29.246	30.434	31.795	35.563	38.885	41.923	42.856	45.642	48.290	54.052	26
30.319	31.528	32.912	36.741	40.113	43.194	44.140	46.963	49.645	55.476	27
31.391	32.620	34.027	37.916	41.337	44.461	45.419	48.278	50.993	56.893	28
32.461	33.711	35.139	39.087	42.557	45.722	46.693	49.588	52.336	58.302	29
33.530	34.800	36.250	40.256	43.773	46.979	47.962	50.892	53.672	59.703	30

Table A-3 t distribution [2, 4]*

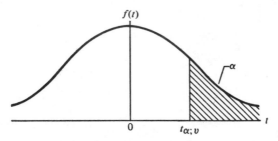

α ν	.40	.30	.20	.10	.050	.025	.010	.005	.001	.0005
1	.325	.727	1.376	3.078	6.314	12.71	31.82	63.66	318.3	636.6
2	.289	.617	1.061	1.886	2.920	4.303	6.965	9.925	22.33	31.60
3	.277	.584	.978	1.638	2.353	3.182	4.541	5.841	10.22	12.94
4	.271	.569	.941	1.533	2.132	2.776	3.747	4.604	7.173	8.610
5	.267	.559	.920	1.476	2.015	2.571	3.365	4.032	5.893	6.859
6	.265	.553	.906	1.440	1.943	2.447	3.143	3.707	5.208	5.959
7	.263	.549	.896	1.415	1.895	2.365	2.998	3.499	4.785	5.405
8	.262	.546	.889	1.397	1.860	2.306	2.896	3.355	4.501	5.041
9	.261	.543	.883	1.383	1.833	2.262	2.821	3.250	4.297	4.781
10	.260	.542	.879	1.372	1.812	2.228	2.764	3.169	4.144	4.587
11	.260	.540	.876	1.363	1.796	2.201	2.718	3.106	4.025	4.437
12	.259	.539	.873	1.356	1.782	2.179	2.681	3.055	3.930	4.318
13	.259	.538	.870	1.350	1.771	2.160	2.650	3.012	3.852	4.221
14	.258	.537	.868	1.345	1.761	2.145	2.624	2.977	3.787	4.140
15	.258	.536	.866	1.341	1.753	2.131	2.602	2.947	3.733	4.073
16	.258	.535	.865	1.337	1.746	2.120	2.583	2.921	3.686	4.015
17	.257	.534	.863	1.333	1.740	2.110	2.567	2.898	3.646	3.965
18	.257	.534	.862	1.330	1.734	2.101	2.552	2.878	3.611	3.922
19	.257	.533	.861	1.328	1.729	2.093	2.539	2.861	3.579	3.883
20	.257	.533	.860	1.325	1.725	2.086	2.528	2.845	3.552	3.850
21	.257	.532	.859	1.323	1.721	2.080	2.518	2.831	3.527	3.819
22	.256	.532	.858	1.321	1.717	2.074	2.508	2.819	3.505	3.792
23	.256	.532	.858	1.319	1.714	2.069	2.500	2.807	3.485	3.767
24	.256	.531	.857	1.318	1.711	2.064	2.492	2.797	3.467	3.745
25	.256	.531	.856	1.316	1.708	2.060	2.485	2.787	3.450	3.725
26	.256	.531	.856	1.315	1.706	2.056	2.479	2.779	3.435	3.707
27	.256	.531	.855	1.314	1.703	2.052	2.473	2.771	3.421	3.690
28	.256	.530	.855	1.313	1.701	2.048	2.467	2.763	3.408	3.674
29	.256	.530	.854	1.311	1.699	2.045	2.462	2.756	3.396	3.659
30	.256	.530	.854	1.310	1.697	2.042	2.457	2.750	3.385	3.646
40	.255	.529	.851	1.303	1.684	2.021	2.423	2.704	3.307	3.551
50	.255	.528	.849	1.298	1.676	2.009	2.403	2.678	3.262	3.495
60	.254	.527	.848	1.296	1.671	2.000	2.390	2.660	3.232	3.460
80	.254	.527	.846	1.292	1.664	1.990	2.374	2.639	3.195	3.415
100	.254	.526	.845	1.290	1.660	1.984	2.365	2.626	3.174	3.389
200	.254	.525	.843	1.286	1.653	1.972	2.345	2.601	3.131	3.339
500	.253	.525	.842	1.283	1.648	1.965	2.334	2.586	3.106	3.310
∞	.253	.524	.842	1.282	1.645	1.960	2.326	2.576	3.090	3.291

* Tabulation of the values of α versus $t_{\alpha;\,\nu}$ for different values of ν.

$$\alpha = P(t > t_{\alpha;\,\nu}) = \int_{t_{\alpha;\,\nu}}^{\infty} f(t)\, dt$$

Table A-4 *F* distribution [4]

Tabulation of the values of $F_{0.10;\, \nu_1;\, \nu_2}$ versus ν_1 and ν_2*

	Degrees of freedom for the numerator (ν_1)																	
	1	2	3	4	5	6	7	8	9	10	15	20	30	50	100	200	500	∞
1	39.9	49.5	53.6	55.8	57.2	58.2	58.9	59.4	59.9	60.2	61.2	61.7	62.3	62.7	63.0	63.2	63.3	63.3
2	8.53	9.00	9.16	9.24	9.29	9.33	9.35	9.37	9.38	9.39	9.42	9.44	9.46	9.47	9.48	9.49	9.49	9.49
3	5.54	5.46	5.39	5.34	5.31	5.28	5.27	5.25	5.24	5.23	5.20	5.18	5.17	5.15	5.14	5.14	5.14	5.13
4	4.54	4.32	4.19	4.11	4.05	4.01	3.98	3.95	3.94	3.92	3.87	3.84	3.82	3.80	3.78	3.77	3.76	3.76
5	4.06	3.78	3.62	3.52	3.45	3.40	3.37	3.34	3.32	3.30	3.24	3.21	3.17	3.15	3.13	3.12	3.11	3.10
6	3.78	3.46	3.29	3.18	3.11	3.05	3.01	2.98	2.96	2.94	2.87	2.84	2.80	2.77	2.75	2.73	2.73	2.72
7	3.59	3.26	3.07	2.96	2.88	2.83	2.78	2.75	2.72	2.70	2.63	2.59	2.56	2.52	2.50	2.48	2.48	2.47
8	3.46	3.11	2.92	2.81	2.73	2.67	2.62	2.59	2.56	2.54	2.46	2.42	2.38	2.35	2.32	2.31	2.30	2.29
9	3.36	2.81	2.81	2.69	2.61	2.55	2.51	2.47	2.44	2.42	2.34	2.30	2.25	2.22	2.19	2.17	2.17	2.16
10	3.28	2.92	2.73	2.61	2.52	2.46	2.41	2.38	2.35	2.32	2.24	2.20	2.16	2.12	2.09	2.07	2.06	2.06
11	3.23	2.86	2.66	2.54	2.45	2.39	2.34	2.30	2.27	2.25	2.17	2.12	2.08	2.04	2.00	1.99	1.98	1.97
12	3.18	2.81	2.61	2.48	2.39	2.33	2.28	2.24	2.21	2.19	2.10	2.06	2.01	1.97	1.94	1.92	1.91	1.90
13	3.14	2.76	2.56	2.43	2.35	2.28	2.23	2.20	2.16	2.14	2.05	2.01	1.96	1.92	1.88	1.86	1.85	1.85
14	3.10	2.73	2.52	2.39	2.31	2.24	2.19	2.15	2.12	2.10	2.01	1.96	1.91	1.87	1.83	1.82	1.80	1.80
15	3.07	2.70	2.49	2.36	2.27	2.21	2.16	2.12	2.09	2.06	1.97	1.92	1.87	1.83	1.79	1.77	1.76	1.76
16	3.05	2.67	2.46	2.33	2.24	2.18	2.13	2.09	2.06	2.03	1.94	1.89	1.84	1.79	1.76	1.74	1.73	1.72
17	3.03	2.64	2.44	2.31	2.22	2.15	2.10	2.06	2.03	2.00	1.91	1.86	1.81	1.76	1.73	1.71	1.69	1.69
18	3.01	2.62	2.42	2.29	2.20	2.13	2.08	2.04	2.00	1.98	1.89	1.84	1.78	1.74	1.70	1.68	1.67	1.66
19	2.99	2.61	2.40	2.27	2.18	2.11	2.06	2.02	1.98	1.96	1.86	1.81	1.76	1.71	1.67	1.65	1.64	1.63
20	2.97	2.59	2.38	2.25	2.16	2.09	2.04	2.00	1.96	1.94	1.84	1.79	1.74	1.69	1.65	1.63	1.62	1.61
22	2.95	2.56	2.35	2.22	2.13	2.06	2.01	1.97	1.93	1.90	1.81	1.76	1.70	1.65	1.61	1.59	1.58	1.57
24	2.93	2.54	2.33	2.19	2.10	2.04	1.98	1.94	1.91	1.88	1.78	1.73	1.67	1.62	1.58	1.56	1.54	1.53
26	2.91	2.52	2.31	2.17	2.08	2.01	1.96	1.92	1.88	1.86	1.76	1.71	1.65	1.59	1.55	1.53	1.51	1.50
28	2.89	2.50	2.29	2.16	2.06	2.00	1.94	1.90	1.87	1.84	1.74	1.69	1.63	1.57	1.53	1.50	1.49	1.48
30	2.88	2.49	2.28	2.14	2.05	1.98	1.93	1.88	1.85	1.82	1.72	1.67	1.61	1.55	1.51	1.48	1.47	1.46
40	2.84	2.44	2.23	2.09	2.00	1.93	1.87	1.83	1.79	1.76	1.66	1.61	1.54	1.48	1.43	1.41	1.39	1.38
50	2.81	2.41	2.20	2.06	1.97	1.90	1.84	1.80	1.76	1.73	1.63	1.57	1.50	1.44	1.39	1.36	1.34	1.33
60	2.79	2.39	2.18	2.04	1.95	1.87	1.82	1.77	1.74	1.71	1.60	1.54	1.48	1.41	1.36	1.33	1.31	1.29
80	2.77	2.37	2.15	2.02	1.92	1.85	1.79	1.75	1.71	1.68	1.57	1.51	1.44	1.38	1.32	1.28	1.26	1.24
100	2.76	2.36	2.14	2.00	1.91	1.83	1.78	1.73	1.70	1.66	1.56	1.49	1.42	1.35	1.29	1.26	1.23	1.21
200	2.73	2.33	2.11	1.97	1.88	1.80	1.75	1.70	1.66	1.63	1.52	1.46	1.38	1.31	1.24	1.20	1.17	1.14
500	2.72	2.31	2.10	1.96	1.86	1.79	1.73	1.68	1.64	1.61	1.50	1.44	1.36	1.28	1.21	1.16	1.12	1.09
∞	2.71	2.30	2.08	1.94	1.85	1.77	1.72	1.67	1.63	1.60	1.49	1.42	1.34	1.26	1.18	1.13	1.08	1.00

Degrees of freedom for the denominator (ν_2)

* $F_{\alpha;\, \nu_1;\, \nu_2}$ and $F_{\alpha;\, \nu_x;\, \nu_y}$ have been used interchangeably in the text.

Table A-5 F distribution

Tabulation of the values of $F_{0.05; \nu_1; \nu_2}$ versus ν_1 and ν_2*

	1	2	3	4	5	6	7	8	9	10	11	12	13	14	15	16	17	18
						Degrees of freedom for the numerator (ν_1)												
1	161	200	216	225	230	234	237	239	241	242	243	244	245	245	246	246	247	247
2	18.5	19.0	19.2	19.2	19.3	19.3	19.4	19.4	19.4	19.4	19.4	19.4	19.4	19.4	19.4	19.4	19.4	19.4
3	10.1	9.55	9.28	9.12	9.01	8.94	8.89	**8.85**	8.81	8.79	8.76	8.74	8.73	8.71	8.70	8.69	8.68	8.67
4	7.71	6.94	6.59	6.39	6.26	6.16	6.09	6.04	6.00	5.96	5.94	5.91	5.82	5.87	5.86	5.84	5.83	5.82
5	6.61	5.79	5.41	5.19	5.05	4.95	4.88	4.82	4.77	4.74	4.70	4.08	4.66	4.64	4.62	4.60	4.59	4.58
6	5.99	5.14	4.76	4.53	4.39	4.28	4.21	4.15	4.10	4.06	4.03	4.00	3.98	3.96	3.94	3.92	3.91	3.90
7	5.59	4.74	4.35	4.12	3.97	3.87	3.79	3.73	3.68	3.64	3.60	3.57	3.55	3.53	3.51	3.49	3.48	3.47
8	5.32	4.46	4.07	3.84	3.69	3.58	3.50	3.44	3.39	3.35	3.31	3.28	3.26	3.24	3.22	3.20	3.19	3.17
9	5.12	4.26	3.86	3.63	3.48	3.37	3.29	3.23	3.18	3.14	3.10	3.07	3.05	3.03	3.01	2.99	2.97	2.96
10	4.96	4.10	3.71	3.48	3.33	3.22	3.14	3.07	3.02	2.98	2.94	2.91	2.89	2.86	2.85	2.83	2.81	2.80
11	4.84	2.98	3.50	3.36	3.20	3.01	2.95	2.90	2.85	2.82	2.82	2.79	2.76	2.74	2.72	2.70	2.69	2.67
12	4.75	3.89	3.49	3.26	3.11	3.00	2.91	2.85	2.80	2.75	2.72	2.69	2.66	2.64	2.62	2.60	2.58	2.57
13	4.67	3.81	3.41	3.18	3.03	2.92	2.83	2.77	2.71	2.67	2.63	2.60	2.58	2.55	2.53	2.51	2.50	2.48
14	4.60	3.74	3.34	3.11	2.96	2.85	2.76	2.70	2.65	2.60	2.57	2.53	2.51	2.48	2.46	2.44	2.43	2.41
15	4.54	3.68	3.29	3.06	2.90	2.79	2.71	2.64	2.59	2.54	2.51	2.48	2.45	2.42	2.40	2.38	2.37	2.35
16	4.49	3.63	3.24	3.01	2.85	2.74	2.66	2.59	2.54	2.49	2.46	2.42	2.40	2.37	2.35	2.33	2.32	2.30
17	4.45	3.59	3.20	2.96	2.81	2.70	2.61	2.55	2.49	2.45	2.41	2.38	2.36	2.33	2.31	2.29	2.27	2.26
18	4.41	3.55	3.16	2.93	2.77	2.66	2.58	2.51	2.46	2.41	2.37	2.34	2.31	2.29	2.27	2.25	2.23	2.22
19	4.38	3.52	3.13	2.90	2.74	2.63	2.54	2.48	2.42	2.38	2.34	2.31	2.28	2.26	2.23	2.21	2.20	2.18
20	4.35	3.49	3.10	2.87	2.71	2.60	2.51	2.45	2.39	2.35	2.31	2.28	2.25	2.22	2.20	2.18	2.17	2.15
21	4.32	3.47	3.07	2.82	2.68	2.57	2.49	2.42	2.37	2.32	2.28	2.25	2.22	2.20	2.18	2.16	2.14	2.12
22	4.30	3.44	3.05	2.84	2.66	2.55	2.46	2.40	2.34	2.30	2.26	2.23	2.20	2.17	2.15	2.13	2.11	2.10
23	4.28	3.42	3.03	2.80	2.64	2.53	2.44	2.37	2.32	2.27	2.23	2.20	2.18	2.15	2.13	2.11	2.09	2.07
24	4.26	3.40	3.01	2.78	2.62	2.51	2.42	2.36	2.30	2.25	2.21	2.18	2.15	2.13	2.11	2.09	2.07	2.05
25	4.24	3.39	2.99	2.76	2.60	2.49	2.40	2.34	2.28	2.24	2.20	2.16	2.14	2.11	2.09	2.07	2.05	2.04
26	4.23	3.37	2.98	2.74	2.59	2.47	2.39	2.32	2.27	2.22	2.18	2.15	2.12	2.09	2.07	2.05	2.03	2.02
27	4.21	3.35	2.96	2.73	2.57	2.46	2.37	2.31	2.25	2.20	2.17	2.13	2.10	2.08	2.06	2.04	2.02	2.00
28	4.20	3.34	2.95	2.71	2.56	2.45	2.36	2.29	2.24	2.19	2.15	2.12	2.09	2.06	2.04	2.02	2.00	1.99
29	4.18	3.33	2.93	2.70	2.55	2.43	2.35	2.28	2.22	2.18	2.14	2.10	2.08	2.05	2.03	2.01	1.99	1.97
30	4.17	3.32	2.92	2.69	2.53	2.42	2.33	2.27	2.21	2.16	2.13	2.09	2.06	2.04	2.01	1.99	1.98	1.96
32	4.15	3.29	2.90	2.67	2.51	2.40	2.31	2.24	2.19	2.14	2.10	2.07	2.04	2.01	1.99	1.97	1.95	1.94
34	4.13	3.28	2.88	2.65	2.49	2.38	2.29	2.23	2.17	2.12	2.08	2.05	2.02	1.99	1.97	1.95	1.93	1.92
36	4.11	3.26	2.87	2.63	2.48	2.36	2.28	2.21	2.15	2.11	2.07	2.03	2.00	1.98	1.95	1.93	1.92	1.90
38	4.10	3.24	2.85	2.62	2.46	2.35	2.26	2.19	2.14	2.09	2.05	2.02	1.99	1.96	1.94	1.92	1.90	1.88
40	4.08	3.23	2.84	2.61	2.45	2.34	2.25	2.18	2.12	2.08	2.04	2.00	1.97	1.95	1.92	1.90	1.89	1.87

Left axis label: Degrees of freedom for the denominator (ν_2)

* $F_{\alpha; \nu_1; \nu_2}$ and $F_{\alpha; \nu_x; \nu_y}$ have been used interchangeably in the text.

Degrees of freedom for the numerator (v_1)																	
19	20	22	24	26	28	30	35	40	45	50	60	80	100	200	500	∞	
248	248	249	249	249	250	250	251	251	251	252	252	252	253	254	254	254	1
19.4	19.4	19.5	19.5	19.5	19.5	19.5	19.5	19.5	19.5	19.5	19.5	19.5	19.5	19.5	19.5	19.5	2
8.67	8.66	8.65	8.64	8.63	8.62	8.62	8.60	8.59	8.59	8.58	8.57	8.56	8.55	8.54	8.53	8.53	3
5.81	5.80	5.79	5.77	5.76	5.75	5.75	5.73	5.72	5.71	5.70	5.69	5.67	5.66	5.65	5.64	5.63	4
4.57	4.56	4.54	4.53	4.52	4.50	4.50	4.48	4.46	4.45	4.44	4.43	4.41	4.41	4.30	4.37	4.37	5
3.88	3.87	3.88	3.84	3.83	3.82	3.81	3.79	3.77	3.76	3.75	3.74	3.72	3.71	3.09	3.63	3.67	6
3.46	3.44	3.43	3.41	3.40	3.39	3.38	3.36	3.34	3.33	3.32	3.30	3.20	3.29	2.25	3.24	3.23	7
3.16	3.15	3.13	3.12	3.10	3.09	3.08	3.06	3.04	3.03	3.02	3.01	2.99	2.97	2.95	2.94	2.93	8
2.95	2.94	2.92	2.90	2.89	2.87	2.86	2.84	2.83	2.81	2.80	2.79	2.77	2.76	2.23	2.72	2.71	9
2.78	2.77	2.75	2.74	2.72	2.71	2.70	2.68	2.66	2.65	2.64	2.62	2.60	2.59	2.55	2.55	2.54	10
2.66	2.65	2.63	2.61	2.59	2.58	2.57	2.55	2.53	2.52	2.51	2.49	2.47	2.46	2.43	2.42	2.40	11
2.56	2.54	2.52	2.51	2.49	2.48	2.47	2.44	2.43	2.41	2.40	2.38	2.36	2.35	2.32	2.31	2.30	12
2.47	2.46	2.44	2.42	2.41	2.39	2.38	2.36	2.34	2.33	2.31	2.30	2.27	2.26	2.23	2.22	2.21	13
2.40	2.39	2.37	2.35	2.33	2.32	2.31	2.28	2.27	2.26	2.24	2.22	2.20	2.19	2.16	2.14	2.13	14
2.34	2.33	2.31	2.29	2.27	2.26	2.25	2.22	2.20	2.19	2.18	2.16	2.14	2.12	2.10	2.08	2.07	15
2.29	2.28	2.25	2.24	2.22	2.21	2.19	2.17	2.15	2.14	2.12	2.11	2.08	2.07	2.04	2.02	2.01	16
2.24	2.23	2.21	2.19	2.17	2.16	2.15	2.12	2.10	2.09	2.08	2.06	2.03	2.02	1.09	1.97	1.96	17
2.20	2.19	2.17	2.15	2.13	2.12	2.11	2.08	2.06	2.05	2.04	2.02	1.99	1.98	1.95	1.93	1.92	18
2.17	2.16	2.13	2.11	2.10	2.08	2.07	2.05	2.03	2.01	2.00	1.98	1.96	1.94	1.91	1.89	1.88	19
2.14	2.12	2.10	2.08	2.07	2.05	2.04	2.01	1.99	1.98	1.97	1.95	1.92	1.91	1.88	1.86	1.84	20
2.11	2.10	2.07	2.05	2.04	2.02	2.01	1.98	1.96	1.95	1.94	1.92	1.89	1.88	1.84	1.82	1.81	21
2.08	2.07	2.03	2.03	2.01	2.00	1.98	1.96	1.94	1.92	1.91	1.89	1.86	1.85	1.82	1.80	1.78	22
2.06	2.05	2.02	2.00	1.99	1.97	1.96	1.93	1.91	1.90	1.88	1.85	1.84	1.82	1.79	1.77	1.76	23
2.04	2.03	2.00	1.98	1.97	1.95	1.94	1.91	1.89	1.88	1.86	1.84	1.82	1.80	1.77	1.75	1.73	24
2.02	2.01	1.98	1.96	1.95	1.93	1.92	1.89	1.87	1.85	1.84	1.82	1.80	1.78	1.75	1.73	1.71	25
2.00	1.99	1.97	1.95	1.93	1.91	1.90	1.87	1.85	1.84	1.82	1.80	1.78	1.76	1.73	1.71	1.69	26
1.99	1.97	1.95	1.93	1.91	1.90	1.88	1.86	1.84	1.82	1.81	1.79	1.76	1.74	1.71	1.69	1.67	27
1.97	1.96	1.93	1.91	1.90	1.88	1.87	1.84	1.82	1.80	1.79	1.77	1.74	1.73	1.60	1.67	1.65	28
1.96	1.94	1.92	1.90	1.88	1.87	1.85	1.83	1.81	1.79	1.77	1.75	1.73	1.71	1.67	1.65	1.64	29
1.95	1.93	1.91	1.89	1.87	1.85	1.84	1.81	1.79	1.77	1.76	1.74	1.71	1.70	1.68	1.64	1.62	30
1.92	1.91	1.88	1.86	1.85	1.83	1.82	1.79	1.77	1.75	1.74	1.71	1.69	1.67	1.63	1.61	1.59	32
1.90	1.89	1.86	1.84	1.82	1.80	1.80	1.77	1.75	1.73	1.71	1.69	1.66	1.65	1.61	1.59	1.57	34
1.88	1.87	1.85	1.82	1.81	1.79	1.78	1.75	1.73	1.71	1.69	1.67	1.64	1.62	1.59	1.56	1.55	36
1.87	1.85	1.83	1.81	1.79	1.77	1.76	1.73	1.71	1.69	1.68	1.65	1.62	1.61	1.57	1.54	1.53	38
1.85	1.84	1.81	1.79	1.77	1.76	1.74	1.72	1.69	1.67	1.66	1.64	1.61	1.50	1.55	1.53	1.51	40

Table A-5 *(continued)*

	Degrees of freedom for the numerator (ν_1)																	
	1	*2*	*3*	*4*	*5*	*6*	*7*	*8*	*9*	*10*	*11*	*12*	*13*	*14*	*15*	*16*	*17*	*18*
42	4.07	3.22	2.83	2.59	2.44	2.32	2.24	2.16	2.11	2.06	2.03	1.99	1.96	1.93	1.91	1.89	1.87	1.86
44	4.06	3.21	2.82	2.58	2.43	2.31	2.23	2.16	2.10	2.05	2.01	1.98	1.95	1.92	1.90	1.88	1.86	1.84
46	4.05	3.20	2.81	2.57	2.42	2.30	2.22	2.15	2.09	2.04	2.00	1.97	1.94	1.91	1.89	1.87	1.85	1.83
48	4.04	3.19	2.80	2.57	2.41	2.29	2.21	2.14	2.08	2.03	1.99	1.96	1.93	1.90	1.88	1.86	1.84	1.82
50	4.03	3.18	2.79	2.56	2.40	2.29	2.20	2.13	2.07	2.03	1.99	1.95	1.92	1.89	1.87	1.85	1.83	1.81
55	4.02	3.16	2.77	2.54	2.38	2.27	2.18	2.11	2.06	2.01	1.97	1.93	1.90	1.88	1.85	1.83	1.81	1.79
60	4.00	3.15	2.76	2.53	2.37	2.25	2.17	2.10	2.04	1.99	1.95	1.92	1.89	1.86	1.84	1.82	1.80	1.78
65	3.99	3.14	2.75	2.51	2.36	2.24	2.15	2.08	2.03	1.98	1.94	1.90	1.87	1.85	1.82	1.80	1.78	1.76
70	3.98	3.13	2.74	2.50	2.35	2.23	2.14	2.07	2.02	1.97	1.93	1.89	1.86	1.84	1.81	1.79	1.77	1.75
80	3.96	3.11	2.73	2.49	2.33	2.21	2.13	2.06	2.00	1.95	1.91	1.88	1.84	1.82	1.79	1.77	1.75	1.73
90	3.95	3.10	2.71	2.47	2.32	2.20	2.11	2.04	1.99	1.94	1.90	1.86	1.83	1.80	1.78	1.76	1.74	1.72
100	3.94	3.09	2.70	2.46	2.31	2.19	2.10	2.03	1.97	1.93	1.89	1.85	1.82	1.79	1.77	1.75	1.73	1.71
125	3.92	3.07	2.68	2.44	2.29	2.17	2.08	2.01	1.96	1.91	1.87	1.83	1.80	1.77	1.76	1.72	1.70	1.69
150	3.90	3.08	2.66	2.43	2.27	2.16	2.07	2.00	1.94	1.89	1.85	1.82	1.79	1.76	1.73	1.71	1.69	1.67
200	3.89	3.04	2.65	2.42	2.26	2.14	2.06	1.98	1.93	1.88	1.84	1.80	1.77	1.74	1.72	1.69	1.67	1.65
300	3.87	3.03	2.63	2.40	2.24	2.13	2.04	1.97	1.91	1.86	1.82	1.78	1.75	1.72	1.70	1.68	1.66	1.64
500	3.86	3.01	2.62	2.39	2.23	2.12	2.03	1.96	1.90	1.85	1.81	1.77	1.74	1.71	1.69	1.66	1.64	1.62
1000	3.85	3.00	2.61	2.38	2.22	2.11	2.02	1.95	1.89	1.84	1.80	1.76	1.73	1.70	1.68	1.65	1.63	1.61
∞	3.84	3.00	2.60	2.37	2.21	2.10	2.01	1.94	1.88	1.83	1.79	1.75	1.72	1.69	1.67	1.64	1.62	1.60

Degrees of freedom for the denominator (ν_2)

		Degrees of freedom for the numerator (ν_1)															
19	20	22	24	26	28	30	35	40	45	50	60	80	100	200	500	∞	
1.84	1.83	1.80	1.78	1.76	1.74	1.73	1.70	1.68	1.66	1.65	1.62	1.59	1.57	1.53	1.51	1.49	42
1.83	1.81	1.79	1.77	1.75	1.73	1.72	1.69	1.67	1.65	1.63	1.61	1.58	1.56	1.52	1.49	1.48	44
1.82	1.80	1.78	1.76	1.74	1.72	1.71	1.68	1.65	1.64	1.62	1.60	1.57	1.55	1.51	1.48	1.46	46
1.81	1.79	1.77	1.75	1.73	1.71	1.70	1.67	1.64	1.62	1.61	1.59	1.56	1.54	1.49	1.47	1.45	48
1.80	1.78	1.76	1.74	1.72	1.70	1.69	1.66	1.63	1.61	1.60	1.58	1.54	1.52	1.48	1.46	1.44	50
1.78	1.76	1.74	1.72	1.70	1.68	1.67	1.64	1.61	1.59	1.58	1.55	1.52	1.50	1.46	1.43	1.41	55
1.76	1.75	1.72	1.70	1.68	1.66	1.65	1.62	1.59	1.57	1.56	1.53	1.50	1.48	1.44	1.41	1.39	60
1.75	1.73	1.71	1.69	1.67	1.65	1.63	1.60	1.58	1.56	1.54	1.52	1.49	1.48	1.42	1.39	1.37	65
1.74	1.72	1.70	1.67	1.65	1.64	1.62	1.59	1.57	1.55	1.53	1.50	1.47	1.45	1.40	1.37	1.35	70
1.72	1.70	1.68	1.65	1.63	1.62	1.60	1.57	1.54	1.52	1.51	1.48	1.45	1.43	1.38	1.35	1.32	80
1.70	1.69	1.66	1.64	1.62	1.60	1.59	1.55	1.53	1.51	1.49	1.46	1.43	1.41	1.36	1.32	1.30	90
1.69	1.68	1.65	1.63	1.61	1.59	1.57	1.54	1.52	1.49	1.48	1.45	1.41	1.39	1.34	1.31	1.28	100
1.67	1.65	1.63	1.60	1.58	1.57	1.55	1.52	1.49	1.47	1.45	1.42	1.39	1.36	1.31	1.27	1.25	125
1.66	1.64	1.61	1.59	1.57	1.55	1.53	1.50	1.48	1.45	1.44	1.41	1.37	1.34	1.29	1.23	1.22	150
1.64	1.62	1.60	1.57	1.55	1.53	1.52	1.48	1.46	1.43	1.41	1.39	1.35	1.32	1.26	1.22	1.19	200
1.62	1.61	1.58	1.55	1.53	1.51	1.50	1.46	1.43	1.41	1.39	1.38	1.32	1.30	1.23	1.19	1.15	300
1.61	1.50	1.56	1.54	1.52	1.50	1.48	1.45	1.42	1.40	1.38	1.34	1.30	1.28	1.21	1.16	1.11	500
1.60	1.58	1.55	1.53	1.51	1.49	1.47	1.44	1.41	1.38	1.36	1.33	1.29	1.26	1.19	1.13	1.08	1000
1.59	1.57	1.54	1.52	1.50	1.48	1.46	1.42	1.39	1.37	1.35	1.32	1.27	1.24	1.17	1.11	1.00	∞

Degrees of freedom for the denominator (ν_2)

Table A-6 *F* distribution

Tabulation of the values of $F_{0.025; \nu_1; \nu_2}$ versus ν_1 and ν_2*

							Degrees of freedom for the numerator (ν_1)											
	1	2	3	4	5	6	7	8	9	10	11	12	13	14	15	16	17	18
1	648	800	864	900	922	937	948	957	963	969	975	977	980	983	985	987	989	990
2	38.5	39.0	39.2	29.2	39.3	39.3	39.4	39.4	39.4	39.4	39.4	39.4	39.4	39.4	39.4	39.4	39.4	39.4
3	17.4	16.0	15.4	15.1	14.9	14.7	14.6	14.5	14.5	14.4	14.4	14.3	14.3	14.3	14.3	14.2	14.2	14.2
4	12.2	10.6	9.98	9.60	9.36	9.20	9.07	8.98	8.90	8.84	8.79	8.75	8.79	8.69	8.66	8.64	8.62	8.60
5	10.0	8.43	7.76	7.39	7.15	6.98	6.85	6.76	8.68	6.62	6.57	6.52	6.49	6.46	6.43	6.41	6.39	6.37
6	8.81	7.28	6.00	6.23	5.90	5.82	5.70	5.60	5.52	5.48	5.41	5.37	5.33	5.30	5.27	5.25	5.23	5.21
7	8.07	6.54	5.89	5.52	5.29	5.12	4.99	4.90	4.82	4.76	4.71	4.67	4.63	4.60	4.57	4.54	4.52	4.50
8	7.57	6.06	5.42	5.05	4.62	4.65	4.53	4.43	4.66	4.30	4.24	4.20	4.16	4.13	4.10	4.06	4.05	4.03
9	7.21	5.71	5.08	4.72	4.48	4.32	4.20	4.10	4.03	3.96	3.91	3.87	3.83	3.80	3.77	3.74	3.72	3.70
10	6.94	5.46	4.83	4.47	4.24	4.07	3.95	3.85	3.78	3.72	3.86	3.67	3.89	3.55	3.52	3.50	3.47	3.45
11	6.72	5.26	4.63	4.28	4.04	3.88	3.76	3.66	3.59	3.53	3.47	3.43	3.39	3.36	3.33	3.30	3.28	3.26
12	6.55	5.10	4.47	4.12	3.89	3.73	3.61	3.51	3.44	3.37	3.32	3.28	3.24	3.21	3.18	3.15	3.13	3.11
13	6.41	4.97	4.35	4.00	3.77	3.70	3.48	3.39	3.31	3.24	3.20	3.14	3.12	3.08	3.05	3.03	3.00	2.98
14	6.30	4.86	4.24	3.89	3.66	3.50	3.38	3.29	3.21	3.15	3.09	3.05	3.01	2.98	2.95	2.92	2.90	2.88
15	6.20	4.76	4.15	3.80	3.58	3.41	3.29	3.20	3.12	3.06	3.01	2.96	2.92	2.89	2.86	2.84	2.81	2.79
16	6.12	4.69	4.08	3.73	3.50	3.34	3.22	3.12	3.05	2.99	2.93	2.89	2.85	2.82	2.79	2.76	2.74	2.72
17	6.04	4.62	4.01	3.66	3.44	3.28	3.16	3.06	2.98	2.92	2.87	2.82	2.79	2.75	2.72	2.70	2.67	2.65
18	5.98	4.56	3.95	3.61	3.38	3.22	3.10	3.01	2.93	2.87	2.81	2.77	2.73	2.70	2.67	2.64	2.62	2.60
19	5.92	4.51	3.90	3.56	3.33	3.17	3.05	2.96	2.88	2.82	2.76	2.72	2.68	2.65	2.62	2.59	2.57	2.55
20	5.87	4.48	3.86	3.51	3.29	3.13	3.01	2.91	2.84	2.77	2.72	2.68	2.64	2.60	2.57	2.55	2.52	2.50
21	5.83	4.42	3.82	3.48	3.25	3.09	2.97	2.87	2.80	2.73	2.68	2.64	2.60	2.56	2.53	2.51	2.48	2.46
22	5.79	4.38	3.78	3.44	3.22	3.05	2.93	2.84	2.76	2.70	2.65	2.60	2.56	2.53	2.50	2.47	2.45	2.43
23	5.75	4.35	3.75	3.41	3.18	3.02	2.90	2.81	2.73	2.67	2.62	2.57	2.53	2.50	2.47	2.46	2.42	2.39
24	5.72	4.32	3.72	3.38	3.15	2.99	2.87	2.78	2.70	2.64	2.59	2.54	2.50	2.47	2.44	2.41	2.39	2.36
25	5.69	4.29	3.69	3.35	3.13	2.97	2.85	2.75	2.68	2.61	2.56	2.51	2.48	2.44	2.41	2.38	2.36	2.34
26	5.66	4.27	3.67	3.33	3.10	2.94	2.82	2.73	2.65	2.59	2.54	2.49	2.45	2.42	2.39	2.36	2.34	2.31
27	5.63	4.24	3.65	3.31	3.08	2.92	2.80	2.71	2.63	2.57	2.51	2.47	2.43	2.39	2.36	2.34	2.31	2.29
28	5.61	4.22	3.63	3.29	3.06	2.90	2.78	2.69	2.61	2.55	2.49	2.45	2.41	2.37	2.34	2.32	2.29	2.27
29	5.59	4.20	3.61	3.27	3.04	2.88	2.76	2.67	2.59	2.53	2.48	2.43	2.39	2.36	2.32	2.30	2.27	2.25
30	5.57	4.18	3.59	3.25	3.03	2.87	2.75	2.65	2.57	2.51	2.46	2.41	2.37	2.34	2.31	2.28	2.26	2.23
32	5.53	4.15	3.56	3.22	3.00	2.84	2.72	2.62	2.54	2.48	2.43	2.38	2.34	2.31	2.28	2.25	2.22	2.20
34	5.50	4.12	3.53	3.19	2.97	2.81	2.69	2.59	2.52	2.45	2.40	2.35	2.31	2.28	2.25	2.22	2.19	2.17
36	5.47	4.09	3.51	3.17	2.94	2.79	2.56	2.57	2.49	2.43	2.37	2.33	2.29	2.25	2.22	2.20	2.17	2.15
38	5.45	4.07	3.48	3.15	2.92	2.76	2.64	2.55	2.47	2.41	2.35	2.31	2.27	2.23	2.20	2.17	2.15	2.13
40	5.42	4.05	3.46	3.13	2.90	2.74	2.62	2.53	2.45	2.39	2.33	2.29	2.25	2.21	2.18	2.15	2.13	2.11

Degrees of freedom for the denominator (ν_2)

* $F_{\alpha; \nu_1; \nu_2}$ and $F_{\alpha; \nu_x; \nu_y}$ have been used interchangeably in the text.

\multicolumn{17}{c}{Degrees of freedom for the numerator (v_1)}	
19	20	22	24	25	28	30	35	40	45	50	60	80	100	200	500	∞	
992	923	995	997	999	1000	1001	1004	1008	1007	1008	1010	1012	1013	1016	1017	1018	1
23.4	30.4	29.5	39.5	39.5	39.5	39.5	39.5	39.5	39.5	39.5	39.5	39.5	39.5	39.5	39.5	39.5	2
14.2	14.2	14.1	14.1	14.1	14.1	14.1	14.1	10.0	14.0	14.0	14.0	14.0	14.0	13.9	13.9	13.9	3
8.53	8.56	8.53	8.51	8.49	8.48	8.48	8.44	8.41	8.39	8.38	8.36	8.33	8.32	8.29	8.27	8.26	4
6.35	6.33	6.30	6.28	6.26	6.24	6.23	6.20	6.18	6.16	6.14	6.12	6.10	6.06	6.05	6.03	6.02	5
5.19	5.1'.	5.14	5.12	5.10	5.08	5.07	5.04	5.01	4.99	4.96	4.96	4.93	4.92	4.88	4.86	4.85	6
4.48	4.47	4.44	4.42	4.92	4.38	4.38	4.33	4.31	4.29	4.28	4.25	4.23	4.21	4.18	4.16	4.14	7
4.02	4.00	3.97	3.95	3.93	3.91	3.90	3.86	3.84	3.82	3.81	3.78	3.76	3.74	3.70	3.78	3.67	8
3.68	3.67	3.64	3.61	3.59	3.58	3.56	3.53	3.51	3.49	3.47	3.45	3.42	3.40	3.37	3.35	3.33	9
3.44	3.42	3.39	3.37	3.34	3.33	3.31	3.28	3.26	3.24	3.22	3.20	3.17	3.15	3.12	3.09	3.08	10
3.24	3.23	3.20	3.17	3.15	3.13	3.12	3.09	3.06	3.04	3.03	3.00	2.97	2.96	2.92	2.90	2.88	11
3.09	3.07	3.04	3.02	3.00	2.98	2.96	2.93	2.91	2.89	2.87	2.85	2.82	2.80	2.76	2.74	2.72	12
2.96	2.95	2.92	2.89	2.87	2.85	2.84	2.80	2.78	2.76	2.74	2.72	2.69	2.67	2.63	2.61	2.60	13
2.86	2.84	2.81	2.79	2.77	2.75	2.73	2.70	2.67	2.65	2.64	2.61	2.58	2.56	2.53	2.50	2.49	14
2.77	2.76	2.73	2.70	2.68	2.66	2.64	2.61	2.58	2.56	2.55	2.52	2.49	2.47	2.44	2.41	2.40	15
2.70	2.68	2.65	2.63	2.60	2.58	2.57	2.53	2.51	2.49	2.47	2.45	2.42	2.40	2.36	2.33	2.32	16
2.63	2.62	2.59	2.56	2.54	2.52	2.50	2.47	2.44	2.42	2.41	2.38	2.35	2.33	2.29	2.28	2.25	17
2.58	2.56	2.53	2.50	2.48	2.46	2.44	2.41	2.38	2.36	2.35	2.32	2.29	2.27	2.23	2.20	2.19	18
2.53	2.51	2.48	2.45	2.43	2.41	2.39	2.36	2.33	2.31	2.30	2.27	2.24	2.22	2.18	2.15	2.13	19
2.43	2.46	2.43	2.41	2.39	2.37	2.35	2.31	2.29	2.27	2.25	2.22	2.19	2.17	2.13	2.10	2.09	20
2.44	2.42	2.39	2.37	2.34	2.33	2.31	2.27	2.25	2.23	2.21	2.18	2.15	2.13	2.09	2.06	2.04	21
2.41	2.39	2.36	2.33	2.31	2.29	2.27	2.24	2.21	2.19	2.17	2.14	2.11	2.09	2.05	2.02	2.00	22
2.37	2.36	2.33	2.30	2.28	2.26	2.24	2.20	2.18	2.15	2.14	2.11	2.08	2.06	2.01	1.99	1.97	23
2.35	2.33	2.30	2.27	2.25	2.23	2.21	2.17	2.15	2.12	2.11	2.08	2.05	2.02	1.98	1.95	1.94	24
2.32	2.30	2.27	2.24	2.22	2.20	2.18	2.13	2.12	2.10	2.08	2.05	2.02	2.00	1.95	1.92	1.91	25
2.29	2.28	2.24	2.22	2.19	2.17	2.16	2.12	2.09	2.07	2.05	2.03	1.99	1.97	1.92	1.90	1.88	26
2.27	2.25	2.22	2.19	2.17	2.15	2.13	2.10	2.07	2.05	2.03	2.00	1.97	1.94	1.90	1.87	1.85	27
2.25	2.23	2.20	2.17	2.15	2.13	2.11	2.08	2.05	2.03	2.01	1.98	1.94	1.92	1.88	1.85	1.83	28
2.23	2.21	2.18	2.15	2.13	2.11	2.09	2.06	2.03	2.01	1.99	1.96	1.92	1.90	1.86	1.83	1.81	29
2.21	2.20	2.16	2.14	2.11	2.08	2.07	2.04	2.01	1.99	1.97	1.94	1.90	1.88	1.84	1.81	1.79	30
2.18	2.16	2.13	2.10	2.08	2.06	2.04	2.00	1.98	1.95	1.93	1.91	1.87	1.85	1.80	1.77	1.75	32
2.15	2.13	2.10	2.07	2.05	2.03	2.01	1.97	1.95	1.92	1.90	1.88	1.84	1.82	1.77	1.74	1.72	34
2.12	2.11	2.08	2.05	2.03	2.00	1.99	1.95	1.92	1.90	1.88	1.85	1.81	1.79	1.74	1.71	1.69	36
2.11	2.09	2.05	2.03	2.00	1.98	1.96	1.93	1.90	1.87	1.85	1.82	1.79	1.76	1.71	1.68	1.66	38
2.08	2.07	2.03	2.01	1.98	1.96	1.94	1.90	1.88	1.85	1.83	1.80	1.76	1.74	1.69	1.66	1.64	40

Degrees of freedom for the denominator (v_2)

Table A-6   (*continued*)

							Degrees of freedom for the numerator (v_1)										
	1	*2*	*3*	*4*	*5*	*6*	*7*	*8*	*9*	*10*	*11*	*12*	*13*	*14*	*15*	*16*	*17*	*18*
42	5.40	4.03	3.45	3.11	2.89	2.73	2.61	2.51	2.44	2.37	2.32	2.27	2.23	2.20	2.16	2.14	2.11	2.09
44	5.39	4.02	3.43	3.09	2.87	2.71	2.59	2.50	2.42	2.36	2.30	2.26	2.21	2.18	2.15	2.12	2.10	2.07
46	5.37	4.00	3.42	3.08	2.88	2.70	2.58	2.48	2.41	2.34	2.29	2.24	2.20	2.17	2.13	2.11	2.08	2.06
48	5.35	3.99	3.40	3.07	2.84	2.69	2.57	2.47	2.39	2.33	2.27	2.23	2.19	2.15	2.12	2.09	2.07	2.05
50	5.34	3.98	3.39	3.06	2.83	2.67	2.55	2.46	2.38	2.32	2.26	2.22	2.18	2.14	2.11	2.06	2.06	2.03
55	5.31	3.95	3.36	3.03	2.81	2.65	2.53	2.43	2.36	2.29	2.24	2.19	2.15	2.11	2.08	2.05	2.03	2.01
60	5.29	3.93	3.34	3.01	2.79	2.63	2.51	2.41	2.33	2.27	2.22	2.17	2.13	2.09	2.06	2.03	2.01	1.98
65	5.27	3.91	3.32	2.99	2.77	2.61	2.49	2.39	2.32	2.25	2.20	2.15	2.11	2.07	2.04	2.01	1.99	1.97
70	5.25	3.89	3.31	2.98	2.75	2.60	2.48	2.38	2.30	2.24	2.18	2.14	2.10	2.06	2.03	2.00	1.97	1.95
80	5.22	3.86	3.28	2.95	2.73	2.57	2.45	2.36	2.28	2.21	2.16	2.11	2.07	2.03	2.00	1.97	1.95	1.93
90	5.20	3.84	3.27	2.93	2.71	2.55	2.43	2.34	2.26	2.19	2.14	2.09	2.05	2.02	1.98	1.95	1.93	1.91
100	5.18	3.83	3.25	2.92	2.70	2.54	2.42	2.32	2.24	2.18	2.12	2.08	2.04	2.00	1.97	1.94	1.91	1.89
125	5.15	3.80	3.22	2.89	2.67	2.51	2.39	2.30	2.22	2.14	2.10	2.05	2.01	1.97	1.94	1.91	1.89	1.86
150	5.13	3.78	3.20	2.87	2.65	2.49	2.37	2.28	2.20	2.13	2.08	2.03	1.99	1.95	1.92	1.89	1.87	1.84
200	5.10	3.76	3.18	2.85	2.63	2.47	2.35	2.26	2.18	2.11	2.06	2.01	1.97	1.93	1.90	1.87	1.84	1.82
300	5.08	3.74	3.16	2.83	2.61	2.45	2.33	2.23	2.16	2.09	2.04	1.99	1.95	1.91	1.88	1.85	1.82	1.80
500	5.05	3.72	3.14	2.81	2.59	2.43	2.31	2.22	2.14	2.07	2.02	1.97	1.93	1.89	1.86	1.83	1.80	1.78
1000	5.04	3.70	3.13	2.80	2.58	2.42	2.30	2.20	2.13	2.06	2.01	1.96	1.92	1.88	1.85	1.82	1.79	1.77
∞	5.02	3.69	3.12	2.79	2.57	2.41	2.29	2.19	2.11	2.05	1.99	1.94	1.90	1.87	1.83	1.80	1.78	1.75

Degrees of freedom for the denominator (v_2)

ν_1: 19	20	22	24	25	28	30	35	40	45	50	60	80	100	200	500	∞	ν_2
2.07	2.05	2.02	1.99	1.96	1.94	1.92	1.89	1.86	1.83	1.81	1.78	1.74	1.72	1.67	1.64	1.62	42
2.05	2.03	2.00	1.97	1.95	1.93	1.91	1.87	1.84	1.82	1.80	1.77	1.73	1.70	1.65	1.62	1.60	44
2.04	2.02	1.99	1.96	1.93	1.91	1.89	1.85	1.82	1.80	1.78	1.75	1.71	1.69	1.63	1.60	1.58	46
2.03	2.01	1.97	1.94	1.92	1.90	1.88	1.84	1.81	1.79	1.77	1.73	1.69	1.67	1.62	1.58	1.56	48
2.01	1.99	1.96	1.93	1.91	1.88	1.87	1.83	1.80	1.77	1.75	1.72	1.68	1.66	1.59	1.57	1.55	50
1.99	1.97	1.93	1.90	1.88	1.86	1.84	1.80	1.77	1.74	1.72	1.69	1.65	1.62	1.57	1.54	1.51	55
1.96	1.94	1.91	1.88	1.86	1.83	1.82	1.78	1.74	1.72	1.70	1.67	1.62	1.60	1.54	1.51	1.48	60
1.95	1.93	1.89	1.86	1.84	1.82	1.80	1.76	1.72	1.70	1.68	1.65	1.60	1.58	1.52	1.48	1.46	65
1.93	1.91	1.88	1.85	1.82	1.80	1.78	1.74	1.71	1.68	1.66	1.63	1.58	1.56	1.50	1.46	1.44	70
1.90	1.88	1.85	1.82	1.79	1.77	1.75	1.71	1.68	1.65	1.63	1.60	1.55	1.53	1.47	1.43	1.40	80
1.88	1.86	1.83	1.80	1.77	1.75	1.73	1.69	1.66	1.63	1.61	1.58	1.53	1.50	1.44	1.40	1.37	90
1.87	1.85	1.81	1.78	1.76	1.74	1.71	1.67	1.64	1.61	1.59	1.56	1.51	1.48	1.42	1.38	1.35	100
1.84	1.82	1.79	1.75	1.73	1.71	1.68	1.64	1.61	1.58	1.56	1.52	1.48	1.45	1.38	1.34	1.30	125
1.82	1.80	1.77	1.74	1.71	1.69	1.67	1.62	1.58	1.56	1.54	1.50	1.45	1.42	1.35	1.31	1.27	150
1.80	1.78	1.74	1.71	1.68	1.66	1.64	1.60	1.56	1.53	1.51	1.47	1.42	1.39	1.32	1.27	1.23	200
1.77	1.75	1.72	1.69	1.66	1.64	1.62	1.57	1.54	1.51	1.48	1.45	1.39	1.36	1.28	1.23	1.18	200
1.78	1.74	1.70	1.67	1.64	1.62	1.60	1.55	1.51	1.49	1.46	1.42	1.37	1.34	1.25	1.19	1.14	500
1.74	1.72	1.69	1.65	1.63	1.60	1.58	1.54	1.50	1.47	1.44	1.41	1.35	1.32	1.23	1.16	1.09	1000
1.73	1.71	1.67	1.64	1.61	1.59	1.57	1.52	1.48	1.45	1.43	1.39	1.23	1.30	1.21	1.13	1.00	∞

Degrees of freedom for the numerator (ν_1)

Degrees of freedom for the denominator (ν_2)

Table A-7 F distribution

Tabulation of the values of $F_{0.01;\,v_1;\,v_2}$ versus v_1 and v_2*

	Degrees of freedom for the numerator (v_1)																	
	1	2	3	4	5	6	7	8	9	10	11	12	13	14	15	16	17	18
	Multiply the numbers of the first row ($v_2 = 1$) by 10																	
1	405	500	540	563	576	596	598	598	602	606	608	611	613	614	616	617	618	619
2	93.5	99.0	99.3	99.3	99.3	99.3	99.4	99.4	99.4	99.4	99.4	99.4	99.4	99.4	99.4	99.4	99.4	99.4
3	34.1	30.8	20.5	28.7	28.2	27.9	27.7	27.5	27.3	27.2	27.1	27.1	27.0	26.9	26.9	25.8	26.8	26.8
4	21.2	18.0	16.7	16.0	15.5	15.2	15.0	14.8	14.7	14.5	14.4	14.4	14.3	14.2	14.2	14.2	14.1	14.1
5	16.8	13.2	12.1	11.4	11.0	10.7	10.5	10.3	10.2	10.1	9.06	9.89	9.82	9.77	9.72	9.68	9.64	9.61
6	13.7	10.9	9.78	9.15	8.75	8.47	8.28	8.10	7.98	7.87	7.79	7.72	7.66	7.60	7.56	7.52	7.48	7.45
7	12.2	9.55	8.45	7.85	7.46	7.19	6.99	6.94	6.72	6.62	6.54	6.47	6.41	6.36	5.31	6.27	6.24	6.21
8	11.3	8.65	7.89	7.01	6.63	6.37	6.18	6.03	5.91	5.81	5.73	5.67	5.61	5.56	5.52	5.48	5.44	5.41
9	10.6	8.02	6.99	6.42	6.06	5.80	5.61	5.47	5.35	5.26	5.13	5.11	5.05	5.00	4.96	4.92	4.89	4.86
10	10.0	7.56	6.55	5.99	5.64	5.39	5.20	5.06	4.94	4.85	4.77	4.71	4.65	4.60	4.56	4.52	4.49	4.46
11	9.65	7.21	6.22	5.67	5.32	5.07	4.89	4.74	4.63	4.54	4.46	4.40	4.34	4.39	4.25	4.21	4.18	4.15
12	9.33	6.93	5.95	5.41	5.06	4.82	4.64	4.50	4.30	4.30	4.22	4.16	4.10	4.05	4.01	3.97	3.94	3.91
13	9.07	6.70	5.74	5.21	4.86	4.62	4.44	4.30	4.19	4.10	4.02	3.96	3.91	3.86	3.82	3.78	3.75	3.72
14	8.86	6.51	5.58	5.04	4.70	4.46	4.28	4.14	4.03	3.94	3.88	3.80	3.75	3.70	3.66	3.62	3.50	3.56
15	8.68	6.26	5.42	4.89	4.56	4.32	4.14	4.00	3.89	3.80	3.72	3.67	3.61	3.56	3.52	3.49	3.45	3.42
16	8.53	6.22	5.29	4.77	4.44	4.20	4.03	3.89	3.78	3.69	3.62	3.55	3.50	3.45	3.41	3.37	3.34	3.31
17	8.60	6.11	5.18	4.67	4.34	4.10	3.93	3.79	3.68	3.59	3.52	3.46	3.40	3.35	3.31	3.27	3.24	3.21
18	8.20	6.01	5.09	4.58	4.25	4.01	3.84	3.71	3.60	3.51	3.43	3.37	3.32	3.27	3.22	3.19	3.16	3.13
19	8.18	5.93	5.01	4.50	4.17	3.94	3.77	3.68	3.52	3.43	3.36	3.30	3.24	3.19	3.15	3.12	3.08	3.05
20	8.10	5.85	4.94	4.43	4.10	3.87	3.70	3.56	3.46	3.37	3.29	3.23	3.18	3.13	3.09	3.05	3.02	2.99
21	8.02	5.78	4.87	4.37	4.04	3.81	3.64	3.51	3.40	3.31	3.24	3.17	3.12	3.07	3.03	2.99	2.96	2.93
22	7.95	5.72	4.82	4.31	3.99	3.76	3.59	3.45	3.35	3.26	3.18	3.12	3.07	3.02	2.98	2.94	2.91	2.88
23	7.86	5.66	4.76	4.26	3.94	3.71	3.54	3.41	3.30	3.21	3.14	3.07	3.02	2.97	2.93	2.89	2.86	2.83
24	7.82	5.61	4.72	4.22	3.90	3.67	3.50	3.36	3.26	3.17	3.09	3.03	2.98	2.93	2.89	2.85	2.83	2.79
25	7.77	5.57	4.68	4.18	3.86	3.63	3.46	3.32	3.22	3.13	3.06	2.99	2.94	2.89	2.85	2.81	2.78	2.75
26	7.72	5.53	4.64	4.14	3.82	3.59	3.42	3.29	3.18	3.09	3.02	2.96	2.90	2.86	2.82	2.78	2.74	2.72
27	7.66	5.49	4.60	4.11	3.78	3.56	3.39	3.26	3.15	3.06	2.99	2.93	2.87	2.82	2.78	2.75	2.71	2.68
28	7.64	5.45	4.57	4.07	3.75	3.53	3.36	3.23	3.12	3.03	2.96	2.90	2.84	2.79	2.75	2.72	2.68	2.65
29	7.60	5.42	4.54	4.04	3.73	3.50	3.33	3.20	3.09	3.00	2.93	2.87	2.81	2.77	2.73	2.69	2.66	2.63
30	7.56	5.39	4.51	4.03	3.70	3.47	3.30	3.17	3.07	2.98	2.91	2.84	2.79	2.74	2.70	2.66	2.63	2.60
32	7.50	5.34	4.46	3.97	3.65	3.43	3.26	3.13	3.02	2.93	2.86	2.80	2.74	2.70	2.66	2.62	2.58	2.55
34	7.44	5.29	4.42	3.93	3.61	3.39	3.23	3.09	2.96	2.89	2.82	2.76	2.70	2.66	2.62	2.58	2.55	2.51
36	7.40	5.25	4.36	3.89	3.57	3.35	3.18	3.05	2.95	2.86	2.79	2.72	2.67	2.62	2.58	2.54	2.51	2.48
38	7.35	5.21	4.34	3.86	3.54	3.32	3.15	3.02	2.92	2.83	2.75	2.69	2.64	2.59	2.55	2.51	2.48	2.45
40	7.31	5.18	4.31	3.83	3.51	3.29	3.12	2.99	2.80	2.80	2.73	2.66	2.61	2.56	2.52	2.48	2.45	2.42

Degrees of freedom for the denominator (v_2)

* $F_{\alpha;\,v_1;\,v_2}$ and $F_{\alpha;\,v_x;\,v_y}$ have been used interchangeably in the text.

Degrees of freedom for the numerator (ν_1)																	
19	20	23	24	26	28	30	35	40	45	50	60	90	100	200	500	∞	
Multiply the numbers of the first row ($\nu_2 = 1$) by 10																	
020	621	623	623	624	625	626	628	629	630	630	631	633	633	635	636	637	1
99.4	99.4	99.5	99.5	99.5	99.5	99.5	99.5	99.5	99.5	99.5	99.5	99.5	99.5	99.5	99.5	99.5	2
26.7	26.7	26.6	26.6	26.6	26.5	26.5	26.5	26.4	26.4	26.4	26.3	26.3	26.2	26.2	26.1	26.1	3
14.0	14.0	14.0	13.9	13.9	13.9	13.8	13.8	13.7	13.7	13.7	13.7	13.6	13.6	13.5	13.5	13.5	4
9.58	9.55	9.51	9.47	9.43	9.40	9.38	9.33	9.39	9.36	9.24	9.20	9.16	9.13	9.08	9.04	9.02	5
7.42	7.40	7.35	7.31	7.28	7.25	7.23	7.18	7.14	7.11	7.09	7.06	7.01	6.99	6.93	6.90	6.88	6
6.18	6.16	6.11	6.07	6.04	6.02	5.99	5.94	5.91	5.88	5.88	5.82	5.78	5.75	5.70	5.67	5.65	7
5.38	5.28	5.32	5.28	5.25	5.22	5.20	5.15	5.12	5.09	5.07	5.03	4.99	4.96	4.91	4.88	4.86	8
4.53	4.81	4.77	4.73	4.70	4.67	4.65	4.60	4.57	4.54	4.52	4.48	4.44	4.42	4.36	4.33	4.31	9
4.43	4.41	4.36	4.33	4.30	4.27	4.25	4.20	4.17	4.14	4.12	4.08	4.04	4.01	3.96	3.93	3.91	10
4.12	4.10	4.06	4.02	3.99	3.95	3.94	3.89	3.88	3.83	3.81	3.78	3.73	3.71	3.66	3.62	3.60	11
3.88	3.86	3.82	3.78	3.75	3.72	3.70	3.65	3.62	3.59	3.57	3.54	3.49	3.47	3.41	3.38	3.36	12
3.69	3.66	3.62	3.59	3.56	3.53	3.51	3.46	3.43	3.40	3.38	3.34	3.30	3.27	3.22	3.19	3.17	13
3.53	3.51	3.46	3.43	3.40	3.37	3.35	3.30	3.27	3.24	3.22	3.18	3.14	3.11	3.06	3.03	3.00	14
3.40	3.37	3.33	3.29	3.26	3.24	3.21	3.17	3.13	3.10	3.08	3.05	3.00	2.98	2.92	2.89	2.87	15
3.28	3.26	3.22	3.18	3.15	3.12	3.10	3.05	3.02	2.99	2.97	2.93	2.89	2.88	2.81	2.78	2.75	16
3.18	3.16	3.12	3.08	3.05	3.03	3.00	2.96	2.92	2.89	2.87	2.83	2.79	2.76	2.71	2.68	2.68	17
3.10	3.08	3.03	3.00	2.97	2.94	2.92	2.87	2.84	2.81	2.78	2.75	2.70	2.68	2.62	2.59	2.57	18
3.03	3.00	2.96	2.92	2.89	2.87	2.84	2.80	2.76	2.73	2.71	2.67	2.63	2.60	2.55	2.51	2.49	19
2.93	2.94	2.90	2.88	2.83	2.80	2.78	2.73	2.69	2.67	2.64	2.61	2.56	2.54	2.48	2.44	2.42	20
2.90	2.88	2.84	2.80	2.77	2.74	2.72	2.67	2.64	2.61	2.58	2.55	2.50	2.46	2.42	2.38	2.36	21
2.85	2.83	2.78	2.75	2.72	2.69	2.67	2.62	2.58	2.55	2.53	2.50	2.45	2.42	2.36	2.33	2.31	22
2.80	2.78	2.74	2.70	2.67	2.64	2.62	2.57	2.54	2.51	2.48	2.45	2.40	2.37	2.32	2.28	2.26	23
2.76	2.74	2.70	2.66	2.63	2.60	2.58	2.53	2.49	2.46	2.44	2.40	2.36	2.33	2.27	2.24	2.21	24
2.72	2.70	2.66	2.62	2.50	2.56	2.54	2.49	2.45	2.42	2.40	2.36	2.32	2.29	2.23	2.19	2.17	25
2.69	2.66	2.62	2.58	2.55	2.53	2.50	2.45	2.42	2.39	2.36	2.33	2.28	2.25	2.19	2.16	2.13	26
2.66	2.63	2.59	2.55	2.52	2.49	2.47	2.42	2.38	2.35	2.33	2.29	2.25	2.22	2.16	2.12	2.10	27
2.63	2.60	2.56	2.52	2.49	2.46	2.44	2.39	2.35	2.32	2.30	2.26	2.22	2.19	2.13	2.09	2.06	28
2.60	2.57	2.53	2.49	2.46	2.44	2.41	2.36	2.33	2.30	2.27	2.23	2.19	2.16	2.10	2.06	2.03	29
2.57	2.55	2.51	2.47	2.44	2.41	2.39	2.34	2.30	2.27	2.24	2.21	2.16	2.13	2.07	2.03	2.01	30
2.53	2.50	2.46	2.42	2.39	2.36	2.34	2.29	2.25	2.22	2.20	2.16	2.11	2.08	2.02	1.98	1.96	32
2.49	2.46	2.42	2.38	2.35	2.32	2.30	2.25	2.21	2.18	2.16	2.12	2.07	2.04	1.98	1.94	1.91	34
2.45	2.43	2.38	2.35	2.32	2.29	2.26	2.21	2.17	2.14	2.12	2.08	2.03	2.00	1.94	1.90	1.87	36
2.42	2.40	2.35	2.32	2.28	2.26	2.23	2.18	2.14	2.11	2.09	2.05	2.00	1.97	1.90	1.86	1.84	38
2.39	2.37	2.33	2.39	2.26	2.23	2.20	2.15	2.11	2.08	2.06	2.02	1.97	1.94	1.87	1.83	1.80	40

Table A-7 (*continued*)

v_2	1	2	3	4	5	6	7	8	9	10	11	12	13	14	15	16	17	18
	Degrees of freedom for the numerator (v_1)																	
	Multiply the numbers of the first row ($v_2 = 1$) *by 10*																	
42	7.28	5.15	4.29	3.80	3.49	3.27	3.10	2.97	2.86	2.78	2.70	2.64	2.59	2.54	2.50	2.46	2.43	2.40
44	7.25	5.12	4.26	3.78	3.47	3.24	3.08	2.95	2.84	2.75	2.68	2.62	2.56	2.52	2.47	2.44	2.40	2.37
46	7.22	5.10	4.24	3.76	3.44	3.22	3.06	2.93	2.82	2.73	2.66	2.60	2.54	2.50	2.45	2.42	2.38	2.35
48	7.19	5.08	4.22	3.74	3.43	3.20	3.04	2.91	2.80	2.72	2.64	2.58	2.53	2.48	2.44	2.40	2.37	2.33
50	7.17	5.06	4.20	3.72	3.41	3.19	3.02	2.89	2.79	2.70	2.63	2.56	2.51	2.46	2.42	2.38	2.35	2.32
55	7.12	5.01	4.16	3.68	3.37	3.15	2.98	2.85	2.75	2.66	2.59	2.53	2.47	2.42	2.38	2.34	2.31	2.28
60	7.08	4.98	4.12	3.65	3.34	3.12	2.95	2.82	2.72	2.63	2.56	2.50	2.44	2.39	2.35	2.31	2.28	2.24
65	7.04	4.95	4.10	3.62	3.31	3.09	2.93	2.80	2.69	2.61	2.53	2.47	2.42	2.37	2.33	2.29	2.26	2.23
70	7.01	4.92	4.08	3.60	3.29	3.07	2.91	2.78	2.67	2.59	2.51	2.45	2.40	2.24	2.31	2.27	2.23	2.20
80	6.98	4.88	4.04	3.56	3.26	3.04	2.87	2.74	2.64	2.55	2.48	2.42	2.36	2.21	2.27	2.23	2.20	2.17
90	6.93	4.85	4.01	3.54	3.23	3.01	2.84	2.72	2.61	2.52	2.45	2.39	2.33	2.29	2.24	2.21	2.17	2.14
100	6.90	4.83	3.96	3.51	3.21	2.99	2.82	2.69	2.59	2.50	2.43	2.37	2.31	2.26	2.22	2.19	2.15	2.12
125	6.84	4.78	3.94	3.47	3.17	2.95	2.79	2.66	2.55	2.47	2.39	2.33	2.28	2.23	2.19	2.15	2.11	2.08
150	6.81	4.75	3.92	3.45	3.14	2.92	2.76	2.63	2.53	2.44	2.37	2.31	2.25	2.20	2.16	2.12	2.09	2.06
200	6.76	4.71	3.88	3.41	3.11	2.89	2.73	2.60	2.50	2.41	2.34	2.27	2.22	2.17	2.13	2.09	2.06	2.02
300	6.72	4.68	3.85	3.38	3.08	2.86	2.70	2.57	2.47	2.36	2.31	2.24	2.19	2.14	2.10	2.06	2.03	1.99
500	6.69	4.65	3.82	3.36	3.05	2.84	2.68	2.55	2.44	2.36	2.28	2.22	2.17	2.12	2.07	2.04	2.00	1.97
1000	6.66	4.63	3.80	3.34	3.04	2.82	2.66	2.53	2.43	2.34	2.27	2.20	2.15	2.10	2.06	2.02	1.98	1.95
∞	6.66	4.61	3.78	3.32	3.02	2.80	2.64	2.51	2.41	2.32	2.25	2.18	2.13	2.08	2.04	2.00	1.97	1.00

Degrees of freedom for the denominator (v_2)

Degrees of freedom for the numerator (v_1)

Multiply the numbers of the first row ($v_2 = 1$) by 10

19	20	23	24	26	28	30	35	40	45	50	60	90	100	200	500	∞	Degrees of freedom for the denominator (v_2)
2.37	2.34	2.30	2.26	2.23	2.20	2.18	2.13	2.09	2.08	2.03	1.99	1.94	1.91	1.85	1.80	1.78	42
2.35	2.32	2.28	2.24	2.21	2.18	2.15	2.10	2.06	2.03	2.01	1.97	1.92	1.89	1.82	1.78	1.75	44
2.33	2.30	2.26	2.22	2.19	2.16	2.13	2.08	2.04	2.01	1.99	1.95	1.90	1.86	1.80	1.75	1.73	46
2.31	2.28	2.24	2.20	2.17	2.14	2.12	2.06	2.02	1.99	1.97	1.93	1.88	1.84	1.78	1.73	1.70	48
2.29	2.27	2.22	2.18	2.15	2.12	2.10	2.05	2.01	1.97	1.95	1.91	1.86	1.82	1.76	1.71	1.68	50
2.25	2.23	2.18	2.15	2.11	2.08	2.06	2.01	1.97	1.92	1.91	1.87	1.81	1.78	1.71	1.67	1.64	55
2.22	2.20	2.15	2.12	2.08	2.05	2.03	1.98	1.94	1.90	1.88	1.84	1.78	1.75	1.68	1.63	1.60	60
2.20	2.17	2.13	2.09	2.06	2.03	2.00	1.95	1.91	1.88	1.85	1.81	1.75	1.72	1.65	1.60	1.57	65
2.18	2.15	2.11	2.07	2.03	2.01	1.98	1.93	1.89	1.85	1.83	1.78	1.73	1.70	1.62	1.57	1.54	70
2.14	2.12	2.07	2.03	2.00	1.97	1.94	1.89	1.85	1.81	1.79	1.75	1.69	1.66	1.58	1.53	1.49	80
2.11	2.09	2.04	2.00	1.97	1.94	1.92	1.86	1.82	1.79	1.76	1.72	1.66	1.62	1.54	1.49	1.46	90
2.09	2.07	2.02	1.98	1.94	1.92	1.89	1.84	1.80	1.76	1.73	1.69	1.63	1.60	1.52	1.47	1.43	100
2.05	2.03	1.98	1.94	1.91	1.88	1.85	1.80	1.79	1.72	1.69	1.65	1.59	1.55	1.47	1.41	1.37	125
2.03	2.00	1.96	1.92	1.88	1.85	1.83	1.77	1.73	1.69	1.66	1.63	1.56	1.52	1.43	1.35	1.33	150
2.00	1.97	1.93	1.89	1.84	1.82	1.79	1.74	1.69	1.66	1.63	1.58	1.52	1.48	1.39	1.33	1.28	200
1.97	1.94	1.89	1.85	1.82	1.79	1.76	1.71	1.68	1.62	1.58	1.55	1.48	1.44	1.35	1.28	1.22	300
1.94	1.92	1.87	1.83	1.79	1.76	1.74	1.68	1.63	1.60	1.55	1.53	1.45	1.41	1.31	1.23	1.16	500
1.92	1.90	1.85	1.81	1.77	1.74	1.72	1.66	1.61	1.57	1.54	1.50	1.43	1.38	1.28	1.19	1.11	1000
1.90	1.86	1.83	1.79	1.76	1.72	1.70	1.64	1.58	1.55	1.52	1.47	1.40	1.36	1.25	1.15	1.00	∞

Table A-8 *F* distribution

Tabulation of the values of $F_{0.005; \nu_1; \nu_2}$ versus ν_1 and ν_2*

		Degrees of freedom for the numerator (ν_1)																
	1	*2*	*3*	*4*	*5*	*6*	*7*	*8*	*9*	*10*	*11*	*12*	*13*	*14*	*15*	*16*	*17*	*18*
	Multiply the numbers of the first row ($\nu_2 = 1$) by 100																	
1	*162*	*200*	*216*	*225*	*231*	*234*	*237*	*239*	*241*	*242*	*243*	*244*	*245*	*246*	*246*	*247*	*247*	*248*
2	198	199	199	199	199	199	199	199	199	199	199	199	199	199	199	199	199	199
3	55.6	49.8	47.5	46.2	45.4	44.3	44.4	44.1	43.9	43.7	43.5	43.4	43.3	43.2	43.1	43.0	42.9	42.9
4	21.2	26.3	24.3	23.2	22.5	22.0	21.6	21.4	21.1	21.0	20.8	20.7	20.6	20.5	20.4	20.4	20.3	20.2
5	22.8	18.2	16.5	15.6	14.3	14.5	14.2	14.0	13.8	13.6	13.5	13.4	13.3	12.2	12.1	12.1	13.0	13.0
6	18.6	14.5	12.9	13.0	11.5	11.1	10.8	10.6	10.4	10.2	10.1	10.0	9.95	9.88	9.81	9.76	9.71	9.08
7	16.2	12.4	10.9	10.0	9.52	9.16	8.89	8.68	8.51	8.38	8.27	8.18	8.10	8.03	7.97	7.93	7.87	7.83
8	14.7	11.0	9.60	8.81	8.30	7.95	7.69	7.50	7.34	7.21	7.10	7.01	6.94	6.37	6.34	6.76	6.72	6.68
9	13.6	10.1	8.72	7.96	7.47	7.13	6.88	6.69	6.54	6.42	6.31	6.23	6.15	6.09	6.03	5.98	5.94	5.90
10	12.8	9.43	8.06	7.34	6.87	6.54	6.30	6.12	5.97	5.85	5.75	5.66	5.50	5.53	5.47	5.42	5.38	5.34
11	12.2	8.91	7.60	6.88	6.42	6.10	5.86	5.68	5.54	5.42	5.32	5.24	5.16	5.10	5.05	5.00	4.96	4.92
12	11.8	8.51	7.23	6.52	6.07	5.76	5.52	5.35	5.20	5.09	4.99	4.91	4.84	4.77	4.72	4.67	4.63	4.50
13	11.4	8.19	6.93	6.23	5.79	5.48	5.25	5.08	4.94	4.82	4.72	4.54	4.57	4.54	4.46	4.41	4.37	4.23
14	11.1	7.92	6.68	6.00	5.56	5.26	5.03	4.86	4.72	4.60	4.51	4.43	4.36	4.30	4.25	4.20	4.18	4.12
15	10.8	7.70	6.48	5.80	5.37	5.07	4.85	4.67	4.54	4.42	4.33	4.25	4.18	4.12	4.07	4.02	3.93	3.95
16	10.6	7.51	6.30	5.64	5.21	4.91	4.69	4.52	4.48	4.27	4.18	4.10	4.03	3.97	3.92	3.87	3.83	3.80
17	10.4	7.35	6.16	5.50	5.07	4.78	4.56	4.39	4.25	4.14	4.05	3.97	3.90	3.84	3.79	3.75	3.71	3.67
18	10.2	7.21	6.03	5.37	4.98	4.66	4.44	4.28	4.14	4.03	3.94	3.86	3.79	3.73	3.68	3.64	3.60	3.56
19	10.1	7.09	5.92	5.27	4.85	4.56	4.34	4.18	4.04	3.93	3.84	3.76	3.70	3.64	3.59	3.54	3.50	3.46
20	9.94	6.99	5.82	5.17	4.76	4.47	4.28	4.09	3.96	3.85	3.76	3.68	3.61	3.55	3.50	3.48	3.42	3.28
21	9.82	6.89	5.73	5.09	4.39	4.63	4.18	4.01	3.88	3.77	3.68	3.60	3.54	3.48	3.43	3.38	3.34	3.31
22	9.73	6.81	5.65	5.02	4.61	4.32	4.11	3.94	3.81	3.70	3.61	3.54	3.47	3.41	3.36	3.31	3.27	3.24
23	9.63	6.72	5.58	4.95	4.54	4.26	4.05	3.85	3.75	3.64	3.55	3.47	3.41	3.35	3.30	3.25	3.21	3.18
24	9.55	6.66	5.52	4.89	4.49	4.20	3.99	3.83	3.69	3.59	3.50	3.42	3.35	3.30	3.25	3.20	3.16	3.12
25	9.48	6.60	5.46	4.84	4.43	4.15	3.94	3.78	3.64	3.54	3.45	3.37	3.30	3.25	3.20	3.15	3.11	3.08
26	9.41	6.54	5.41	4.79	4.36	4.10	3.89	3.73	3.60	3.49	3.40	3.33	3.26	3.20	3.15	3.17	3.07	3.03
27	9.34	6.49	5.36	4.74	3.44	4.06	3.85	3.69	3.56	3.45	3.36	3.28	3.22	3.16	3.11	3.07	3.03	2.99
28	9.28	6.44	5.32	4.70	4.30	4.02	3.81	3.65	3.52	3.41	3.32	3.25	3.18	3.12	3.07	2.08	2.99	2.95
29	9.23	6.40	5.26	4.66	4.26	3.98	3.77	3.61	3.48	3.38	3.29	3.21	3.15	3.09	3.04	2.09	2.95	2.92
30	9.18	6.35	5.24	4.62	4.23	3.95	3.74	3.58	3.45	3.34	3.25	3.18	3.11	3.06	3.01	2.98	2.92	2.89
32	9.09	6.28	5.17	4.56	4.17	3.89	3.68	3.52	3.39	3.29	3.20	3.12	3.06	3.00	2.95	2.90	2.86	2.83
34	9.01	6.22	5.11	4.50	4.11	3.84	3.63	3.47	3.34	3.24	3.15	3.07	3.01	2.95	2.90	2.85	2.81	2.78
36	8.94	6.15	5.06	4.46	4.06	3.79	3.58	3.42	3.30	3.19	3.10	3.03	2.96	2.90	2.85	2.81	2.77	2.73
38	8.88	6.11	5.02	4.41	4.02	3.75	3.54	3.39	3.25	3.15	3.06	2.99	2.92	2.87	2.82	2.77	2.73	2.70
40	8.83	6.07	4.98	4.37	3.99	3.71	3.51	3.35	3.22	3.12	3.03	2.95	2.90	2.83	2.78	2.74	2.70	2.66

Degrees of freedom for the denominator (ν_2)

* $F_{\alpha; \nu_1; \nu_2}$ and $F_{\alpha; \nu_x; \nu_y}$ have been used interchangeably in the text.

Table
Tabul

Degrees of freedom for the numerator (v_1)

19	20	22	24	26	28	30	35	40	45	50	60	80	100	200	500	∞	

Multiply the numbers of the first row ($v_2 = 1$) by 100

248	248	249	249	250	250	250	251	251	252	252	253	253	253	254	254	255	1
199	199	199	199	199	199	199	199	199	199	199	199	199	199	199	200	200	2
42.8	42.8	42.7	42.6	42.6	42.5	42.5	42.4	42.3	42.3	42.2	42.1	42.1	42.0	41.9	41.9	41.8	3
20.2	20.2	20.1	20.0	20.0	19.9	19.9	19.8	19.8	19.7	19.7	19.6	19.5	19.5	19.4	19.4	19.3	4
12.9	12.9	12.8	12.8	12.7	12.6	12.7	12.6	12.5	12.5	12.5	12.4	12.3	12.3	12.2	12.2	12.1	5
9.62	9.59	9.53	9.47	9.43	9.39	9.36	9.29	9.24	9.20	9.17	9.12	9.06	9.03	8.95	8.91	8.83	6
7.79	7.75	7.69	7.64	7.60	7.57	7.53	7.47	7.42	7.38	7.25	7.21	7.25	7.22	7.15	7.10	7.08	7
6.64	6.61	6.55	6.50	6.46	6.43	6.40	6.33	6.29	6.25	6.22	6.18	6.12	6.09	6.02	5.98	5.95	8
5.86	5.83	5.78	5.73	5.69	5.65	5.62	5.56	5.52	5.48	5.45	5.41	5.36	5.32	5.26	5.21	5.19	9
5.30	5.27	5.22	5.17	5.13	5.10	5.07	5.01	4.97	4.93	4.90	4.98	4.80	4.77	4.71	4.67	4.64	10
4.89	4.86	4.80	4.76	4.72	4.68	4.65	4.60	4.55	4.52	4.49	4.44	4.39	4.36	4.20	4.25	4.23	11
4.56	4.53	4.48	4.43	4.39	4.36	4.33	4.27	4.23	4.19	4.17	4.12	4.07	4.04	3.97	3.93	3.90	12
4.30	4.27	4.22	4.17	4.13	4.10	4.07	4.01	3.97	3.94	3.91	3.87	3.81	3.78	3.71	3.67	3.65	13
4.09	4.08	4.01	3.96	3.92	3.89	3.86	3.80	3.76	3.73	3.70	3.66	3.60	3.57	3.50	3.46	3.44	14
3.91	3.88	3.83	3.79	3.75	3.72	3.69	3.63	3.58	3.55	3.52	3.43	3.43	3.30	3.33	3.29	3.28	15
3.76	3.73	3.68	3.64	3.60	3.57	3.54	3.48	3.44	3.40	3.37	3.33	3.28	3.25	3.18	3.14	3.11	16
3.64	3.61	3.56	3.51	3.47	3.44	3.41	3.35	3.31	3.28	3.25	3.21	3.15	3.12	3.05	3.01	2.98	17
3.53	3.50	3.45	3.40	3.36	3.33	3.33	3.25	3.20	3.17	3.14	3.10	3.04	3.01	2.94	2.90	2.87	18
3.43	3.40	3.35	3.31	3.27	3.24	3.21	3.15	3.11	3.07	3.04	3.00	2.95	2.91	2.85	2.80	2.78	19
3.35	3.32	3.27	3.22	3.18	3.15	3.12	3.07	3.02	2.99	2.96	2.92	2.86	2.83	2.76	2.72	2.69	20
3.27	3.24	3.19	3.15	3.11	3.08	3.05	2.99	2.95	2.91	2.88	2.84	2.78	2.75	2.68	2.64	2.61	21
3.20	3.18	3.12	3.08	3.04	3.01	2.98	2.92	2.88	2.84	2.82	2.77	2.72	2.69	2.62	2.57	2.55	22
3.15	3.12	3.06	3.02	2.98	2.95	2.92	2.86	2.82	2.78	2.76	2.71	2.66	2.62	2.56	2.51	2.48	23
3.09	3.06	3.01	2.97	2.93	2.90	2.87	2.81	2.77	2.73	2.70	2.68	2.60	2.57	2.50	2.46	2.43	24
3.04	3.01	2.96	2.92	2.88	2.85	2.82	2.76	2.72	2.68	2.65	2.61	2.55	2.52	2.45	2.41	2.38	25
3.00	2.97	2.92	2.87	2.83	2.80	2.77	2.72	2.67	2.64	2.61	2.56	2.51	2.47	2.40	2.36	2.33	26
2.96	2.93	2.88	2.83	2.79	2.76	2.73	2.67	2.63	2.58	2.57	2.52	2.47	2.43	2.36	2.32	2.29	27
2.92	2.89	2.84	2.79	2.76	2.72	2.69	2.64	2.59	2.56	2.53	2.48	2.43	2.39	2.32	2.28	2.25	28
2.88	2.86	2.80	2.76	2.72	2.69	2.66	2.60	2.56	2.52	2.49	2.45	2.39	2.36	2.28	2.24	2.21	29
2.85	2.82	2.77	2.73	2.69	2.66	2.63	2.57	2.52	2.49	2.46	2.42	2.36	2.32	2.25	2.21	2.18	30
2.80	2.77	2.71	2.67	2.63	2.60	2.57	2.51	2.47	2.43	2.40	2.36	2.30	2.26	2.19	2.15	2.11	32
2.75	2.72	2.66	2.62	2.58	2.55	2.52	2.46	2.42	2.38	2.35	2.30	2.25	2.21	2.14	2.09	2.06	34
2.70	2.67	2.62	2.58	2.54	2.50	2.48	2.42	2.37	2.33	2.30	2.26	2.20	2.17	2.09	2.04	2.01	36
2.66	2.63	2.58	2.54	2.50	2.47	2.44	2.38	2.33	2.29	2.27	2.22	2.16	2.12	2.05	2.00	1.97	38
2.63	2.60	2.55	2.50	2.46	2.43	2.40	2.34	2.30	2.28	2.23	2.18	2.12	2.09	2.01	1.96	1.93	40

Degrees of freedom for the denominator (v_2)

* F::

Table A-8 (*continued*)

ν_2	1	2	3	4	5	6	7	8	9	10	11	12	13	14	15	16	17	18
42	8.78	6.03	4.94	4.34	3.95	3.68	3.48	3.32	3.19	3.09	3.00	2.92	2.86	2.80	2.75	2.71	2.67	2.63
44	8.74	5.99	4.91	4.31	3.92	3.65	3.45	3.29	3.16	3.06	2.97	2.89	2.83	2.77	2.72	2.68	2.64	2.60
46	8.70	5.96	4.88	4.28	3.90	3.62	3.42	3.26	3.14	3.03	2.94	2.87	2.80	2.75	2.70	2.65	2.61	2.58
48	8.66	5.93	4.85	4.25	3.87	3.60	3.40	3.24	3.11	3.01	2.92	2.85	2.78	2.72	2.67	2.63	2.59	2.55
50	8.63	5.90	4.83	4.23	3.85	3.58	3.38	3.22	3.09	2.99	2.90	2.82	2.76	2.70	2.65	2.61	2.57	2.53
55	8.55	5.84	4.77	4.18	3.80	3.53	3.33	3.17	3.05	2.94	2.85	2.78	2.71	2.66	2.61	2.56	2.52	2.49
60	8.49	5.80	4.73	4.14	3.76	3.49	3.29	3.12	3.01	2.90	2.82	2.74	2.68	2.62	2.57	2.53	2.49	2.45
65	8.44	5.75	4.68	4.11	3.73	3.45	3.26	3.10	2.98	2.87	2.79	2.71	2.63	2.59	2.54	2.49	2.45	2.42
70	8.40	5.72	4.65	4.08	3.70	3.43	3.23	3.08	2.95	2.85	2.76	2.68	2.62	2.56	2.51	2.47	2.43	2.39
80	8.33	5.67	4.61	4.03	3.65	3.39	3.19	3.03	2.91	2.80	2.72	2.64	2.58	2.52	2.47	2.43	2.39	2.35
90	8.28	5.62	4.57	3.99	3.62	3.35	3.15	3.00	2.87	2.77	2.68	2.61	2.54	2.49	2.44	2.39	2.35	2.32
100	8.24	5.59	4.54	3.96	3.59	3.33	3.13	2.97	2.85	2.74	2.66	2.58	2.52	2.46	2.41	2.37	2.33	2.29
125	8.17	5.53	4.49	3.91	3.54	3.28	3.08	2.93	2.80	2.70	2.61	2.54	2.47	2.42	2.37	2.32	2.28	2.24
150	8.12	5.49	4.43	3.88	3.51	3.25	3.05	2.89	2.77	2.67	2.58	2.51	2.44	2.38	2.33	2.29	2.25	2.21
200	8.06	5.44	4.41	3.84	3.47	3.21	3.01	2.85	2.73	2.63	2.54	2.47	2.40	2.35	2.30	2.26	2.21	2.18
300	8.00	5.39	4.37	3.80	3.43	3.17	2.97	2.81	2.69	2.59	2.51	2.43	2.37	2.31	2.26	2.21	2.17	2.14
500	7.95	5.36	4.33	3.76	3.40	3.14	2.94	2.79	2.66	2.56	2.48	2.40	2.34	2.28	2.23	2.19	2.14	2.11
1000	7.92	5.33	4.31	3.74	3.37	3.11	2.92	2.77	2.64	2.54	2.45	2.38	2.32	2.28	2.21	2.16	2.12	2.09
∞	7.88	5.30	4.28	3.72	3.35	2.09	2.90	2.74	2.62	2.52	2.43	2.36	2.29	2.24	2.19	2.14	2.10	2.08

						Degrees of freedom for the numerator (ν_1)										
19	*20*	*22*	*24*	*26*	*28*	*30*	*35*	*40*	*45*	*50*	*60*	*80*	*100*	*200*	*500*	*∞*
					Multipl. the numbers of the first row ($\nu_2 = 1$) *by 100*											
2.60	2.57	2.52	2.47	2.43	2.40	2.37	2.31	2.26	2.23	2.20	2.15	2.09	2.06	1.98	1.93	1.90
2.57	2.54	2.49	2.44	2.40	2.37	2.34	2.28	2.24	2.20	2.17	2.12	2.06	2.03	1.95	1.90	1.87
2.54	2.51	2.46	2.42	2.38	2.34	2.32	2.26	2.21	2.17	2.14	2.10	2.04	2.00	1.92	1.87	1.84
2.52	2.49	2.44	2.39	2.36	2.32	2.29	2.23	2.19	2.15	2.12	2.07	2.01	1.97	1.89	1.84	1.81
2.50	2.47	2.42	2.37	2.33	2.30	2.27	2.21	2.16	2.13	2.10	2.05	1.99	1.95	1.87	1.82	1.79
2.45	2.42	2.37	2.33	2.29	2.26	2.23	2.16	2.12	2.08	2.05	2.00	1.94	1.90	1.82	1.77	1.73
2.42	2.39	2.33	2.29	2.24	2.22	2.19	2.13	2.08	2.04	2.01	1.98	1.90	1.86	1.78	1.73	1.69
2.39	2.36	2.30	2.26	2.22	2.19	2.16	2.09	2.05	2.01	1.98	1.93	1.87	1.83	1.74	1.69	1.65
2.36	2.33	2.28	2.23	2.19	2.16	2.13	2.07	2.02	1.98	1.93	1.90	1.84	1.80	1.71	1.65	1.62
2.32	2.29	2.23	2.19	2.15	2.11	2.08	2.02	1.97	1.93	1.90	1.85	1.79	1.75	1.66	1.60	1.56
2.28	2.25	2.20	2.15	2.12	2.08	2.05	1.99	1.94	1.90	1.87	1.82	1.75	1.71	1.62	1.56	1.52
2.26	2.23	2.17	2.13	2.09	2.05	2.02	1.96	1.91	1.87	1.84	1.79	1.72	1.68	1.59	1.53	1.49
2.21	2.18	2.13	2.08	2.04	2.01	1.98	1.91	1.86	1.82	1.79	1.74	1.67	1.63	1.53	1.47	1.42
2.18	2.15	2.10	2.05	2.01	1.98	1.94	1.88	1.83	1.79	1.76	1.70	1.63	1.59	1.49	1.42	1.37
2.14	2.11	2.06	2.01	1.97	1.94	1.91	1.84	1.79	1.75	1.71	1.66	1.58	1.54	1.44	1.37	1.31
2.10	2.07	2.02	1.97	1.93	1.90	1.87	1.80	1.75	1.71	1.67	1.61	1.54	1.50	1.39	1.31	1.25
2.07	2.04	1.99	1.94	1.90	1.87	1.84	1.77	1.72	1.67	1.64	1.58	1.51	1.46	1.35	1.26	1.18
2.05	2.02	1.97	1.92	1.88	1.84	1.81	1.75	1.69	1.65	1.61	1.56	1.48	1.43	1.31	1.22	1.12
2.03	2.00	1.95	1.90	1.86	1.82	1.79	1.72	1.67	1.63	1.59	1.53	1.45	1.40	1.28	1.17	1.00

Degrees of freedom for the denominator (ν_2): 42, 44, 46, 48, 50, 55, 60, 65, 70, 80, 90, 100, 125, 150, 200, 300, 500, 1000, ∞

Table A-9   *F* distribution

Tabulation of the values of $F_{0.001; \nu_1; \nu_2}$ versus ν_1 and ν_2*

		1	2	3	4	5	6	7	8	9	10	15	20	30	50	100	200	500	∞
		\multicolumn{18}{c}{*Degrees of freedom for the numerator* (ν_1)}

Degrees of freedom for the numerator (ν_1)

Multiply the numbers of the first row ($\nu_2 = 1$) *by 1000*

ν_2	1	2	3	4	5	6	7	8	9	10	15	20	30	50	100	200	500	∞
1	405	500	540	562	576	586	593	598	602	606	616	621	626	630	633	635	636	637
2	998	999	999	999	999	999	999	999	999	999	999	999	999	999	999	999	999	999
3	168	148	141	137	135	133	132	131	130	129	127	126	125	125	124	124	124	124
4	74.1	61.2	56.2	53.4	51.7	50.5	49.7	49.0	48.5	48.0	46.8	46.1	45.4	44.9	44.5	44.3	44.1	44.0
5	47.0	36.6	33.2	31.1	29.8	28.8	28.2	27.6	27.2	26.9	25.9	25.4	24.9	24.4	24.1	23.9	23.8	23.8
6	35.5	27.0	23.7	21.9	20.8	20.0	19.5	19.0	18.7	18.4	17.6	17.1	16.7	16.3	16.0	15.9	15.8	15.8
7	29.2	21.7	18.8	17.2	16.2	15.5	15.0	14.6	14.3	14.1	13.3	12.9	12.5	12.2	11.9	11.8	11.7	11.7
8	25.4	18.5	15.8	14.4	13.5	12.9	12.4	12.0	11.8	11.5	10.8	10.5	10.1	9.80	9.57	9.46	9.39	9.34
9	22.9	16.4	13.9	12.6	11.7	11.1	10.7	10.4	10.1	9.89	9.24	8.90	8.55	8.26	8.04	7.93	7.86	7.81
10	21.0	14.9	12.6	11.3	10.5	9.92	9.52	9.20	8.96	8.75	8.13	7.80	7.47	7.19	6.98	6.87	6.81	6.76
11	19.7	13.8	11.6	10.4	9.58	9.05	8.66	8.35	8.12	7.92	7.32	7.01	6.68	6.41	6.21	6.10	6.04	6.00
12	18.6	13.0	10.8	9.63	8.89	8.38	8.00	7.71	7.48	7.29	6.71	6.40	6.09	5.83	5.63	5.52	5.46	5.42
13	17.8	12.3	10.2	9.07	8.35	7.86	7.49	7.21	6.98	6.80	6.23	5.93	5.62	5.37	5.17	5.07	5.01	4.97
14	17.1	11.8	9.73	8.62	7.92	7.43	7.08	6.80	6.58	6.40	5.85	5.56	5.25	5.00	4.80	4.70	4.64	4.60
15	16.6	11.3	9.34	8.25	7.57	7.09	6.74	6.47	6.26	6.08	5.53	5.25	4.95	4.70	4.51	4.41	4.35	4.31
16	16.1	11.0	9.00	7.94	7.27	6.81	6.46	6.19	5.98	5.81	5.27	4.99	4.70	4.45	4.26	4.16	4.10	4.06
17	15.7	10.7	8.73	7.68	7.02	6.56	6.22	5.96	5.75	5.58	5.05	4.78	4.48	4.24	4.05	3.95	3.89	3.85
18	15.4	10.4	8.49	7.46	6.81	6.35	6.02	5.76	5.56	5.39	4.87	4.59	4.30	4.06	3.87	3.77	3.71	3.67
19	15.1	10.2	8.28	7.26	6.61	6.18	5.84	5.59	5.39	5.22	4.70	4.43	4.14	3.90	3.71	3.61	3.55	3.51
20	14.8	9.95	8.10	7.10	6.46	6.02	5.69	5.44	5.24	5.08	4.56	4.29	4.01	3.77	3.58	3.48	3.42	3.38
22	14.4	9.61	7.80	6.81	6.19	5.76	5.44	5.19	4.99	4.83	4.32	4.06	3.77	3.53	3.34	3.25	3.19	3.15
24	14.0	9.34	7.55	6.59	5.98	5.55	5.23	4.99	4.80	4.64	4.14	3.87	3.59	3.35	3.16	3.07	3.01	2.97
26	13.7	9.12	7.36	6.41	5.80	5.38	5.07	4.83	4.64	4.48	3.99	3.72	3.45	3.20	3.01	2.92	2.86	2.82
28	13.5	8.93	7.19	6.25	5.66	5.24	4.93	4.69	4.50	4.35	3.86	3.60	3.32	3.08	2.89	2.79	2.73	2.70
30	13.3	8.77	7.05	6.12	5.53	5.12	4.82	4.58	4.39	4.24	3.75	3.49	3.22	2.98	2.79	2.69	2.63	2.59
40	12.6	8.25	6.60	5.70	5.13	4.73	4.43	4.21	4.02	3.87	3.40	3.15	2.87	2.64	2.44	2.34	2.28	2.23
50	12.2	7.95	6.34	5.46	4.90	4.51	4.22	4.00	3.82	3.67	3.20	2.95	2.68	2.44	2.24	2.14	2.07	2.03
60	12.0	7.76	6.17	5.31	4.76	4.37	4.09	3.87	3.69	3.54	3.08	2.83	2.56	2.31	2.11	2.01	1.93	1.89
80	11.7	7.54	5.97	5.13	4.58	4.21	3.92	3.70	3.53	3.39	2.93	2.68	2.40	2.16	1.95	1.84	1.77	1.72
100	11.5	7.41	5.85	5.01	4.48	4.11	3.83	3.61	3.44	3.30	2.84	2.59	2.32	2.07	1.87	1.75	1.68	1.62
200	11.2	7.15	5.64	4.81	4.29	3.92	3.65	3.43	3.26	3.12	2.67	2.42	2.15	1.90	1.68	1.55	1.46	1.39
500	11.0	7.01	5.51	4.69	4.18	3.82	3.54	3.33	3.16	3.02	2.58	2.33	2.05	1.80	1.57	1.43	1.32	1.23
∞	10.8	6.91	5.42	4.62	4.10	3.74	3.47	3.27	3.10	2.96	2.51	2.27	1.99	1.73	1.49	1.34	1.21	1.00

Degrees of freedom for the denominator (ν_2)

* $F_{\alpha; \nu_1; \nu_2}$ and $F_{\alpha; \nu_x; \nu_y}$ have been used interchangeably in the text.

Table A-10　Gamma function

Tabulation of values of $\Gamma(n)$ versus n

n	$\Gamma(n)$	n	$\Gamma(n)$	n	$\Gamma(n)$	n	$\Gamma(n)$
1.00	1.00000	1.25	.90640	1.50	.88623	1.75	.91906
1.01	.99433	1.26	.90440	1.51	.88659	1.76	.92137
1.02	.98884	1.27	.90250	1.52	.88704	1.77	.92376
1.03	.98355	1.28	.90072	1.53	.88757	1.78	.92623
1.04	.97844	1.29	.89904	1.54	.88818	1.79	.92877
1.05	.97350	1.30	.89747	1.55	.88887	1.80	.93138
1.06	.96874	1.31	.89600	1.56	.88964	1.81	.93408
1.07	.96415	1.32	.89464	1.57	.89049	1.82	.93685
1.08	.95973	1.33	.89338	1.58	.89142	1.83	.93969
1.09	.95546	1.34	.89222	1.59	.89243	1.84	.94261
1.10	.95135	1.35	.89115	1.60	.89352	1.85	.94561
1.11	.94739	1.36	.89018	1.61	.89468	1.86	.94869
1.12	.94359	1.37	.88931	1.62	.89592	1.87	.95184
1.13	.93993	1.38	.88854	1.63	.89724	1.88	.95507
1.14	.93642	1.39	.88785	1.64	.89864	1.89	.95838
1.15	.93304	1.40	.88726	1.65	.90012	1.90	.96177
1.16	.92980	1.41	.88676	1.66	.90167	1.91	.96523
1.17	.92670	1.42	.88636	1.67	.90330	1.92	.96878
1.18	.92373	1.43	.88604	1.68	.90500	1.93	.97240
1.19	.92088	1.44	.88580	1.69	.90678	1.94	.97610
1.20	.91817	1.45	.88565	1.70	.90864	1.95	.97988
1.21	.91558	1.46	.88560	1.71	.91057	1.96	.98374
1.22	.91311	1.47	.88563	1.72	.91258	1.97	.98768
1.23	.91075	1.48	.88575	1.73	.91466	1.98	.99171
1.24	.90852	1.49	.88595	1.74	.91683	1.99	.99581
						2.00	1.00000

$$\Gamma(n) = \int_0^\infty e^{-x} x^{n-1}\, dx$$

$$\Gamma(n+1) = n\Gamma(n)$$

$$\Gamma(1) = 1$$

$$\Gamma\left(\frac{1}{2}\right) = \sqrt{\pi}$$

$$\Gamma\left(\frac{n}{2}\right) = \left(\frac{n}{2}-1\right)! = \begin{cases} \left(\frac{n}{2}-1\right)\left(\frac{n}{2}-2\right)\cdots(3)\cdot(2)\cdot(1) & \text{for } n \text{ even and } n > 2 \\ \left(\frac{n}{2}-1\right)\left(\frac{n}{2}-2\right)\cdots\left(\frac{3}{2}\right)\left(\frac{1}{2}\right)\sqrt{\pi} & \text{for } n \text{ odd and } n > 2 \end{cases}$$

Table A-11   Median ranks [5]

j*	1	2	3	4	Sample size n 5	6	7	8	9	10
1	.5000	.2929	.2063	.1591	.1294	.1091	.0943	.0830	.0741	.0670
2		.7071	.5000	.3864	.3147	.2655	.2295	.2021	.1806	.1632
3			.7937	.6136	.5000	.4218	.3648	.3213	.2871	.2594
4				.8409	.6853	.5782	.5000	.4404	.3935	.3557
5					.8706	.7345	.6352	.5596	.5000	.4519
6						.8909	.7705	.6787	.6065	.5481
7							.9057	.7979	.7129	.6443
8								.9170	.8194	.7406
9									.9259	.8368
10										.9330

j*	11	12	13	14	Sample size n 15	16	17	18	19	20
1	.0611	.0561	.0519	.0483	.0452	.0424	.0400	.0378	.0358	.0341
2	.1489	.1368	.1266	.1178	.1101	.1034	.0975	.0922	.0874	.0831
3	.2366	.2175	.2013	.1873	.1751	.1644	.1550	.1465	.1390	.1322
4	.3244	.2982	.2760	.2568	.2401	.2254	.2125	.2009	.1905	.1812
5	.4122	.3789	.3506	.3263	.3051	.2865	.2700	.2553	.2421	.2302
6	.5000	.4596	.4253	.3958	.3700	.3475	.3275	.3097	.2937	.2793
7	.5878	.5404	.5000	.4653	.4350	.4085	.3850	.3641	.3453	.3283
8	.6756	.6211	.5747	.5347	.5000	.4695	.4425	.4184	.3968	.3774
9	.7634	.7018	.6494	.6042	.5650	.5305	.5000	.4728	.4484	.4264
10	.8511	.7825	.7240	.6737	.6300	.5915	.5575	.5272	.5000	.4755
11	.9389	.8632	.7987	.7432	.6949	.6525	.6150	.5816	.5516	.5245
12		.9439	8734	.8127	.7599	.7135	.6725	.6359	.6032	.5736
13			.9481	.8822	.8249	.7746	.7300	.6903	.6547	.6226
14				.9517	.8899	.8356	.7875	.7447	.7063	.6717
15					.9548	.8966	.8450	.7991	.7579	.7207
16						.9576	.9025	.8535	.8095	.7698
17							.9600	.9078	.8610	.8188
18								.9622	.9126	.8678
19									.9642	.9169
20										.9659

* Order number

					Sample size n					
j*	21	22	23	24	25	26	27	28	29	30
1	.0330	.0315	.0301	.0288	.0277	.0266	.0256	.0247	.0239	.0231
2	.0797	.0761	.0728	.0698	.0670	.0645	.0621	.0599	.0579	.0559
3	.1264	.1207	.1155	.1108	.1064	.1023	.0986	.0951	.0919	.0888
4	.1731	.1653	.1582	.1517	.1457	.1402	.1351	.1303	.1259	.1217
5	.2198	.2099	.2009	.1927	.1851	.1781	.1716	.1655	.1599	.1546
6	.2665	.2545	.2437	.2337	.2245	.2159	.2081	.2007	.1939	.1875
7	.3132	.2992	.2864	.2746	.2638	.2538	.2445	.2359	.2279	.2004
8	.3599	.3438	.3291	.3156	.3032	.2917	.2810	.2711	.2619	.2533
9	.4066	.3884	.3718	.3566	.3425	.3295	.3175	.3063	.2959	.2862
10	.4533	.4330	.4145	.3975	.3819	.3674	.3540	.3415	.3299	.3191
11	.5000	.4776	.4572	.4385	.4212	.4053	.3905	.3767	.3639	.3519
12	.5466	.5223	.5000	.4795	.4606	.4431	.4270	.4119	.3979	.3848
13	.5933	.5669	.5427	.5204	.5000	.4810	.4635	.4471	.4319	.4177
14	.6400	.6115	.5854	.5614	.5393	.5189	.5000	.4823	.4659	.4506
15	.6867	.6561	.6281	.6024	.5787	.5568	.5364	.5176	.5000	.4835
16	.7334	.7007	.6708	.6433	.6180	.5946	.5729	.5528	5340	.5164
17	.7801	.7454	.7135	.6843	.6574	.6325	.6094	.5880	,680	.5493
18	.8268	.7900	.7562	.7253	.6967	.6704	.6459	.6232	6020	.5822
19	.8735	.8346	.7990	.7662	.7361	.7082	.6824	.6584	.6360	.6151
20	.9202	.8792	.8417	.8072	.7754	.7461	.7189	.6936	.6700	.6480
21	.9669	.9238	.8844	.8482	.8148	.7840	.7554	.7288	.7040	.6808
22		.9684	.9271	.8891	.8542	.8218	.7918	.7640	.7380	7137
23			.9698	.9301	.8935	.8597	.8283	.7992	.7720	7465
24				.9711	.9329	.8976	.8648	.8344	.8060	7795
25					.9722	.9354	.9013	.8696	.8400	8124
26						.9733	.9378	.9048	.8740	8453
27							.9743	.9400	.9080	.8782
28								.9752	.9420	.9111
29									.9760	.9440
30										.9768

					Sample size n					
j*	31	32	33	34	35	36	37	38	39	40
1	.0223	.0216	.0210	.0203	.0198	.0192	.0187	.0182	.0177	.0173
2	.0542	.0525	.0509	.0494	.0480	.0467	.0454	.0442	.0431	.0420
3	.0860	.0833	.0808	.0785	.0763	.0741	.0722	.0703	.0685	.0668
4	.1178	.1142	.1108	.1075	.1043	.1016	.0989	.0963	.0939	.0915
5	.1497	.1451	.1407	.1366	.1327	.1291	.1256	.1274	.1192	.1163
6	.1815	.1759	.1706	.1657	.1610	.1566	.1524	.1484	.1446	.1410
7	.2134	.2068	.2006	.1947	.1892	.1840	.1791	.1744	.1700	.1658
8	.2452	.2376	.2305	.2238	.2175	.2115	.2058	.2005	.1954	.1905
9	.2771	.2685	.2605	.2529	.2457	.2390	.2326	.2265	.2206	.2153
10	.3089	.2994	.2904	.2819	.2740	.2664	.2593	.2526	.2461	.2401
11	.3407	.3302	.3203	.3110	.3022	.2939	.2861	.2786	.2715	.2648
12	.3726	.3611	.3503	.3401	.3305	.3214	.3128	.3046	.2969	.2896
13	.4044	.3919	.3802	.3691	.3587	.3489	.3395	.3307	.3223	.3143
14	.4363	.4228	.4101	.3982	.3870	.3763	.3663	.3567	.3477	.3391
15	.4681	.4537	.4401	.4273	.4152	.4038	.3930	.3828	.3730	.3638
16	.5000	.4845	.4700	.4563	.4435	.4313	.4197	.4088	.3984	.3886
17	.5318	.5154	.5000	.4854	.4717	.4587	.4465	.4348	.4238	.4133
18	.5636	.5462	.5299	.5145	.5000	.4862	.4732	.4609	.4492	.4381
19	.5955	.5771	.5598	.5436	.5282	.5137	.5000	.4869	.4746	.4628
20	.6273	.6080	.5898	.5726	.5564	.5412	.5267	.5130	.5000	.4876
21	.6592	.6388	.6197	.6017	.5847	.5686	.5534	.5390	.5253	.5123
22	.6910	.6697	.6496	.6308	.6129	.5961	.5802	.5651	.5507	.5371
23	.7228	.7005	.6796	.6598	.6412	.6236	.6069	.5911	.5761	.5618
24	.7547	.7314	.7095	.6889	.6694	.6510	.6336	.6171	.6015	.5866
25	.7865	.7623	.7394	.7180	.6977	.6785	.6604	.6432	.6269	.6113
26	.8184	.7931	.7694	.7470	.7259	.7060	.6871	.6692	.6522	.6361
27	.8502	.8240	.7993	.7761	.7542	.7335	.7138	.6953	.6776	.6608
28	.8821	.8548	.8293	.8052	.7824	.7609	.7406	.7213	.7030	.6856
29	.9139	.8857	.8592	.8342	.8107	.7884	.7673	.7473	.7284	.7103
30	.9457	.9166	.8891	.8633	.8389	.8159	.7941	.7734	.7538	.7351
31	.9776	.9474	.9191	.8924	.8672	.8433	.8208	.7994	.7791	.7599
32		.9783	.9490	.9214	.8954	.8708	.8475	.8255	.8045	.7846
33			.9789	.9505	.9237	.8983	.8743	.8515	.8299	.8094
34				.9796	.9519	.9258	.9010	.8775	.8553	.8341
35					.9801	.9532	.9277	.9036	.8807	.8589
36						.9807	.9545	.9296	.9060	.8836
37							.9812	.9557	.9314	.9084
38								.9817	.9568	.9331
39									.9822	.9579
40										.9826
41										
42										
43										
44										
45										
46										
47										
48										
49										
50										

* Order number.

				Sample	size n				
41	42	43	44	45	46	47	48	49	50
.0169	.0165	.0161	.0157	.0154	.0150	.0147	.0144	.0141	.0138
.0410	.0400	.0391	.0382	.0374	.0366	.0358	.0351	.0343	.0337
.0652	.0636	.0622	.0608	.0594	.0581	.0569	.0557	.0546	.0535
.0893	.0872	.0852	.0833	.0814	.0797	.0780	.0764	.0748	.0733
.1135	.1108	.1082	.1058	.1035	.1012	.0991	.0970	.0951	.0932
.1376	.1344	.1313	.1283	.1255	.1228	.1202	.1177	.1153	.1130
.1618	.1580	.1543	.1508	.1475	.1443	.1413	.1384	.1356	.1329
.1859	.1816	.1774	.1734	.1695	.1659	.1654	.1590	.1558	.1527
.2101	.2051	.2004	.1959	.1916	.1874	.1835	.1797	.1760	.1726
.2342	.2287	.2234	.2184	.2136	.2090	.2046	.2004	.1963	.1924
.2584	.2523	.2465	.2409	.2356	.2305	.2257	.2210	.2165	.2122
.2826	.2759	.2695	.2635	.2577	.2521	.2463	.2417	.2368	.2321
.3067	.2995	.2926	.2860	.2797	.2736	.2679	.2623	.2570	.2519
.3309	.3231	.3156	.3085	.3017	.2952	.2890	.2830	.2773	.2718
.3550	.3466	.3387	.3310	.3237	.3168	.3101	.3037	.2975	.2916
.3792	.3702	.3617	.3536	.3458	.3383	.3312	.3243	.3178	.3114
.4033	.3938	.3847	.3761	.3678	.3599	.3523	.3450	.3380	.3313
.4275	.4174	.4078	.3986	.3898	.3814	.3734	.3656	.3582	.3511
.4516	.4410	.4308	.4211	.4118	.4030	.3945	.3863	.3785	.3710
.4758	.4646	.4539	.4436	.4339	.4245	.4156	.4070	.3987	.3908
.5000	.4842	.4769	.4662	.4559	.4461	.4367	.4276	.4190	.4107
.5241	.5117	.5000	.4887	.4779	.4676	.4578	.4483	.4392	.4305
.5483	.5353	.5230	.5112	.5000	.4893	.4789	.4690	.4595	.4503
.5724	.5589	.5460	.5337	.5220	.5107	.5000	.4896	.4797	.4702
.5966	.5825	.5691	.5563	.5440	.5323	.5210	.5103	.5000	.4900
.6207	.6061	.5921	.5788	.5660	.5538	.5421	.5309	.5202	.5099
.6449	.6297	.6152	.6013	.5881	.5754	.5632	.5516	.5404	.5297
.6690	.6533	.6382	.6238	.6101	.5969	.5843	.5723	.5607	.5496
.6932	.6768	.6612	.6464	.6321	.6185	.6054	.5929	.5809	.5694
.7173	.7004	.6843	.6689	.6541	.6400	.6265	.6136	.6012	.5892
.7415	.7240	.7073	.6914	.6762	.6616	.6476	.6343	.6214	.6091
.7657	.7476	.7304	.7139	.6982	.6831	.6687	.6549	.6417	.6289
.7898	.7712	.7534	.7364	.7202	.7047	.6898	.6756	.6619	.6488
.8140	.7948	.7765	.7590	.7422	.7263	.7109	.6962	.6821	.6686
.8381	.8184	.7995	.7815	.7643	.7478	.7320	.7169	.7024	.6885
.8623	.8419	.8315	.8040	.7863	.7693	.7531	.7376	.7226	.7083
.8864	.8655	.8456	.8265	.8083	.7909	.7742	.7582	.7429	.7281
.9106	.8891	.8686	.8491	.8304	.8125	.7953	.7789	.7631	.7480
.9347	.9127	.8917	.8716	.8524	.8340	.8164	.7916	.7834	.7678
.9589	.9363	.9147	.8941	.8744	.8556	.8375	.8202	.8036	.7877
.9830	.9599	.9377	.9166	.8964	.8771	.8586	.8409	.8239	.8075
	.9834	.9608	.9392	.9185	.8987	.8797	.8615	.8441	.8273
		.9838	.9617	.9405	.9202	.9008	.8822	.8643	.8472
			.9842	.9625	.9418	.9219	.9029	.8846	.8670
				.9845	.9633	.9430	.9235	.9048	.8869
					.9849	.9641	.9442	.9251	.9067
						.9852	.9648	.9433	.9266
							.9855	.9656	.9464
								.9858	.9662
									.9861

Table A-13   Median ranks

j*	Sample size n									
	51	52	53	54	55	56	57	58	59	60
1	.0135	.0133	.0130	.0128	.0126	.0123	.0121	.0119	.0117	.0115
2	.0330	.0324	.0318	.0312	.0306	.0301	.0295	.0290	.0285	.0281
3	.0525	.0515	.0505	.0496	.0487	.0478	.0470	.0462	.0454	.0446
4	.0719	.0705	.0692	.0679	.0667	.0655	.0644	.0633	.0622	.0612
5	.0914	.0896	.0879	.0863	.0848	.0833	.0818	.0804	.0790	.0777
6	.1108	.1087	.1067	.1047	.1028	.1010	.0992	.0975	.0959	.0943
7	.1303	.1278	.1254	.1231	.1208	.1187	.1166	.1146	.1127	.1108
8	.1497	.1469	.1441	.1415	.1389	.1365	.1341	.1318	.1296	.1274
9	.1692	.1660	.1629	.1599	.1570	.1542	.1515	.1489	.1464	.1440
10	.1886	.1850	.1816	.1782	.1750	.1719	.1689	.1660	.1632	.1605
11	.2081	.2041	.2003	.1966	.1931	.1896	.1863	.1831	.1801	.1771
12	.2276	.2232	.2190	.2150	.2111	.2074	.2038	.2003	.1969	.1936
13	.2470	.2423	.2378	.2334	.2292	.2251	.2212	.2174	.2137	.2102
14	.2665	.2614	.2565	.2518	.2472	.2428	.2386	.2345	.2306	.2268
15	.2859	.2805	.2752	.2702	.2653	.2606	.2560	.2516	.2474	.2433
16	.3054	.2996	.2939	.2885	.2833	.2783	.2735	.2688	.2642	.2599
17	.3248	.3186	.3127	.3069	.3014	.2960	.2909	.2859	.2811	.2764
18	.3443	.3377	.3314	.3253	.3194	.3138	.3083	.3030	.2979	.2930
19	.3638	.3568	.3501	.3437	.3375	.3315	.3257	.3201	.3148	.3095
20	.3832	.3759	.3689	.3621	.3555	.3492	.3431	.3373	.3316	.3261
21	.4027	.3950	.3876	.3805	.3736	.3670	.3606	.3544	.3484	.3427
22	.4221	.4141	.4063	.3988	.3916	.3847	.3780	.3715	.3653	.3592
23	.4416	.4332	.4250	.4172	.4097	.4024	.3954	.3886	.3821	.3758
24	.4610	.4522	.4438	.4356	.4277	.4202	.4128	.4058	.3989	.3923
25	.4805	.4713	.4625	.4540	.4458	.4379	.4303	.4229	.4158	.4089
26	.5000	.4904	.4812	.4724	.4638	.4556	.4477	.4400	.4326	.4254
27	.5194	.5095	.5000	.4908	.4819	.4734	.4651	.4571	.4494	.4420
28	.5389	.5286	.5187	.5091	.5000	.4911	.4825	.4743	.4663	.4589
29	.5583	.5477	.5374	.5275	.5180	.5088	.5000	.4914	.4831	.4751
30	.5778	.5667	.5561	.5459	.5361	.5265	.5174	.5085	.5000	.4917
31	.5972	.5858	.5749	.5643	.5541	.5443	.5348	.5256	.5168	.5082
32	.6167	.6049	.5936	.5827	.5722	.5620	.5522	.5428	.5336	.5248
33	.6361	.6240	.6123	.6011	.5902	.5797	.5696	.5599	.5505	.5413
34	.6556	.6431	.6310	.6194	.6083	.5975	.5871	.5770	.5673	.5579
35	.6751	.6622	.6498	.6378	.6263	.6152	.6045	.5941	.5841	.5746
36	.6945	.6813	.6685	.6562	.6444	.6329	.6219	.6113	.6010	.5910
37	.7140	.7003	.6872	.6746	.6624	.6507	.6393	.6284	.6178	.6076
38	.7334	.7194	.7060	.6930	.6805	.6684	.6568	.6455	.6346	.6241
39	.7529	.7385	.7247	.7114	.6985	.6861	.6742	.6626	.6515	.6407
40	.7723	.7576	.7434	.7297	.7166	.7039	.6915	.6798	.6683	.6572
41	.7918	.7767	.7621	.7481	.7346	.7216	.7090	.6969	.6851	.6738
42	.8113	.7958	.7809	.7665	.7527	.7393	.7264	.7140	.7020	.6904
43	.8307	.8149	.7996	.7849	.7707	.7571	.7439	.7311	.7188	.7069
44	.8502	.8339	.8183	.8033	.7888	.7748	.7613	.7483	.7357	.7235

* Order number.

					Sample size n					
j*	51	52	53	54	55	56	57	58	59	60
45	.8696	.8530	.8371	.8217	.8068	.7925	.7787	.7654	.7525	.7400
46	.8891	.8721	.8558	.8400	.8249	.8103	.7961	.7825	.7693	.7566
47	.9085	.8912	.8745	.8584	.8429	.8280	.8136	.7996	.7862	.7731
48	.9280	.9103	.8932	.8768	.8610	.8457	.8310	.8168	.8030	.7897
49	.9474	.9294	.9120	.8952	.8790	.8635	.8484	.8339	.8198	.8063
50	.9669	.9485	.9307	.9136	.8971	.8812	.8658	.8510	.8367	.8228
51	.9864	.9675	.9494	.9320	.9151	.8989	.8833	.8681	.8535	.8394
52		.9866	.9681	.9503	.9332	.9166	.9007	.8853	.8703	.8559
53			.9869	.9687	.9512	.9344	.9181	.9024	.8872	.8725
54				.9871	.9693	.9521	.9355	.9195	.9040	.8891
55					.9873	.9698	.9529	.9366	.9209	.9056
56						.9876	.9704	.9538	.9377	.9222
57							.9878	.9709	.9545	.9387
58								.9880	.9714	.9553
59									9882	.9718
60										.9884

Table of

j*	Sample size n									
	1	2	3	4	5	6	7	8	9	10
1	.0500	.0253	.0170	.0127	.0102	.0085	.0074	.0065	.0057	.0051
2		.2236	.1354	.0976	.0764	.0629	.0534	.0468	.0410	.0368
3			.3684	.2486	.1893	.1532	.1287	.1111	.0978	.0873
4				.4729	.3426	.2713	.2253	.1929	.1688	.1500
5					.5493	.4182	.3413	.2892	.2514	.2224
6						.6070	.4793	.4003	.3449	.3035
7							.6518	.5293	.4504	.3934
8								.6877	.5709	.4931
9									.7169	.6058
10										.7411

Table of

j*	Sample size n									
	1	2	3	4	5	6	7	8	9	10
1	.9500	.7764	.6316	.5271	.4507	.3930	.3482	.3123	.2831	.2589
2		.9747	.8646	.7514	.6574	.5818	.5207	.4707	.4291	.3942
3			.9830	.9024	.8107	.7287	.6587	.5997	.5496	.5069
4				.9873	.9236	.8468	.7747	.7108	.6551	.6076
5					.9898	.9371	.8713	.8071	.7436	.6965
6						.9915	.9466	.8889	.8312	.7776
7							.9926	.9532	.9032	.8500
8								.9935	.9590	.9127
9									.9943	.9632
10										.9949

* Order number.

5% ranks

j*	11	12	13	14	15	16	17	18	19	20
					Sample size n					
1	.0047	.0043	.0040	.0037	.0034	.0032	.0030	.0029	.0028	.0026
2	.0333	.0307	.0281	.0263	.0245	.0227	.0216	.0205	.0194	.0183
3	.0800	.0719	.0665	.0611	.0574	.0536	.0499	.0476	.0452	.0429
4	.1363	.1245	.1127	.1047	.0967	.0910	.0854	.0797	.0761	.0725
5	.2007	.1824	.1671	.1527	.1424	.1321	.1247	.1173	.1099	.1051
6	.2713	.2465	.2255	.2082	.1909	.1786	.1664	.1575	.1485	.1396
7	.3498	.3152	.2883	.2652	.2459	.2267	.2128	.1990	.1887	.1785
8	.4356	.3909	.3548	.3263	.3016	.2805	.2601	.2449	.2298	.2183
9	.5299	.4727	.4274	.3904	.3608	.3350	.3131	.2912	.2749	.2587
10	.6356	.5619	.5054	.4600	.4226	.3922	.3542	.3429	.3201	.3029
11	.7616	.6613	.5899	.5343	.4893	.4517	.4208	.3937	.3703	.3469
12		.7791	.6837	.6146	.5602	.5156	.4781	.4460	.4196	.3957
13			.7942	.7033	.6366	.5834	.5395	.5022	.4711	.4434
14				.8074	.7206	.6562	.6044	.5611	.5242	.4932
15					.8190	.7360	.6738	.6233	.5809	.5444
16						.8274	.7475	.6871	.6379	.5964
17							.8358	.7589	.7005	.6525
18								.8441	.7704	.7138
19									.8525	.7818
20										.8609

95% ranks

j*	11	12	13	14	15	16	17	18	19	20
					Sample size n					
1	.2384	.2209	.2058	.1926	.1810	.1726	.1642	.1559	.1475	.1391
2	.3644	.3387	.3163	.2967	.2794	.2640	.2525	.2411	.2296	.2182
3	.4701	.4381	.4101	.3854	.3634	.3438	.3262	.3129	.2995	.2862
4	.5644	.5273	.4946	.4657	.4398	.4166	.3956	.3767	.3621	.3475
5	.6502	.6091	.5726	.5400	.5107	.4844	.4605	.4389	.4191	.4036
6	.7287	.6848	.6452	.6096	.5774	.5483	.5219	.4978	.4758	.4556
7	.7993	.7535	.7117	.6737	.6392	.6078	.5792	.5540	.5289	.5068
8	.8637	.8176	.7745	.7348	.6984	.6650	.6458	.6063	.5804	.5666
9	.9200	.8755	.8329	.7918	.7541	.7195	.6869	.6571	.6297	.6043
10	.9667	.9281	.8873	.8473	.8091	.7733	.7399	.7088	.6799	.6531
11	.9953	.9693	.9335	.8953	.8576	.8214	.7872	.7551	.7251	.6971
12		.9957	.9719	.9389	.9033	.8679	.8336	.8010	.7702	.7413
13			.9960	.9737	.9426	.9090	.8753	.8425	.8113	.7817
14				.9963	.9755	.9464	.9146	.8827	.8525	.8215
15					.9966	.9773	.9501	.9203	.8901	.8604
16						.9968	.9784	.9534	.9239	.8949
17							.9970	.9795	.9548	.9275
18								.9971	.9806	.9571
19									.9972	.9817
20										.9974

Table A-15 2.5% and 97.5% ranks

j*	1	2	3	4	Sample size n 5	6	7	8	9	10
1	.025	.0126	.0084	.0063	.0050	.0042	.0036	.0032	.0028	.0025
2		.1581	.9043	.0676	.0527	.0433	.0367	.0318	.0281	.0252
3			.2924	.1941	.1466	.1181	.0990	.0852	.0748	.0667
4				.3976	.2836	.2228	.1840	.1570	.1370	.1215
5					.4782	.3588	.2904	.2449	.2120	.1871
6						.5407	.4213	.3491	.2993	.2624
7							.5904	.4735	.3999	.3475
8								.6306	.5175	.4439
9									.6637	.5550
10										.6915

j*	1	2	3	4	Sample size n 5	6	7	8	9	10
1	.9750	.8419	.7076	.6024	.5218	.4593	.4096	.3694	.3363	.3085
2		.9874	.9057	.8059	.7164	.6412	.5787	.5265	.4825	.4450
3			.9916	.9324	.8534	.7772	.7096	.6509	.6001	.5561
4				.9937	.9473	.8819	.8160	.7551	.7007	.6525
5					.9950	.9567	.9010	.8430	.7880	.7376
6						.9958	.9633	.9148	.8630	.8129
7							.9964	.9682	.9252	.8785
8								.9968	.9619	.9333
9									.9972	.9748
10										9975

* Order number

2.5% ranks

j*	|	11	12	13	14	Sample size n 15	16	17	18	19	20
1		.0023	.0021	.0019	.0018	.0017	.0016	.0015	.0014	.0013	.0013
2		.0228	.0209	.0192	.0178	.0166	.0155	.0146	.0137	.0130	.0123
3		.0602	.0549	.0504	.0466	.0432	.0405	.0380	.0358	.0338	.0321
4		.1325	.0992	.0909	.0839	.0779	.0727	.0681	.0640	.0605	.0572
5		.1675	.1516	.1386	.1276	.1182	.1102	.1031	.0969	.0915	.0866
6		.2338	.2109	.1922	.1766	.1634	.1520	.1421	.1334	.1258	.1189
7		.3079	.2767	.2513	.2304	.2127	.1975	.1844	.1730	.1629	.1539
8		.3903	.3489	.3158	.2886	.2659	.2465	.2298	.2153	.2025	.1912
9		.4822	.4281	.3857	.3513	.3229	.2988	.2781	.2602	.2445	.2316
10		.5872	.5159	.4619	.4190	.3838	.3543	.3291	.3076	.2886	.2720
11		.7151	.6151	.5455	.4920	.4490	.4134	.3832	.3574	.3350	.3153
12			.7353	.6397	.5719	.5191	.4762	.4404	.4099	.3836	.3605
13				.7529	.6613	.5954	.5435	.5010	.4652	.4345	.4078
14					.7684	.6820	.6165	.5657	.5236	.4880	.4572
15						.7820	.6977	.6356	.5858	.5443	.5089
16							.7941	.7131	.6529	.6042	.5634
17								.8049	.7271	.6686	.6211
18									.8147	.7397	.6829
19										.8235	.7513
20											.8316

97.5% ranks

j*	|	11	12	13	14	Sample size n 15	16	17	18	19	20
1		.2849	.2647	.2471	.2316	.2180	.2059	.1951	.1853	.1765	.1684
2		.4128	.3848	.3603	.3387	.3195	.3023	.2869	.2729	.2603	.2487
3		.5178	.4841	.4545	.4281	.4046	.3835	.3644	.3471	.3314	.3170
4		.6097	.5719	.5381	.5080	.4809	.4565	.4343	.4142	.3958	.3789
5		.6921	.6511	.6143	.5810	.5510	.5238	.4990	.4764	.4557	.4366
6		.7662	.7233	.6842	.6486	.6162	.5866	.5596	.5348	.5120	.4911
7		.8325	.7891	.7487	.7114	.6771	.6457	.6167	.5901	.5655	.5428
8		.8907	.8484	.8078	.7696	.7341	.7012	.6708	.6426	.6164	.5922
9		.9389	.9008	.8614	.8234	.7873	.7535	.7219	.6924	.6650	.6395
10		.9772	.9451	.9091	.8724	.8366	.8025	.7702	.7398	.7114	.6847
11		.9977	.9791	.9496	.9161	.8818	.8480	.8156	.7847	.7555	.7280
12			.9979	.9808	.9534	.9221	.8898	.8579	.8270	.7975	.7694
13				.9981	.9822	.9567	.9273	.8969	.8666	.8371	.8088
14					.9982	.9834	.9505	.9319	.9031	.8742	.8461
15						.9983	.9845	.9620	.9360	.9085	.8811
16							.9984	.9854	.9642	.9395	.9134
17								.9985	.9863	.9662	.9427
18									.9986	.9870	.9679
19										.9987	.9877
20											.9987

Table of

					Sample size n					
j*	1	2	3	4	5	6	7	8	9	10
1	.0050	.0025	.0017	.0013	.0010	.0008	.0007	.0006	.0006	.0005
2		.0707	.0414	.0294	.0229	.0187	.0158	.0139	.0123	.0110
3			.1710	.1109	.0828	.0663	.0553	.0475	.0419	.0374
4				.2659	.1851	.1436	.1177	.0999	.0868	.0777
5					.3466	.2540	.2030	.1698	.1461	.1281
6						.4135	.3151	.2578	.2191	.1909
7							.4691	.3685	.3074	.2649
8								.5157	.4150	.3518
9									.5551	.4557
10										.5887

Table of

					Sample size n					
j*	1	2	3	4	5	6	7	8	9	10
1	.9950	.9293	.8290	.7341	.6534	.5865	.5309	.4843	.4449	.4113
2		.9975	.9586	.8891	.8149	.7460	.6849	.6315	.5850	.5443
3			.9983	.9706	.9172	.8564	.7970	.7422	.6926	.6482
4				.9987	.9771	.9337	.8823	.8302	.7809	.7351
5					.9990	.9813	.9447	.9001	.8539	.8091
6						.9992	.9842	.9525	.9132	.8717
7							.9993	.9861	.9581	.9223
8								.9994	.9877	.9626
9									.9994	.9890
10										.9995

* Order number.

0.5% ranks

j*	11	12	13	14	15	16	17	18	19	20
					Sample size n					
1	.0005	.0004	.0004	.0004	.0003	.0003	.0003	.0003	.0003	.0002
2	.0098	.0090	.0083	.0077	.0072	.0067	.0063	.0059	.0056	.0053
3	.0338	.0303	.0280	.0257	.0241	.0225	.0209	.0198	.0189	.0280
4	.0695	.0633	.0571	.0529	.0488	.0458	.0429	.0400	.0381	.0364
5	.1152	.1042	.0954	.0866	.0806	.0745	.0702	.0659	.0617	.0509
6	.1692	.1531	.1393	.1282	.1170	.1092	.1014	.0958	.0802	.0846
7	.2332	.2085	.1895	.1736	.1603	.1471	.1377	.1284	.1215	.1147
8	.3067	.2725	.2454	.2290	.2064	.1914	.1764	.1657	.1549	.1469
9	.3915	.3448	.3087	.2799	.2572	.2376	.2211	.2047	.1926	.1807
10	.4914	.4271	.3794	.3421	.3118	.2998	.2670	.2493	.2316	.2185
11	.6178	.5230	.4590	.4108	.3727	.3415	.3170	.2946	.2759	.2572
12		.6431	.5510	.4877	.4395	.4009	.3690	.3431	.3205	.3010
13			.6653	.5760	.5137	.4656	.4268	.3945	.3680	.3448
14				.6849	.5984	.5372	.4896	.4508	.4182	.3913
15					.7024	.6186	.5587	.5116	.4729	.4412
16						.7181	.6340	.5754	.5290	.4807
17							.7322	.6494	.5921	.5465
18								.7450	.6649	.6089
19									.7566	.6803
20										.7673

99.5% ranks

j*	11	12	13	14	15	16	17	18	19	20
					Sample size n					
1	.3822	.3569	.3347	.3151	.2976	.2819	.2678	.2550	.2434	.2327
2	.5086	.4770	.4490	.4240	.4016	.3814	.3660	.3506	.3351	.3197
3	.6085	.5729	.5410	.5123	.4863	.4628	.4413	.4246	.4079	.3911
4	.6933	.6552	.6206	.5892	.5605	.5344	.5104	.4884	.4710	.4535
5	.7668	.7275	.6913	.6579	.6273	.5991	.5732	.5492	.5271	.5193
6	.8307	.7915	.7546	.7201	.6882	.6585	.6310	.6055	.5818	.5598
7	.8848	.8469	.8104	.7710	.7428	.7002	.6836	.6569	.6320	.6087
8	.9305	.8958	.8607	.8264	.7936	.7624	.7330	.7054	.6795	.6552
9	.9661	.9367	.9046	.8718	.8397	.8086	.7789	.7507	.7241	.6990
10	.9902	.9697	.9429	.9134	.8830	.8529	.8236	.7953	.7684	.7428
11	.9995	.9910	.9720	.9471	.9194	.8908	.8623	.8343	.8074	.7815
12		.9996	.9917	.9743	.9512	.9255	.8986	.8716	.8451	.8193
13			.9996	.9923	.9759	.9542	.9298	.9042	.8785	.8531
14				9996	.9928	.9775	.9571	.9341	.9198	.8853
15					.9997	.9933	9791	.9600	.9383	.9154
16						.9997	.9937	.9801	.9618	.9491
17							.9997	.9941	.9811	.9636
18								.9997	.9944	.9820
19									.9997	.9947
20										.9998

Table A-17 Position of the Weibull mean

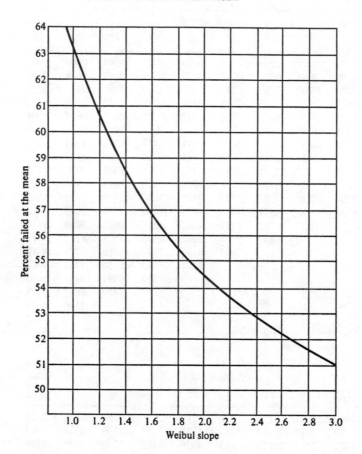

Table A-18   Weibull slope error—90% confidence interval

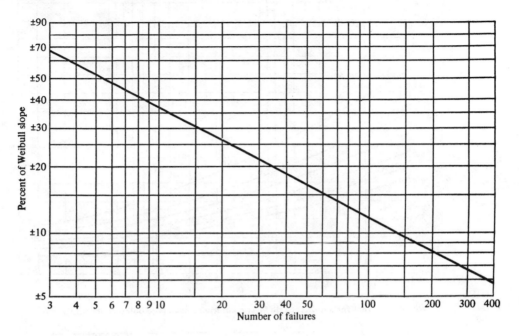

Table A-19   Weibull slope error—50% confidence interval

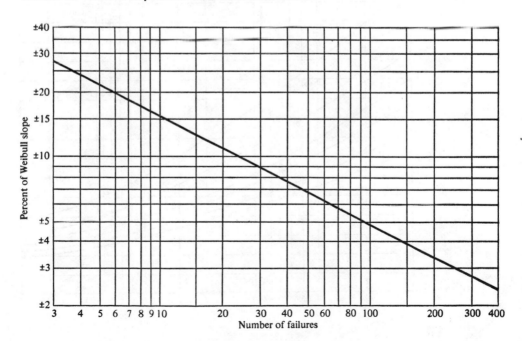

Table A-20 Test for significant difference in mean lives (Weibull distribution)

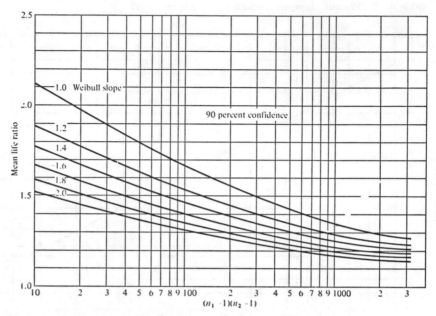

Table A-21 Test for significant difference in mean lives (Weibull distribution)

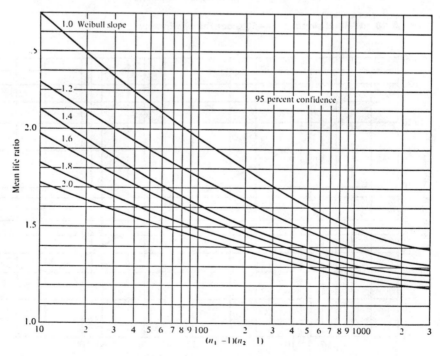

Table A-22   Test for significant difference in mean lives (Weibull distribution)

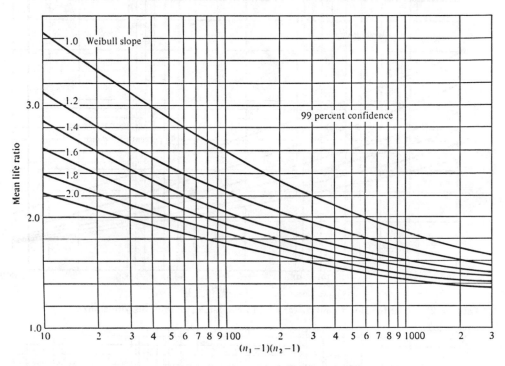

Table A-23   Test for significant difference in B_{10} lives (Weibull distribution)

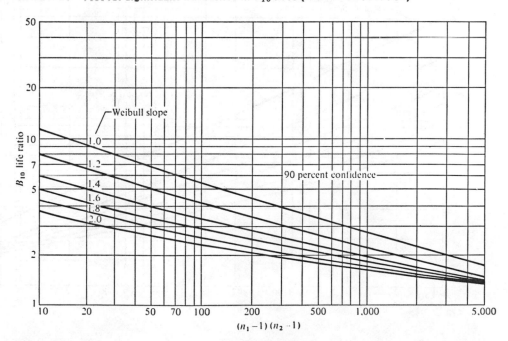

Table A-24   Test for significant difference in B_{10} lives (Weibull distribution)

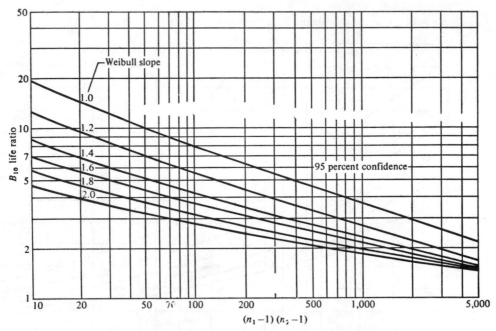

Table A-25   Test for significant difference in B_{10} lives (Weibull distribution)

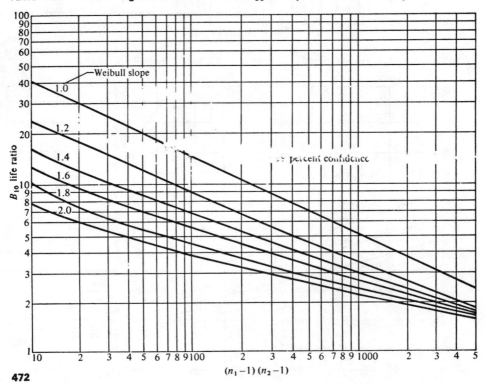

Table A-26 Values of assurance that the true mean value to failure exceeds the desired mean value to failure

Reduced sample size—normal distribution
(N = statistical sample size; k = number of failures)

$N=1$		$N=2$			$N=3$			
$\dfrac{W_0-W_D*}{\sigma}$	k 0	$\dfrac{W_0-W_D*}{\sigma}$	k		$\dfrac{W_0-W_D*}{\sigma}$	k		
			0	1		0	1	2
−4.0	0.000	−4.0	0.000	0.000	−4.0	0.000	0.000	0.000
−3.6	0.000	−3.6	0.000	0.000	−3.6	0.000	0.000	0.000
−3.2	0.001	−3.2	0.001	0.000	−3.2	0.002	0.000	0.000
−2.8	0.003	−2.8	0.005	0.000	−2.8	0.008	0.000	0.000
−2.4	0.008	−2.4	0.016	0.000	−2.4	0.024	0.000	0.000
−2.0	0.023	−2.0	0.045	0.001	−2.0	0.067	0.002	0.000
−1.8	0.036	−1.8	0.071	0.001	−1.8	0.104	0.004	0.000
−1.6	0.055	−1.6	0.107	0.003	−1.6	0.156	0.009	0.000
−1.4	0.081	−1.4	0.155	0.007	−1.4	0.223	0.019	0.001
−1.2	0.115	−1.2	0.217	0.013	−1.2	0.307	0.037	0.002
−1.0	0.159	−1.0	0.292	0.025	−1.0	0.405	0.068	0.004
−0.9	0.184	−0.9	0.334	0.034	−0.9	0.457	0.089	0.006
−0.8	0.212	−0.8	0.379	0.045	−0.8	0.511	0.116	0.010
−0.7	0.242	−0.7	0.425	0.059	−0.7	0.564	0.147	0.014
−0.6	0.274	−0.6	0.473	0.075	−0.6	0.618	0.184	0.021
−0.5	0.308	−0.5	0.522	0.095	−0.5	0.669	0.227	0.029
−0.4	0.345	−0.4	0.570	0.119	−0.4	0.718	0.274	0.041
−0.3	0.382	−0.3	0.618	0.146	−0.3	0.764	0.326	0.056
−0.2	0.421	−0.2	0.664	0.177	−0.2	0.806	0.382	0.074
−0.1	0.460	−0.1	0.709	0.212	−0.1	0.843	0.440	0.097
0.0	0.500	0.0	0.750	0.250	0.0	0.875	0.500	0.125
0.1	0.540	0.1	0.788	0.291	0.1	0.903	0.560	0.157
0.2	0.579	0.2	0.823	0.336	0.2	0.926	0.618	0.194
0.3	0.618	0.3	0.854	0.382	0.3	0.944	0.674	0.236
0.4	0.655	0.4	0.881	0.430	0.4	0.959	0.726	0.282
0.5	0.692	0.5	0.905	0.478	0.5	0.971	0.773	0.331
0.6	0.726	0.6	0.925	0.527	0.6	0.979	0.816	0.382
0.7	0.758	0.7	0.941	0.575	0.7	0.986	0.853	0.436
0.8	0.788	0.8	0.955	0.621	0.8	0.990	0.884	0.489
0.9	0.816	0.9	0.966	0.666	0.9	0.994	0.911	0.543
1.0	0.841	1.0	0.975	0.708	1.0	0.996	0.932	0.595
1.2	0.885	1.2	0.987	0.783	1.2	0.998	0.963	0.693
1.4	0.919	1.4	0.993	0.845	1.4	0.999	0.981	0.777
1.6	0.945	1.6	0.997	0.893	1.6	1.000	0.991	0.844
1.8	0.964	1.8	0.999	0.929	1.8	1.000	0.996	0.896
2.0	0.977	2.0	0.999	0.955	2.0	1.000	0.998	0.933
2.4	0.992	2.4	1.000	0.984	2.4	1.000	1.000	0.976
2.8	0.997	2.8	1.000	0.995	2.8	1.000	1.000	0.992
3.2	0.999	3.2	1.000	0.999	3.2	1.000	1.000	0.998
3.6	1.000	3.6	1.000	1.000	3.6	1.000	1.000	1.000
4.0	1.000	4.0	1.000	1.000	4.0	1.000	1.000	1.000

$$* \ \frac{W_0-W_D}{\sigma} = \frac{W_0/W_D - 1}{\sigma/W_D}.$$

Table A-27 Values of assurance that the true mean value to failure exceeds the desired mean value to failure

Reduced sample size—normal distribution
($N=$ statistical sample size; $k=$ number of failures)

$\dfrac{W_0 - W_D{}^*}{\sigma}$	k				$\dfrac{W_0 - W_D{}^*}{\sigma}$	k				
N=4	0	1	2	3	**N=5**	0	1	2	3	4
−4.0	0.000	0.000	0.000	0.000	−4.0	0.000	0.000	0.000	0.000	0.000
−3.6	0.001	0.000	0.000	0.000	−3.6	0.001	0.000	0.000	0.000	0.000
−3.2	0.003	0.000	0.000	0.000	−3.2	0.003	0.000	0.000	0.000	0.000
−2.8	0.010	0.000	0.000	0.000	−2.8	0.013	0.000	0.000	0.000	0.000
−2.4	0.032	0.000	0.000	0.000	−2.4	0.040	0.001	0.000	0.000	0.000
−2.0	0.088	0.003	0.000	0.000	−2.0	0.109	0.005	0.000	0.000	0.000
−1.8	0.136	0.007	0.000	0.000	−1.8	0.167	0.012	0.000	0.000	0.000
−1.6	0.202	0.017	0.001	0.000	−1.6	0.246	0.027	0.002	0.000	0.000
−1.4	0.286	0.035	0.002	0.000	−1.4	0.344	0.055	0.005	0.000	0.000
−1.2	0.387	0.068	0.006	0.000	−1.2	0.457	0.105	0.013	0.001	0.000
−1.0	0.499	0.121	0.014	0.001	−1.0	0.579	0.181	0.031	0.003	0.000
−0.9	0.557	0.157	0.022	0.001	−0.9	0.638	0.231	0.046	0.005	0.000
−0.8	0.614	0.199	0.032	0.002	−0.8	0.696	0.287	0.067	0.008	0.000
−0.7	0.670	0.248	0.046	0.003	−0.7	0.750	0.350	0.095	0.014	0.001
−0.6	0.723	0.303	0.066	0.006	−0.6	0.799	0.418	0.131	0.022	0.002
−0.5	0.771	0.363	0.090	0.009	−0.5	0.842	0.489	0.175	0.034	0.003
−0.4	0.815	0.427	0.121	0.014	−0.4	0.879	0.561	0.227	0.051	0.005
−0.3	0.854	0.494	0.159	0.021	−0.3	0.910	0.631	0.287	0.074	0.008
−0.2	0.887	0.560	0.204	0.031	−0.2	0.935	0.698	0.354	0.104	0.013
−0.1	0.915	0.626	0.255	0.045	−0.1	0.954	0.759	0.426	0.142	0.021
0.0	0.938	0.688	0.313	0.063	0.0	0.969	0.813	0.500	0.188	0.031
0.1	0.955	0.745	0.374	0.085	0.1	0.979	0.858	0.574	0.241	0.046
0.2	0.969	0.796	0.440	0.113	0.2	0.987	0.896	0.646	0.302	0.065
0.3	0.979	0.841	0.506	0.146	0.3	0.992	0.926	0.713	0.369	0.090
0.4	0.986	0.879	0.573	0.185	0.4	0.995	0.949	0.773	0.439	0.121
0.5	0.991	0.910	0.637	0.229	0.5	0.997	0.966	0.825	0.511	0.158
0.6	0.994	0.934	0.697	0.277	0.6	0.998	0.978	0.869	0.582	0.201
0.7	0.997	0.954	0.752	0.330	0.7	0.999	0.986	0.905	0.650	0.250
0.8	0.998	0.968	0.801	0.386	0.8	1.000	0.992	0.933	0.713	0.304
0.9	0.999	0.978	0.843	0.443	0.9	1.000	0.995	0.954	0.769	0.362
1.0	0.999	0.986	0.879	0.501	1.0	1.000	0.997	0.979	0.819	0.421
1.2	1.000	0.994	0.932	0.613	1.2	1.000	0.999	0.987	0.895	0.543
1.4	1.000	0.998	0.965	0.714	1.4	1.000	1.000	0.995	0.945	0.656
1.6	1.000	0.999	0.983	0.798	1.6	1.000	1.000	0.998	0.973	0.754
1.8	1.000	1.000	0.993	0.864	1.8	1.000	1.000	1.000	0.988	0.833
2.0	1.000	1.000	0.997	0.912	2.0	1.000	1.000	1.000	0.995	0.891
2.4	1.000	1.000	1.000	0.968	2.4	1.000	1.000	1.000	0.999	0.960
2.8	1.000	1.000	1.000	0.990	2.8	1.000	1.000	1.000	1.000	0.987
3.2	1.000	1.000	1.000	0.997	3.2	1.000	1.000	1.000	1.000	0.997
3.6	1.000	1.000	1.000	0.999	3.6	1.000	1.000	1.000	1.000	0.999
4.0	1.000	1.000	1.000	1.000	4.0	1.000	1.000	1.000	1.000	1.000

$$^* \; \frac{W_0 - W_D}{\sigma} = \frac{W_0/W_D - 1}{\sigma/W_D}.$$

Table A-28 Values of assurance that the true mean value to failure exceeds the desired mean value to failure

Reduced sample size—normal distribution
(N = statistical sample size; k = number of failures)

$\dfrac{W_0 - W_D*}{\sigma}$	$N=6$					
	k					
	0	1	2	3	4	5
−4.0	0.000	0.000	0.000	0.000	0.000	0.000
−3.6	0.001	0.000	0.000	0.000	0.000	0.000
−3.2	0.004	0.000	0.000	0.000	0.000	0.000
−2.8	0.015	0.000	0.000	0.000	0.000	0.000
−2.4	0.048	0.001	0.000	0.000	0.000	0.000
−2.0	0.129	0.007	0.000	0.000	0.000	0.000
−1.8	0.197	0.018	0.001	0.000	0.000	0.000
−1.6	0.287	0.039	0.003	0.000	0.000	0.000
−1.4	0.397	0.079	0.009	0.001	0.000	0.000
−1.2	0.520	0.145	0.023	0.002	0.000	0.000
−1.0	0.645	0.244	0.055	0.007	0.001	0.000
−0.9	0.705	0.306	0.080	0.013	0.001	0.000
−0.8	0.760	0.374	0.114	0.021	0.002	0.000
−0.7	0.810	0.447	0.157	0.034	0.004	0.000
−0.6	0.854	0.523	0.210	0.052	0.007	0.000
−0.5	0.891	0.598	0.272	0.077	0.012	0.001
−0.4	0.921	0.671	0.342	0.112	0.021	0.002
−0.3	0.944	0.738	0.419	0.155	0.033	0.003
−0.2	0.962	0.798	0.499	0.209	0.051	0.006
−0.1	0.975	0.849	0.579	0.272	0.076	0.009
0.0	0.984	0.891	0.656	0.344	0.109	0.016
0.1	0.991	0.924	0.728	0.421	0.151	0.025
0.2	0.994	0.949	0.791	0.501	0.202	0.038
0.3	0.997	0.967	0.845	0.581	0.262	0.056
0.4	0.998	0.979	0.888	0.658	0.329	0.079
0.5	0.999	0.988	0.923	0.728	0.402	0.109
0.6	1.000	0.993	0.948	0.790	0.477	0.146
0.7	1.000	0.996	0.966	0.843	0.553	0.190
0.8	1.000	0.998	0.979	0.886	0.626	0.240
0.9	1.000	0.999	0.987	0.920	0.694	0.295
1.0	1.000	0.999	0.993	0.945	0.756	0.355
1.2	1.000	1.000	0.998	0.977	0.855	0.480
1.4	1.000	1.000	0.999	0.991	0.921	0.603
1.6	1.000	1.000	1.000	0.997	0.961	0.713
1.8	1.000	1.000	1.000	0.999	0.982	0.803
2.0	1.000	1.000	1.000	1.000	0.993	0.871
2.4	1.000	1.000	1.000	1.000	0.999	0.952
2.8	1.000	1.000	1.000	1.000	1.000	0.985
3.2	1.000	1.000	1.000	1.000	1.000	0.996
3.6	1.000	1.000	..000	1.000	1.000	0.999
4.0	1.000	1.000	1.000	1.000	1.000	1.000

$$* \ \frac{W_0 - W_D}{\sigma} = \frac{W_0/W_D - 1}{\sigma/W_D}$$

Table A-29 Values of assurance that the true mean value to failure exceeds the desired mean value to failure

Reduced sample size—normal distribution
(N = statistical sample size; k = number of failures)

$\dfrac{W_0 - W_D{}^*}{\sigma}$	k						
	0	1	2	3	4	5	6
−4.0	0.000	0.000	0.000	0.000	0.000	0.000	0.000
−3.6	0.001	0.000	0.000	0.000	0.000	0.000	0.000
−3.2	0.005	0.000	0.000	0.000	0.000	0.000	0.000
−2.8	0.018	0.000	0.000	0.000	0.000	0.000	0.000
−2.4	0.056	0.001	0.000	0.000	0.000	0.000	0.000
−2.0	0.149	0.010	0.000	0.000	0.000	0.000	0.000
−1.8	0.226	0.024	0.001	0.000	0.000	0.000	0.000
−1.6	0.326	0.052	0.005	0.000	0.000	0.000	0.000
−1.4	0.446	0.104	0.014	0.001	0.000	0.000	0.000
−1.2	0.575	0.188	0.037	0.005	0.000	0.000	0.000
−1.0	0.702	0.308	0.085	0.015	0.002	0.000	0.000
−0.9	0.759	0.379	0.122	0.025	0.003	0.000	0.000
−0.8	0.811	0.456	0.169	0.041	0.006	0.001	0.000
−0.7	0.856	0.535	0.227	0.063	0.011	0.001	0.000
−0.6	0.894	0.614	0.296	0.095	0.019	0.002	0.000
−0.5	0.924	0.688	0.372	0.137	0.032	0.004	0.000
−0.4	0.948	0.757	0.455	0.191	0.052	0.008	0.001
−0.3	0.966	0.817	0.541	0.256	0.080	0.015	0.001
−0.2	0.978	0.867	0.624	0.331	0.118	0.025	0.002
−0.1	0.987	0.907	0.703	0.413	0.167	0.040	0.004
0.0	0.992	0.938	0.773	0.500	0.227	0.063	0.008
0.1	0.996	0.960	0.833	0.587	0.297	0.093	0.013
0.2	0.998	0.975	0.882	0.669	0.376	0.133	0.022
0.3	0.999	0.985	0.920	0.744	0.459	0.183	0.034
0.4	0.999	0.992	0.948	0.809	0.545	0.243	0.052
0.5	1.000	0.996	0.968	0.863	0.623	0.312	0.076
0.6	1.000	0.998	0.981	0.905	0.704	0.386	0.106
0.7	1.000	0.999	0.989	0.937	0.773	0.465	0.144
0.8	1.000	0.999	0.994	0.959	0.831	0.544	0.189
0.9	1.000	1.000	0.997	0.975	0.878	0.621	0.241
1.0	1.000	1.000	0.998	0.985	0.915	0.692	0.298
1.2	1.000	1.000	1.000	0.995	0.963	0.812	0.425
1.4	1.000	1.000	1.000	0.999	0.986	0.896	0.554
1.6	1.000	1.000	1.000	1.000	0.995	0.948	0.674
1.8	1.000	1.000	1.000	1.000	0.999	0.976	0.774
2.0	1.000	1.000	1.000	1.000	1.000	0.990	0.851
2.4	1.000	1.000	1.000	1.000	1.000	0.999	0.944
2.8	1.000	1.000	1.000	1.000	1.000	1.000	0.982
3.2	1.000	1.000	1.000	1.000	1.000	1.000	0.995
3.6	1.000	1.000	1.000	1.000	1.000	1.000	0.999
4.0	1.000	1.000	1.000	1.000	1.000	1.000	1.000

$N = 7$

$$* \ \frac{W_0 - W_D}{\sigma} = \frac{W_0/W_D - 1}{\sigma/W_D}.$$

Table A-30 Values of assurance that the true mean value to failure exceeds the desired mean value to failure

Reduced sample size—normal distribution
(N = statistical sample size; k = number of failures)

$\dfrac{W_0 - W_D{}^*}{\sigma}$	$N = 8$							
	k							
	0	1	2	3	4	5	6	7
−4.0	0.000	0.000	0.000	0.000	0.000	0.000	0.000	0.000
−3.6	0.001	0.000	0.000	0.000	0.000	0.000	0.000	0.000
−3.2	0.005	0.000	0.000	0.000	0.000	0.000	0.000	0.000
−2.8	0.020	0.000	0.000	0.000	0.000	0.000	0.000	0.000
−2.4	0.064	0.002	0.000	0.000	0.000	0.000	0.000	0.000
−2.0	0.168	0.013	0.001	0.000	0.000	0.000	0.000	0.000
−1.8	0.254	0.031	0.002	0.000	0.000	0.000	0.000	0.000
−1.6	0.363	0.067	0.007	0.001	0.000	0.000	0.000	0.000
−1.4	0.490	0.132	0.022	0.002	0.000	0.000	0.000	0.000
−1.2	0.624	0.233	0.055	0.008	0.001	0.000	0.000	0.000
−1.0	0.749	0.370	0.120	0.026	0.004	0.000	0.000	0.000
−0.9	0.804	0.449	0.169	0.043	0.007	0.001	0.000	0.000
−0.8	0.851	0.531	0.230	0.068	0.013	0.002	0.000	0.000
−0.7	0.891	0.613	0.302	0.103	0.024	0.004	0.000	0.000
−0.6	0.923	0.690	0.383	0.150	0.040	0.007	0.001	0.000
−0.5	0.948	0.761	0.470	0.210	0.065	0.013	0.002	0.000
−0.4	0.966	0.823	0.559	0.282	0.100	0.023	0.003	0.000
−0.3	0.979	0.874	0.646	0.365	0.147	0.040	0.006	0.000
−0.2	0.987	0.914	0.726	0.454	0.207	0.064	0.012	0.001
−0.1	0.993	0.944	0.797	0.547	0.280	0.098	0.021	0.002
0.0	0.996	0.965	0.855	0.637	0.363	0.145	0.035	0.004
0.1	0.998	0.979	0.902	0.720	0.453	0.203	0.056	0.007
0.2	0.999	0.988	0.936	0.793	0.546	0.274	0.086	0.013
0.3	1.000	0.994	0.960	0.853	0.635	0.354	0.126	0.021
0.4	1.000	0.997	0.977	0.900	0.718	0.441	0.177	0.034
0.5	1.000	0.998	0.987	0.935	0.790	0.530	0.239	0.052
0.6	1.000	0.999	0.993	0.960	0.850	0.617	0.310	0.077
0.7	1.000	1.000	0.996	0.976	0.897	0.698	0.387	0.109
0.8	1.000	1.000	0.998	0.987	0.932	0.770	0.469	0.149
0.9	1.000	1.000	0.999	0.993	0.957	0.831	0.551	0.196
1.0	1.000	1.000	1.000	0.996	0.974	0.880	0.630	0.251
1.2	1.000	1.000	1.000	0.999	0.992	0.945	0.767	0.376
1.4	1.000	1.000	1.000	1.000	0.998	0.978	0.868	0.510
1.6	1.000	1.000	1.000	1.000	0.999	0.993	0.933	0.637
1.8	1.000	1.000	1.000	1.000	1.000	0.998	0.969	0.746
2.0	1.000	1.000	1.000	1.000	1.000	0.999	0.987	0.832
2.4	1.000	1.000	1.000	1.000	1.000	1.000	0.998	0.936
2.8	1.000	1.000	1.000	1.000	1.000	1.000	1.000	0.980
3.2	1.000	1.000	1.000	1.000	1.000	1.000	1.000	0.995
3.6	1.000	1.000	1.000	1.000	1.000	1.000	1.000	0.999
4.0	1.000	1.000	1.000	1.000	1.000	1.000	1.000	1.000

$$* \quad \frac{W_0 - W_D}{\sigma} = \frac{W_0/W_D - 1}{\sigma/W_D}.$$

Table A-31 Values of assurance that the true mean value to failure exceeds the desired mean value to failure

Reduced sample size—normal distribution
(N = statistical sample size; k = number of failures)

$\dfrac{W_0 - W_D{}^*}{\sigma}$	k								
	0	1	2	3	4	5	6	7	8
−4.0	0.000	0.000	0.000	0.000	0.000	0.000	0.000	0.000	0.000
−3.6	0.001	0.000	0.000	0.000	0.000	0.000	0.000	0.000	0.000
−3.2	0.006	0.000	0.000	0.000	0.000	0.000	0.000	0.000	0.000
−2.8	0.023	0.000	0.000	0.000	0.000	0.000	0.000	0.000	0.000
−2.4	0.071	0.002	0.000	0.000	0.000	0.000	0.000	0.000	0.000
−2.0	0.187	0.017	0.001	0.000	0.000	0.000	0.000	0.000	0.000
−1.8	0.280	0.039	0.003	0.000	0.000	0.000	0.000	0.000	0.000
−1.6	0.398	0.084	0.011	0.001	0.000	0.000	0.000	0.000	0.000
−1.4	0.532	0.161	0.031	0.004	0.000	0.000	0.000	0.000	0.000
−1.2	0.667	0.278	0.075	0.014	0.002	0.000	0.000	0.000	0.000
−1.0	0.789	0.430	0.160	0.041	0.007	0.001	0.000	0.000	0.000
−0.9	0.840	0.514	0.221	0.066	0.014	0.002	0.000	0.000	0.000
−0.8	0.883	0.599	0.294	0.102	0.025	0.004	0.000	0.000	0.000
−0.7	0.917	0.680	0.377	0.151	0.043	0.008	0.001	0.000	0.000
−0.6	0.944	0.754	0.467	0.214	0.070	0.016	0.002	0.000	0.000
−0.5	0.964	0.819	0.560	0.290	0.110	0.029	0.005	0.001	0.000
−0.4	0.978	0.872	0.650	0.378	0.163	0.050	0.010	0.001	0.000
−0.3	0.987	0.914	0.733	0.472	0.230	0.081	0.019	0.003	0.000
−0.2	0.993	0.945	0.805	0.569	0.311	0.124	0.034	0.006	0.000
−0.1	0.996	0.966	0.864	0.662	0.403	0.182	0.057	0.011	0.001
0.0	0.998	0.980	0.910	0.746	0.500	0.254	0.090	0.020	0.002
0.1	0.999	0.989	0.943	0.818	0.597	0.338	0.136	0.034	0.004
0.2	1.000	0.994	0.966	0.876	0.689	0.431	0.195	0.055	0.007
0.3	1.000	0.997	0.981	0.919	0.770	0.528	0.267	0.086	0.013
0.4	1.000	0.999	0.990	0.950	0.837	0.622	0.350	0.128	0.022
0.5	1.000	0.999	0.995	0.971	0.890	0.710	0.440	0.181	0.036
0.6	1.000	1.000	0.998	0.984	0.930	0.786	0.533	0.246	0.056
0.7	1.000	1.000	0.999	0.992	0.957	0.849	0.623	0.320	0.083
0.8	1.000	1.000	1.000	0.996	0.975	0.898	0.706	0.401	0.117
0.9	1.000	1.000	1.000	0.998	0.986	0.934	0.779	0.486	0.160
1.0	1.000	1.000	1.000	0.999	0.993	0.959	0.840	0.570	0.211
1.2	1.000	1.000	1.000	1.000	0.998	0.986	0.925	0.722	0.333
1.4	1.000	1.000	1.000	1.000	1.000	0.996	0.969	0.839	0.468
1.6	1.000	1.000	1.000	1.000	1.000	0.999	0.989	0.916	0.602
1.8	1.000	1.000	1.000	1.000	1.000	1.000	0.997	0.961	0.720
2.0	1.000	1.000	1.000	1.000	1.000	1.000	0.999	0.983	0.813
2.4	1.000	1.000	1.000	1.000	1.000	1.000	1.000	0.998	0.929
2.8	1.000	1.000	1.000	1.000	1.000	1.000	1.000	1.000	0.977
3.2	1.000	1.000	1.000	1.000	1.000	1.000	1.000	1.000	0.894
3.6	1.000	1.000	1.000	1.000	1.000	1.000	1.000	1.000	0.999
4.0	1.000	1.000	1.000	1.000	1.000	1.000	1.000	1.000	1.000

$N = 9$

* $\dfrac{W_0 - W_D}{\sigma} = \dfrac{W_0/W_D - 1}{\sigma/W_D}.$

Table A-32 Values of assurance that the true mean value to failure exceeds the desired mean value to failure

Reduced sample size—normal distribution
(N = statistical sample size; k = number of failures)

					$N=10$					
$\dfrac{W_0 - W_D*}{\sigma}$						k				
	0	1	2	3	4	5	6	7	8	9
−4.0	0.000	0.000	0.000	0.000	0.000	0.000	0.000	0.000	0.000	0.000
−3.6	0.002	0.000	0.000	0.000	0.000	0.000	0.000	0.000	0.000	0.000
−3.2	0.007	0.000	0.000	0.000	0.000	0.000	0.000	0.000	0.000	0.000
−2.8	0.025	0.000	0.000	0.000	0.000	0.000	0.000	0.000	0.000	0.000
−2.4	0.079	0.003	0.000	0.000	0.000	0.000	0.000	0.000	0.000	0.000
−2.0	0.206	0.021	0.001	0.000	0.000	0.000	0.000	0.000	0.000	0.000
−1.8	0.306	0.048	0.005	0.000	0.000	0.000	0.000	0.000	0.000	0.000
−1.6	0.431	0.101	0.015	0.001	0.000	0.000	0.000	0.000	0.000	0.000
−1.4	0.569	0.191	0.041	0.006	0.001	0.000	0.000	0.000	0.000	0.000
−1.2	0.706	0.323	0.099	0.021	0.003	0.000	0.000	0.000	0.000	0.000
−1.0	0.822	0.487	0.203	0.060	0.013	0.002	0.000	0.000	0.000	0.000
−0.9	0.869	0.574	0.275	0.095	0.023	0.004	0.001	0.000	0.000	0.000
−0.8	0.908	0.659	0.358	0.143	0.041	0.009	0.001	0.000	0.000	0.000
−0.7	0.937	0.737	0.450	0.206	0.069	0.017	0.003	0.000	0.000	0.000
−0.6	0.959	0.806	0.546	0.283	0.110	0 031	0.006	0.001	0.000	0.000
−0.5	0.975	0.863	0.640	0.373	0.165	0.054	0.013	0.002	0.000	0.000
−0.4	0.985	0.908	0.727	0.471	0.237	0.089	0.024	0.004	0.000	0.000
−0.3	0.992	0.942	0.802	0.572	0.323	0.138	0.043	0.009	0.001	0.000
−0.2	0.996	0.965	0.864	0.668	0.420	0.203	0.072	0.017	0.003	0.000
−0.1	0.998	0.980	0.911	0.755	0.522	0.284	0.114	0.032	0.005	0.000
0.0	0.999	0.989	0.945	0.828	0.623	0.377	0.172	0.055	0.011	0.001
0.1	1.000	0.995	0.968	0.886	0.716	0.478	0.245	0.089	0.020	0.002
0.2	1.000	0.997	0.983	0.928	0.797	0.580	0.332	0.136	0.035	0.004
0.3	1.000	0.995	0.991	0.957	0.862	0.677	0.428	0.198	0.058	0.008
0.4	1.000	1.000	0.996	0.976	0.911	0.763	0.529	0.273	0.092	0.015
0.5	1.000	1.000	0.998	0.987	0.946	0.835	0.627	0.360	0.137	0.025
0.6	1.000	1.000	0.999	0.994	0.969	0.890	0.717	0.454	0 194	0.041
0.7	1.000	1.000	1.000	0.997	0.983	0.931	0.794	0.550	0.263	0.063
0.8	1.000	1.000	1.000	0.999	0.991	0.959	0.857	0.642	0.341	0.092
0.9	1.000	1.000	1.000	0.999	0.996	0.977	0.905	0.725	0.426	0.131
1.0	1.000	1.000	1.000	1.000	0.998	0.987	0.940	0.797	0.513	0.178
1.2	1.000	1.000	1.000	1.000	1.000	0.997	0.979	0.901	0.677	0.294
1.4	1.000	1.000	1.000	1.000	1.000	0.999	0.994	0.959	0.809	0.431
1.6	1.000	1.000	1.000	1.000	1.000	1.000	0.999	0.985	0.899	0.569
1.8	1.000	1.000	1.000	1.000	1.000	1.000	1.000	0.995	0.952	0.694
2.0	1.000	1.000	1.000	1.000	1.000	1.000	1.000	0.999	0.979	0.794
2.4	1.000	1.000	1.000	1.000	1.000	1.000	1.000	1.000	0.997	0.921
2.8	1.000	1.000	1.000	1.000	1.000	1.000	1.000	1.000	1.000	0.975
3.2	1.000	1.000	1.000	1.000	1.000	1.000	1.000	1.000	1.000	0.993
3.6	1.000	1.000	1.000	1.000	1.000	1.000	1.000	1.000	1.000	0.998
4.0	1.000	1.000	1.000	1.000	1.000	1.000	1.000	1.000	1.000	1.000

$$* \quad \frac{W_0 - W_D}{\sigma} = \frac{W_0/W_D - 1}{\sigma/W_D}.$$

Table A-33 Values of assurance that the true mean (or characteristic) time to failure exceeds the desired mean (or characteristic) time to failure

Reduced sample size—exponential and Weibull distributions
(N = statistical sample size; k = number of failures)

$N = 1$		$N = 2$			$N = 3$			
T_r^*	k	T_r^*	k		T_r^*	k		
	0		0	1		0	1	2
0.1	0.095	0.1	0.181	0.009	0.1	0.259	0.025	0.001
0.2	0.181	0.2	0.330	0.033	0.2	0.451	0.087	0.006
0.3	0.259	0.3	0.451	0.067	0.3	0.593	0.167	0.017
0.4	0.330	0.4	0.551	0.109	0.4	0.699	0.254	0.036
0.5	0.393	0.5	0.632	0.155	0.5	0.777	0.343	0.061
0.6	0.451	0.6	0.699	0.204	0.6	0.835	0.427	0.092
0.7	0.503	0.7	0.753	0.253	0.7	0.878	0.505	0.128
0.8	0.551	0.8	0.798	0.303	0.8	0.909	0.576	0.167
0.9	0.593	0.9	0.835	0.352	0.9	0.933	0.639	0.209
1.0	0.632	1.0	0.865	0.400	1.0	0.950	0.694	0.253
1.1	0.667	1.1	0.889	0.445	1.1	0.963	0.741	0.297
1.2	0.699	1.2	0.909	0.488	1.2	0.973	0.782	0.341
1.3	0.727	1.3	0.926	0.529	1.3	0.980	0.818	0.385
1.4	0.753	1.4	0.939	0.568	1.4	0.985	0.848	0.428
1.5	0.777	1.5	0.950	0.604	1.5	0.989	0.873	0.469
1.6	0.798	1.6	0.959	0.637	1.6	0.992	0.894	0.508
1.7	0.817	1.7	0.967	0.668	1.7	0.994	0.912	0.546
1.8	0.835	1.8	0.973	0.697	1.8	0.995	0.927	0.582
1.9	0.850	1.9	0.978	0.723	1.9	0.997	0.940	0.615
2.0	0.865	2.0	0.982	0.748	2.0	0.998	0.950	0.646
2.2	0.889	2.2	0.988	0.791	2.2	0.999	0.966	0.703
2.4	0.909	2.4	0.992	0.827	2.4	0.999	0.977	0.752
2.6	0.926	2.6	0.994	0.857	2.6	1.000	0.984	0.793
2.8	0.939	2.8	0.996	0.882	2.8	1.000	0.989	0.828
3.0	0.950	3.0	0.998	0.903	3.0	1.000	0.993	0.858
3.2	0.959	3.2	0.998	0.920	3.2	1.000	0.995	0.883
3.4	0.967	3.4	0.999	0.934	3.4	1.000	0.997	0.903
3.6	0.973	3.6	0.999	0.946	3.6	1.000	0.998	0.920
3.8	0.978	3.8	0.999	0.956	3.8	1.000	0.999	0.934
4.0	0.982	4.0	1.000	0.964	4.0	1.000	0.999	0.946
4.5	0.989	4.5	1.000	0.978	4.5	1.000	1.000	0.967
5.0	0.993	5.0	1.000	0.987	5.0	1.000	1.000	0.980
5.5	0.996	5.5	1.000	0.992	5.5	1.000	1.000	0.988
6.0	0.998	6.0	1.000	0.995	6.0	1.000	1.000	0.993
6.5	0.998	6.5	1.000	0.997	6.5	1.000	1.000	0.995
7.0	0.999	7.0	1.000	0.998	7.0	1.000	1.000	0.997
7.5	0.999	7.5	1.000	0.999	7.5	1.000	1.000	0.998
8.0	1.000	8.0	1.000	0.999	8.0	1.000	1.000	0.999
8.5	1.000	8.5	1.000	1.000	8.5	1.000	1.000	0.999
9.0	1.000	9.0	1.000	1.000	9.0	1.000	1.000	1.000
9.5	1.000	9.5	1.000	1.000	9.5	1.000	1.000	1.000
10.0	1.000	10.0	1.000	1.000	10.0	1.000	1.000	1.000

* $T_r = \dfrac{T_0}{T_D}$ for exponential distribution; $T_r = \left(\dfrac{T_0}{\theta}\right)^b$ for Weibull distribution.

Table A-34 Values of assurance that the true mean (or characteristic) time to failure exceeds the desired mean (or characteristic) time to failure

Reduced sample size—exponential and Weibull distributions
($N =$ statistical sample size; $k =$ number of failures)

T_r^*	N=4 k=0	1	2	3	T_r^*	N=5 k=0	1	2	3	4
0.1	0.330	0.048	0.003	0.000	0.1	0.393	0.075	0.007	0.000	0.000
0.2	0.551	0.153	0.021	0.001	0.2	0.632	0.225	0.045	0.005	0.000
0.3	0.699	0.277	0.056	0.005	0.3	0.777	0.387	0.113	0.018	0.001
0.4	0.798	0.401	0.108	0.012	0.4	0.865	0.532	0.204	0.043	0.004
0.5	0.865	0.513	0.172	0.024	0.5	0.918	0.652	0.306	0.082	0.009
0.6	0.909	0.611	0.243	0.041	0.6	0.950	0.746	0.409	0.132	0.019
0.7	0.939	0.693	0.318	0.064	0.7	0.970	0.817	0.506	0.192	0.032
0.8	0.959	0.759	0.392	0.092	0.8	0.982	0.869	0.594	0.257	0.051
0.9	0.973	0.813	0.464	0.124	0.9	0.989	0.908	0.671	0.326	0.074
1.0	0.982	0.856	0.531	0.160	1.0	0.993	0.935	0.736	0.395	0.101
1.1	0.988	0.889	0.593	0.198	1.1	0.996	0.955	0.791	0.462	0.132
1.2	0.992	0.915	0.650	0.238	1.2	0.998	0.969	0.835	0.526	0.167
1.3	0.994	0.936	0.700	0.280	1.3	0.998	0.978	0.871	0.585	0.204
1.4	0.996	0.951	0.744	0.322	1.4	0.999	0.985	0.900	0.640	0.243
1.5	0.998	0.963	0.783	0.364	1.5	0.999	0.990	0.923	0.689	0.283
1.6	0.998	0.972	0.816	0.406	1.6	1.000	0.993	0.941	0.733	0.324
1.7	0.999	0.979	0.845	0.446	1.7	1.000	0.995	0.955	0.772	0.365
1.8	0.999	0.984	0.870	0.485	1.8	1.000	0.997	0.965	0.806	0.405
1.9	0.999	0.988	0.891	0.523	1.9	1.000	0.998	0.974	0.836	0.445
2.0	1.000	0.991	0.909	0.559	2.0	1.000	0.999	0.980	0.862	0.483
2.2	1.000	0.995	0.937	0.625	2.2	1.000	0.999	0.989	0.902	0.556
2.4	1.000	0.997	0.956	0.684	2.4	1.000	1.000	0.994	0.932	0.622
2.6	1.000	0.998	0.970	0.734	2.6	1.000	1.000	0.996	0.953	0.680
2.8	1.000	0.999	0.980	0.778	2.8	1.000	1.000	0.998	0.967	0.731
3.0	1.000	1.000	0.986	0.815	3.0	1.000	1.000	0.999	0.978	0.775
3.2	1.000	1.000	0.991	0.847	3.2	1.000	1.000	0.999	0.985	0.812
3.4	1.000	1.000	0.994	0.873	3.4	1.000	1.000	1.000	0.990	0.844
3.6	1.000	1.000	0.996	0.895	3.6	1.000	1.000	1.000	0.993	0.871
3.8	1.000	1.000	0.997	0.913	3.8	1.000	1.000	1.000	0.995	0.893
4.0	1.000	1.000	0.998	0.929	4.0	1.000	1.000	1.000	0.997	0.912
4.5	1.000	1.000	0.999	0.956	4.5	1.000	1.000	1.000	0.999	0.946
5.0	1.000	1.000	1.000	0.973	5.0	1.000	1.000	1.000	1.000	0.967
5.5	1.000	1.000	1.000	0.984	5.5	1.000	1.000	1.000	1.000	0.980
6.0	1.000	1.000	1.000	0.990	6.0	1.000	1.000	1.000	1.000	0.988
6.5	1.000	1.000	1.000	0.994	6.5	1.000	1.000	1.000	1.000	0.993
7.0	1.000	1.000	1.000	0.996	7.0	1.000	1.000	1.000	1.000	0.995
7.5	1.000	1.000	1.000	0.998	7.5	1.000	1.000	1.000	1.000	0.997
8.0	1.000	1.000	1.000	0.999	8.0	1.000	1.000	1.000	1.000	0.998
8.5	1.000	1.000	1.000	0.999	8.5	1.000	1.000	1.000	1.000	0.999
9.0	1.000	1.000	1.000	1.000	9.0	1.000	1.000	1.000	1.000	0.999
9.5	1.000	1.000	1.000	1.000	9.5	1.000	1.000	1.000	1.000	1.000
10.0	1.000	1.000	1.000	1.000	10.0	1.000	1.000	1.000	1.000	1.000

* $T_r = \dfrac{T_0}{T_D}$ for exponential distribution; $T_r = \left(\dfrac{T_0}{\theta}\right)^b$ for Weibull distribution.

Table A-35   Values of assurance that the true mean (or characteristic) time to failure exceeds the desired mean (or characteristic) time to failure

Reduced sample size—exponential and Weibull distributions
(N = statistical sample size; k = number of failures)

T_r^*	$N = 6$					
	k					
	0	1	2	3	4	5
0.1	0.451	0.105	0.014	0.001	0.000	0.000
0.2	0.699	0.299	0.077	0.012	0.001	0.000
0.3	0.835	0.488	0.184	0.043	0.006	0.000
0.4	0.909	0.642	0.312	0.097	0.017	0.001
0.5	0.950	0.756	0.442	0.170	0.038	0.004
0.6	0.973	0.838	0.561	0.257	0.070	0.008
0.7	0.985	0.894	0.663	0.350	0.113	0.016
0.8	0.992	0.931	0.746	0.443	0.164	0.028
0.9	0.995	0.956	0.812	0.531	0.223	0.044
1.0	0.998	0.972	0.862	0.611	0.287	0.064
1.1	0.999	0.982	0.900	0.681	0.352	0.088
1.2	0.999	0.989	0.929	0.742	0.418	0.116
1.3	1.000	0.993	0.949	0.793	0.481	0.148
1.4	1.000	0.996	0.964	0.836	0.542	0.183
1.5	1.000	0.997	0.975	0.871	0.599	0.220
1.6	1.000	0.998	0.982	0.899	0.651	0.258
1.7	1.000	0.999	0.988	0.921	0.698	0.298
1.8	1.000	0.999	0.992	0.939	0.740	0.338
1.9	1.000	1.000	0.994	0.952	0.777	0.378
2.0	1.000	1.000	0.996	0.964	0.810	0.418
2.2	1.000	1.000	0.998	0.979	0.864	0.494
2.4	1.000	1.000	0.999	0.988	0.904	0.565
2.6	1.000	1.000	1.000	0.993	0.932	0.629
2.8	1.000	1.000	1.000	0.996	0.953	0.686
3.0	1.000	1.000	1.000	0.998	0.967	0.736
3.2	1.000	1.000	1.000	0.999	0.978	0.779
3.4	1.000	1.000	1.000	0.999	0.985	0.816
3.6	1.000	1.000	1.000	1.000	0.990	0.847
3.8	1.000	1.000	1.000	1.000	0.993	0.873
4.0	1.000	1.000	1.000	1.000	0.995	0.895
4.5	1.000	1.000	1.000	1.000	0.998	0.935
5.0	1.000	1.000	1.000	1.000	0.999	0.960
5.5	1.000	1.000	1.000	1.000	1.000	0.976
6.0	1.000	1.000	1.000	1.000	1.000	0.985
6.5	1.000	1.000	1.000	1.000	1.000	0.991
7.0	1.000	1.000	1.000	1.000	1.000	0.995
7.5	1.000	1.000	1.000	1.000	1.000	0.997
8.0	1.000	1.000	1.000	1.000	1.000	0.998
8.5	1.000	1.000	1.000	1.000	1.000	0.999
9.0	1.000	1.000	1.000	1.000	1.000	0.999
9.5	1.000	1.000	1.000	1.000	1.000	1.000
10.0	1.000	1.000	1.000	1.000	1.000	1.000

* $T_r = \dfrac{T_0}{T_D}$ for exponential distribution; $T_r = \left(\dfrac{T_0}{\theta}\right)^b$ for Weibull distribution.

Table A-36 Values of assurance that the true mean (or characteristic) time to failure exceeds the desired mean (or characteristic) time to failure

Reduced sample size—exponential and Weibull distributions
(N = statistical sample size; k = number of failures)

				$N = 7$			
T_r^*				k			
	0	1	2	3	4	5	6
0.1	0.503	0.138	0.022	0.002	0.000	0.000	0.000
0.2	0.753	0.371	0.117	0.024	0.003	0.000	0.000
0.3	0.878	0.578	0.263	0.079	0.015	0.002	0.000
0.4	0.939	0.730	0.421	0.168	0.043	0.006	0.000
0.5	0.970	0.833	0.566	0.277	0.090	0.017	0.001
0.6	0.985	0.899	0.686	0.394	0.154	0.036	0.004
0.7	0.993	0.940	0.779	0.507	0.232	0.065	0.008
0.8	0.996	0.965	0.848	0.610	0.318	0.103	0.015
0.9	0.998	0.979	0.897	0.697	0.406	0.150	0.026
1.0	0.999	0.988	0.932	0.770	0.491	0.205	0.040
1.1	1.000	0.993	0.955	0.827	0.572	0.264	0.059
1.2	1.000	0.996	0.971	0.872	0.644	0.327	0.081
1.3	1.000	0.998	0.981	0.907	0.708	0.391	0.108
1.4	1.000	0.999	0.988	0.933	0.763	0.453	0.138
1.5	1.000	0.999	0.992	0.952	0.810	0.514	0.171
1.6	1.000	1.000	0.995	0.966	0.849	0.572	0.206
1.7	1.000	1.000	0.997	0.976	0.880	0.625	0.244
1.8	1.000	1.000	0.998	0.983	0.906	0.674	0.282
1.9	1.000	1.000	0.999	0.988	0.927	0.718	0.322
2.0	1.000	1.000	0.999	0.992	0.943	0.757	0.361
2.2	1.000	1.000	1.000	0.996	0.996	0.823	0.440
2.4	1.000	1.000	1.000	0.998	0.980	0.873	0.514
2.6	1.000	1.000	1.000	0.999	0.989	0.910	0.583
2.8	1.000	1.000	1.000	1.000	0.993	0.937	0.645
3.0	1.000	1.000	1.000	1.000	0.996	0.956	0.699
3.2	1.000	1.000	1.000	1.000	0.998	0.970	0.747
3.4	1.000	1.000	1.000	1.000	0.999	0.979	0.789
3.6	1.000	1.000	1.000	1.000	0.999	0.986	0.824
3.8	1.000	1.000	1.000	1.000	1.000	0.990	0.854
4.0	1.000	1.000	1.000	1.000	1.000	0.993	0.879
4.5	1.000	1.000	1.000	1.000	1.000	0.998	0.925
5.0	1.000	1.000	1.000	1.000	1.000	0.999	0.954
5.5	1.000	1.000	1.000	1.000	1.000	1.000	0.972
6.0	1.000	1.000	1.000	1.000	1.000	1.000	0.983
6.5	1.000	1.000	1.000	1.000	1.000	1.000	0.990
7.0	1.000	1.000	1.000	1.000	1.000	1.000	0.994
7.5	1.000	1.000	1.000	1.000	1.000	1.000	0.996
8.0	1.000	1.000	1.000	1.000	1.000	1.000	0.998
8.5	1.000	1.000	1.000	1.000	1.000	1.000	0.999
9.0	1.000	1.000	1.000	1.000	1.000	1.000	0.999
9.5	1.000	1.000	1.000	1.000	1.000	1.000	0.999
10.0	1.000	1.000	1.000	1.000	1.000	1.000	1.000

$* \ T_r = \dfrac{T_0}{T_D}$ for exponential distribution; $T_r = \left(\dfrac{T_0}{\theta}\right)^b$ for Weibull distribution.

Table A-37 Values of assurance that the true mean (or characteristic) time to failure exceeds the desired mean (or characteristic) time to failure

Reduced sample size—exponential and Weibull distributions
(N = statistical sample size; k = number of failures)

	$N=8$							
	k							
T_r*	0	1	2	3	4	5	6	7
0.1	0.551	0.173	0.033	0.004	0.000	0.000	0.000	0.000
0.2	0.798	0.440	0.163	0.041	0.007	0.001	0.000	0.000
0.3	0.909	0.655	0.344	0.127	0.032	0.005	0.000	0.000
0.4	0.959	0.799	0.523	0.251	0.084	0.019	0.002	0.000
0.5	0.982	0.887	0.671	0.391	0.164	0.046	0.008	0.001
0.6	0.992	0.938	0.782	0.526	0.263	0.090	0.018	0.002
0.7	0.996	0.966	0.860	0.644	0.371	0.149	0.037	0.004
0.8	0.998	0.982	0.912	0.741	0.479	0.221	0.064	0.008
0.9	0.999	0.991	0.946	0.816	0.579	0.302	0.100	0.015
1.0	1.000	0.995	0.967	0.872	0.667	0.386	0.144	0.025
1.1	1.000	0.997	0.980	0.913	0.742	0.469	0.196	0.039
1.2	1.000	0.999	0.988	0.941	0.804	0.549	0.253	0.057
1.3	1.000	0.999	0.993	0.961	0.853	0.622	0.314	0.078
1.4	1.000	1.000	0.996	0.974	0.891	0.687	0.376	0.104
1.5	1.000	1.000	0.998	0.983	0.920	0.744	0.438	0.133
1.6	1.000	1.000	0.999	0.989	0.942	0.793	0.498	0.165
1.7	1.000	1.000	0.999	0.993	0.958	0.834	0.555	0.199
1.8	1.000	1.000	1.000	0.996	0.970	0.868	0.609	0.236
1.9	1.000	1.000	1.000	0.997	0.979	0.896	0.659	0.274
2.0	1.000	1.000	1.000	0.998	0.985	0.918	0.704	0.312
2.2	1.000	1.000	1.000	0.999	0.993	0.950	0.780	0.391
2.4	1.000	1.000	1.000	1.000	0.996	0.970	0.840	0.467
2.6	1.000	1.000	1.000	1.000	0.998	0.983	0.886	0.539
2.8	1.000	1.000	1.000	1.000	0.999	0.990	0.919	0.605
3.0	1.000	1.000	1.000	1.000	1.000	0.994	0.943	0.665
3.2	1.000	1.000	1.000	1.000	1.000	0.997	0.961	0.717
3.4	1.000	1.000	1.000	1.000	1.000	0.998	0.973	0.762
3.6	1.000	1.000	1.000	1.000	1.000	0.999	0.981	0.801
3.8	1.000	1.000	1.000	1.000	1.000	0.999	0.987	0.834
4.0	1.000	1.000	1.000	1.000	1.000	1.000	0.991	0.863
4.5	1.000	1.000	1.000	1.000	1.000	1.000	0.997	0.915
5.0	1.000	1.000	1.000	1.000	1.000	1.000	0.999	0.947
5.5	1.000	1.000	1.000	1.000	1.000	1.000	1.000	0.968
6.0	1.000	1.000	1.000	1.000	1.000	1.000	1.000	0.980
6.5	1.000	1.000	1.000	1.000	1.000	1.000	1.000	0.988
7.0	1.000	1.000	1.000	1.000	1.000	1.000	1.000	0.993
7.5	1.000	1.000	1.000	1.000	1.000	1.000	1.000	0.996
8.0	1.000	1.000	1.000	1.000	1.000	1.000	1.000	0.997
8.5	1.000	1.000	1.000	1.000	1.000	1.000	1.000	0.998
9.0	1.000	1.000	1.000	1.000	1.000	1.000	1.000	0.999
9.5	1.000	1.000	1.000	1.000	1.000	1.000	1.000	0.999
10.0	1.000	1.000	1.000	1.000	1.000	1.000	1.000	1.000

* $T_r = \dfrac{T_0}{T_D}$ for exponential distribution; $T_r = \left(\dfrac{T_0}{\theta}\right)^b$ for Weibull distribution.

Table A-38 Values of assurance that the true mean (or characteristic) time to failure exceeds the desired mean (or characteristic) time to failure

Reduced sample size—exponential and Weibull distributions
(N = statistical sample size; k = number of failures)

T_r^*					k				
	0	1	2	3	4	5	6	7	8
0.1	0.593	0.209	0.047	0.007	0.001	0.000	0.000	0.000	0.000
0.2	0.835	0.505	0.214	0.063	0.013	0.002	0.000	0.000	0.000
0.3	0.933	0.721	0.425	0.183	0.056	0.012	0.002	0.000	0.000
0.4	0.973	0.852	0.614	0.341	0.139	0.040	0.008	0.001	0.000
0.5	0.989	0.924	0.756	0.501	0.253	0.092	0.023	0.003	0.000
0.6	0.995	0.962	0.852	0.641	0.381	0.168	0.051	0.009	0.001
0.7	0.998	0.981	0.913	0.753	0.508	0.261	0.093	0.021	0.002
0.8	0.999	0.991	0.951	0.835	0.623	0.363	0.150	0.039	0.005
0.9	1.000	0.996	0.972	0.893	0.720	0.466	0.220	0.065	0.009
1.0	1.000	0.998	0.985	0.932	0.797	0.564	0.297	0.101	0.016
1.1	1.000	0.999	0.992	0.958	0.856	0.651	0.378	0.144	0.026
1.2	1.000	1.000	0.996	0.974	0.900	0.727	0.460	0.194	0.040
1.3	1.000	1.000	0.998	0.984	0.931	0.790	0.538	0.249	0.057
1.4	1.000	1.000	0.999	0.991	0.954	0.841	0.610	0.309	0.078
1.5	1.000	1.000	0.999	0.994	0.969	0.881	0.676	0.370	0.103
1.6	1.000	1.000	1.000	0.997	0.980	0.912	0.733	0.431	0.131
1.7	1.000	1.000	1.000	0.998	0.987	0.936	0.783	0.490	0.163
1.8	1.000	1.000	1.000	0.999	0.991	0.953	0.825	0.547	0.197
1.9	1.000	1.000	1.000	0.999	0.994	0.966	0.860	0.601	0.233
2.0	1.000	1.000	1.000	1.000	0.996	0.976	0.889	0.651	0.270
2.2	1.000	1.000	1.000	1.000	0.999	0.988	0.932	0.737	0.348
2.4	1.000	1.000	1.000	1.000	0.999	0.994	0.959	0.806	0.425
2.6	1.000	1.000	1.000	1.000	1.000	0.997	0.976	0.860	0.499
2.8	1.000	1.000	1.000	1.000	1.000	0.999	0.986	0.900	0.569
3.0	1.000	1.000	1.000	1.000	1.000	0.999	0.992	0.929	0.632
3.2	1.000	1.000	1.000	1.000	1.000	1.000	0.995	0.951	0.688
3.4	1.000	1.000	1.000	1.000	1.000	1.000	0.997	0.966	0.737
3.6	1.000	1.000	1.000	1.000	1.000	1.000	0.998	0.976	0.779
3.8	1.000	1.000	1.000	1.000	1.000	1.000	0.999	0.984	0.816
4.0	1.000	1.000	1.000	1.000	1.000	1.000	1.000	0.989	0.847
4.5	1.000	1.000	1.000	1.000	1.000	1.000	1.000	0.996	0.904
5.0	1.000	1.000	1.000	1.000	1.000	1.000	1.000	0.998	0.941
5.5	1.000	1.000	1.000	1.000	1.000	1.000	1.000	0.999	0.964
6.0	1.000	1.000	1.000	1.000	1.000	1.000	1.000	1.000	0.978
6.5	1.000	1.000	1.000	1.000	1.000	1.000	1.000	1.000	0.987
7.0	1.000	1.000	1.000	1.000	1.000	1.000	1.000	1.000	0.992
7.5	1.000	1.000	1.000	1.000	1.000	1.000	1.000	1.000	0.995
8.0	1.000	1.000	1.000	1.000	1.000	1.000	1 000	1.000	0.997
8.5	1.000	1.000	1.000	1.000	1.000	1.000	1 000	1.000	0.998
9.0	1.000	1.000	1.000	1.000	1.000	1.000	1 000	1.000	0.999
9.5	1.000	1.000	1.000	1.000	1.000	1.000	1.000	1.000	0.999
10.0	1.000	1.000	1.000	1.000	1.000	1.000	1.000	1.000	1.000

$N = 9$

* $T_r = \dfrac{T_0}{T_D}$ for exponential distribution; $T_r = \left(\dfrac{T_0}{\theta}\right)^b$ for Weibull distribution.

Table A-39 Values of assurance that the true mean (or characteristic) time to failure exceeds the desired mean (or characteristic) time to failure

Reduced sample size—exponential and Weibull distributions
(N = statistical sample size; k = number of failures)

					$N = 10$					
T_r*					k					
	0	1	2	3	4	5	6	7	8	9
0.1	0.632	0.245	0.062	0.011	0.001	0.000	0.000	0.000	0.000	0.000
0.2	0.865	0.565	0.266	0.090	0.022	0.004	0.000	0.000	0.000	0.000
0.3	0.950	0.776	0.502	0.246	0.089	0.024	0.004	0.001	0.000	0.000
0.4	0.982	0.892	0.692	0.431	0.206	0.073	0.018	0.003	0.000	0.000
0.5	0.993	0.950	0.822	0.601	0.351	0.156	0.050	0.011	0.001	0.000
0.6	0.998	0.977	0.902	0.736	0.499	0.264	0.103	0.028	0.005	0.000
0.7	0.999	0.990	0.948	0.834	0.631	0.385	0.178	0.057	0.011	0.001
0.8	1.000	0.996	0.973	0.899	0.740	0.506	0.267	0.100	0.023	0.003
0.9	1.000	0.998	0.986	0.940	0.823	0.617	0.366	0.157	0.043	0.005
1.0	1.000	0.999	0.993	0.966	0.882	0.711	0.466	0.225	0.069	0.010
1.1	1.000	1.000	0.997	0.980	0.924	0.788	0.561	0.300	0.105	0.017
1.2	1.000	1.000	0.998	0.989	0.952	0.848	0.646	0.380	0.147	0.028
1.3	1.000	1.000	0.999	0.994	0.970	0.893	0.721	0.459	0.197	0.042
1.4	1.000	1.000	1.000	0.997	0.982	0.926	0.784	0.536	0.252	0.059
1.5	1.000	1.000	1.000	0.998	0.989	0.949	0.835	0.607	0.310	0.080
1.6	1.000	1.000	1.000	0.999	0.993	0.966	0.876	0.672	0.370	0.105
1.7	1.000	1.000	1.000	1.000	0.996	0.977	0.908	0.729	0.430	0.133
1.8	1.000	1.000	1.000	1.000	0.998	0.985	0.932	0.779	0.489	0.164
1.9	1.000	1.000	1.000	1.000	0.999	0.990	0.950	0.821	0.546	0.198
2.0	1.000	1.000	1.000	1.000	0.999	0.994	0.964	0.857	0.599	0.234
2.2	1.000	1.000	1.000	1.000	1.000	0.997	0.982	0.910	0.694	0.309
2.4	1.000	1.000	1.000	1.000	1.000	0.999	0.991	0.945	0.772	0.386
2.6	1.000	1.000	1.000	1.000	1.000	1.000	0.996	0.967	0.833	0.462
2.8	1.000	1.000	1.000	1.000	1.000	1.000	0.998	0.980	0.880	0.534
3.0	1.000	1.000	1.000	1.000	1.000	1.000	0.999	0.989	0.914	0.600
3.2	1.000	1.000	1.000	1.000	1.000	1.000	1.000	0.993	0.940	0.660
3.4	1.000	1.000	1.000	1.000	1.000	1.000	1.000	0.996	0.958	0.712
3.6	1.000	1.000	1.000	1.000	1.000	1.000	1.000	0.998	0.971	0.758
3.8	1.000	1.000	1.000	1.000	1.000	1.000	1.000	0.999	0.980	0.798
4.0	1.000	1.000	1.000	1.000	1.000	1.000	1.000	0.999	0.986	0.831
4.5	1.000	1.000	1.000	1.000	1.000	1.000	1.000	1.000	0.995	0.894
5.0	1.000	1.000	1.000	1.000	1.000	1.000	1.000	1.000	0.998	0.935
5.5	1.000	1.000	1.000	1.000	1.000	1.000	1.000	1.000	0.999	0.960
6.0	1.000	1.000	1.000	1.000	1.000	1.000	1.000	1.000	1.000	0.975
6.5	1.000	1.000	1.000	1.000	1.000	1.000	1.000	1.000	1.000	0.985
7.0	1.000	1.000	1.000	1.000	1.000	1.000	1.000	1.000	1.000	0.991
7.5	1.000	1.000	1.000	1.000	1.000	1.000	1.000	1.000	1.000	0.994
8.0	1.000	1.000	1.000	1.000	1.000	1.000	1.000	1.000	1.000	0.997
8.5	1.000	1.000	1.000	1.000	1.000	1.000	1.000	1.000	1.000	0.998
9.0	1.000	1.000	1.000	1.000	1.000	1.000	1.000	1.000	1.000	0.999
9.5	1.000	1.000	1.000	1.000	1.000	1.000	1.000	1.000	1.000	0.999
10.0	1.000	1.000	1.000	1.000	1.000	1.000	1.000	1.000	1.000	1.000

* $T_r = \dfrac{T_0}{T_D}$ for exponential distribution; $T_r = \left(\dfrac{T_0}{\theta}\right)^b$ for Weibull distribution.

Table A-40 Nomograph for calculating one-sided lower estimates of reliability at several confidence levels

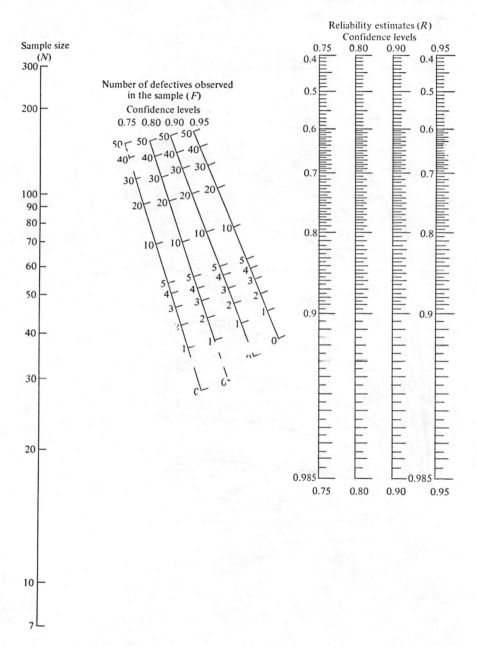

Table A-41 Lower-limit estimates of reliability at 90% confidence—binomial distribution

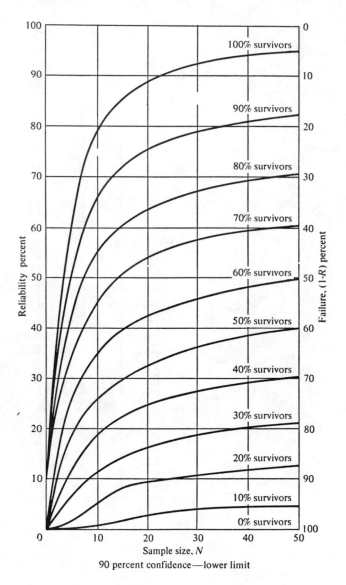

90 percent confidence—lower limit

Table A-42 Upper-limit estimates of reliability at 90% confidence—binomial distribution

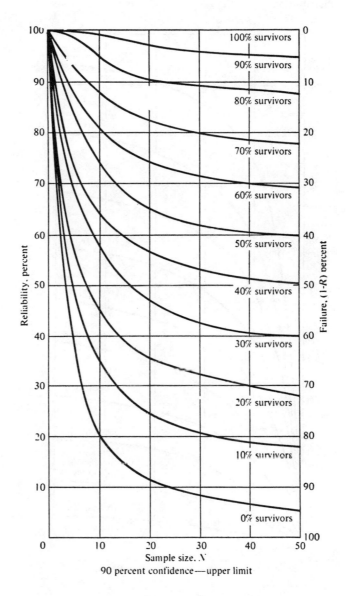

90 percent confidence—upper limit

Table A-43 Lower-limit estimates of reliability at 95% confidence—binomial distribution

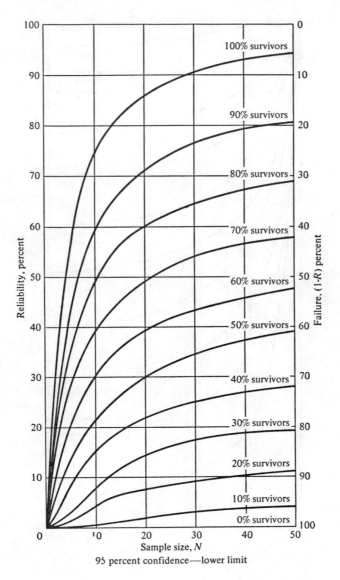

95 percent confidence—lower limit

Table A-44 Upper-limit estimates of reliability at 95% confidence—binomial distribution

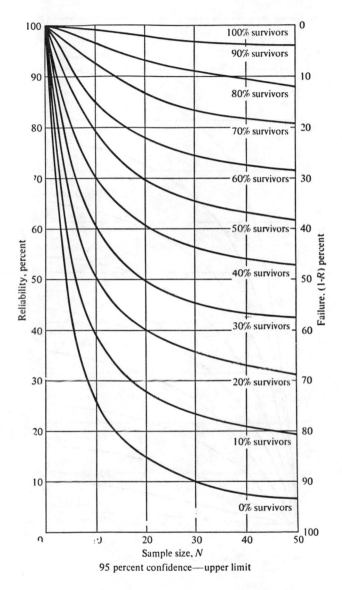

Sample size, N

95 percent confidence—upper limit

Table A-45   Lower-limit estimates of reliability at 99% confidence—binomial distribution

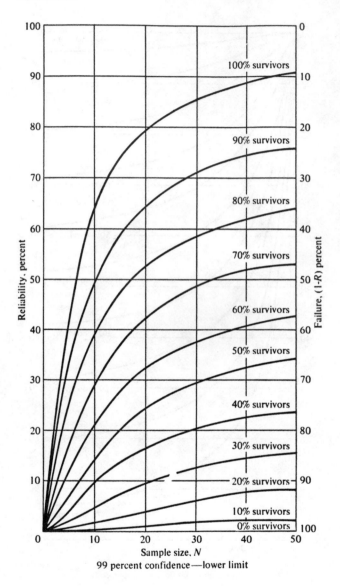

Sample size, N

99 percent confidence—lower limit

Table A-46   Upper-limit estimates of reliability at 99% confidence—binomial distribution

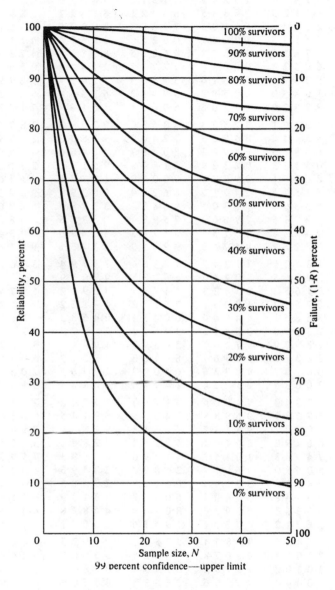

99 percent confidence—upper limit

Table A-47   Random numbers (constant probability)

3 8 0 0 1	8 8 9 7 7	1 1 3 5 4	5 1 2 8 1	4 4 9 7 2
3 7 4 0 2	1 5 2 4 3	3 1 3 1 2	5 6 0 2 0	6 9 7 9 7
9 7 1 2 5	2 4 3 3 5	6 9 9 2 1	6 1 4 2 7	3 2 9 0 5
2 1 8 2 6	6 1 1 0 5	7 9 8 8 8	8 0 8 7 5	9 3 5 4 1
7 3 1 3 5	1 9 0 8 7	0 6 2 5 6	3 8 6 2 9	0 1 1 3 4
0 7 6 3 8	4 2 6 7 8	4 6 0 6 5	8 6 8 3 6	4 7 3 9 0
6 0 5 2 8	9 8 0 8 6	5 2 7 7 7	2 1 4 2 4	0 5 6 2 2
8 3 5 9 6	9 4 6 1 4	5 4 5 6 3	5 2 4 9 7	5 2 7 9 4
1 0 8 5 0	0 0 5 8 2	5 9 9 5 2	6 8 5 6 8	5 6 0 5 8
3 9 8 2 0	9 7 7 0 3	5 0 6 9 1	8 5 1 9 9	7 1 3 3 6
4 0 7 9 1	3 2 5 3 3	7 8 4 3 0	9 2 9 9 7	2 8 7 2 5
5 5 4 4 4	0 4 8 0 5	7 7 4 0 0	9 7 8 5 2	1 0 3 9 3
6 3 3 1 5	6 8 9 5 3	8 0 4 5 7	8 8 4 6 6	4 3 8 0 6
0 3 1 3 3	0 2 5 2 9	5 1 8 7 8	9 5 1 1 1	2 7 1 5 1
8 6 9 6 1	9 9 9 7 0	9 0 0 7 0	1 3 7 7 1	9 1 3 6 9
5 4 5 2 5	7 4 7 1 7	6 6 2 0 9	9 0 3 9 8	9 3 0 3 9
6 3 3 9 0	1 0 8 0 5	8 6 8 8 2	3 6 7 8 0	6 2 8 6 5
3 2 9 1 5	7 7 6 0 2	7 5 9 7 4	0 8 7 6 6	0 0 8 0 0
0 4 2 3 0	3 2 1 3 5	9 3 7 8 3	7 6 6 8 6	6 4 3 4 0
8 8 2 3 2	4 5 7 5 3	9 2 4 1 9	1 1 2 3 1	9 2 1 6 8
1 0 7 2 1	3 1 3 4 7	4 2 8 5 4	7 3 7 9 4	3 4 6 7 7
3 9 7 5 5	3 0 2 4 0	6 6 1 3 8	9 9 0 2 9	4 5 3 0 5
3 1 6 5 2	2 3 8 2 3	5 4 7 5 3	4 8 9 2 2	5 9 7 4 7
8 7 6 6 2	1 9 0 5 1	7 5 5 3 2	0 8 1 2 2	1 6 5 2 0
8 3 6 5 1	4 4 6 4 0	6 0 4 1 3	0 4 5 3 3	6 8 6 5 2
2 3 7 9 0	0 0 8 1 2	9 9 6 7 0	2 1 0 2 2	7 9 3 7 5
1 8 3 7 0	9 7 2 0 7	2 6 5 9 3	1 9 4 4 1	3 3 5 2 1
8 8 3 1 8	2 4 7 6 7	6 9 1 0 8	5 9 2 1 7	5 9 5 8 9
0 0 1 5 7	4 8 3 3 6	6 4 0 0 2	1 7 3 6 5	2 0 5 5 4
3 0 6 3 5	3 1 2 2 4	1 4 7 1 0	5 8 8 3 3	5 9 4 0 4
1 7 3 4 0	4 4 9 0 6	4 1 4 6 8	8 9 7 4 2	2 7 2 1 2
3 7 5 8 9	9 6 9 8 8	0 4 5 5 9	6 9 1 4 5	0 3 8 6 7
7 0 3 2 2	7 5 1 7 2	4 1 6 1 5	9 1 8 6 4	7 9 8 9 5
6 6 4 9 2	2 6 4 0 1	5 9 2 7 3	4 2 8 8 6	5 5 2 5 6
7 4 0 8 3	1 0 1 5 7	4 1 3 9 6	0 5 6 9 3	7 5 9 3 7
8 0 6 8 0	2 6 0 1 7	2 5 8 0 7	2 3 5 9 6	8 0 2 3 6
7 4 5 0 8	3 3 1 9 3	0 6 1 7 0	2 7 3 0 3	3 0 3 4 5
0 9 9 3 8	0 4 4 0 3	6 0 8 0 8	3 3 1 3 9	9 5 5 8 5
9 5 3 6 3	3 0 9 3 7	8 0 9 4 0	4 4 8 9 8	4 4 1 3 7
9 6 7 1 2	4 7 3 9 1	1 9 5 1 6	8 3 0 0 1	2 7 4 1 4
4 5 4 4 5	3 3 8 3 4	0 7 5 1 0	0 6 0 7 3	8 0 6 9 2
2 1 0 6 7	9 8 0 7 8	5 8 5 2 4	5 9 0 0 8	0 2 6 8 0
3 0 3 0 2	7 0 7 8 2	5 1 9 3 4	0 8 8 4 4	8 6 9 8 9
7 0 0 4 0	7 3 3 0 6	8 4 8 3 3	8 8 5 0 5	9 4 9 9 4
1 3 3 5 1	6 7 4 4 0	4 3 5 4 6	7 3 2 5 9	5 9 8 1 7
0 4 7 1 6	0 2 0 4 3	4 6 1 8 1	8 1 5 5 0	1 3 0 6 2
5 7 6 5 7	1 3 0 4 7	3 4 3 2 1	5 5 8 3 1	2 5 3 4 3
7 2 7 6 4	7 9 6 3 0	7 1 2 0 4	3 9 4 6 5	4 2 0 9 3
4 5 2 2 8	7 6 4 1 7	2 7 2 9 8	6 6 4 1 5	3 6 6 3 6
6 4 6 6 5	3 0 7 5 7	5 6 3 2 2	7 5 8 8 9	1 4 4 8 6

Table A-48   Random normal deviates [6]

0.527	1.007	2.351	0.847	1.947	0.545—	0.490	1.091	1.389	0.103
0.655	0.571	2.605	0.457—	3.037—	0.001	0.127	0.946	1.962	1.158
0.160—	0.011	0.567	1.658—	0.356	2.245—	0.579—	0.232	2.122	0.380—
0.596	0.121—	0.591—	2.333	1.242—	1.144	0.350—	1.403	0.493	0.317—
0.170—	0.024—	0.051—	1.395—	1.176	0.168—	0.825	0.412—	0.951	0.569
0.422	0.163	0.054—	1.155—	0.446—	0.641—	0.243	1.850—	3.003—	0.801—
0.771—	0.223	0.746	1.001—	0.641—	1.024—	0.490	0.337	1.142	0.704—
0.790	0.329—	0.170—	0.811—	0.900—	0.843—	0.502—	0.705—	1.063—	0.328—
0.214	0.820—	1.125—	0.424	0.624—	0.872—	0.895—	0.053—	0.151—	0.353—
2.475—	1.584—	0.781	3.218	0.449—	1.115	1.828—	3.083—	0.113	1.156—
0.843	0.351	0.641—	0.202—	0.598	0.873	0.090	0.194—	0.015—	0.230
0.095	1.162	1.281—	1.685	0.417	0.433—	1.217	1.204—	0.773—	1.696—
1.222	0.604	0.185—	0.566—	1.594—	2.350—	1.105—	0.227	1.266	2.668
0.728	1.088	0.706—	0.173—	0.449	2.111—	0.315	0.798—	0.386	0.260—
1.191—	0.945	1.855	1.477	1.228—	1.464—	0.490—	0.552—	0.274—	0.738
1.285	0.334—	1.357	0.252	0.232	0.082—	0.210	1.169	1.036	0.357
0.539—	0.970—	0.776	0.823—	0.042—	0.322—	0.464	0.778	1.044	0.319—
1.554	0.644—	0.557	1.349—	1.343	0.019—	1.787	1.104	1.207—	0.243
0.839	0.476	0.319—	0.531	0.612—	1.456—	0.894—	1.157	1.206	0.560—
0.824—	0.406	0.485—	0.475	0.029	1.773—	0.427—	0.864	1.488	1.194—
0.176—	1.089—	0.994—	0.515—	0.005—	2.765	1.286—	0.232—	0.006	0.894—
0.121—	1.227	1.727—	1.255	0.736	0.359—	1.540—	0.448—	0.898	0.714—
0.001—	0.233—	1.632	0.129—	0.329	0.265—	0.174—	0.589	0.326—	0.139
0.311	1.305	0.763	0.557—	1.056—	1.293	0.706—	0.693	0.824—	0.428
0.619	0.942—	0.784—	0.172—	0.098—	1.424—	0.642	0.542	1.890—	0.743
1.215—	0.403	0.831—	1.139	0.237	0.108—	0.484—	0.888—	0.491—	0.020
2.021—	0.671	0.117	0.873—	0.777—	0.594—	0.408	1.582—	1.938	0.164—
0.618	0.013—	0.585—	1.060—	0.221—	0.359—	0.256	0.587—	2.461—	1.251
0.039—	0.614	0.226	0.993—	0.305	1.196—	1.017	0.051	0.442	0.218—
0.108—	0.768—	0.088—	2.563 —	0.303	1.758—	0.620—	1.147—	0.107	0.205—
0.374	0.084	0.575—	0.683	0.350	0.078—	0.958—	0.787	0.644—	0.466—
0.349—	1.932—	0.681—	0.438	1.214	0.795	1.742	0.603—	2.538	1.243—
0.321—	0.747	1.026—	1.451	0.383	2.195	0.646—	1.146	0.672—	0.761—
0.602—	1.920—	0.381—	0.008	1.342	1.701	0.370—	0.444	0.011	0.966—
0.945—	1.450	0.701—	0.938—	0.643—	0.410—	0.825	0.864	0.133—	1.295—
0.039—	0.565—	0.849—	0.121	1.820—	1.697—	0.259	0.429—	0.958	1.139—
1.356—	0.249—	0.630—	0.395	0.124—	1.040	0.859	0.133—	0.369—	1.106—
0.809	1.509	1.428	1.987	1.152—	1.267	0.428—	0.544	0.588	0.321—
0.615	1.192	0.632—	0.021—	0.874—	0.721	0.408	1.543	1.163	1.657
0.040—	1.834—	0.865	0.188—	1.339—	0.051	0.042—	0.799	0.668—	1.713
2.208—	0.926	0.518—	0.904—	1.532	1.070	0.993—	0.106—	0.733—	1.058—
0.302—	0.092—	0.696—	0.373	1.174	1.504—	0.190	0.011—	0.328—	1.075—
0.600—	1.241—	0.916	0.317—	0.711—	2.028—	0.119—	0.218	1.825—	0.241—
0.042—	0.989	0.092—	0.631	0.495—	1.065	0.142	0.444—	0.210	0.187
0.284—	0.548—	0.774	1.780	0.677	0.231	0.203	1.221—	1.657	0.847
1.170	0.386	2.184—	1.067	0.873—	0.437—	0.531	2.506—	0.302—	0.601—
1.194—	0.026	0.127	0.979—	0.025	1.009	1.569	1.328—	0.227—	1.518—
0.169—	0.136	1.323—	0.851	1.272	1.010	0.929—	0.451	1.025	0.368
0.880—	1.077	0.269	0.576—	0.262—	0.266	0.561—	0.412	0.917	1.067
1.645—	1.687	0.412	0.992—	0.965—	0.388-	0.034	1.140	0.122	0.258—

Table A-49 Values of correlation coefficient r [2]

ν	95% Confidence level				99% Confidence level				ν
	Total number of variables				Total number of variables				
	2	3	4	5	2	3	4	5	
1	.997	.999	.999	.999	1.000	1.000	1.000	1.000	1
2	.950	.975	.983	.987	.990	.995	.997	.998	2
3	.878	.930	.950	.961	.959	.976	.983	.987	3
4	.811	.881	.912	.930	.917	.949	.962	.970	4
5	.754	.836	.874	.898	.874	.917	.937	.949	5
6	.707	.795	.839	.867	.834	.886	.911	.927	6
7	.666	.758	.807	.838	.798	.855	.885	.904	7
8	.632	.726	777	.811	.765	.827	.860	.882	8
9	.602	.697	.750	.786	.735	.800	.836	.861	9
10	.576	.671	.726	.763	.708	.776	.814	.840	10
11	.553	.648	.703	.741	.684	.753	.793	.821	11
12	.532	.627	.683	.722	.661	.732	.773	.802	12
13	.514	.608	.664	.703	.641	.712	.755	.785	13
14	.497	.590	.646	.686	.623	.694	.737	.768	14
15	.482	.574	.630	.670	.606	.677	.721	.752	15
16	.468	.559	.615	.655	.590	.662	.706	.738	16
17	.456	.545	.601	.641	.575	.647	.691	.724	17
18	.444	.532	.587	.628	.561	.633	.678	.710	18
19	.433	.520	.575	.615	.549	.620	.665	.698	19
20	.423	.509	.563	.604	.537	.608	.652	.685	20
21	.413	.498	.552	.592	.526	.596	.641	.674	21
22	.404	.488	.542	.582	.515	.585	.630	.663	22
23	.396	.479	.532	.572	.505	.574	.619	.652	23
24	.388	.470	.523	.562	.496	.565	.609	.642	24
25	.381	.462	.514	.553	.487	.555	.600	.633	25
26	.374	.454	.506	.545	.478	.546	.590	.624	26
27	.367	.446	.498	.536	.470	.538	.582	.615	27
28	.361	.439	.490	.529	.463	.530	.573	.606	28
29	.355	.432	.482	.521	.456	.522	.565	.598	29
30	.349	.426	.476	.514	.449	.514	.558	.591	30
35	.325	.397	.445	.482	.418	.481	.523	.556	35
40	.304	.373	.419	.455	.393	.454	.494	.526	40
45	.288	.353	.397	.432	.372	.430	.470	.501	45
50	.273	.336	.379	.412	.354	.410	.449	.479	50
60	.250	.308	.348	.380	.325	.377	.414	.442	60
70	.232	.286	.324	.354	.302	.351	.386	.413	70
80	.217	.269	.304	.332	.283	.330	.362	.389	80
90	.205	.254	.288	.315	.267	.312	.343	.368	90
100	.195	.241	.274	.300	.254	.297	.327	.351	100
125	.174	.216	.246	.269	.228	.266	.294	.316	125
150	.159	.198	.225	.247	.208	.244	.270	.290	150
200	.138	.172	.196	.215	.181	.212	.234	.253	200
300	.113	.141	.160	.176	.148	.174	.192	.208	300
400	.098	.122	.139	.153	.128	.151	.167	.180	400
500	.088	.109	.124	.137	.115	.135	.150	.162	500
1,000	.062	.077	.088	.097	.081	.096	.106	.116	1,000

Table A-50   Natural logarithms

To find the natural logarithm of a number which is 1/10, 1/100, 1/1000, etc., of a number whose logarithm is given, subtract from the given logarithm $\log_e$ 10, 2 $\log_e$ 10, 3 $\log_e$ 10, etc.

To find the natural logarithm of a number which is 10, 100, 1000, etc., times a number whose logarithm is given, add to the given logarithm $\log_e$ 10, 2 $\log_e$ 10, 3 $\log_e$ 10, etc.

$\log_e$ 10 = 2.30253 50930	6 $\log_e$ 10 = 13.31551 05530
2 $\log_e$ 10 = 4.60517 01960	7 $\log_e$ 10 = 16.11809 56510
3 $\log_e$ 10 = 6.90775 52790	8 $\log_e$ 10 = 18.42068 07440
4 $\log_e$ 10 = 9.21034 03720	9 $\log_e$ 10 = 20.72326 58369
5 $\log_e$ 10 = 11.51292 54540	10 $\log_e$ 10 = 23.02585 09298

1.00–4.99

N	0	1	2	3	4	5	6	7	8	9
1.0	0.00000	.00995	.01980	.02956	.03922	.04879	.05827	.06766	.07696	.08618
.1	.09531	.10436	.11333	.12222	.13103	.13976	.14842	.15700	.16551	.17395
.2	.18232	.19062	.19885	.20701	.21511	.22314	.23111	.23902	.23686	.25464
.3	.26236	.27003	.27763	.28518	.29267	.30010	.30748	.31481	.32208	.32930
.4	.33647	.34359	.35066	.35767	.36464	.37156	.37844	.38526	.39204	.39878
.5	.40547	.41211	.41871	.42527	.43178	.43825	.44469	.45108	.45742	.46373
.6	.47000	.47623	.48243	.48858	.49470	.50078	.50682	.51282	.51879	.52473
.7	.53063	.53649	.54232	.54812	.55389	.55962	.56531	.57098	.57661	.58222
.8	.55779	.59333	.59884	.60432	.60977	.61519	.62058	.62594	.63127	.63658
.9	.04185	.64710	.65233	.65752	.66269	.66783	.67294	.67803	.68310	.68812
2.0	0.69315	.69813	.70310	.70804	.71295	.71784	.72271	.72755	.73237	.73716
.1	.72194	.74669	.75142	.75612	.76081	.76547	.77011	.77473	.77932	.78390
.2	.78846	.79299	.79751	.80200	.80648	.81093	.81536	.81978	.82418	.82855
.3	.83291	.83725	.84157	.84587	.85015	.85442	.85866	.86289	.86710	.87129
.4	.87547	.87963	.88377	.88789	.89200	.89809	.90016	.90422	.90826	.91228
.5	.91629	.92028	.92426	.92822	.93216	.93609	.94001	.94391	.94779	.95166
.6	.95551	.95935	.96317	.96698	.97078	.97456	.97833	.98208	.98582	.98954
.7	.99325	.99695	.00063	.00430	.00796	.01160	.01523	.01885	.02245	.02604
.8	1.02962	.03318	.03674	.04028	.04380	.04732	.05082	.05431	.05779	.06126
9	.06471	.06815	.07158	.07500	.07841	.08181	.08519	.08856	.09192	.09527
3.0	1.09861	.10194	.10526	.10856	.11186	.11514	.11841	.12168	.12493	.12817
.1	.13140	.13462	.13783	.14103	.14422	.14740	.15057	.15373	.15688	.16002
.2	.16315	.16627	.16938	.17248	.17557	.17865	.18173	.18479	.18784	.19089
.3	.19392	.19695	.19996	.20297	.26597	.20896	.21194	.21491	.21788	.22083
.4	.22378	.22671	.22964	.32356	.23547	.23837	.24127	.24415	.24703	.24990
.5	.25276	.25562	.25846	.26130	.26413	.26695	.26976	.27257	.27536	.27815
.6	.28893	.28371	.28647	.28923	.29198	.29473	.29746	.30019	.30291	.30563
.7	.30833	.31103	.31372	.31641	.31909	.32176	.32442	.32708	.32972	.33237
.8	.33500	.33763	.34025	.34286	.34547	.34807	.35067	.35325	.35584	.35841
9	.36098	.38354	.36609	.36864	.37118	.37372	.37624	.37877	.38128	.38379

Table A-50 (*continued*)

4.00–7.49

N	0	1	2	3	4	5	6	7	8	9
4.0	1.28629	.38879	.39128	.39377	.39624	.39872	.40118	.40364	.40610	.40854
.1	.41099	.41342	.41585	.41828	.42070	.42311	.42552	.42792	.43041	.43270
.2	.43508	.43746	.43984	.44220	.44456	.44692	.44927	.45161	.45395	.45629
.3	.45862	.46094	.46326	.46557	.46787	.47018	.47247	.47476	.47705	.47933
.4	.48160	.48387	.48614	.48840	.49085	.49290	.49515	.49739	.49962	.50135
.5	.50408	.50630	.50851	.51072	.51293	.51513	.51732	.51951	.52170	.52386
.6	.52606	.52823	.53039	.53256	.53471	.53687	.53902	.54116	.54330	.54543
.7	.54758	.54969	.55181	.55393	.55604	.55814	.56025	.56235	.56444	.56653
.8	.56862	.57070	.57277	.57485	.57691	.57808	.58104	.58309	.58515	.58719
.9	.58924	.59127	.59331	.59534	.59737	.59939	.60141	.60342	.60543	.60744
5.0	1.60944	.61144	.61343	.61542	.61741	.61939	.62137	.62334	.62531	.62726
.1	.62924	.63120	.63315	.63511	.63705	.63900	.64094	.64287	.64491	.64673
.2	.64866	.65058	.65250	.65441	.65632	.65823	.66013	.66203	.66388	.66582
.3	.66771	.66959	.67147	.67335	.67523	.67710	.67896	.68083	.68289	.68455
.4	.68640	.68825	.69010	.69164	.69378	.69562	.69745	.69928	.70111	.70293
.5	.70475	.70656	.70838	.71019	.71199	.71380	.71560	.71740	.71919	.72098
.6	.72277	.72455	.72633	.72811	.72988	.73166	.73342	.73519	.73695	.73670
.7	.74047	.74222	.74397	.74572	.74746	.74920	.75094	.75267	.75440	.75613
.8	.75786	.75958	.76130	.76302	.76473	.76644	.76715	.76985	.77156	.77326
.9	.77495	.77665	.77834	.78002	.78171	.78339	.78339	.78675	.78842	.79009
6.0	1.79176	.79342	.79509	.79675	.79840	.80006	.80171	.80336	.80500	.80665
.1	.80829	.80993	.81156	.81319	.81482	.81645	.81808	.81970	.82132	.82294
.2	.82455	.82616	.82777	.82938	.83098	.83258	.83418	.83578	.83737	.83869
.3	.84055	.84214	.84372	.84530	.84688	.84845	.85003	.85160	.85317	.85473
.4	.85630	.85786	.85942	.86097	.86253	.86408	.86563	.86718	.86872	.87026
.5	.87180	.87334	.87487	.87641	.87794	.87947	.88099	.88251	.88403	.88555
.6	.88707	.88858	.89010	.89160	.89311	.89462	.89612	.89762	.89912	.90061
.7	.90211	.90360	.90509	.90658	.90806	.90954	.91102	.91250	.91398	.91545
.8	.91692	.91839	.91986	.92132	.92279	.92425	.92571	.92716	.92862	.93007
.9	.93152	.93297	.93442	.93586	.93730	.93874	.94018	.94162	.94305	.94488
7.0	1.94591	.94734	.94876	.95019	.95161	.95303	.95445	.95586	.95727	.95969
.1	.96009	.96150	.96291	.96431	.96571	.96711	.96851	.96991	.97130	.97269
.2	.97408	.97547	.97685	.97824	.97962	.98100	.98238	.98376	.98513	.98650
.3	.98787	.98924	.99061	.99198	.99334	.99470	.99606	.99742	.99877	.00013
.4	2.00148	.00283	.00418	.00553	.00687	.00821	.00956	.01089	.01223	.01357

7.50–9.99

N	0	1	2	3	4	5	6	7	8	9
.5	.01490	.01624	.01757	.01890	.02022	.02155	.02287	.02419	.02551	.02683
.6	.02815	.02946	.03078	.03209	.03340	.03471	.03601	.03732	.03982	.03992
.7	.04122	.04252	.04381	.04511	.04640	.04769	.04898	.05027	.05156	.05284
.8	.05412	.05540	.05668	.05796	.05924	.06051	06179	.06306	.06433	.06560
.9	.06686	.06813	.06939	.07065	.07191	.07317	.07143	.07568	.07694	.07819
8.0	2.07944	.08069	.08194	.08318	.08443	.08567	.08791	.08815	.08939	.09061
.1	.09186	.09310	.09433	.09556	.09679	.09802	.09924	.10047	.10169	.10291
.2	.10413	.10535	.10657	.10779	.10900	.11021	.11142	.11263	.11384	.11505
.3	.11626	.11746	.11866	.11986	.12106	.12226	.12346	.12465	.12585	.12704
.4	.12823	.12942	.13061	.13180	.13298	.13417	.13535	.13653	.13771	.13889
.5	.14007	.14124	.14242	.14359	.14476	.14593	.14710	.14827	.14943	.15060
.6	.15176	.15292	.15409	.15524	.15640	.15745	.15871	.15987	.16102	.16217
.7	.16332	.16447	.16562	.16677	.16791	.16905	.17020	.17134	.17248	.17361
.8	.17475	.17589	.17702	.17816	.17929	.18042	.18155	.18267	.18380	.18493
.9	.18605	.18717	.18830	.18942	.19054	.19165	.19277	.19389	.19500	.19611
9.0	2.19722	.19834	.19944	.20055	.20166	.20276	.20387	.20497	.20607	.20717
.1	.20827	.20937	.21047	.21157	.21266	.21375	.21485	.21594	.21703	.21812
.2	.21920	.22029	.22138	.22246	.22354	.22462	.22570	.22678	.22796	.22894
.3	.23001	.23109	.23216	.23324	.23431	.23538	.23654	.23751	.23858	.23965
.4	.24071	.24177	.24284	.24390	.24496	.24601	.24707	.24813	.24918	.25024
.5	.25129	.25234	.25339	.25444	.25549	.25654	.25759	.25963	.25968	.26072
.6	.26176	.26280	.26384	.26488	.26592	.26696	.26799	.26903	.27008	.27109
.7	.27213	.27316	.27419	.27521	.27624	.27727	.27829	.27932	.28034	.28136
.8	.28238	.28340	.28442	.28544	.28646	.28747	.28849	.28950	.29051	29152
.9	.29253	.29354	.29455	.29556	.29657	.29757	.29852	.29958	.30058	30158

Table A-51 Base 10 logarithms (100–999)

N	0	1	2	3	4	5	6	7	8	9	*Proportional Parts*								
											1	2	3	4	5	6	7	8	9
10	0000	0043	0086	0128	0170	0212	0253	0294	0334	0274	*4	8	12	17	21	25	29	33	37
11	0414	0453	0492	0531	0589	0607	0845	0383	0719	0755	4	8	11	15	19	23	26	30	34
12	0792	0828	0864	0899	0934	0969	1004	1038	1072	1106	3	7	10	14	17	21	24	28	31
13	1139	1173	1206	1239	1271	1303	1335	1367	1399	1430	3	6	10	13	16	19	23	26	29
14	1461	1492	1523	1553	1584	1614	1644	1673	1703	1732	3	6	9	12	15	18	21	24	27
15	1761	1790	1818	1847	1875	1903	1931	1959	1987	2014	*3	6	8	11	14	17	20	22	25
16	2041	2068	2095	2122	2148	2175	2201	2227	2253	2279	3	5	8	11	13	16	18	21	24
17	2304	2330	2355	2380	2405	2430	2455	2480	2504	2529	2	5	7	10	12	15	17	20	22
18	2553	2577	2601	2625	2648	2672	2695	2718	2742	2765	2	5	7	9	12	14	16	19	21
19	2788	2810	2833	2856	2878	2900	2923	2945	2967	2989	2	4	7	9	11	13	16	18	20
20	3010	3032	3054	3075	3096	3118	3139	3160	3181	3201	2	4	6	8	11	13	15	17	19
21	3222	3243	3263	3284	3304	3324	3345	3365	3385	3404	2	4	6	8	10	12	14	16	18
22	3424	3444	3464	3483	3502	3522	3541	3560	3579	3598	2	4	6	8	10	12	14	15	17
23	3617	3636	3655	3674	3692	3711	3729	3747	3766	3784	2	4	6	7	9	11	13	15	17
24	3802	3820	3838	3856	3874	3892	3909	3927	3945	3962	2	4	5	7	9	11	12	14	16
25	3979	3997	4014	4031	4048	4065	4082	4099	4116	4133	2	3	5	7	9	10	12	14	15
26	4150	4166	4183	4200	4216	4232	4249	4265	4281	4298	2	3	5	7	8	10	11	13	15
27	4314	4330	4346	4362	4378	4393	4409	4425	4440	4456	2	3	5	6	8	9	11	13	14
28	4472	4487	4502	4518	4533	4548	4564	4579	4594	4609	2	3	5	6	8	9	11	12	14
29	4624	4639	4654	4669	4683	4608	4713	4728	4742	4747	1	3	4	6	7	9	10	12	13
30	4771	4786	4800	4814	4829	4843	4857	4871	4886	4900	1	3	4	6	7	9	10	11	13
31	4914	4928	4942	4955	4969	4983	4997	5011	5024	5038	1	3	4	6	7	8	10	11	12
32	5051	5065	5079	5092	5105	5119	5132	5145	5159	5172	1	3	4	5	7	8	9	11	12
33	5185	5198	5211	5224	5237	5250	5263	5276	5289	5302	1	3	4	5	6	8	9	10	12
34	5315	5328	5340	5353	5366	5378	5391	5403	5416	5428	1	3	4	5	6	8	9	10	11
35	5441	5453	5465	5478	5490	5502	5514	5527	5539	5551	1	2	4	5	6	7	9	10	11
36	5563	5575	5587	5599	5611	5623	5635	5647	5658	5670	1	2	4	5	6	7	8	10	11
37	5682	5694	5705	5717	5729	5740	5752	5763	5775	5786	1	2	3	5	6	7	8	9	10
38	5798	5809	5821	5832	5843	5855	5866	5877	5888	5899	1	2	3	5	6	7	8	9	10
39	5911	5922	5933	5944	5955	5966	5977	5988	5999	6010	1	2	3	4	5	7	8	9	10
40	6021	6031	6042	6053	6064	6075	6035	6096	6107	6117	1	2	3	4	5	6	8	9	10
41	6128	6138	6149	6160	6170	6180	6191	6201	6212	6222	1	2	3	4	5	6	7	8	9
42	6232	6243	6253	6263	6274	6284	6294	6304	6314	6325	1	2	3	4	5	6	7	8	9
43	6335	6345	6355	6365	6375	6385	6395	6405	6415	6425	1	2	3	4	5	6	7	8	9
44	6435	6444	6454	6464	6474	6484	6493	6503	6513	6522	1	2	3	4	5	6	7	8	9
45	6532	6542	6551	6561	6571	6580	6590	6599	6609	6618	1	2	3	4	5	6	7	8	9
46	6828	6637	6646	6656	6665	6675	6684	6693	6702	6712	1	2	3	4	5	6	7	7	8
47	6721	6730	6739	6749	6758	6767	6776	6785	6794	6803	1	2	3	4	5	5	6	7	8
48	6812	6821	6830	6839	6848	6857	6866	6875	6884	6893	1	2	3	4	5	5	6	7	8
49	6902	6911	6920	6928	6937	6946	6955	6964	6972	6981	1	2	3	4	4	5	6	7	8
50	6990	6998	7007	7016	7024	7033	7042	7050	7059	7067	1	2	3	3	4	5	6	7	8
51	7076	7084	7093	7101	7110	7118	7126	7135	7143	7152	1	2	3	3	4	5	6	7	8
52	7160	7168	7177	7185	7193	7202	7210	7218	7226	7235	1	2	2	3	4	5	6	7	7
53	7243	7251	7259	7267	7275	7284	7292	7300	7308	7316	1	2	2	3	4	5	6	6	7
54	7324	7332	7340	7348	7356	7364	7372	7380	7388	7396	1	2	2	3	4	5	6	6	7
N	0	1	2	3	4	5	6	7	8	9	1	2	3	4	5	6	7	8	9

* Interpolation in this section of the table is inaccurate.

N	0	1	2	3	4	5	6	7	8	9	Proportional Parts								
											1	2	3	4	5	6	7	8	9
55	7404	7412	7419	7427	7435	7443	7451	7459	7466	7474	1	2	2	3	4	5	5	6	7
56	7482	7490	7497	7505	7513	7520	7528	7536	7543	7551	1	2	2	3	4	5	5	6	7
57	7559	7566	7574	7582	7589	7597	7604	7612	7619	7627	1	2	2	3	4	5	5	6	7
58	7634	7642	7649	7657	7664	7672	7679	7686	7694	7701	1	1	2	3	4	4	5	6	7
59	7709	7716	7723	7731	7738	7745	7752	7760	7767	7774	1	1	2	3	4	4	5	6	7
60	7782	7789	7796	7803	7810	7818	7825	7832	7839	7846	1	1	2	3	4	4	5	6	6
61	7853	7860	7868	7875	7882	7889	7896	7903	7910	7917	1	1	2	3	4	4	5	6	6
62	7924	7931	7938	7945	7952	7950	7966	7973	7980	7987	1	1	2	3	3	4	5	6	6
63	7993	8000	8007	8014	8021	8028	8035	8041	8048	8055	1	1	2	3	3	4	5	5	6
64	8062	8069	8075	8082	8089	8096	8102	8109	8116	8122	1	1	2	3	3	4	5	5	6
65	8129	8136	8142	8149	8156	8162	8169	8176	8182	8189	1	1	2	3	3	4	5	5	6
66	8195	8202	8209	8215	8222	8228	8235	8241	8248	8254	1	1	2	3	3	4	5	5	6
67	8261	8267	8274	8280	8287	8293	8299	8306	8312	8319	1	1	2	3	3	4	5	5	6
68	8325	8331	8338	8344	8351	8357	8363	8370	8376	8382	1	1	2	3	3	4	4	5	6
69	8388	8395	8401	8407	8414	8420	8428	8432	8439	8445	1	1	2	2	3	4	4	5	6
70	8451	8457	8463	8470	8476	8482	8488	8494	8500	8506	1	1	2	2	3	4	4	5	6
71	8513	8519	8525	8531	8537	8543	8549	8555	8561	8567	1	1	2	2	3	4	4	5	5
72	8573	8579	8585	8591	8597	8603	8609	8615	8621	8627	1	1	2	2	3	4	4	5	6
73	8633	8639	8645	8651	8657	8663	8669	8675	8681	8686	1	1	2	2	3	4	4	5	5
74	8692	8698	8704	8710	8716	8722	8727	8733	8739	8745	1	1	2	2	3	4	4	5	5
75	8751	8756	8762	8768	8774	8779	8785	8791	8797	8802	1	1	2	2	3	3	4	5	5
76	8808	8814	8820	8825	8831	8837	8842	8848	8854	8859	1	1	2	2	3	3	4	5	5
77	8855	8871	8876	8882	8887	8893	8899	8904	8910	8915	1	1	2	2	3	3	4	4	5
78	8921	8927	8932	8938	8943	8949	8954	8960	8965	8971	1	1	2	2	3	3	4	4	5
79	8976	8982	8987	8993	8998	9004	9009	9015	9020	9025	1	1	2	2	3	3	4	4	5
80	9031	9036	9042	9047	9053	9058	9063	9069	9074	9079	1	1	2	2	3	3	4	4	5
81	9085	9090	9096	9101	9106	9112	9117	9122	9128	9133	1	1	2	2	3	3	4	4	5
82	9138	9143	9149	9154	9159	9165	9170	9175	9180	9186	1	1	2	2	3	3	4	4	5
83	9191	9196	9201	9206	9212	9217	9222	9227	9232	9238	1	1	2	2	3	3	4	4	5
84	9243	9248	9253	9258	9263	9269	9274	9279	9284	9289	1	1	2	2	3	3	4	4	5
85	9294	9299	9304	9309	9315	9320	9325	9330	9335	9340	1	1	2	2	3	3	4	4	5
86	9345	9350	9355	9360	9365	9370	9375	9380	9385	9390	1	1	2	2	3	3	4	4	5
87	9395	9400	9405	9410	9415	9420	9425	9430	9435	9440	0	1	1	2	2	3	3	4	4
88	9445	9450	9455	9460	9465	9469	9474	9479	9484	9489	0	1	1	2	2	3	3	4	4
89	9494	9499	9504	9509	9513	9518	9523	9528	9533	9538	0	1	1	2	2	3	3	4	4
90	9542	9547	9552	9557	9562	9566	9571	9576	9581	9586	0	1	1	2	2	3	3	4	4
91	9500	9595	9600	9605	9609	9613	9619	9624	9628	9633	0	1	1	2	2	3	3	4	4
92	9638	9643	9647	9652	9657	9661	9666	9671	9675	9680	0	1	1	2	2	3	3	4	4
93	9685	9689	9694	9699	9703	9703	9713	9717	9722	9727	0	1	1	2	2	3	3	4	4
94	9731	9736	9741	9745	9750	9754	9759	9763	9763	9773	0	1	1	2	2	3	3	4	4
95	9777	9782	9786	9791	9795	9800	9805	9809	9814	9818	0	1	1	2	2	3	3	4	4
96	9628	9827	9332	9836	9841	9845	9850	9854	9859	9863	0	1	1	2	2	3	3	4	4
97	9868	9872	9877	9881	9886	9890	9894	9899	9903	9908	0	1	1	2	2	3	3	4	4
98	9912	9917	9921	9926	9930	9934	9939	9943	9948	9952	0	1	1	2	2	3	3	4	4
99	9956	9961	9965	9989	9974	9978	9983	9987	9991	9996	0	1	1	2	2	3	3	3	4
N	0	1	2	3	4	5	6	7	8	9	1	2	3	4	5	6	7	8	9

Table A-52   Base 10 antilogarithms (0.000–0.999)

	0	1	2	3	4	5	6	7	8	9	Proportional Parts 1	2	3	4	5	6	7	8	9
.00	1000	1002	1005	1007	1009	1012	1014	1016	1019	1021	0	0	1	1	1	1	2	2	2
.01	1023	1026	1028	1030	1033	1035	1038	1040	1043	1045	0	0	1	1	1	1	2	2	2
.02	1047	1050	1052	1054	1057	1059	1062	1064	1067	1069	0	0	1	1	1	1	2	2	2
.03	1072	1074	1076	1079	1081	1084	1086	1089	1091	1094	0	0	1	1	1	1	2	2	2
04	1096	1099	1102	1104	1107	1109	1112	1114	1117	1119	0	1	1	1	1	2	2	2	2
.05	1122	1125	1127	1130	1132	1135	1138	1140	1143	1146	0	1	1	1	1	2	2	2	2
.06	1148	1151	1153	1156	1159	1161	1164	1167	1169	1172	0	1	1	1	1	2	2	2	2
.07	1175	1178	1180	1183	1186	1189	1191	1194	1197	1199	0	1	1	1	1	2	2	2	2
.08	1202	1205	1208	1211	1213	1216	1219	1222	1225	1227	0	1	1	1	1	2	2	2	2
.09	1230	1233	1236	1239	1242	1245	1247	1250	1253	1256	0	1	1	1	1	2	2	2	2
.10	1259	1262	1265	1268	1271	1274	1276	1279	1282	1285	0	1	1	1	1	2	2	2	3
.11	1288	1291	1294	1297	1300	1303	1306	1309	1312	1315	0	1	1	1	2	2	2	3	3
.12	1318	1321	1324	1327	1330	1334	1337	1340	1343	1346	0	1	1	1	2	2	2	2	3
.13	1349	1352	1355	1358	1361	1365	1368	1371	1374	1377	0	1	1	1	2	2	2	3	3
.14	1380	1384	1387	1390	1393	1396	1400	1403	1406	1409	0	1	1	1	2	2	2	3	3
.15	1413	1416	1419	1422	1426	1429	1432	1435	1439	1442	0	1	1	1	2	2	2	3	3
.16	1445	1449	1452	1455	1459	1462	1466	1469	1472	1476	0	1	1	1	2	2	2	3	3
.17	1479	1483	1486	1489	1493	1496	1500	1503	1507	1510	0	1	1	1	2	2	3	3	3
.18	1514	1517	1521	1524	1528	1531	1535	1538	1542	1545	0	1	1	1	2	2	2	3	3
.19	1549	1552	1556	1560	1563	1567	1570	1574	1578	1581	0	1	1	1	2	2	3	3	3
.20	1585	1589	1592	1596	1600	1603	1607	1611	1614	1618	0	1	1	1	2	2	3	3	3
.21	1622	1626	1629	1633	1637	1641	1644	1648	1652	1656	0	1	1	2	2	2	3	3	3
.22	1660	1663	1667	1671	1675	1679	1683	1687	1690	1694	0	1	1	2	2	2	3	3	3
.23	1698	1702	1706	1710	1714	1718	1722	1726	1730	1734	0	1	1	2	2	2	3	3	4
.24	1738	1742	1746	1750	1754	1758	1762	1766	1770	1774	0	1	1	2	2	2	3	3	4
.25	1778	1782	1786	1791	1795	1799	1803	1807	1811	1816	0	1	1	2	2	2	3	3	4
.26	1820	1824	1828	1832	1837	1841	1845	1849	1854	1858	0	1	1	2	2	3	3	3	4
.27	1862	1866	1871	1875	1879	1884	1888	1892	1897	1901	0	1	1	2	2	3	3	3	4
.28	1905	1910	1914	1919	1923	1928	1932	1936	1941	1945	0	1	1	2	2	3	3	4	4
.29	1950	1954	1959	1963	1968	1972	1977	1982	1986	1991	0	1	1	2	2	3	3	4	4
.30	1995	2000	2004	2009	2014	2018	2023	2028	2032	2037	0	1	1	2	2	3	3	4	4
.31	2042	2046	2051	2056	2061	2065	2070	2075	2080	2084	0	1	1	2	2	3	3	4	4
.32	2089	2094	2099	2104	2109	2113	2118	2123	2128	2133	0	1	1	2	2	3	3	4	4
.33	2138	2143	2148	2153	2158	2163	2168	2173	2178	2183	0	1	1	2	2	3	3	4	4
.34	2188	2193	2198	2203	2208	2213	2218	2223	2228	2234	1	1	2	2	3	3	4	4	5
	0	1	2	3	4	5	6	7	8	9	1	2	3	4	5	6	7	8	9

	0	1	2	3	4	5	6	7	8	9	Proportional Parts								
											1	2	3	4	5	6	7	8	9
.35	2239	2244	2249	2254	2259	2265	2270	2275	2280	2286	1	1	2	2	3	3	4	4	5
.36	2291	2296	2301	2307	2312	2317	2323	2328	2333	2339	1	1	2	2	3	3	4	4	5
.37	2344	2350	2355	2360	2366	2371	2377	2382	2388	2393	1	1	2	2	3	3	4	4	5
.38	2399	2404	2410	2415	2421	2427	2432	2438	2443	2449	1	1	2	2	3	3	4	4	5
.39	2455	2460	2466	2472	2477	2483	2489	2495	2500	2506	1	1	2	2	3	3	4	5	5
.40	2512	2518	2523	2529	2535	2541	2547	2553	2559	2564	1	1	2	2	3	4	4	5	5
.41	2570	2576	2582	2588	2594	2600	2606	2612	2618	2624	1	1	2	2	3	4	4	5	5
.42	2630	2636	2642	2649	2655	2661	2667	2673	2679	2685	1	1	2	2	3	4	4	5	6
.43	2692	2698	2704	2710	2716	2723	2729	2735	2742	2748	1	1	2	3	3	4	4	5	6
.44	2754	2761	2767	2773	2780	2786	2793	2799	2805	2812	1	1	2	3	3	4	4	5	6
.45	2818	2825	2831	2838	2844	2851	2858	2864	2871	2877	1	1	2	3	3	4	5	5	6
.46	2884	2891	2897	2904	2911	2917	2924	2931	2938	2944	1	1	2	3	3	4	5	5	6
.47	2951	2958	2965	2972	2979	2985	2992	2999	3006	3013	1	1	2	3	3	4	5	5	6
.48	3020	3027	3034	3041	3048	3055	3062	3069	3076	3083	1	1	2	3	4	4	5	6	6
.49	3090	3097	3105	3112	3119	3126	3133	3141	3148	3155	1	1	2	3	4	4	5	6	6
.50	3162	3170	3177	3184	3192	3199	3206	3214	3221	3228	1	1	2	3	4	4	5	6	7
.51	3236	3243	3251	3258	3266	3273	3281	3289	3296	3304	1	2	2	3	4	5	5	6	7
.52	3311	3319	3327	3334	3342	3350	3357	3365	3373	3381	1	2	2	3	4	5	5	6	7
.53	3388	3396	3404	3412	3420	3428	3436	3443	3451	3459	1	2	2	3	4	5	6	6	7
54	3467	3475	3483	3491	3499	3508	3516	3524	3532	3540	1	2	2	3	4	5	6	6	7
.55	3548	3556	3565	3573	3581	3589	3597	3606	3614	3622	1	2	2	3	4	5	6	7	7
.56	3631	3639	3648	3656	3664	3673	3681	3690	3698	3707	1	2	3	3	4	5	6	7	8
.57	3715	3724	3733	3741	3750	3758	3767	3776	3784	3793	1	2	3	3	4	5	6	7	8
.58	3802	3811	3819	3828	3837	3846	3855	3864	3873	3882	1	2	3	4	4	5	6	7	8
.59	3890	3899	3908	3917	3926	3936	3945	3954	3963	3972	1	2	3	4	5	5	6	7	8
.60	3981	3990	3999	4009	4018	4027	2036	4046	4055	4064	1	2	3	4	5	6	6	7	8
.61	4074	4083	4093	4102	4111	4121	4130	4140	4150	4159	1	2	3	4	5	6	7	8	9
.62	4169	4178	4188	4198	4207	4217	4227	4236	4246	4256	1	2	3	4	5	6	7	8	9
.63	4266	4276	4285	4295	4305	4315	4325	4335	4345	4355	1	2	3	4	5	6	7	8	9
.64	4365	4375	4385	4395	4406	4416	4426	4436	4446	4457	1	2	3	4	5	6	7	8	9
.65	4467	4477	4487	4498	4508	4519	4529	4539	4550	4560	1	2	3	4	5	6	7	8	9
.66	4571	4581	4592	4603	4613	4624	4634	4645	4656	4667	1	2	3	4	5	6	7	9	10
.67	4677	4688	4699	4710	4721	4732	4742	4753	4764	4775	1	2	3	4	5	7	8	9	10
.68	4786	4797	4808	4819	4831	4842	4853	4864	4875	4887	1	2	3	4	6	7	8	9	10
.69	4898	4909	4920	4932	4943	4955	4966	4977	4989	5000	1	2	3	5	6	7	8	9	10
	0	1	2	3	4	5	6	7	8	9	1	2	3	4	5	6	7	8	9

Table A-52 (*continued*)

	0	1	2	3	4	5	6	7	8	9	Proportional Parts 1	2	3	4	5	6	7	8	9
.70	5012	5023	5035	5047	5058	5070	5082	5093	5105	5117	1	2	4	5	6	7	8	9	11
71	5120	5140	5152	5164	5176	5188	5200	5212	5224	5236	1	2	4	5	6	7	8	10	11
.72	5248	5260	5272	5284	5297	5309	5321	5333	5346	5358	1	2	4	5	6	7	9	10	11
73	5370	5383	5395	5408	5420	5433	5445	5458	5470	5483	1	3	4	5	6	8	9	10	11
.74	5495	5508	5521	5534	5546	5559	5572	5585	5598	5610	1	3	4	5	6	8	9	10	12
.75	5623	5636	5649	5662	5675	5689	5702	5715	5728	5741	1	3	4	5	7	8	9	10	12
.76	5754	5768	5781	5794	5808	5821	5834	5848	5861	5875	1	3	4	5	7	8	9	11	12
77	5888	5902	5916	5929	5943	5957	5970	5984	5998	6012	1	3	4	5	7	8	10	11	12
78	6026	6039	6053	6067	6081	6095	6109	6124	6138	6152	1	3	4	6	7	8	10	11	13
79	6166	6180	6194	6209	6223	6237	6252	6266	6281	6295	1	3	4	6	7	9	10	11	13
80	6310	6324	6339	6353	6368	6383	6397	6412	6427	6442	1	3	4	6	7	9	10	12	13
81	6457	6471	6486	6501	6516	6531	6546	6561	6577	6592	2	3	5	6	8	9	11	12	14
.82	6607	6622	6637	6653	6668	6683	6699	6714	6730	6745	2	3	5	6	8	9	11	12	14
83	6761	6776	6792	6808	6823	6839	6855	6871	6887	6902	2	3	5	6	8	9	11	13	14
84	6978	6934	6950	6966	6982	6998	7015	7031	7047	7063	2	3	5	6	8	10	11	13	15
85	7079	7096	7112	7129	7145	7161	7178	7194	7211	7228	2	3	5	7	8	10	12	13	15
.86	7244	7261	7278	7295	7311	7328	7345	7362	7379	7396	2	3	5	7	8	10	12	13	15
.87	7413	7430	7447	7464	7482	7499	7516	7534	7551	7568	2	3	5	7	9	10	12	14	16
88	7586	7603	7621	7638	7656	7674	7691	7709	7727	7745	2	4	5	7	9	11	12	14	16
89	7762	7780	7798	7816	7834	7852	7870	7889	7907	7925	2	4	5	7	9	11	13	14	16
.90	7943	7962	7980	7998	8017	8035	8054	8072	8091	8110	2	4	6	7	9	11	13	15	17
.91	8128	8147	8166	8185	8204	8222	8241	8260	8270	8299	2	4	6	8	9	11	13	15	17
.92	8318	8337	8356	8375	8395	8414	8433	8453	8472	8492	2	4	6	8	10	12	14	15	17
.93	8511	8531	8551	8570	8590	8610	8630	8650	8670	8690	2	4	6	8	10	12	14	16	18
.94	8710	8730	8750	8770	8790	8810	8831	8851	8872	8892	2	4	6	8	10	12	14	16	18
.95	8913	8933	8954	8974	8995	9016	9036	9057	9078	9099	2	4	6	8	10	12	15	17	19
.96	9120	9141	9162	9183	9204	9226	9247	9268	9290	9311	2	4	6	8	11	13	15	17	19
.97	9333	9354	9376	9397	9419	9441	9462	9484	9506	9528	2	4	7	9	11	13	15	17	20
.98	9550	9572	9594	9616	9638	9661	9683	9705	9727	9750	2	4	7	9	11	13	16	18	20
.99	9772	9795	9817	9840	9863	9868	9908	9931	9934	9977	2	5	7	9	11	14	16	18	20
	0	1	2	3	4	5	6	7	8	9	1	2	3	4	5	6	7	8	9

Table A-53 Base 10 logarithms (0.100–0.999); four-place common logarithms of decimal fractions*

N	0	1	2	3	4	5	6	7	8	9
.10	−1.000	−.9957	−.9914	−.9872	−.9830	−.9788	−.9747	−.9706	−.9666	−.9626
.11	−.9586	−.9547	−.9506	−.9469	−.9431	−.9303	−.9355	−.9318	−.9281	−.9245
.12	−.9208	−.9172	−.9136	−.9101	−.9066	−.9031	−.8996	−.8962	−.8928	−.8894
.13	−.8861	−.8827	−.8794	−.8761	−.8729	−.8697	−.8665	−.8633	−.8601	−.8570
.14	−.8539	−.8508	−.8477	−.8447	−.8416	−.8136	−.8356	−.8327	−.8297	−.8268
.15	−.8239	−.8210	−.8182	−.8153	−.8125	−.8007	−.8069	−.8041	−.8013	−.7986
.16	−.7959	−.7932	−.7905	−.7878	−.7852	−.7825	−.7799	−.7773	−.7747	−.7721
.17	−.7696	−.7670	−.7645	−.7620	−.7595	−.7570	−.7545	−.7520	−.7496	−.7471
.18	−.7447	−.7423	−.7399	−.7375	−.7352	−.7328	−.7305	−.7282	−.7258	−.7235
.19	−.7212	−.7190	−.7167	−.7144	−.7122	−.7100	−.7077	−.7055	−.7033	−.7011
.20	−.6990	−.6968	−.6946	−.6925	−.6904	−.6882	−.6861	−.6840	−.6819	−.6799
.21	−.6778	−.6757	−.6737	−.6716	−.6606	−.6676	−.6655	−.6635	−.6615	−.6596
.22	−.6576	−.6558	−.6536	−.6517	−.6498	−.6478	−.6459	−.6440	−.6423	−.6402
23	−.6383	−.6364	−.6345	−.6326	−.6308	−.6289	−.6271	−.6253	−.6234	−.6216
24	−.6198	−.6180	−.6162	−.6144	−.6126	−.6108	−.6091	−.6073	−.6055	−.6038
.25	−.6021	−.6003	−.5986	−.5909	−.5952	−.5935	−.5918	−.5901	−.5884	−.5867
.26	−.5850	−.5834	−.5817	−.5800	−.5784	−.5768	−.5751	−.5735	−.5719	−.5702
.27	−.5686	−.5670	−.5654	−.5638	−.5622	−.5607	−.5591	−.5575	−.5560	−.5544
.28	−.5528	−.5513	−.5498	−.5482	−.5467	−.5452	−.5436	−.5421	−.5406	−.5391
.29	−.5376	−.5361	−.5346	−.5331	−.5317	−.5302	−.5287	−.5272	−.5258	−.5243
30	−.5229	−.5214	−.5200	−.5186	−.5171	−.5157	−.5143	−.5129	−.5114	−.5100
.31	−.5086	−.5072	−.5058	−.5045	−.5031	−.5017	−.5003	−.4989	−.4976	−.4962
.32	−.4949	−.4935	−.4921	−.4908	−.4895	−.4881	−.4868	−.4855	−.4841	−.4828
.33	−.4815	.4802	−.4789	−.4776	−.4763	−.4750	−.4737	−.4724	−.4711	−.4698
.34	−.4685	−.4672	−.4660	−.4647	−.4634	−.4622	−.4609	−.4597	−.4584	−.4572
.35	−.4559	−.4547	−.4535	−.4522	−.4510	−.4498	−.4486	−.4473	−.4461	−.4449
.36	−.4437	−.4425	−.4413	−.4401	−.4389	−.4377	−.4365	−.4353	−.4342	−.4330
.37	−.4318	−.4306	−.4295	−.4283	−.4271	−.4260	−.4248	−.4237	−.4225	−.4214
.38	−.4202	−.4191	−.4179	−.4186	−.4157	−.4115	−.4134	−.4123	−.4112	−.4101
.39	−.4089	−.4078	−.4067	−.4056	−.4045	−.4034	−.4023	−.4012	−.4001	−.3990
.40	−.3979	−.3969	−.3958	−.3947	−.3936	−.3925	−.3915	−.3904	−.3893	−.3883
.41	−.3872	−.3862	−.3851	−.3840	−.3830	−.3820	−.3809	−.3799	−.3788	−.3778
.42	−.3768	−.3757	−.3747	−.3737	−.3726	−.3716	−.3706	−.3696	−.3686	−.3675
.43	−.3665	−.3655	−.3645	−.3635	−.3625	−.3615	−.3605	−.3595	−.3585	−.3575
.44	−.3565	−.3556	−.3546	−.3536	−.3526	−.3516	−.3507	−.3497	−.3487	−.3478
.45	−.3468	−.3458	−.3449	−.3439	−.3429	−.3420	−.3410	−.3401	−.3391	−.3382
.46	−.3372	−.3363	−.3354	−.3344	−.3335	−.3325	−.3316	−.3307	−.3298	−.3288
.47	−.3279	−.3270	−.3261	−.3251	−.3242	−.3233	−.3224	−.3215	−.3206	−.3197
.48	−.3188	−.3179	−.3170	−.3161	−.3152	−.3143	−.3134	−.3125	−.3116	−.3107
.49	−.3098	−.3089	−.3080	−.3072	−.3063	−.3054	−.3045	−.3036	−.3028	−.3019
.50	−.3010	−.3002	−.2993	−.2984	−.2976	−.2967	−.2958	−.2950	−.2941	−.2933
.51	−.2924	−.2916	−.2907	−.2899	−.2890	−.2882	−.2874	−.2865	−.2857	−.2848
.52	−.2840	−.2832	−.2823	−.2815	−.2807	−.2798	−.2790	−.2782	−.2774	−.2765
.53	−.2757	−.2749	−.2741	−.2733	−.2725	−.2716	−.2708	−.2700	−.2692	−.2684
.54	−.2676	−.2668	−.2660	−.2652	−.2644	−.2636	−.2620	−.2620	−.2612	−.2604

* This table can be used conveniently for finding the cologarithms of the decimal numbers, since the colog $N = -\log N$. For example, colog 0.61 = 0.2147.

Table A-53 (*continued*)

N	0	1	2	3	4	5	6	7	8	9
.55	−.2596	−.2588	−.2581	−.2573	−.2565	−.2557	−.2549	−.2541	−.2534	−.2526
.56	−.2518	−.2510	−.2503	−.2495	−.2487	−.2480	−.2472	−.2464	−.2457	−.2449
.57	−.2441	−.2434	−.2426	−.2418	−.2411	−.2403	−.2396	−.2388	−.2381	−.2373
.58	−.2366	−.2358	−.2351	−.2343	−.2336	−.2328	−.2321	−.2314	−.2306	−.2299
.59	−.2291	−.2284	−.2277	−.2269	−.2262	−.2255	−.2248	−.2240	−.2233	−.2226
.60	−.2218	−.2211	−.2204	−.2197	−.2190	−.2182	−.2175	−.2168	−.2161	−.2154
.61	−.2147	−.2140	−.2132	−.2125	−.2118	−.2111	−.2104	−.2097	−.2090	−.2083
.62	−.2076	−.2069	−.2062	−.2055	−.2048	−.2041	−.2034	−.2027	−.2020	−.2013
.63	−.2007	−.2000	−.1993	−.1986	−.1979	−.1972	−.1965	−.1959	−.1952	−.1945
.64	−.1938	−.1931	−.1925	−.1918	−.1911	−.1904	-.1898	−.1891	−.1884	−.1878
.65	−.1871	−.1864	−.1858	−.1851	−.1844	−.1838	−.1831	−.1824	−.1818	−.1811
.66	−.1805	−.1798	−.1791	−.1785	−.1778	−.1772	−.1765	−.1750	−.1752	−.1746
.67	−.1730	−.1733	−.1726	−.1720	−.1713	−.1707	−.1701	−.1694	−.1686	−.1681
.68	−.1675	−.1669	−.1662	−.1656	−.1649	−.1643	−.1637	−.1630	−.1624	−.1618
.69	−.1612	−.1605	−.1599	−.1592	−.1586	−.1580	−.1574	−.1568	−.1561	−.1555
70	−.1549	−.1543	−.1537	−.1530	−.1524	−.1518	−.1512	−.1506	−.1500	−.1494
.71	−.1487	−.1481	−.1475	−.1469	−.1463	−.1457	−.1451	−.1445	−.1439	−.1433
.72	−.1427	−.1421	−.1415	−.1409	−.1403	−.1397	−.1391	−.1385	−.1379	−.1373
.73	−.1367	−.1361	−.1355	−.1349	−.1343	−.1337	−.1331	−.1325	−.1319	−.1314
.74	−.1308	−.1302	−.1296	−.1290	−.1284	−.1278	−.1273	−.1267	−.1261	−.1255
.75	−.1249	−.1244	−.1238	−.1232	−.1226	−.1221	−.1215	−.1209	−.1203	−.1198
.76	−.1192	−.1186	−.1180	−.1175	−.1169	−.1163	−.1158	−.1152	−.1146	−.1141
.77	−.1135	−.1129	−.1124	−.1118	−.1113	−.1107	−.1101	−.1096	−.1090	−.1085
.78	−.1079	−.1073	−.1068	−.1062	−.1057	−.1051	−.1046	−.1040	−.1035	−.1029
.79	−.1024	−.1018	−.1013	−.1007	−.1002	−.0996	−.0991	−.0985	−.0980	−.0975
.80	−.0969	−.0964	−.0958	−.0953	−.0947	−.0942	−.0937	−.0931	−.0926	−.0921
.81	−.0915	−.0910	−.0904	−.0899	−.0894	−.0888	−.0883	−.0878	−.0872	−.0867
.82	−.0862	−.0857	−.0851	−.0846	−.0841	−.0835	−.0830	−.0825	−.0820	−.0814
.83	−.0809	−.0804	−.0799	−.0794	−.0788	−.0783	−.0778	−.0773	−.0768	−.0762
.84	−.0757	−.0752	−.0747	−.0742	−.0737	−.0731	−.0726	−.0721	−.0716	−.0711
.85	−.0706	−.0701	−.0696	−.0691	−.0685	−.0680	−.0675	−.0670	−.0665	−.0660
.86	−.0655	−.0650	−.0645	−.0640	−.0635	−.0630	−.0625	−.0620	−.0615	−.0610
.87	−.0605	−.0600	−.0595	−.0590	−.0585	−.0580	−.0575	−.0570	−.0565	−.0560
.88	−.0555	−.0550	−.0545	−.0540	−.0535	−.0531	−.0526	−.0521	−.0516	−.0511
.89	−.0506	−.0501	−.0496	−.0491	−.0487	−.0482	−.0477	−.0472	−.0467	−.0462
.90	−.0458	−.0453	−.0448	−.0443	−.0438	−.0434	−.0429	−.0424	−.0419	−.0414
.91	−.0410	−.0405	−.0400	−.0395	−.0391	−.0386	−.0381	−.0376	−.0372	−.0367
.92	−.0362	−.0357	−.0353	−.0348	−.0343	−.0339	−.0334	−.0329	−.0325	−.0320
.93	−.0315	−.0311	−.0306	−.0301	−.0297	−.0292	−.0287	−.0283	−.0278	−.0273
.94	−.0269	−.0264	−.0259	−.0255	−.0250	−.0246	−.0241	−.0237	−.0232	−.0227
.95	−.0223	−.0218	−.0214	−.0209	−.0205	−.0200	−.0195	−.0191	−.0186	−.0182
.96	−.0177	−.0173	−.0168	−.0164	−.0159	−.0155	−.0150	−.0146	−.0141	−.0137
.97	−.0132	−.0128	−.0123	−.0119	−.0114	−.0110	−.0106	−.0101	−.0097	−.0092
.98	−.0088	−.0083	−.0079	−.0074	−.0070	−.0066	−.0061	−.0057	−.0052	−.0043
.99	−.0044	−.0039	−.0035	−.0031	−.0026	−.0022	−.0017	−.0013	−.0009	−.0004

Table A-54 Logarithms of the factorial

n	$\log \lfloor n$	n	$\log \lfloor n$	n	$\log \lfloor n$	n	$\log \lfloor n$
1	.000 0000	51	66.190 0480	101	159.974 3280	151	264.934 9704
2	.301 0300	52	67.808 6484	102	161.938 9858	152	267.117 7139
3	.778 1613	53	69.630 9243	103	163.995 7634	153	269.302 4054
4	1.380 2112	54	71.363 3180	104	166.018 7948	154	271.489 9281
5	2.079 1812	55	73.103 6807	105	168.088 9851	155	273.680 2570
6	2.857 3325	56	74.651 8837	106	170.080 2809	156	275.873 2824
7	3.702 4305	57	76.807 7456	107	172.056 6747	157	278.089 2820
8	4.605 5205	58	78.371 1716	108	174.123 0286	158	280.287 9301
9	5.559 7630	59	80.148 0236	109	176.123 5250	159	282.499 3243
10	6.550 7630	60	81.920 1748	110	179.200 9176	160	284.972 4563
11	7.601 1567	61	83.706 5047	111	180.248 2408	161	286.000 2321
12	8.680 3370	62	85.497 8094	112	182.295 4589	162	289.089 7971
13	9.794 2803	63	87.297 2309	113	184.343 5371	163	291.201 9847
14	10.940 4084	64	89.103 4169	114	186.403 4410	164	293.516 8296
15	12.116 4995	65	90.916 3308	115	189.403 1388	165	294.734 3125
16	13.320 6196	66	92.735 8742	116	190.530 5978	166	297.944 6203
17	14.551 0885	67	94.581 3480	117	192.598 7836	167	300.177 1371
18	15.808 3410	68	98.294 4579	118	194.670 6656	168	302.402 4464
19	17.085 0916	69	96.223 3070	119	195.748 2199	169	304.430 2231
20	18.286 1248	70	100.078 4060	120	196.026 3636	170	306.390 7830
21	19.708 3439	71	101.979 6834	121	200.908 1792	171	309 083 7781
22	21.050 7686	72	103.768 9950	122	202.224 5320	172	311.329 3063
23	22.412 4914	73	105.650 3187	123	206.084 6443	173	313.567 3567
24	23.792 7097	74	107.519 5505	124	207.177 6558	174	315.807 2019
25	25.190 6457	75	109.294 6117	125	208.274 7769	175	318.080 9400
26	26.605 6180	76	111.275 4253	126	211.375 1464	176	320.296 4526
27	28.036 9828	77	113.161 9160	127	213.478 9501	177	322.544 4259
28	29.484 1408	78	115.054 0108	128	215.556 1801	178	324.794 8459
29	30.946 5388	79	116.851 6377	129	217.998 7498	179	327.047 0829
30	32.423 6601	80	118.654 7877	130	219.801 6838	180	329.202 9714
31	33.915 0216	81	120.763 2187	131	221.927 9865	181	331.560 6300
32	35.420 1717	82	122.877 0266	132	224.065 6484	182	333.820 7214
33	36.938 6857	83	124.596 1047	133	226.172 3200	185	336.053 1725
34	38.470 1646	84	126.520 2340	134	228.289 4948	184	338.347 9903
35	40.014 2326	85	128.449 2020	135	230.429 6226	185	340.918 1620
36	41.570 5351	86	130.284 2013	136	233.563 3876	186	342.384 6750
37	43.138 7369	87	132.323 8206	137	234.700 0861	187	345.168 5168
38	44.719 5205	88	134.268 3033	138	236.826 2272	188	347.420 6744
39	46.309 5851	89	136.217 6933	139	239.268 2620	189	349.707 1343
40	47.911 6451	90	138.171 8358	140	241.129 1100	190	351.964 4498
41	49.524 4289	91	140.120 9772	141	243.978 3201	191	354.228 9232
42	51.147 6782	92	143.084 7650	142	245.420 6176	192	356.550 2244
43	52.781 1487	93	145.083 2480	143	247.165 2835	193	358.825 7817
44	54.434 3993	94	146.033 3758	144	248.744 3100	194	361.123 8635
45	56.077 6119	95	148.014 0394	145	251.206 0960	195	363.413 6181
46	57.740 5697	96	149.986 3707	146	254.070 0626	196	366.708 8748
47	59.412 4876	97	151.982 1484	147	253.257 2642	197	268.000 3404
48	61.083 8053	98	153.974 3685	148	258.407 6160	198	370.297 0058
49	63.794 1049	99	155.970 0097	149	260.580 3032	199	372.525 8500
50	64.482 0749	100	157.970 0087	150	261.726 5934	200	374.806 8956

n	log |n	n	log |n	n	log |n	n	log |n
201	377.200 0847	251	494.909 2901	301	616.964 3825	351	743.037 2313
202	379.505 4361	252	497.310 6607	302	619.444 3766	352	745.123 2240
203	381.812 9321	253	499.713 7812	303	621.926 4191	353	747.791 5867
204	384.122 5623	254	502.110 6149	304	624.408 0227	354	750.560 6080
205	386.434 3161	255	504.525 1561	305	628.828 9326	355	753.830 6308
206	388.748 1834	256	508.933 3850	306	630.378 7140	356	755.828 2303
207	391.064 1537	257	509.343 3268	307	631.036 6523	357	757.634 9495
208	393.382 2170	258	511.764 2479	308	634.364 4031	358	760.483 2310
209	395.702 3633	259	514.106 2474	309	638.844 2818	359	763.043 9260
210	398.024 5826	260	518.683 2310	310	639.235 7238	360	765.800 2083
211	400.348 8651	261	519.489 5314	311	641.323 4826	361	768.157 7367
212	402.675 2009	262	521.413 1028	312	644.223 6253	362	770.716 4643
213	405.003 6805	263	523.838 1105	313	648.818 1825	363	773.278 2509
214	407.333 9943	264	528.259 7225	314	649.315 1123	364	775.637 4623
215	409.666 4328	265	529.682 9823	315	651.813 4227	365	778.350 7663
216	412.000 8865	266	531.107 6500	316	654.213 1088	366	780.283 3203
217	414.337 3463	267	533.634 3818	317	658.814 1091	367	783.407 8268
218	416.657 8027	268	535.838 4980	318	659.316 5968	368	785.033 7601
219	419.016 2469	269	538.398 2463	319	661.820 2869	369	788.620 7694
220	421.358 6695	270	540.823 6121	320	664.225 6269	370	791.668 9422
221	423.703 0618	271	543.256 0914	321	666.838 0419	371	793.793 3481
222	426.049 4148	272	545.091 1603	322	669.339 8978	372	796.302 9261
223	428.397 7197	273	548.127 3129	323	671.569 1003	373	798.960 5860
224	430.747 9677	274	550.564 0536	324	674.359 6453	374	801.613 4025
225	433.100 4502	275	553.004 3883	325	678.871 6287	375	804.087 4028
226	435.454 2586	276	566.445 2068	326	679.254 7483	376	808.028 0248
227	437.410 2845	277	557.087 7880	327	681.899 3440	377	809.189 0880
228	440.168 8193	278	560.231 8226	328	684.415 1679	378	811.816 8178
229	442.528 0568	279	562.777 4340	329	688.933 3826	379	814.286 1670
230	444.889 7827	280	564 184 5820	330	689.450 8777	380	816.974 9408
231	447.253 2946	281	567.579 8894	331	691.970 7057	381	819.555 0055
232	449.618 8826	282	570.129 2476	332	694.491 2438	382	822.137 9429
233	451.966 2385	283	572.575 2339	333	697.014 2980	383	824.781 1277
234	452.928 2984	284	575.588 0679	334	699.923 1720	384	827.404 7077
235	456.726 5223	285	577.423 4971	335	702.063 0703	385	829.820 9198
236	459.099 4343	286	579.999 8331	336	704.589 4166	386	832.477 8039
237	461.474 1826	287	582.397 7450	337	707.117 0485	387	835.085 8179
238	463.850 7596	288	584.257 1376	338	709.645 9852	388	837.634 0498
239	466.229 1575	289	587.318 0354	339	712.176 1849	389	840.243 8091
240	468.609 3687	290	589.780 4334	340	714.707 6438	390	842.824 0626
241	470.991 3857	291	592.244 2884	341	717.840 2882	391	845.427 2406
242	473.375 3011	292	594.709 7092	342	719.774 4342	392	848.020 5287
243	475.780 8074	293	597.178 5706	343	722.209 7194	393	850.614 9192
244	478.148 1972	294	599.644 9943	344	724.348 2768	394	853.310 4154
245	480.537 3633	295	602.114 7463	345	727.384 0069	395	855.207 0125
246	482.926 2984	296	604.588 0079	346	729.923 1720	396	856.404 7077
247	485.320 9954	297	607.068 7942	347	732.463 6015	397	861.003 4968
248	487.715 4470	298	609.523 0106	348	735.005 0907	398	863.603 3813
249	490.111 6464	299	612.008 6818	349	737.547 9068	399	869.204 2534
250	498.509 5854	300	614.465 2020	350	740.091 9741	400	898.808 4143

REFERENCES

1. Croxton, Frederick E.: "Elementary Statistics with Applications in Medicines," Prentice-Hall, Inc., Englewood Cliffs, N.J., 1953.
2. Fisher, R. A., and F. Yates: "Statistical Tables for Biological, Agricultural, and Medical Research," Oliver & Boyd, Ltd., Edinburgh, 1957.
3. Thomson, Catherine M.: Table of Percentage Points of the χ^2 Distribution, *Biometrika*, vol. 32, part II, October, 1941.
4. Hald, A.: "Statistical Tables and Formulas," John Wiley & Sons, Inc., New York.
5. Johnson, L. G.: "The Statistical Treatment of Fatigue Experiments," Elsevier Publishing Company, Amsterdam, 1964.
6. Rand Corporation: "A Million Random Digits with 100,000 Normal Deviates," The Free Press of Glencoe, Ill., Chicago, 1955.

Glossary

Accelerated experiment The experiment employed to reduce the testing time or the sample size to less than what it would normally take to reach meaningful conclusions.

Acceptance region The region of values for which the null hypothesis is accepted.

Alpha (α) error The probability that the null hypothesis H_0 is rejected when it is true. Also called type I error.

Alternate hypothesis H_A The assumption that the null hypothesis is not true.

Average See *mean of a sample.*

Beta (β) error The probability of accepting the null hypothesis H_0 when it is false. Also called type II error.

B_q life Life by which q percent of the items would fail.

Characteristic life Life by which 63.2 percent of the items would fail, for Weibull distribution.

Confidence band The area enclosed by two curves passing through the upper and lower limits of a series of confidence intervals. Usually these curves are drawn on either side of a median-rank line.

Confidence level $(1 - \alpha)$ The probability that a random variable x lies between the interval $(x - C_{\alpha/2})$ to $(x + C_{\alpha/2})$.

Confidence limits The two values that define the confidence interval $(x - C_{\alpha/2})$ to $(x + C_{\alpha/2})$.

Continuous random variable A random variable which can attain any value continuously in some specified interval.

Correlation analysis A method for establishing a degree of association between variables, continuous or discrete.

Correlation coefficient r A quantitative measure of association between variables. A measure of how well a curve fits the test data. A value of 1.0 indicates a perfect relationship, and zero indicates no relationship.

Cumulative distribution function F(a) The function which yields the probability that x is less than or equal to a. Mathematically,

$$F(a) = \begin{cases} \displaystyle\sum_{i=-\infty}^{a} p_{x_i} & x_i \leq a \text{ for discrete } x \\[2ex] \displaystyle\int_{-\infty}^{a} f(x)\,dx & \text{for continuous } x \end{cases}$$

Cutoff point The point which separates the acceptance region from the rejection region.

Degrees of freedom The number of independent measurements available for estimating a population parameter.

Density function f(x) The function which yields the probability that the random variable takes on any one of its admissible values.

Design hypothesis H_D An assurance that the design detects or avoids some specific mean value.

Discrete random variable A random variable which can attain values only from a definite number of discrete values.

Durability The useful life of a part, assembly, or system. This life may be an objective, a firm requirement, or a measured life value.

Expected value See *mean of a population*.

Experiment An operation under controlled conditions to determine an unknown effect; to illustrate or verify a known law; to test or establish a hypothesis.

Experimental error Variation in observations made under identical test conditions. Also called residual error.

Factorial experiment An experiment which extracts information on several design factors more efficiently than the traditional test.

Failure A component or a system is said to have failed if it ceases to perform satisfactorily the function for which it is intended.

Fatigue The process of progressive localized permanent structural change occurring in a material subjected to fluctuating loading.

Fatigue life   The number of cycles of stress or strain that a given specimen or part sustains before failure.

Hazard rate λ   The probability that a given item on test will fail between X and $X + dX$ time when it has already survived the time X.

Histogram   A graphical representation of a frequency distribution by a series of rectangles, where the width of the rectangle represents the range of a variable and the height represents the frequency of occurrence.

Interaction   If the effects of a factor A are not the same at all levels of another factor B, then A and B are said to interact.

Least-square line   A line fitted to a series of test points such that the sum of the squares of the deviations of the test points from that line is minimum. Theoretically, the best-fitting line for given points.

Location parameter x_0   The minimum value of a random variable x in Weibull distribution.

Mean of a population (μ)   A value for a continuous variable x, with a distribution function $f(x)$, given by

$$\mu = E(x) = \int_{-\infty}^{\infty} xf(x)\, dx$$

and for a discrete variable

$$\mu = E(x) = \sum_{i=1}^{n} x_i p_{x_i}$$

where x_i's are the values of the discrete random variable and p_{x_i} is the probability of occurrence associated with x_i.

Mean of a sample $(\bar{x})$ A value obtained by dividing the sum of observations by their total number.   Also called average.

Mean-life ratio   The ratio of the mean lives of two different populations.

Mean time between failures $(MTBF)$ The limit of the ratio of operating time to the number of observed failures as the number of failures approaches infinity.

$$\text{MTBF} = \lim_{n \to \infty} \frac{T}{n}$$

Mean time to failure $(MTTF)$ The operating time of a part divided by a total number of failures of that part during a given period of time.

Median of a population   The value of x at which the cumulative distribution function $f(x)$ is 0.5.

Median of a sample   The number in the middle, when all the observations are ranked in their order of magnitude.

Median rank   A rank of test observations, usually failures, used for constructing a cumulative distribution plot for estimating the population; a 50 percent rank.

Mode of a population   The value of x corresponding to the peak value of the probability of occurrence of any given continuous or discrete distribution.

Mode of a sample   The value of x corresponding to the maximum frequency on the distribution curve, or the highest probability in the case of a discrete random variable.

Multiple regression analysis   Same as regression analysis, except that it involves more than two variables.

Nonparametric experiment   An experiment that does not depend on the assumption of a particular distribution of test data.

Null hypothesis H_0   An assumption that the mean $\bar{x}$ or the variance s^2 of a sample did come from the population whose mean was μ_0 or variance $\sigma_0{}^2$.

One-sided alternative   The case where the value of the parameter under consideration has either an upper bound or a lower bound, but not both.

Parameter   A constant defining some property of the density function of a variable, such as a mean or a standard deviation.

Population   A group of similar items from which a sample is drawn for test purposes; usually assumed to be infinitely large.

Power efficiency of an experiment   The ratio of a parametric test sample size to the nonparametric test sample size for both tests to give the same power, when used with a normally distributed population.

Power of an experiment   The probability of rejecting the null hypothesis H_0 when it is false and accepting the alternate hypothesis H_A when it is true.

$$\text{Power} = 1 - \beta$$

Probability of occurrence   The number of successful occurrences divided by the total number of trials.

Probability paper   Graph paper constructed so that cumulative distribution curves plot as straight lines.

Random variable   A variable which can assume any value from a set of possible values.

Ranks   Values assigned to items in a sample to determine their relative occurrence in a population.

Regression analysis   A method for establishing the functional relationship between two variables, where the variation in one measurement is considered while the other is held fixed. This relation can be linear or nonlinear.

Reliability   The reliability of a part, an assembly, or a system is the probability that it will perform satisfactorily for a specified period of time under specified operating conditions.

Replication   Observations made under identical test conditions.

Residual error   *Experimental error.*

Sample   A random selection of items from a lot or population, usually made for evaluating the characteristics of the lot or population

Scale parameter   See *characteristic life.*

Sequential experiment   A procedure in which items are tested in sequence, the test results are reviewed at the end of each test, and tests of significance are applied to the data accumulated up to that time.

Shape parameter   See *Weibull slope.*

Standard deviation of a population (σ)   Square root of the variance of a population.

Standard deviation of a sample (s)   Square root of the variance of a sample; a measure of the scatter of test data about the mean.

Sudden-death test   A test on groups of items which is discontinued as soon as one item in each group fails; requires only a fraction of the time necessary for a full failure test.

Suspended item   An item removed from test prior to failure.

Test of significance   A procedure to determine whether a quantity subjected to random variation differs from a postulated value by an amount greater than that due to random variation alone.

Two-sided alternative   The case when the value of the parameter under consideration falls within a range and is thus bounded on both the upper and the lower end.

Type I error   *alpha* (α) *error.*

Type II error   *beta* (β) *error.*

Variance of a population (σ^2)   A value for a variable x with a distribution function $f(x)$, given by

$$\sigma^2 = E[(x - \mu)^2] = \int_{\infty}^{\infty} (x - \mu)^2 f(x)\, dx$$

Variance of a sample (s^2)   The sum of squares of the differences between each observation and the sample average divided by the sample size minus 1.

Variation analysis   A procedure used to isolate one type of variation from another (for example, separating product variation from test variation).

Weibull slope (b)   A slope of a straight line drawn through the test points plotted on the Weibull probability paper.

Index